Building Regulations in Brief

Eighth edition

Ray Tricker and Samantha Alford

Routledge
Taylor & Francis Group

LONDON AND NEW YORK

This eighth edition published 2014
by Routledge
2 Park Square, Milton Park, Abingdon, Oxon, OX14 4RN

Simultaneously published in the USA and Canada
by Routledge
711 Third Avenue, New York, NY 10017

First edition 2003 (Reprinted twice 2003)
Second edition 2004 (Reprinted 2004)
Third edition 2005
Fourth edition 2006
Fifth edition 2007
Sixth edition 2011
Seventh edition 2012

Routledge is an imprint of the Taylor & Francis Group, an informa business

British Library Cataloguing in Publication Data
A catalogue record for this book is available from the British Library

Library of Congress Cataloging-in-Publication Data
Tricker, Ray author.
   Building regulations in brief / Ray Tricker and Samantha Alford.—Eighth edition.
   pages cm
   Includes bibliographical references and index.
   ISBN 978-0-415-72171-4 (pbk. : alk. paper)—ISBN 978-1-315-79446-4 (ebook)
   1. Building laws—England. I. Alford, Samantha, author. II. Title.
   KD1140.T75 2014
   343.41'07869—dc23
   2013040150

ISBN13: 978–0–415–72171–4 (pbk)
ISBN13: 978–1–315–79446–4 (ebk)

Typeset in Times New Roman
by Swales & Willis Ltd, Exeter, Devon
Printed and bound by CPI Group (UK) Ltd, Croydon, CR0 4YY

Building Regulations in Brief

This eighth edition of the most popular and trusted guide to the building regulations is the most comprehensive revision yet. It reflects all the latest amendments to Building Regulations, Planning Permission and the Approved Documents A, B, C, H, K, P and Regulation 7, incorporating all amendments up to December 2013 (including the changes to Leaflets L1A and L2A which come into effect April 2014).

This new edition also contains details of the new national planning guidance system and initiatives to speed up the planning process, such as the new online planning application process. It contains an updated list of fees for planning consents and provides guidance on the changes to permitted development rights in Agricultural, Business and Residential buildings which came into force on 1 October 2013.

Giving practical information throughout on how to work with (and within) the regulations, this book enables compliance in the simplest and most cost-effective manner possible. The no-nonsense approach of *Building Regulations in Brief* cuts through the confusion and explains the meaning of the regulations; consequently it has become a favourite for anyone involved in the building industry as well as those planning to have work carried out in their home.

Ray Tricker is the Senior Consultant (Management Systems) of Herne European Consultancy Ltd (a company offering organizations access to highly skilled and specialist consultants to help them enhance their business performance), and is also an established author, with over 40 titles published. He served with the Royal Corps of Signals (for a total of 37 years), during which time he held various managerial posts, culminating in being appointed as the Chief Engineer of NATO's Communications Security Agency (ACE COMSEC).

Samantha Alford served with the Royal Air Force for 17 years, where she held various logistics and managerial posts, the majority of which were in the operations and planning fields. She has a wealth of experience in the business management, scheduling and presentation fields. She has an M.Sc. in Logistics Management, is an expert in integrated (i.e. quality, environmental and safety) management and is a qualified internal, external and third-party quality auditor, an experienced instructor and a published author.

Contents

Preface

As you will have noticed, this eighth edition of *Building Regulations in Brief* is back in black and white, in an effort to keep costs down for our readers. We have continued to provide a simplified explanation of the numerous regulations which relate to buildings that our readers have told us they value. We have worked hard to ensure that this book is as up to date as we can possibly make it, and it even contains the changes to Approved Document L, which only came into effect in April 2014. We believe that it provides a comprehensive guide to a complicated area of legislation which has changed greatly over the last 18 months. Indeed, we are consistently told that there is no other book on the market that provides the detail that we do in such an easy-to-use format.

We recognize that it is only 18 months since the last edition of this book was published, but again much has changed with the source legislation, and the planning system in the UK has been simplified and completely overhauled. We have therefore found it necessary to produce a new edition. Since the issue of the Building Regulations 2010 there have been a number of changes to the regulations. In particular, Approved Documents A, B, C, H, K, L, M and P and Regulation 7 have been revised, and leaflet N has been removed. A range of initiatives were introduced in March 2012 in an effort to speed up the planning process, and there have been major wholesale changes to the planning regulations as a result of the Coalition Government's initiative to reduce red tape and the Localism Act 2011. The previous technical guidance relating to planning permission has been substantially reduced, and all of these changes have been included up to the date of publication (April 2014). It is anticipated that there will be further changes to the Planning Act 2008 in due course.

At the time of writing a new streamlined planning process is being introduced which the government hope will provide support for growth and create jobs and homes in the UK. Better protection for the natural and historic environment and more community power to make local decisions have been the result. The planning website (the Planning Portal) has been updated, a centralized online application process has been introduced for all local authorities in England, and new pages set out guidance on a range of issues, including:

- a new affordability test for determining how many homes should be built;
- discouragement of any plans to introduce a new parking tax on people's driveways and parking spaces;
- encouragement of more town centre parking spaces and an end of aggressive 'anti-car' traffic calming measures like speed bumps;
- introduction of new neighbourhood planning guidance to help communities with their own plans;

- new local green space guidance to help councils and local communities to plan for open spaces and to protect local green spaces which are special to local people;
- new neighbourhood planning guidance to help more communities start their own plans;
- provision of housing for older people – to encourage councils to build more bungalows and plan positively for an ageing population;
- the opening up of planning appeal hearings to filming.

The National Planning Policy Framework was published on 27 March 2012 and is a major part of the reforms to simplify the planning system. It sets out the planning policies for England and states how the policies are to be applied. The Framework provides guidance for local planning authorities and decision takers, both in drawing up plans and in making decisions about planning applications. There are no specific policies for nationally significant infrastructure projects in the Framework, as these are determined by the Secretary of State on a case-by-case basis. The Framework is the result of extensive consultation and is based upon the presumption in favour of sustainable development.

During May 2013 the UK Parliament agreed to increase the size of single-storey rear extensions which can be built under permitted development and brought into force the associated mandatory neighbour consultation scheme. This applies only for a period of three years, between 30 May 2013 and 30 May 2016, but the extension sizes under permitted development have been doubled during this period from $4\,\mathrm{m}^2$ to $8\,\mathrm{m}^2$ for detached houses, and from $3\,\mathrm{m}^2$ to $6\,\mathrm{m}^2$ for all other houses.

Although the regulations themselves are comparatively short, they rely on their technical detail being available in a series of Approved Documents and a vast number of British, European and international standards, codes of practice, drafts for development, published documents and other non-statutory guidance documents. The main problem, from the point of view of the average builder and DIY enthusiast, is that the Building Regulations are far too professional for their purposes. Because they cover every aspect of building, they are often far too detailed and contain far too many options. All that builders or DIY people really require is sufficient information to enable them to comply with the regulations in the simplest and most cost-effective manner possible.

Building inspectors, acting on behalf of local authorities, are primarily concerned with whether a building complies with the requirements of the Building Regulations, and, to assess this, they need to 'see the calculations'. But how do DIY enthusiasts and/or builders obtain these calculations? Where can they find, for instance, the policy and requirements for the load-bearing elements of a structure?

Builders, through experience, are normally aware of the overall requirements for building projects such as laying drains, building walls or installing central heating or glazing. However, they still need a reminder when they come across a different situation for the first time (e.g. how deep will the foundations have to be if the builder is going to construct a building on soft soil?).

On the other hand, DIY enthusiasts, keen on building their own extension, conservatory, garage or workshop, usually have no past experience, and need the relevant information – but in a form that they can easily understand without having had the advantage of many years of experience. In fact, what they really need is a rule-of-thumb guide to the basic requirements.

We know from a number of surveys that the majority of builders and virtually all DIY enthusiasts are self-taught, and most of their knowledge is gained through experience. When they hit a problem, it is usually discussed over a drink with friends in the building trade, as opposed to seeking professional help. What they really need is a reference book to enable them to understand (or remind themselves of) the official requirements.

The aim of this book, therefore, is to provide the reader with a brief guide that can act as an *aide-mémoire* to the current requirements of the Building Regulations. Intended readers are primarily builders and the DIY community (who need to know the regulations but do not require the detail), but the book, with its ready reference and no-nonsense approach, will be equally useful to students, architects, designers, building surveyors and inspectors.

The structure of this book

In essence, this book is in two parts. The first part (Chapters 1 to 5) provides a user-friendly introduction to the Building Act 1984 and its associated Building Regulations. It explains the meaning of the Building Regulations, their current status, requirements, associated documentation and how local authorities and councils view their importance. We also describe the content of the guidance documents (such as the overview of the key elements of each Approved Document contained in Chapter 3), and provide details of how to get planning permission and how much it will cost. This first part concludes with a series of 'What if?' scenarios, providing answers to the most common questions that DIY enthusiasts and builders might ask concerning building projects (for ease of reference this list is alphabetical). The second part is the bit of the book of which we are most proud, and the part about which we get the most comments. It contains Chapter 6, 'Meeting the requirements of the Building Regulations'. In essence, this part provides a foundations-up approach, detailing all the regulations relating to specific projects (e.g. building walls). We have developed Chapter 6 extensively over the 11 years of writing previous versions of this book.

In summary, the chapters in this book are:

Chapter 1 – The Building Act 1984
Chapter 2 – The Building Regulations 2010
Chapter 3 – The requirements of the Building Regulations
Chapter 4 – Planning permission

Chapter 5 – Requirements for planning permission and Building Regulations approval
Chapter 6 – Meeting the requirements of the Building Regulations

Chapter 6 is further supported by appendices which cover specific areas in greater detail as entities in their own right, rather than how they apply to a particular project.

Appendix A – Entrance and access
Appendix B – Conservation of fuel and power
Appendix C – Fire and escape
Appendix D – Electrical safety (dwellings)
Appendix E – Fire resistance
Appendix F – Means of escape
Appendix G – Entrance and access

All the appendices are available at www.routledge.com/9780415721714, and they are also available in the printed version.

The book concludes with a list of acronyms, a bibliography and a full index.

The following symbols will help you get the most out of this book. In the margins you will find:

An important requirement or point

A good idea or suggestion

and within the text:

Notes: These are used to provide further amplification or information.

Italic text, which indicates a direct quote from the relevant Act, Regulation or Approved Document

Shaded boxes are used in Chapter 6 to show either the full text of the Building Regulation's *legal requirements* or a paraphrased version of these requirements.

The current legislation is the Building Regulations 2010. It is made by the Secretary of State for Environment, Food and Rural Affairs under powers delegated by Parliament under the Building Act 1984. These Regulations have replaced **all** former regulations and Statutory Instruments.

The Building Act 1984

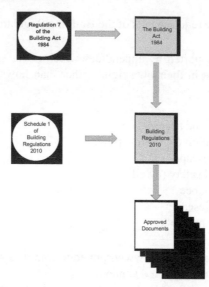

The Secretary of State ensures that the health, welfare and convenience of persons living in or working in (or near) buildings are secured through the Building Act 1984. One of the Act's prime purposes is to assist in the conservation of fuel and power, and prevent waste, undue consumption and the misuse and contamination of water.

The Act imposes on owners and occupiers of buildings a set of requirements concerning the design and construction of buildings, and the provision of services, fittings and equipment used in (or in connection with) buildings.

 The Building Act 1984 does not apply to Scotland or to Northern Ireland. The Building Act 1984 has five parts:

Part 1 – The Building Regulations
Part 2 – Supervision of building work etc. other than by a local authority
Part 3 – Other provisions about buildings
Part 4 – General
Part 5 – Supplementary

Part 5 then contains seven schedules, the prime function of which is to list the principal areas requiring regulation and to show how the Building Regulations are to be controlled by local authorities. These schedules are:

Schedule 1 – Building Regulations
Schedule 2 – Relaxation of Building Regulations
Schedule 3 – Inner London

Schedule 4 – Provisions consequential upon public body's notice
Schedule 5 – Transitional provisions
Schedule 6 – Consequential amendments
Schedule 7 – Repeals

Schedule 1 is the most important (from the point of view of builders), as it shows, in general terms, how the Building Regulations are to be administered by local authorities, the approved methods of construction and the approved types of materials that are to be used in (or in connection with) buildings.

Regulation 7 of the Building Act 1984 covers materials and workmanship and states that building work shall be carried out:

(a) with adequate and proper materials which

    (i) are appropriate for the circumstances in which they are used,
    (ii) are adequately mixed or prepared, and
    (iii) are applied, used or fixed so as adequately to perform the functions for which they are designed; and

(b) in a workmanlike manner.

The Building Regulations

Building Regulations 2010 (SI 2010/2214) form Schedule 1 of the Building Act and are a set of minimum requirements and basic performance standards designed to secure the health, safety and welfare of people in and around buildings and to conserve fuel and energy in England and Wales.

The Building Regulations describe the mandatory requirements for completing **all** building work, including:

- accommodation for specific purposes (e.g. for disabled persons);
- air pressure plants;
- cesspools (and other methods for treating and disposing of foul matter);
- dimensions of rooms and other spaces (inside buildings);
- drainage (including waste disposal units);
- emission of smoke, gases, fumes, grit or dust (or other noxious or offensive substances);
- fire precautions (services, fittings and equipment, means of escape);
- lifts (escalators, hoists, conveyors and moving footways);
- materials and components (suitability, durability and use);
- means of access to and egress from buildings;
- natural lighting and ventilation of buildings;
- open spaces around buildings;
- prevention of infestation;
- provision of power outlets;
- resistance to moisture and decay;
- site preparation;

- solid fuel, oil, gas, electricity installations (including appliances, storage tanks, heat exchangers, ducts, fans and other equipment);
- standards of heating, artificial lighting, mechanical ventilation and air-conditioning;
- structural strength and stability (overloading, impact and explosion, underpinning, safeguarding of adjacent buildings);
- telecommunications services (wiring installations for telephones, radio and television);
- third-party liability (danger and obstruction to persons working or passing by building work);
- transmission of heat;
- transmission of sound;
- waste (storage, treatment and removal);
- water services, fittings and fixed equipment (including wells and bore-holes for supplying water); and
- matters connected with (or ancillary to) any of the foregoing matters.

They are legal requirements laid down by Parliament and are based on the Building Act 1984. The Building Regulations:

- are approved by Parliament;
- are designed to ensure structural stability;
- contribute to meeting the needs of disabled people;
- deal with the minimum standards of design and building work for the construction of domestic, commercial and industrial buildings;
- ensure the health and safety of people in and around buildings (by providing functional requirements for building design and construction);
- promote energy efficiency in buildings;
- promote the use of suitable materials to provide adequate durability, fire and weather resistance, and the prevention of damp;
- set out the procedure for ensuring that building work meets the standards laid down;
- stipulate the minimum amount of ventilation and natural light to be provided for habitable rooms.

The level of safety and standards acceptable are set out as guidance in **Approved Documents**, which are discussed below. Compliance with the detailed guidance of the Approved Documents is usually considered as evidence that the Building Regulations themselves have been complied with.

Approved Documents

The Building Regulations are supported by separate documents which correspond to the different areas covered by the regulations. These are called *Approved Documents*, and they contain practical and technical guidance on ways in which the requirements of Schedule 1 and Regulation 7 of the Building Act 1984 can be met.

Each Approved Document reproduces the actual *requirements* contained in the Building Regulations relevant to the subject area. These are then followed by *practical and technical guidance* (together with examples) showing how the requirements can be met in some of the more common building situations. There may, however, be alternative ways of complying with the requirements to those shown in the Approved Documents, and you are, therefore, under no obligation to adopt any particular solution contained in an Approved Document if you prefer to meet the requirement(s) in some other way.

If you intend to carry out building work you should always check with the building control body, either the local authority or an approved inspector, that your proposals comply with Building Regulations.

The current set of Approved Documents are in 13 parts, A to P (less I, N and O), and consist of:

A   Structure
B   Fire safety
C   Site preparation and resistance to contaminants and moisture
D   Toxic substances
E   Resistance to the passage of sound
F   Ventilation
G   Sanitation, hot water safety and water efficiency
H   Drainage and waste disposal
J   Combustion appliances and fuel storage systems
K   Protection from falling, collision and impact
L   Conservation of fuel and power
M   Access to and use of buildings
P   Electrical safety – dwellings

The Planning Portal actually shows these as different titles.

In accordance with Regulation 8 of the Building Regulations, the requirements in Parts A to D and F to K (except for paragraphs G2 and J6) of Schedule 1 to the Building Regulations do not require anything to be done except for the purpose of securing reasonable standards of health and safety for persons in or about buildings (and any others who may be affected by buildings or matters connected with buildings).

Notes:

(1) Paragraphs H2 and J6 are excluded from Regulation 8 because they deal directly with the prevention of contamination of water.
(2) Parts E and M (which deal, respectively, with resistance to the passage of sound, and access to and use of buildings) are excluded from Regulation 8 because they address the welfare and convenience of building users.

(3) Part L is excluded from Regulation 8 because it addresses the conservation of fuel and power.

Planning permission

Planning permission is the single biggest hurdle for anyone who has acquired land on which to build a house, or wants to extend or carry out other building work on a property. There is never a guarantee that permission will be given, and without that permission no project can start. The planning system itself is not at all user-friendly. There is a bewildering array of formalities to go through and ever more stringent requirements to satisfy. Planning permission has never been more difficult to get, nor so sought after. Every year over half a million applications are made, and the number is rising. The majority of applications are now made using the Planning Portal's online system.

The purpose of the planning system is to protect the environment as well as public amenities and facilities. It is **not** designed to protect the interests of one person over another. Within the framework of legislation approved by Parliament, councils are tasked to ensure that development is allowed where it is needed, while ensuring that the character and amenity of the area are not adversely affected by new buildings or changes in the use of existing buildings and/or land.

Provided that the work you are completing does not affect the external appearance of the building, you are allowed to make certain changes to your home without having to apply to the local council for permission. These are called *permitted development rights*. These rights were relaxed during the summer of 2013, and the details are contained in Chapter 4. However, you will still find that 'building work' will probably require you to have planning permission – so be warned!

The actual details of planning requirements are complex but, for most domestic developments, the planning authority is only really concerned with construction work such as an extension to a house (e.g. a conservatory) or the provision of a new garage or new outbuildings. Structures such as walls, fences and decking also need to be considered, because their height or siting might well infringe the rights of neighbours and other members of the community. The planning authority will also want to approve any change of use, such as converting a house into flats or running a business from premises previously occupied as a dwelling only.

The voluntary Code for Sustainable Homes was launched as part of a package of measures towards zero-carbon development in 2006. Full technical guidance on how to comply with the code was published in April 2007. Further details of the code can be found in Chapter 4.

Main changes in this new edition of *Building Regulations in Brief*

For this new edition we have incorporated the amendments to Building Regulations 2010 up to October 2013. We include details of initiatives introduced to speed up the planning process and the changes to the planning regulations as a result. As the previous technical guidance relating to planning permission has been substantially reduced and much more guidance is available on the Planning Portal, we have included all the information we think will be useful to our readers. A number of changes to permitted development rights to agricultural, business and residential buildings which took effect on 30 May 2013 and came into force on 1 October 2013 have also been included. We also provide information on the Community Right to Build order, which allows certain community organizations to put forward smaller-scale developments on a specific site, without the need for planning permission.

The main changes to the Approved Documents (and dates they came into effect) are as set out below.

Regulation 7 – July 2013

There is no change to the actual regulation, but the guidance has changed in accordance with EU regulations coming into force in July, and the regulation is in a new format. In essence changes are:

*   construction products now have to be CE marked;
*   examples of materials susceptible to changes in their properties have been removed;
*   guidance on resistance to moisture and substances in the subsoil have been moved from this document, and now there is merely a cross-reference to Approved Document C;
*   the section containing reference to the environmental impact of building work has been deleted.

These changes have caused consequential amendments to all of the Approved Documents. Therefore the section on materials and workmanship, independent certification schemes and technical specifications in **every** Approved Document has changed.

Approved Document B – April 2013

The document has the following changes:

*   Appendix F contains new standard numbers;
*   changes have been made to the section on independent certification schemes;
*   the Construction Products Directive section has been removed;
*   fire door standards have been updated;
*   fire doors and European standards have changed;
*   fire resistance standards have been updated;
*   para B1(a) is changed;
*   the standard for smoke alarms has been updated (BSEN 14604:2005).

Approved Document H – April and July 2013

In the main, the changes to Document H are new standards, and regulation numbers have changed. Other changes include:

- a new table at Appendix C;
- changes to the adoption of sewers;
- minor amendments to H5 – limits on application;
- new standards numbers;
- references to more 'other publications' than in the previous document.

Approved Document K – April 2013

This document has been completely rewritten in a new format, and it has been expanded to take in Approved Document N. The language is simplified, as are the diagrams, but there are no new technical requirements. Changes are as follows:

- changes to the maximum gradient for landings at the bottom of flights;
- a new format;
- introduction of the new terms of 'utility stair' and 'general access stair';
- more information on the appearance and construction of ramps;
- more information on landings for buildings other than dwellings;
- new examples of tread profiles;
- new key dimensions diagram for stairs in buildings other than dwellings;
- new requirements for handrails in buildings other than dwellings;
- new scope for ramps;
- a new section on handrails for dwellings and a diagram (omitted in the last Approved Document);
- a new width for ramps in buildings other than dwellings;
- changes to some figures and section titles;
- the introduction of maximum headroom for alternating tread stairs, and a cross-reference to comply with the new construction section;
- changes to the minimum and maximum rise and going, which are now in tabular form;
- diagrams which show the requirements for buildings other than dwellings;
- new figures;
- a new section on the length of stairs for buildings other than dwellings;
- a new section on stepped gangways in assembly buildings (football stadiums);
- more information on external steps and school building going;
- a new section on the width of flights of stairs in buildings other than dwellings, as well as in dwellings, and a change to the size of divisions of flights of stairs;
- a requirement for ramps to have handrails.

Approved Document L – October 2013

As well as the new materials and workmanship changes there are new regulation numbers and changes to the requirements in all four parts of Approved

Document L. Some of the wording has changed in the standard, but the meaning is the same. A summary of changes is as follows:

- new regulation numbers and text;
- a new definition of the energy performance certificate;
- new information on fixed building services;
- all new dwellings now need to achieve a better fabric efficiency target as well as to deliver a 6% CO_2 saving;
- all new non-domestic buildings now need to achieve a better fabric efficiency target as well as to deliver a 9% CO_2 saving;
- high-efficiency alternative systems for new buildings;
- new materials and workmanship text;
- new paragraphs 4.17 B, C and D;
- new definitions of high-efficiency alternative systems, major renovations and recommendation reports;
- new information on replacement of thermal elements;
- a new definition of major renovations.

Approved Document M – changed in January 2013

The Equality Act 2010 replaces the Disability Discrimination Act 1995 and the Equality Act 2006, and other changes include:

- all guidance on stairs and ramps (other than those that are part of entrances) has moved to Document K;
- all guidance on guarding and handrails has moved to Document K;
- all guidance on glass doors and screens has been updated and moved to Document K;
- guidance on access statements has been updated to a new title, 'Access strategies', and there is additional narrative;
- guidance on the forces required to open doors has been updated;
- there are new technical references for WC pans and suites;
- there is a new section on changing places and toilets;
- there are new definitions for contrast visibility, general access stair, light reflectance and utility stair.

Approved Document N – repealed in 2013

Approved Document P – changed in April 2013

The range of electrical installation work that is notifiable has been reduced. An installer can get a registered third party to certify notifiable work. The document now refers to BS7671:2008. Much of the supporting documentation has been removed.

Foreword

In my role, quick reference to the Approved Documents is sometimes essential on-site, and with so much useful information contained in one clear volume, I find this book invaluable.

This simple approach to the regulations, without delving too deeply into detail, gives a perfect guide for the DIY person, student, site operative and architect, as well as other professionals across the board; the key being that it helps you to locate the vital information. It is also extremely cost-effective compared with the full documents, which are not easy to read or as accessible due to the multiple volumes.

A wide range of angles is covered, including Statutory Appeals and what you can do if you do not agree with the process. There is a guide to what requirements you need for planning, which could well save the reader a great deal of time and cost. The book clearly identifies a range of common issues, and this can be reassuring at times for the individual reading it, where comparisons can be made when making final decisions. For example, it advises that there may be variations in the planning requirements and to some extent the Building Regulations from one area of the country to another, and that this can sometimes be misunderstood.

Even the more experienced personnel refer back to the Approved Documents, but in this edition the references are contained in one handy volume, enabling you to easily keep up-to-date with the latest regulations as well.

This book helps you to understand what to expect from a legal point of view, the hurdles you might encounter along the way, and the interface of the governing bodies. With this complexity in mind, this book is essential for understanding the Building Regulations and the requirements of the Building Act.

Appendix B Conservation of Fuel and Power typifies the writers' simplicity in explaining the additions to Part L, whereas the actual Approved Documents take some reading to understand the sections completely.

In each of the sections, the book gives you guidelines for meeting the requirements for Building Regulations and the Act, assisting everyone from the DIY enthusiast to the professional, and providing an excellent platform for the legal requirements and where to begin in achieving compliance.

This is an essential reference book for understanding the Building Regulations, and it is cleverly designed to put the building regulations in brief by giving you all the information needed to gain compliance, but at the same time simplifying the requirements. The book is a very cost-effective option rather than buying the full volumes and, in my opinion, is excellent reading.

A must for all people involved in Building Regulations and construction.

Mark Heggs MBEng, MIHEEM, MICWCI
University of Leicester

About the authors

Ray Tricker (MSc, IEng, FIET, FCMI, FCQI, FIRSE) is a senior consultant with over 50 years' continuous service in quality, safety and environmental management, project management, communication electronics, railway command, control and signalling systems, information technology and the development of molecular nanotechnology.

He served with the Royal Corps of Signals (for a total of 37 years) during which time he held various managerial posts culminating in being appointed as the Chief Engineer of NATO's Communication Security Agency (ACE COMSEC).

Most of Ray's work since leaving the services has centred on the European Railways. He has held a number of posts with the Union Internationale des Chemins de fer (UIC) (e.g. Quality Manager of the European Train Control System (ETCS)) and with the European Union (EU) Commission (e.g. T500 Review Team Leader, European Rail Traffic Management System (ERTMS) Users Group Project Coordinator, HEROE Project Coordinator) and currently (as well as writing books on diverse subjects such as ISO 9001:2008 and building, wiring and water regulations for Routledge, Butterworth-Heinemann and Newnes) he is busy assisting small businesses from around the world (usually on a no-cost basis) to produce their own auditable quality and/or integrated management systems to meet the requirements of ISO 9001, ISO 14001 and OHSAS 18001. He is a UKAS assessor (for the assessment of certification bodies for the harmonization of the trans-European high-speed railway network), and recently he was the Quality, Safety and Environmental Manager for the project management consultant overseeing the multibillion-dollar Trinidad Rapid Rail System.

Currently he is working as the Senior Management Consultant for Herne European Consultancy Ltd (www.herne.org.uk), a company specializing in offering organizations access to a range of highly skilled and specialist consultants to help these companies enhance their business performance.

One day he might retire!

Ray is once more joined by **Samantha Alford** (MSc, MCIPS), who is an established technical author, instructor and business management specialist. She has over 25 years' experience in capability planning, governance, oversight and event management and is a director and owner of Professional Procurement & Project Management Ltd (www.pppmanagement. co.uk), a company offering outstanding support and advice on strategic and operational procurement. Sam has a strong supply chain, planning, business and performance management background, which was gained working on projects for the Royal Air Force, Manchester Business School, Pearson Education, and Taylor & Francis. She has assisted a variety of organizations with advisory services for fundraising, business planning and process documentation. She has provided these services in both the private and not-for-profit sectors. She is a school governor and has been the chairman of a combined secondary and primary school parents' association since 2012.

1

The Building Act 1984

The Building Act 1984 applies in England and Wales and is the United Kingdom statute under which the Building Regulations have been made.

1.1 What is the Building Act 1984? (Building Act 1984 Section 1)

The Building Act 1984 is the mechanism by which the Secretary of State ensures that the health, welfare and convenience of persons living in or working in or near buildings is secured.

The primary purpose of the Building Act 1984 is to assist in the conservation of fuel and power, to prevent waste, undue consumption, misuse or contamination of water and to ensure that those who are in or about buildings are kept safe. The Building Act 1984 imposes a set of requirements on owners and occupiers of buildings which cover the design and construction of buildings, and the provision of services, fittings and equipment used in (or in connection with) buildings. These involve and cover:

- a method of controlling (inspecting and reporting) buildings;
- how services, fittings and equipment may be used;
- the inspection and maintenance of any service, fitting or equipment used.

1.1.1 What about the rest of the United Kingdom?

The Building Act 1984 only applies to England and Wales. Separate acts and regulations apply in Scotland or Northern Ireland; these are shown in Table 1.1.

Scotland

In Scotland, the requirements for buildings are controlled by the Building (Scotland) Act 2003 and the Building (Scotland) Regulations 2004 as amended by the Building (Miscellaneous Amendments) (Scotland) Regulations 2013, which came into effect in October 2013. The methods for implementing these requirements are similar to those for England and Wales, except that the guidance documents for achieving compliance are contained in two Technical

Handbooks (2013), one for domestic work and one for non-domestic work, and other associated guidance.

The main procedural difference between the Scottish system and the others is that a building warrant is **still** required before work can start in Scotland.

Table 1.1 Building legislation within the United Kingdom

	Act	Regulations	Implementation
England and Wales	Building Act 1984	Building Regulations 2010	Approved Documents
Scotland	Building (Scotland) Act 2003	Building (Scotland) Regulations 2004 (as amended)	Technical Handbooks
Northern Ireland	Building Act (Northern Ireland) Order 1979	Building Regulations (Northern Ireland) 2000 (as amended)	Technical Booklets

Northern Ireland

On the other hand, in Northern Ireland the primary legislation for buildings is the Building Regulations (Northern Ireland) Order 1979, under which Building Regulations (Northern Ireland) 2000 (as amended) are made. These regulations revoked and replaced with amendments the previous 1994 Building Regulations; they have since been amended by the Building (Amendment) Regulations (Northern Ireland) 2012. The Principal Regulations comprise 15 parts and are supported by Technical Booklets, which are then used to ensure that the requirements are implemented.

Wales

The changes to the Approved Documents (K and N) which came into effect in England in 2013 have not yet been ratified by the Welsh Assembly; therefore Approved Documents K and N **still** apply in Wales.

Table 1.2 provides a summary of the titles of the sections of Building Regulations that apply in the United Kingdom.

1.2 What does the Building Act 1984 contain?

The Building Act 1984 is made up of five parts:

Part 1	The Building Regulations
Part 2	Supervision of building work, etc. other than by a local authority
Part 3	Other provisions about buildings
Part 4	General
Part 5	Supplementary

These parts are then broken down into a number of sections and subsections as shown in Appendix 1A at the end of this chapter.

Table 1.2 Building Regulations for the United Kingdom

England and Wales		Scotland	Northern Ireland
Approved Document A	Structure	Technical Handbook 2013 – Structure	Technical Booklet D: 2012 – Structure
Approved Document B	Fire safety	Technical Handbook 2013 – Fire	Technical Booklet E: 2012 – Fire safety
Approved Document C	Site preparation and resistance to contaminants and water	Technical Handbook 2013 – Environment	Technical Booklet C: 2012 – Site preparation and resistance to contaminants and moisture
Approved Document D	Toxic substances	Technical Handbook 2013 – Environment	Technical Booklet B: 2012 – Materials and workmanship
Approved Document E	Resistance to the passage of sound	Technical Handbook 2013 – Noise	Technical Booklet G: 2012 – Resistance to the passage of sound
Approved Document F	Ventilation	Technical Handbook 2013 – Environment	Technical Booklet K: 2012 – Ventilation
Approved Document G	Sanitation, hot water safety and water efficiency	Technical Handbook 2013 Section 4 – Environment	Technical Booklet P: 2012 – Sanitary appliances and unvented hot water storage systems and reducing the risk of scalding
Approved Document H	Drainage and waste disposal	Technical Handbook 2013 – Environment	Technical Booklet N: 2012 – Drainage
Approved Document J	Combustion appliances and fuel storage systems	Technical Handbook 2013 – Safety	Technical Booklet L: 2012 – Combustion appliances and fuel storage systems
Approved Document K	Protection from falling, collision and impact	Technical Handbook 2013 – Safety Technical Handbook 2013 – Energy	Technical Booklet H: 2012 – Stairs, ramps, guarding and protection from impact Technical Booklet V: 2012 – Glazing
Approved Document L	Conservation of fuel and power	Technical Handbook 2013 – Energy	Technical Booklet F: 2012 – Conservation of fuel and power
Approved Document M	Access to and use of buildings	Technical Handbook 2013 – Safety	Technical Booklet R: 2012 – Access to and use of buildings
Approved Document P	Electrical safety	Technical Handbook 2013 – Safety	

1.3 What are the Supplementary Regulations?

The Supplementary Regulations make up Part 5 of the Building Act. They comprise seven schedules, whose function is to list the principal areas requiring regulation and to show how the Building Regulations are to be controlled by local authorities. These schedules are:

Schedule 1	Building Regulations
Schedule 2	Relaxation of Building Regulations for existing work
Schedule 3	Inner London
Schedule 4	Provisions consequential upon public body's notice
Schedule 5	Transitional provisions
Schedule 6	Consequential amendments
Schedule 7	Repeals

The details of what is contained within each of these schedules are shown in alphabetical order in Appendix 1B at the end of this chapter. We also discuss each area on the following pages.

1.4 What are 'Approved Documents'?
(Building Act 1984 Section 6)

The Secretary of State makes available a series of documents (called '**Approved Documents**') which are intended to provide practical guidance with respect to the requirements of the Building Regulations. More details about Approved Documents are contained in Chapter 3 and on the Planning Portal (http://www.planningportal.gov.uk).

1.5 How are buildings classified? (Building Act 1984 Section 34)

For the purpose of the Building Act, the six normal classifications for buildings are:

* by reference to size;
* by description;
* by design;
* by purpose;
* by location; or
* by 'any other characteristic whatsoever'!

1.6 Who polices the Building Act?

Under the terms of the Building Act 1984, local authorities are responsible for ensuring that any building work that takes place in their area conforms to the requirements of the associated Building Regulations. They have the authority to:

- make you take down and remove or rebuild anything that contravenes a regulation;
- make you complete alterations so that your work complies with the Building Regulations;
- employ a third party (and then send you the bill!) to take down and rebuild non-conforming buildings or parts of buildings.

You can be prosecuted or ordered to carry out remedial work on a property whether you are the owner or merely the occupier.

Local Authorities can, in certain circumstances, even take you to court and have you fined – especially if you fail to complete the removal or rebuilding of the non-conforming work.

1.7 How is my building work evaluated for conformance with the Building Regulations?
(Building Act 1984 Section 33)

Part of the local authority's duty is to make regular checks that all building work being completed is in conformance with the approved plan and the Building Regulations. These checks would normally be completed at certain stages of the work (e.g. the excavation of foundations), and tests will include:

- tests of the soil or subsoil of the site of the building;
- tests of any material, component or combination of components that has been, is being or is proposed to be used in the construction of a building;
- tests of any service, fitting or equipment that has been, is being or is proposed to be provided in or in connection with a building.

The cost of carrying out these tests will normally be charged to the owner or occupier of the building.

The local authority has the power to ask the person responsible for the building work to complete some of these tests on its behalf.

1.8 What are the duties of the local authority?
(Building Act 1984 Section 91)

It is the duty of local authorities to ensure that the requirements of the Building Act 1984 are carried out and that the appropriate associated Building Regulations are enforced, subject to:

- the provisions of Part I of the Public Health Act 1936 (relating to united districts and joint boards);
- Section 151 of the Local Government, Planning and Land Act 1980 (relating to urban development areas);
- Section 1(3) of the Public Health (Control of Disease) Act 1984 (relating to port health authorities).

Limitations are placed on a Local Authority's ability to enforce Building Regulations for work on buildings which are owned by exempt bodies, work that is subject to an initial notice or work relating to a plans certificate which was issued or accepted before an initial notice ceases to be in force.

1.8.1 What document controls must local authorities have in place? (Building Act 1984 Sections 92 and 93)

The Secretary of State mandates a written format for all notices, applications, orders, consents, demands and other documents that are required by this act or by a local authority (or an officer of a local authority).

All the documents that a local authority provides under the Building Act 1984 must be correctly authenticated and signed by:

- the proper officer for the authority;
- the district surveyor (for documents relating to matters within his or her specialization);
- an officer authorized by the authority to sign documents (of a particular kind).

 If a document bears the signature of an officer of the local authority it is deemed to have been issued by that authority.

1.8.2 How do local authorities 'serve' notices and documents? (Building Act 1984 Section 94)

Notices, orders, consents, demands or other documents which are authorized or required by the Building Act 1984 may be given or served to a person in the following ways:

- by delivering it to the person concerned or by sending it to the person concerned at his or her last known residence;

- by leaving it, or sending it in a pre-paid letter addressed to an officer of a local authority, at his or her office;
- by delivering or sending it to the secretary or clerk of an incorporated company at its registered or principal office;
- by leaving it, or sending it addressed to a person who receives the rack rent for any premises, at his or her place of business;
- where it is not possible to ascertain the name and address of the person on whom the documentation is to be served it may be addressed to the 'owner' or 'occupier' of the premises and delivered to a person on the premises or affixed to a conspicuous part of the premises.

Notices for an officer of the local authority are considered to be 'served' if they are left or sent in a prepaid letter addressed to the officer at his or her office.

If it is not possible to ascertain the name and address of the person to or on whom it should be given or served (or if the premises are unoccupied) then the notice, order, consent, demand or other document can be addressed to the 'owner' or 'occupier' of the premises (naming them) and delivered to 'some person on the premises' or, if there is not anyone at the premises to whom it can be delivered, a copy of the document can be fixed to a conspicuous part of the premises.

1.9 What are the powers of the local authority? (Building Act 1984 Sections 97–101)

The local authority has the following powers under the Building Act 1984 and its associated Building Regulations:

- overall responsibility for the construction and maintenance of sewers and drains and the laying and maintenance of water mains and pipes;
- the ability to sell any materials that have been removed, by it, from any premises when executing works under this act (paying all proceeds, less expenses, from this sale to the owner or occupier);
- the authority to complete remedial and essential work itself (on repayment of expenses) if the owner or occupier refuses to do this work him- or herself;
- the authority to make the owner or occupier of any premises complete essential and remedial work in connection with the Building Act 1984 (particularly with respect to the construction, laying, alteration or repair of a sewer or drain).

This does not apply to any refuse that is, or has been, removed by the local authority.

1.9.1 Has the local authority any power to enter premises?
(Building Act 1984 Section 95)

An authorized officer of a local authority has a right to enter any premises, at all 'reasonable hours', for the following reasons:

* to carry out the local authority's function as a local authority;
* to carry out work or take any action that is authorized or required by the Building Act, or by Building Regulations;
* to check if there is (or has been) a contravention of the Building Act or of Building Regulations that the local authority has a duty to enforce;
* to see if there are any circumstances that would require local authority action or the local authority to carry out any work.

 If the premises are a factory or workplace the local authority must give 24 hours' notice of its intention to enter the property.

If the local authority is refused admission to any premises (or the premises are unoccupied) then the local authority can apply to a Justice of the Peace for a warrant authorizing entry.

1.10 Who are approved inspectors? (Building Act 1984 Section 49)

An approved inspector is a person who is approved by the Secretary of State (or a body such as a local authority or county council designated by the Secretary of State) to inspect, supervise and authorize building work. Lists of approved inspectors are available from all local authorities. The Construction Industry Council maintains and operates the Register of Approved Inspectors (http://approvedinspectors.org.uk/members-directory/).

 If an approved inspector gives a notice or certificate that falsely claims to comply with the Building Regulations and/or the Building Act then he or she is liable to prosecution.

1.10.1 What happens when an initial notice is presented by an approved inspector? (Building Act 1984 Section 47)

An approved inspector and the person intending to carry out the work will have to present an initial notice and plan of work to the local authority. Once accepted, the approved inspector is authorized to inspect and supervise all work being completed and to provide certificates and notices. Acceptance of an initial notice by a local authority is treated as 'depositing plans of work'.

Under Section 47 of the Building Act, the local authority is required to accept all certificates and notices, unless the initial notice and plans contravene a local ruling. Whilst the initial notice continues to be in force the local authority is not allowed to give a notice in relation to any of the work being carried out or take any action for a contravention of Building Regulations.

If the local authority rejects the initial notice for any reason, then the approved inspector can appeal to a magistrates' court for a ruling. If still dissatisfied, the approved inspector can appeal to the Crown Court.

Cancellation of the initial notice (Building Act 1984 Sections 52 and 53)

If an approved inspector is unable to carry out or complete his or her functions, or is of the opinion that there is a contravention of the Building Regulations, then he or she can cancel the initial notice lodged with the local authority.

Equally, if the person carrying out the work has good reason to consider that the approved inspector is unable (or unwilling) to carry out his or her functions, then that person can cancel the initial notice given to the local authority. In this instance if the person fails to cancel the initial notice with the local authority he or she is liable on summary conviction to a fine.

Any cancellation of an initial notice is effective from the date on which the notice is given. The fact that the initial notice has ceased to be in force does not affect the right of an approved inspector to give a new initial notice relating to any of the work that was previously specified in the original notice.

 If it appears to the local authority that the work to which an initial notice relates has not commenced within three years of the date on which it accepted the initial notice it may cancel the notice.

1.10.2 What are plans certificates? (Building Act 1984 Section 50)

When an approved inspector has inspected and is satisfied that the plans of work specified in the initial notice do not contravene the Building Regulations in any way, he or she will provide a certificate (referred to as a 'plans certificate') to the local authority. This plans certificate:

- can relate to the whole or part of the work specified in the initial notice;
- does not have any effect unless the local authority accepts it;
- may only be rejected by the local authority 'on prescribed grounds'.

1.10.3 What are final certificates? (Building Act 1984 Section 51)

Once the approved inspector is satisfied that all work has been completed in accordance with the work specified in the initial notice, he or she will provide a certificate (referred to as a 'final certificate') to the local authority and the person who carried out the work. This certificate will detail the approved inspector's acceptance of the work and, once it is acknowledged by the local authority, the approved inspector's job will have been completed and (from the point of view of the local authority) the approved inspector will have been considered 'to have discharged his duties'.

 If work has not commenced within three years the local authority can cancel the initial notice.

1.10.4 Who retains all these records? (Building Act 1984 Section 56)

Local authorities are required to keep a register of all initial notices, amendment notices, public body's notices and certificates (including plans certificates, final certificates, and public body's final certificates) which have been given to them. The information retained includes information relating to the insurance cover provided with respect to the work to be carried out. The local authority is required to make this register available for public inspection during normal working hours.

1.10.5 Can public bodies supervise their own work? (Building Act 1984 Section 54)

If a public body (e.g. local authority or county council) is of the opinion that building work that is to be completed on one of its own buildings can be adequately supervised by one of its employees and/or agents, then it can provide the local authority with a notice (referred to as a 'public body's notice') together with its plan of work.

Once it is accepted by the local authority, the public body is authorized to inspect and supervise all work being completed and to provide certificates and notices. Acceptance by a local authority of a public body's notice is treated as 'depositing plans of work'.

If the local authority rejects the public body's notice for any reason, then the public body can appeal to a magistrates' court for a ruling. If still dissatisfied, it can appeal to the Crown Court.

1.11 Can I appeal against a local authority's ruling? (Building Act 1984 Sections 36, 39, 40, 41, 42, 55, 81, 83, 86, 102 and 103)

You may appeal against a local authority's ruling in the following circumstances:

- if the local authority fails to notify you within two months of the application (or an extended period as agreed by both parties) of its decision to reject an application to dispense with or relax a requirement in the Building Regulations;

- if the local authority rejects an application to dispense with or relax a requirement in the Building Regulations;
- if you disagree with a local authority notice requiring works or demolition;
- if you disagree with a notice to remove or alter offending work which has been issued by the local authority after the work has been completed;
- if you are aggrieved by the local authority's rejection of any of the following:

  - an initial notice;
  - an amendment notice;
  - a public body's notice;
  - a plans certificate;
  - a final certificate;
  - a public body's certificate;
  - a public body's final certificate.

You may appeal against the local authority's decision to the magistrates' court in the area where the work is to be carried out. If your appeal is unsuccessful at the magistrates' court you may appeal to the Crown Court. Where the Secretary of State has given a ruling, this ruling may, in certain circumstances, be appealed to the High Court on a point of law.

There are many conditions which relate to the appeals procedures. These include strict timescales which must be adhered to and grounds on which you may or may not appeal. You are recommended to study the relevant sections of the act listed above for further guidance and to seek professional advice before embarking on what could be a costly legal case.

1.11.1 What about compensation? (Building Act 1984 Sections 106, 107 and 108)

If an owner or occupier considers that a ruling obtained from the local authority is incorrect, he or she can appeal (in the first case) to the local magistrates' court. If, on appeal, the magistrates rule against the local authority, then the owner or occupier of the building concerned is entitled to compensation from the local authority (any compensation over £50 is subject to arbitration). If, on the other hand, the magistrates rule in favour of the local authority, then the local authority is entitled to recover any expenses and interest that it has incurred. In this case the monies can be paid in instalments over a period up to 30 years.

Be sure of your facts before you ask a magistrates' court for a ruling!

1.12 Are there any exemptions from Building Regulations? (Building Act 1984 Sections 3, 4 and 5)

The following are exempt from the Building Regulations:

- a 'public body' (i.e. local authorities, county councils and any other body 'that acts under an enactment for public purposes and not for its own profit') – this can be rather a grey area and it is best to seek advice if you think that you come under this category;
- buildings or classes of buildings at a particular location that have been declared exempt by the Secretary of State;
- educational buildings and buildings belonging to 'statutory undertakers', the UK Atomic Energy Authority or the Civil Aviation Authority;
- prescribed classes of buildings, services, fittings or equipment.

1.12.1 What about Crown buildings? (Building Act 1984 Sections 44 and 87)

Although the majority of the requirements of the Building Regulations are applicable to Crown buildings (i.e. a building in which there is a Crown or Duchy of Lancaster or Duchy of Cornwall interest) or government buildings (held in trust for Her Majesty) there are occasional deviations, and therefore you should seek the advice of the Treasury before submitting plans for work on a Crown building.

1.12.2 What about buildings in inner London? (Building Act 1984 Sections 44, 46 and 88)

You will find that the majority of the requirements found in the Building Regulations are also applicable to buildings in inner London boroughs. However, certain exemptions from the regulations are made for buildings in Inner Temple and Middle Temple. There are, however, some important deviations for buildings in inner London, which are contained in Schedule 3 of Part 5 and reproduced in Appendix 1B at the end of this chapter. You should seek the advice of the local authority concerned before submitting plans.

1.12.3 What about the United Kingdom Atomic Energy Authority? (Building Act 1984 Section 45)

The Building Regulations do **not** apply to buildings belonging to or occupied by the United Kingdom Atomic Energy Authority (UKAEA), unless they are dwelling houses or offices.

1.13 Can I apply for a relaxation in certain circumstances? (Building Act 1984 Sections 8–11 and 39)

The Building Act allows the local authority to dispense with, or relax, a Building Regulation if it believes that that requirement is unreasonable in relation to a particular type of work being carried out.

The local authority may charge a fee for reviewing and deciding on these matters.

In the majority of cases, applications to dispense with or relax Building Regulations can be settled locally. In more complicated cases the local authority can seek guidance from the Secretary of State, who will give a direction as to whether the requirement may be relaxed or dispensed with (either unconditionally or subject to certain conditions).

If a question arises between the local authority and the person who has executed (or proposes to execute) any work regarding:

* the application of Building Regulations;
* whether the plans are in conformity with the Building Regulations;
* whether the work has been executed in conformance with these plans

then the question can be referred to the Secretary of State for determination. In these cases, the Secretary of State's decision will be deemed final. The proposal to relax Building Regulations must be published in a local newspaper not less than 21 days before work commences.

Note: Schedule 2 of the Building Act 1984 provides guidance and rules for the application of Building Regulations to work that has been carried out prior to the local authority (under the Building Act 1984 Section 36) dispensing with, or relaxing, some of the requirements contained in the Building Regulations. This schedule is quite difficult to understand and, if it affects you, you are strongly advised to discuss it with the local authority before proceeding any further.

1.14 What is 'type approval'? (Building Act 1984 Sections 12 and 13)

Type approval is where the Secretary of State is empowered to approve a particular type of building matter as complying, either generally or specifically, with a particular requirement of the Building Regulations. This power of approval is normally delegated by the Secretary of State to the local council or other nominated public body.

1.15 What causes some plans for building work to be rejected? (Building Act 1984 Sections 16 and 17)

If a plan for proposed building work is accompanied by a certificate from a person or persons approved by the Secretary of State then only in extreme circumstances can the local authority reject the plans.

The local authority will reject all plans for building work that are defective or that contravene any of the Building Regulations.

In all cases the local authority will advise the person putting forward the plans why they have been rejected (giving details of the relevant regulation or section) and, where possible, indicate what amendments and/or modifications will have to be made in order to get them approved. This will be provided within two months of the plans being submitted. The person who initially put forward the plans is then responsible for making amendments or alterations and resubmitting them for approval.

1.16 Must I complete the approved work in a certain time? (Building Act 1984 Section 32)

Once a building plan has been passed by the local authority, then '**work must commence**' within three years from the date that it was approved. Failure to do so could result in the local authority cancelling the approved plans, and you will have to resubmit them if you want to carry on with your project. If the local authority advises you that the deposit of plans is of no effect it is as if the plans had never been deposited.

Note: The meaning of the term 'work must commence' can vary, but normally it means physically laying foundations of the building (in some cases it could mean more work has to be completed). It is always best to check with the local authority for clarification when your plans are first approved.

1.17 What happens if I contravene any of these requirements? (Building Act 1984 Sections 2, 7, 35, 36 and 112)

If you contravene the Building Regulations or wilfully obstruct a person acting 'in the execution of the Building Act 1984 or of its associated Building Regulations', then on summary conviction you could be liable to a fine including the recovery of costs or, in exceptional circumstances, even a short holiday in one of HM Prisons!

1.18 What about civil liability? (Building Act 1984 Section 38)

It is an aim of the Building Act 1984 that all building work is completed safely and without risk to people employed on the site or visiting the site. Any contravention of the Building Regulations that causes injury (or death) to any person is liable to prosecution in the normal way.

1.19 What is the Building Regulations Advisory Committee? (Building Act 1984 Section 14)

The Building Act allows the Secretary of State to appoint a committee (known as the Building Regulations Advisory Committee) to review, amend, improve and produce new Building Regulations and associated documentation (e.g. such as Approved Documents – see earlier).

1.20 Does the fire authority have any say in Building Regulations? (Building Act 1984 Section 15)

When a requirement 'encroaches' on something that is normally handled by the fire authority, such as the provision of means of escape and structural fire precautions, the local authority **must** consult the fire authority before making any decision.

1.21 Can I change a plan of work once it has been approved? (Building Act 1984 Section 31)

It is possible for the person intending to carry out building work, who has had his or her plan (or plans) passed by the local authority, to change it. In this case the person will have to submit a set of revised plans to the local authority, showing precisely how he or she plans to deviate from the approved plan and ask for the local authority's approval. If the deviation or change is a small one this can usually be achieved by talking to the local planning officer, but if it is a major change it could result in the resubmission of a complete plan of the revised building work.

1.22 What about dangerous buildings?
(Building Act 1984 Sections 77 and 78)

The local authority can make an order restricting the use of a dangerous building until such time as a magistrates' court is satisfied that all necessary works have been completed.

If a building, or part of a building or structure, is in such a dangerous condition (or is used to carry loads which would make it dangerous) then the local authority may apply to a magistrates' court to make an order requiring the owner:

*   to carry out work to avert the danger;
*   to demolish the building or structure, or any dangerous part of it, and remove any rubbish resulting from the demolition.

Before the authority exercises its powers the owner must be informed.

1.22.1 Emergency measures

In emergencies, the local authority can make the owner take immediate action to remove the danger or it can complete the necessary action itself. In these cases, the local authority is entitled to recover from the owner such expenses as are reasonably incurred by it, for example:

*   arranging for the building or structure to be monitored;
*   fencing off the building or structure.

1.22.2 Can I demolish a dangerous building?
(Building Act 1984 Section 80)

Be careful; penalties can be very severe for demolishing something illegally!

You must have good reasons for knocking down a building, such as making way for rebuilding or improvement (which in most cases would be incorporated in the same planning application). You are not allowed to begin any demolition work (even on a dangerous building) unless you have given the local authority notice of your intention **and** either this has been acknowledged by the local authority or the relevant notification period has expired. In this notice you will have to:

*   show how you intend to demolish the building;
*   specify the building to be demolished;
*   state the reason(s) for wanting to demolish it.

    Copies of this notice will have to be sent to:

*   any public electricity supplier in whose area the building is;
*   any public gas supplier in whose area the building is;

- the local authority;
- the occupier of any building adjacent to the building in question.

Note: This regulation does not apply to the demolition of an internal part of an occupied building, or a greenhouse, conservatory, shed or prefabricated garage (that forms part of that building) or an agricultural building.

1.22.3 Can I be made to demolish a dangerous building?
(Building Act 1984 Sections 81, 82 and 83)

If the local authority considers that a building is so dangerous that it should be demolished, it is also entitled to issue a notice to the owner requiring him or her:

- in accordance with the Housing Act 1985 to arrange with the relevant statutory undertakers for the disconnection of gas, electricity and water supplies to the building;
- to disconnect, seal and remove any sewer or drain in or under the building;
- to leave the site in a satisfactory condition following completion of all demolition work;
- to make good the surface of the ground that has been disturbed in connection with this removal of drains;
- to remove material or rubbish resulting from the demolition and clear the site;
- to repair and make good any damage to an adjacent building caused by the demolition or by the negligent act or omission of any person engaged in it;
- to shore up any building adjacent to the building to which the notice relates;
- to weatherproof any surfaces of an adjacent building that are exposed by the demolition.

Before complying with this notice, the owner must give the local authority 48 hours' notice of commencement.

Note: In certain circumstances, the owner of an adjacent building may be liable to assist in the cost of shoring up his or her part of the building and waterproofing the surfaces. It could be worth checking this point with the local authority!

1.23 What about defective buildings? (Building Act 1984 Sections 76, 79 and 80)

If a building or structure is, because of its ruinous or dilapidated condition, liable to cause damage to (or be a nuisance to) the amenities of the neighbourhood, then the local authority can require the owner:

- to carry out necessary repairs and/or restoration; or
- to demolish the building or structure (or any part of it) and to remove all the rubbish or other material resulting from this demolition.

If, however, the building or structure is in a defective state and remedial action (envisaged under Section 93 of the Environmental Protection Act 1990) would cause an unreasonable delay, then the local authority can serve an abatement notice stating that within nine days **it** intends to complete such works as it deems necessary to remedy the defective state and recover the 'expenses reasonably incurred in so doing' from the person to whom the notice was sent.

If appropriate, the owner can (within seven days), after the local authority's notice has been served, serve a counter-notice stating that he or she intends to remedy the defects specified in the first-mentioned notice him- or herself.

 Note: A local authority is not entitled to serve a notice, or commence any work in accordance with a notice that it has served, if the execution of the works would (to its knowledge) be in contravention of a building preservation order that has been made under Section 29 of the Town and Country Planning Act.

1.24 What are the rights of the owner or occupier of the premises? (Building Act 1984 Sections 102–107)

When a person has been given a notice by a local authority to complete work, he or she has the right to appeal to a magistrates' court on any of the following grounds:

- that the notice or requirement is not justified by the terms of the provision under which it purports to have been given;
- that there has been some informality, defect or error in (or in connection with) the notice;
- that the authority has refused (unreasonably) to approve completion of alternative works, or that the works required by the notice to be executed are unreasonable or unnecessary;
- that the time limit set to complete the work is insufficient;
- that the notice should lawfully have been served on the occupier of the premises in question instead of on the owner (or vice versa);
- that some other person (who is likely to benefit from completion of the work) should share in the expense of the works.

Appendix 1A Contents of the Building Act 1984

Subsection title	Part	Subsection	Section title
Advertisement of proposal for relaxation of Building Regulations	1	10	Relaxation of Building Regulations
Appeal against notice requiring works	4	102	Execution of works
Appeal against notice under Section 81	3	83	Defective premises, demolition, etc.
Appeal against refusal, etc. to relax Building Regulations	1	39	Appeals in certain cases
Appeal against Section 36 notice	1	40	Appeals in certain cases
Appeal and statement of case to High Court in certain cases	1	42	Appeals in certain cases
Appeal to Crown Court	1	41	Appeals in certain cases
Appeal to Crown Court	3	86	Appeal to Crown Court
Appeals	2	55	Supplementary
Application for relaxation	1	9	Relaxation of Building Regulations
Application of provisions to Crown property	3	87	Application of provisions to Crown property
Application to Crown	1	44	Application of Building Regulations to Crown, etc.
Application to United Kingdom Atomic Energy Authority	1	45	Application of Building Regulations to Crown, etc.
Approval of documents for purposes of Building Regulations	1	6	Approved Documents
Approval of persons to give certificates, etc.	1	17	Passing of plans
Approved inspectors	2	49	Supervision of plans and work by approved inspectors
Arbitration	4	111	Compensation and recovery of sums
Authentication of documents	4	93	Documents
Breaking open of streets	4	101	Execution of works
Building over sewer	1	18	Passing of plans
Building Regulations	5	1	Schedule 1
Cancellation of initial notice	2	52	Supervision of plans and work by approved inspectors

Appendix 1A (Continued)

Subsection title	Part	Subsection	Section title
Cellars and rooms below subsoil water level	3	74	Buildings
Change of person intending to carry out work	2	51c	Supervision of plans and work by approved inspectors
Civil liability	1	38	Breach of Building Regulations
Classification of buildings	1	34	Classification of buildings
Commencement	5	134	Supplementary
Compensation for damage	4	106	Compensation and recovery of sums
Compliance or non-compliance with Approved Documents	1	7	Approved Documents
Consents under Section 74	3	75	Buildings
Consequential amendments	5	6	Schedule 6
Consequential amendments and repeals	5	133	Supplementary
Construction and availability of sewers	4	125	Interpretation
Construction of certain references concerning Temples	4	127	Interpretation
Construction of Part 11	2	58	Supplementary
Consultation with Building Regulations Advisory Committee and other bodies	1	14	Consultation
Consultation with fire authority	1	15	Consultation
Content and enforcement of notice requiring works	4	99	Execution of works
Continuing offences	4	114	Prosecutions
Continuing requirements	1	2	Power to make Building Regulations
Dangerous building	3	77	Defective premises, demolition, etc.
Dangerous building – emergency measures	3	78	Defective premises, demolition, etc.
Default powers of Secretary of State	4	116	Default powers
Defective premises	3	76	Defective premises, demolition, etc.
Delegation of power to approve	1	13	Type approval of building matter

Appendix 1A (Continued)

Subsection title	Part	Subsection	Section title
Plans certificates	2	50	Supervision of plans and work by approved inspectors
Power of Secretary of State to approve type of building matter	1	12	Type approval of building matter
Power to enter premises	4	95	Entry on premises
Power to execute work	4	97	Execution of works
Power to make Building Regulations	1	1	Power to make Building Regulations
Power to require occupier to permit work	4	98	Execution of works
Procedure on appeal or application to magistrates' court	4	103	General provisions about appeals and applications
Procedure on appeal to Secretary of State on certain matters	1	43	Appeals in certain cases
Proposed departure from plans	1	31	Proposed departure from plans
Prosecution of offences	4	113	Prosecutions
Protection for dock and railway undertakings	4	128	Savings
Protection of members, etc. of authorities	4	115	Protection of members, etc. of authorities
Provision of closets in building	3	64	Provision of sanitary conveniences
Provision of drainage	1	21	Passing of plans
Provision of exits, etc.	1	24	Passing of plans
Provision of facilities for refuse	1	23	Passing of plans
Provision of food storage accommodation in house	3	70	Buildings
Provision of sanitary conveniences in workplace	3	65	Provision of sanitary conveniences
Provision of water supply	1	25	Passing of plans
Provision of water supply in occupied house	3	69	Buildings
Provisions consequential upon public body's notice	5		Schedule 4
Raising of chimney	3	73	Buildings
Recording and furnishing of information	2	56	Supplementary
Recovery of expenses, etc.	4	107	Compensation and recovery of sums

Appendix 1A (Continued)

Subsection title	Part	Subsection	Section title
References in acts to building by-laws	3	89	Miscellaneous
Relaxation of Building Regulations	1	8	Relaxation of Building Regulations
Relaxation of Building Regulations	5		Schedule 2
Removal or alteration of offending work	1	36	Breach of Building Regulations
Repair, etc. of drain	3	61	Drainage
Repeals	5		Schedule 7
Replacement of earth closets, etc.	3	66	Provision of sanitary conveniences
Restriction of application of Part IV to Schedule 3	4	131	Savings
Ruinous and dilapidated buildings and neglected sites	3	79	Defective premises, demolition, etc.
Sale of materials	4	100	Execution of works
Saving for Local Land Charges Act 1975	4	129	Savings
Saving for other laws	4	130	Savings
Service of documents	4	94	Documents
Short title and extent	5	135	Supplementary
Supplementary provisions as to entry	4	96	Entry on premises
Tests for conformity with Building Regulations	1	33	Tests for conformity with Building Regulations
Transitional provisions	5		Schedule 5
Transitional provisions	5	132	Supplementary
Type relaxation of Building Regulations	1	11	Relaxation of Building Regulations
Use and ventilation of soil pipes	3	60	Drainage
Use of materials unsuitable for permanent building	1	20	Passing of plans
Use of short-lived materials	1	19	Passing of plans
Variation of work to which initial notice relates	2	51a	Supervision of plans and work by approved inspectors
Variation or revocation of order transferring powers	4	118	Default powers

Appendix 1B Contents of the schedules forming Part 5 of the Building Act 1984

What does Schedule 1 of the Building Act 1984 contain?

Building Regulations also apply to alterations and extensions being completed on buildings erected before the date on which the regulations came into force.

Schedule 1 of Part 5 of the Building Act 1984 contains Building Regulations gives details of how the generic requirements of the Building Act are to be met. Compliance with Building Regulations is required for **all**:

*   alterations and extensions of buildings (including services, fixtures and fittings);
*   alterations or extensions including new, altered or extended services, fittings or equipment;
*   building construction;
*   material changes of use;
*   provision of new services, fittings or equipment.

Schedule 1 of the Building Act 1984 shows, in general terms, how the Building Regulations are to be administered by local authorities, the approved methods of construction and the approved types of materials that are to be used in (or in connection with) buildings.

How are the Building Regulations controlled?

To assist local authorities, Section 1 shows:

*   how certificates signifying compliance with the Building Regulations are to be issued;
*   how copies of deposited plans are administered and retained;
*   how documents are to be controlled;
*   how local authorities can accept certificates from a person (or persons) nominated to act on their behalf;
*   how local authorities can seek external expertise to assist them in their duties;
*   how notices are given;
*   how plans of proposed work (or work already executed) are deposited;
*   how proposed work can be prohibited;
*   how samples are taken;
*   how work is inspected and tested;
*   what fees (and what level of fees) local authorities can charge;
*   when a dispute arises, how local authorities can refer the matter to the Secretary of State.

What are the requirements of the Building Regulations?

Schedule 1 describes the mandatory requirements for completing **all** building work. These include:

- accommodation for specific purposes;
- air pressure plants;
- cesspools (and other methods for reception, treatment and disposal of foul matter);
- drainage (including waste disposal units);
- emission of smoke, gases, fumes, grit or dust (or other noxious and/or offensive substances);
- fire precautions (services, fittings and equipment, means of escape);
- lifts (escalators, hoists, conveyors and moving footways);
- materials and components (suitability, durability and use);
- means of access to and egress from;
- natural lighting and ventilation of buildings;
- open spaces around buildings;
- prevention of infestation;
- resistance to moisture and decay;
- site preparation;
- solid fuel, oil, gas and electricity installations (including appliances, storage tanks, heat exchangers, ducts, fans and other equipment);
- standards of heating, artificial lighting, mechanical ventilation and air-conditioning and provision of power outlets;
- structural strength and stability (overloading, impact and explosion, underpinning, safeguarding of adjacent buildings);
- telecommunications services (including telephones and radio and television wiring installations);
- third-party liability (danger and obstruction to persons working or passing by building work);
- transmission of heat;
- transmission of sound;
- waste (storage, treatment and removal);
- water services, fittings and fixed equipment (including wells and boreholes for supplying water);

and matters connected with (or ancillary to) any of the foregoing matters.

What is Schedule 2 of the Building Act 1984?

This schedule provides guidance in connection with work that has been carried out prior to a local authority (under the Building Act 1984 Section 36) dispensing with or relaxing some of the requirements contained in the Building Regulations.

Schedule 2 is quite difficult to understand, and if it affects you then we would strongly advise that you discuss it with the local authority before proceeding any further.

What is Schedule 3 of the Building Act 1984?

Schedule 3 applies to how Building Regulations are to be used in inner London. As well as ruling which sections of the act may be omitted, this schedule provides details of how by-laws concerning the demolition of buildings in inner London may be made.

What sections of the Building Act 1984 are not applied to inner London?

In inner London, because of its existing and changed circumstances (compared to other cities in England and Wales), certain sections of the Building Act are inappropriate (these are listed in Tables 1.3 and 1.4). In this case additional requirements, which are applicable to inner London **only**, have been approved instead. These primarily cover the demolition of buildings.

What about inner London's by-laws?

By authority of the Building Act 1984, the council of any inner London borough may make by-laws in relation to the demolition of buildings in the borough, requiring:

*   regulating the hours during which ceilings may be broken down and mortar may be shot, or be allowed to fall, into any lower floor;
*   any person proposing to demolish a building to give to the borough council such notice of his or her intention to do so as may be specified in the by-laws;
*   the demolition of internal parts of buildings before any external walls are taken down;
*   the fixing of floor-level fans on buildings undergoing demolition;
*   the hoarding up of windows in a building where all the sashes and glass have been removed;
*   the placing of screens or mats, the use of water or the taking of other precautions to prevent nuisances arising from dust.

Table 1.3 Sections inapplicable to inner London

Section	Subsection
Buildings	<ul><li>Provision of exits, etc. (except fire escape)</li><li>Provision of water supply</li><li>Entrances, exits, etc. to be required in certain cases</li><li>Means of escape from fire</li><li>Raising of chimney</li><li>Cellars and rooms below subsoil water level</li><li>Consents under Section 74</li></ul>
Defective premises, demolition, etc.	<ul><li>Dangerous building</li><li>Dangerous building – emergency measures</li><li>Ruinous and dilapidated buildings and neglected sites</li><li>Notice to local authority of intended demolition</li><li>Local authority's power to serve notice about demolition</li><li>Notices under Section 81</li><li>Appeal against notice under Section 81</li></ul>

Table 1.4 Sections inapplicable to Temples

Section	Subsection
Drainage	• Drainage of building • Use and ventilation of soil pipes • Repair, etc. of drains

What is Schedule 4 of the Building Act 1984?

Schedule 4 of the Building Act 1984 concerns the authority and ruling of a public body's notices and certificates.

What is a public body's plans certificate?

When a public body is satisfied that the work specified in its (as well as another) public body's notices has been completed as detailed (and in full accordance with the Building Regulations) then that public body will give that local authority a certificate of completion.

This certificate is called a 'public body's plans certificate' and can relate either to the whole or to part of the work specified in the public body's notice. Acceptance by the local authority signifies satisfactory completion of the planned work and the public body's notice ceases to apply to that work.

What is a public body's final certificate?

When a public body is satisfied that all work specified in its (or another's) public body's notice has been completed in compliance with the Building Regulations then that public body will give the local authority a certificate of completion. This is referred to as a 'final certificate'.

How long is the duration of a public body's notice?

A public body's notice comes into force when it is accepted by the local authority and continues in force until the expiry of an agreed period of time.

Local authorities are authorized by the Building Regulations to extend the notice in certain circumstances.

What is Schedule 5 of the Building Act 1984?

Schedule 5 of Part 5 lists the transitional effect of the Building Act 1984 on the following acts:

- the Public Health Act 1936;
- the Clean Air Act 1956;
- the Housing Act 1957;
- the Public Health Act 1961;
- the London Government Act 1963;
- the Local Government Act 1972;

- the Health and Safety at Work, etc. Act 1974;
- the Local Government (Miscellaneous Provisions) Act 1982.

What is Schedule 6 of the Building Act 1984?

Schedule 6 lists the consequential amendments that will have to be made to existing acts of Parliament owing to the acceptance of the Building Act 1984. These amendments concern:

- the Restriction of Ribbon Development Act 1935;
- the Public Health Act 1936;
- the Atomic Energy Authority Act 1954;
- the Clean Air Act 1956;
- the Housing Act 1957;
- the Radioactive Substances Act 1960;
- the Public Health Act 1961;
- the London Government Act 1963;
- the Offices, Shops and Railway Premises Act 1963;
- the Faculty Jurisdiction Measure 1964;
- the Fire Precautions Act 1971;
- the Local Government Act 1972;
- the Local Land Charges Act 1975;
- the Safety of Sports Grounds Act 1975;
- the Development of Rural Wales Act 1976;
- the Local Government (Miscellaneous Provisions) Act 1976;
- the Interpretation Act 1978;
- the Highways Act 1980;
- the New Towns Act 1981;
- the Public Health (Control of Disease) Act 1984.

What is Schedule 7 of the Building Act 1984?

Schedule 7 lists the cancellation (repeal) of some sections of existing acts of Parliament, owing to acceptance of the Building Act 1984. These cancellations concern:

- the Public Health Act 1936;
- the Education Act 1944;
- the Water Act 1945;
- the Atomic Energy Authority Act 1954;
- the Radioactive Substances Act 1960;
- the Public Health Act 1961;
- the London Government Act 1963;
- the Greater London Council (General Powers) Act 1967;
- the Fire Precautions Act 1971;
- the Local Government Act 1972;

- the Water Act 1973;
- the Control of Pollution Act 1974;
- the Health and Safety at Work, etc. Act 1974;
- the Airports Authority Act 1975;
- the Local Government (Miscellaneous Provisions) Act 1976;
- the City of London (Various Powers) Act 1977;
- the Criminal Law Act 1977;
- the Education Act 1980;
- the Highways Act 1980;
- the Water Act 1981;
- the Civil Aviation Act 1982;
- the Local Government (Miscellaneous Provisions) Act 1982;
- the Housing and Building Control Act 1984.

2

The Building Regulations 2010

The Building Regulations are national standards that apply to all types of building and set standards for design and construction including foundations, damp-proofing and the overall stability of new and altered buildings. To ensure that buildings are safe and accessible and to limit any waste or environmental damage the regulations also include requirements relating to insulation, ventilation, heating, fire protection and means of escape as well as ensuring that appropriate facilities are provided for people, including those with disabilities. Those who intend to carry out building work must arrange for their work to be checked to ensure that the work meets the required standards by an independent third party.

The Building Regulations contain both procedural regulations and technical requirements. The former state what work needs Building Regulations approval and how to get it, while the latter set the standards for the work being undertaken.

Since the issue of the Building Regulations 2010 there have been a number of changes to all the regulations. In particular leaflets B, H, K, L, M and P and Regulation 7 have been revised and leaflet N has been removed.

 Building Regulations approval is **separate** from planning permission. Receiving planning permission is not the same as taking action to ensure that the building works comply with the Building Regulations. Even when planning permission is not required, most building works, including alterations to existing structures, are subject to minimum standards of construction (Building Regulations) to safeguard public health and safety.

2.1 What is the purpose of the Building Regulations?

The Building Regulations are legal requirements laid down by Parliament, based on the Building Act 1984. They are approved by Parliament and deal with the **minimum** standards of design and building work for the construction of domestic, commercial and industrial buildings.

Building Regulations ensure that new developments or alterations and/or extensions to buildings are all carried out to an agreed standard that protects the health and safety of people in and around the building. Builders and developers are required by law to obtain building control approval, which is an independent check that the Building Regulations have been complied with.

The responsibility for checking the Building Regulations have been met falls to building control bodies (BCBs) – either local authority building control or private sector approved inspector building control.

2.2 Why do we need the Building Regulations?

The Great Fire of London in 1666 was the single most significant event to have shaped today's legislation. The rapid growth of the fire through adjoining timber buildings highlighted the need for builders to consider the possible spread of fire between properties when rebuilding work commenced. This resulted in the first building construction legislation that required all buildings to have some form of fire resistance.

During the Industrial Revolution (200 years after the Great Fire) poor living and working conditions in ever expanding, densely populated urban areas caused outbreaks of cholera and other serious diseases. Poor sanitation, damp conditions and lack of ventilation forced the government to take action, and building control took on the greater role of health and safety through the first Public Health Act of 1875. This act had two major revisions in 1936 and 1961, and led to the first set of national building standards – the Building Regulations 1965.

The current legislation is the Building Regulations 2010 (Statutory Instrument No. 2214) and the Building (Approved Inspectors etc.) Regulations 2010 (Statutory Instrument No. 2215), which were made by the Secretary of State for Environment, Food and Rural Affairs under powers delegated by Parliament under the Building Act of 1984.

The Building Regulations are basic performance standards, and the level of safety and acceptable standards are set out as guidance in the Approved Documents (which are quite frequently referred to as 'parts' of the Building Regulations). Compliance with the detailed guidance of the Approved Documents is usually considered as evidence that the regulations themselves have been complied with.

2.3 What building work is covered by the Building Regulations?

The Building Regulations cover all new building work. This means that, if you want to put up a new building, extend or alter an existing one, or provide new and/or additional fittings in a building such as drains or heat-producing appliances, washing and sanitary facilities and hot water storage (particularly unvented hot water systems), the Building Regulations will probably apply.

It should be remembered that, although it may appear that the regulations do not apply to some of the work you wish to undertake, the end result of doing that work could well lead to you contravening some of the regulations. You should also recognize that some work – whether it is controlled or not – could have implications for an adjacent property. In such cases it would be advisable to take professional advice and consult the local authority or an approved inspector. Some examples are:

- building parapets, which may increase snow accumulation and lead to an excessive increase in loading on a shared roof;
- removing a buttressed support to a party wall;
- removing a tree close to a wall of an adjoining property;
- underpinning part of a building.

Some building work may also be subject to other statutory requirements such as planning permission, fire precautions, water regulations or licensing/registration.

2.4 What are the requirements associated with the Building Regulations?

The Building Regulations contain a list of requirements (referred to as 'Schedule 1') which are designed to ensure the health and safety of people in and around buildings, to promote energy conservation and to provide access and facilities for disabled people. In total there are 13 parts (A–P less I, N and O) to these requirements, and these cover subjects such as structure, fire and electrical safety, ventilation and drainage.

The requirements are expressed in broad, functional terms in order to give designers and builders the maximum flexibility in preparing their plans.

2.5 What are the Approved Documents?

If guidance in an Approved Document is followed, there will be a presumption of compliance with the requirement(s) covered by the guidance. However, this presumption is not conclusive, so simply following guidance does not actually guarantee you compliance in an individual case!

Approved Documents contain practical and technical guidance on ways in which the requirements of each part of the Building Regulations can be met. Each Approved Document reproduces the *requirements* contained in the Building Regulations relevant to the subject area. This is then followed by *practical and technical guidance*, with examples, on how the requirements can be met in some of the more common building situations.

There may be alternative ways of complying with the requirements to those shown in the Approved Documents and you are, therefore, under no obligation to adopt any particular solution in an Approved Document if you prefer to meet the relevant requirement(s) in some other way.

If you are intending to carry out building work, it is best always to check with your building control body (BCB) or approved inspector to ensure that the proposals comply with Building Regulations. There are two types of BCB – a local authority building control (LABC) and a private sector approved inspector building control (AIBC). Customers are free to choose which type of building control body they use on their project.

In accordance with Regulation 8 of the Building Regulations, the requirements in Approved Documents A–D, F–K and P (except for paragraphs G2, H2 and J7) of Schedule 1 to the Building Regulations do not require anything specific to be done except those things which are needed to secure reasonable standards of health and safety for persons in or about buildings (and any others who may be affected by buildings or matters connected with buildings).

The Building Regulations are constantly reviewed to meet the growing demand for better, safer and more accessible buildings as well as the need to reflect emerging harmonized European Standards. Where there are any issues common to one or more parts (such as the guidance on airtightness in Approved Document L corresponding to the requirements for ventilation in Approved Document F) these have been taken into consideration. Any changes necessary are brought into operation after consultation with all interested parties.

The Approved Documents are in 13 parts (A–P less I, N and O) and are listed in Table 2.1. You can download pdf copies of the Building Act and statutory instruments from the UK government website (http://www.legislation.gov.uk/), and the Approved Documents can be downloaded from the Planning Portal or viewed online (http://www.planningportal.gov.uk/buildingregulations/approveddocuments).

Alternatively you can buy a copy of the Approved Documents (and the Building Act 1984 if you wish) from The Stationery Office (TSO) or some bookshops. Occasionally they are available from libraries.

2.5.1 Approved Document A Structure

The requirements cover the three main structural areas of loading, ground movement and collapse, to ensure that:

(1)   all structural elements of a building can safely carry the loads expected to be placed on them;
(2)   foundations are adequate for any movement of the ground (e.g. caused by landslip or subsidence);
(3)   large buildings are strong enough to withstand an accident without collapsing.

Table 2.1 List of Approved Documents

Section	Title	Edition	Latest amendment
A	Structure	2004	2010 and 2013
B	Fire safety	2006	2010 and 2013
C	Site preparation and resistance to contaminants and moisture	2004	2010 and 2013
D	Toxic substances	1992	2002, 2010 and 2013
E	Resistance to the passage of sound	2003	2004, 2010 and 2013
F	Ventilation	2010	2013
G	Sanitation, hot water safety and water efficiency	2010	2010 and 2013
H	Drainage and waste disposal	2002	2010 and 2013
J	Combustion appliances and fuel storage systems	2010	2013
K	Protection from falling, collision and impact	2013	
L	Conservation of fuel and power (L1A and L2A)	2013	2013
	Conservation of fuel and power (L1B and L2B)	2010	
M	Access to and use of buildings	2004	2010 and 2013
P	Electrical safety	2013	
	Approved Document to support Regulation 7 – Materials and workmanship	2013	

2.5.2 Approved Document B Fire safety

There are nine aspects of fire safety in the construction of buildings: both dwelling houses and other buildings. These are that:

(1) the buildings shall have means of early warning and escape in the event of a fire;

(2) the internal spread of fire should be inhibited within the building by ensuring linings adequately resist the spread of flame over their surface and have a rate of heat release or fire growth that is reasonable;

(3) the building shall be designed and constructed so that, in the event of fire, its stability will be maintained for a reasonable period;

(4) walls common to two or more dwellings are designed and constructed to adequately resist the spread of fire between those buildings;

(5) where a building is subdivided, fire spread shall be inhibited through the use of fire-resisting materials or fire suppression systems;

(6) the building is constructed and designed so that the unseen spread of fire and smoke within concealed spaces in its structure is inhibited;

(7)   external walls of a building are able to resist the spread of fire from one building to another and roofs are able to resist the spread of fire from one room to another;
(8)   the roof of the building can resist the spread of fire over the roof and from one building to another;
(9)   buildings are designed and constructed so as to provide reasonable assistance to firefighters in the protection of life and to enable fire appliances to gain access to the building.

2.5.3 Approved Document C Site preparation and resistance to contaminants and moisture

There are four requirements to this part:

(1)   that, before any building works commence, all vegetation and topsoil are removed;
(2)   that any contaminated ground is treated, neutralized or removed before a building is erected;
(3)   that subsoil drainage is provided to waterlogged sites;
(4)   that **all** floors, walls and roof of a building should not be adversely affected by ground moisture, precipitation, interstitial and surface condensation, and the spillage of water from sanitary fixings.

2.5.4 Approved Document D Toxic substances

This part requires walls to be constructed in such a way that reasonable precautions have been taken to prevent the permeation of toxic fumes emanating from any insulating material that is inserted into cavities (including cavity walls).

2.5.5 Approved Document E Resistance to passage of sound

This part has four main requirements:

(1)   that dwellings shall provide reasonable resistance to sound from other parts of the same building and/or from adjoining buildings;
(2)   that internal walls and floors of dwellings shall provide reasonable resistance to sound;
(3)   that the construction of common internal parts of buildings (containing flats or rooms for residential purposes) shall prevent unreasonable reverberation;
(4)   that school rooms shall be acoustically insulated against noise.

2.5.6 Approved Document F Ventilation

This part has four main requirements.

(1) that people in the building are provided with an adequate means of ventilation;
(2) that mechanical ventilation systems are suitably commissioned;
(3) that the person carrying out the work is required to give the building owner sufficient information about the ventilation system and its maintenance so that it may be operated effectively;
(4) that the person carrying out the work should ensure that a test of the air flow rate is carried out and the local authority advised of the results of this testing.

2.5.7 Approved Document G Hygiene

There are 12 elements in this part:

(1) that wholesome water has to be supplied to any place where drinking water is drawn off and to any sink provided in any area where food is prepared;
(2) that wholesome water or softened wholesome water shall be supplied to any washbasin or bidet, fixed bath or shower in a bathroom or in/adjacent to a room containing a sanitary convenience;
(3) that there is a suitable installation for the provision of water of suitable quality to any sanitary convenience fitted with a flushing device;
(4) that, for the prevention of undue consumption of water, all fittings and fixed appliances must use water efficiently;
(5) that the potential consumption of wholesome water by persons occupying a dwelling must not exceed 125 litres per person per day;
(6) that heated wholesome water must be provided to any washbasin or bidet provided in or adjacent to a room containing a sanitary convenience, or in a bathroom, and any sink provided in any area where food is prepared;
(7) that hot water storage vessels shall prevent the temperature of the water stored from exceeding 100° C;
(8) that a discharge from a safety device shall be safely conveyed to where it is visible – without causing danger to persons in or about the building;
(9) that the hot water supply to a bath shall not exceed 48° C;
(10) that buildings are required to have satisfactory sanitary conveniences and washing facilities;
(11) that all dwellings are required to have a bathroom containing a washbasin and either a fixed bath or a shower with hot and cold water;
(12) that unvented hot water systems over a certain size are required to have safety provisions to prevent explosion.

2.5.8 Approved Document H Drainage and waste disposal

 H1 does not apply to the diversion of water which has been used for personal washing or for the washing of clothes, linen or other articles to collection systems for reuse. Requirement H3 does not apply to the gathering of rainwater for reuse.

There are six aspects of this part:

(1) that adequate drains are provided to take foul water from within buildings into a public or private sewer, septic tank or cesspool;

(2) that, where no public sewer is available, a suitable wastewater treatment system should be made available which is affixed with a durable notice containing details of continuing maintenance required;

(3) that adequate provision is made to take rainwater from roofs of buildings and paved areas to an appropriate soakaway, watercourse or sewer;

(4) that building work should not be detrimental to the continued mainte-nance of a drain, sewer or disposal main;

(5) that a system for discharging water to a sewer as described in paragraph 3 above shall be separate from that provided to carry foul water from the building;

(6) that adequate provision is made for the storage of solid waste and access provided for building occupants to the place of storage and from the place of storage to a refuse collection point.

2.5.9 Approved Document J Combustion appliances and fuel storage systems

Requirements J1, J2 and J4 only apply to fixed combustion appliances (includ-ing incinerators). Requirement J3 only applies to fixed combustion appliances in dwellings. Requirement J5 applies only to fixed oil storage tanks with a capacity greater than 90 litres, LPG storage tanks with a capacity greater than 150 litres and their associated pipes.

There are nine main aspects to this part:

(1) that combustion appliances are provided with a supply of fresh air to pre-vent overheating and for efficient working of any flue;

(2) that adequate provision is made to discharge the products of combustion from a combustion appliance to the outside air;

(3) that there should be a means of warning of the release of carbon monox-ide from a combustion appliance;

(4) that combustion appliances, flue pipes, chimneys and fireplaces are con-structed in such a manner as to reduce the risk of people suffering burns or the building catching fire;

(5) that, where a hearth, fireplace, flue or chimney is provided or extended, a durable notice should be placed at a suitable place in the building provid-ing information on the performance capabilities of the facility;

(6) that liquid fuel storage systems and the pipes connecting them to com-bustion appliances are separated from buildings and the boundary of the premises to reduce the risk of the fuel igniting in the event of a fire;

(7) that liquid fuel storage tanks and pipes are constructed in a manner to minimize the risk of fuel escaping;

(8)   that liquid fuel storage tanks are separated from buildings to reduce the risk of fuel igniting in the event of a fire in the adjacent building;

(9)   that a notice is provided giving details of how to respond in the event of an oil escape.

2.5.10 Approved Document K Protection from falling, collision and impact

Note: Approved Document K has been completely rewritten and came into effect in 2013. Is in a new, easier-to-read format and now includes the information on glazing safety previously included in Approved Document N.

There are ten main aspects to this part:

(1) that stairs, ladders and ramps which form part of the building are designed in a manner that allows the safe passage of people between levels in or about the building;

(2) that, to avoid persons falling off stairwells, balconies, floors, some roofs, light wells and basement areas (or similar sunken areas) connected to a building, they need to be suitably guarded according to the building's use;

(3) that, to avoid vehicles falling off buildings, car park floors, ramps and other raised areas, they need to be provided with vehicle barriers;

(4) that, to avoid collision with vehicles in loading bays, refuges or exits shall be provided;

(5) that suitable protection be provided against impact with glazing;

(6) that, to avoid danger to people from colliding with an open window, skylight or ventilator, some form of guarding may be needed;

(7) that transparent glazing with which people are likely to come into contact while moving in or about the building shall incorporate features which make it apparent;

(8) that windows, skylights and ventilators which can be opened by people in or about the building are constructed or equipped so that they may be opened, closed or adjusted safely;

(9) that provision is made for any windows or skylights or any transparent or translucent walls, ceilings or roofs to be safely accessible for cleaning;

(10) that measures are taken to avoid injury to people by the falling on to them of a door or gate that opens or slides upwards, and that opening powered doors and gates do not trap people and are capable of being opened in the event of a power failure.

2.5.11 Approved Document L Conservation of fuel and power

This part is split into four separate parts, which cover the conservation of fuel and power in:

(1)   new dwellings (L1A);
(2)   existing dwellings (L1B);
(3)   new buildings other than dwellings (L2A);
(4)   existing buildings other than dwellings (L2B).

Note: The requirements of Parts L1A and L2A have been amended completely since the last version of this book and came into effect in April 2014. They are in the new format for Approved Documents that comprises five sections and appendices.

L1A

There are ten main aspects to this part, covering the conservation of fuel and power in new dwellings. This part requires:

(1) that fuel and power are conserved by limiting heat gains and losses;
(2) that the Secretary of State approves the methodology of calculation of the energy performance;
(3) that new buildings meet minimum energy performance requirements;
(4) that the building does not exceed the CO_2 limits set for it;
(5) that fabric energy-efficiency rates are provided for new dwellings;
(6) that the person carrying out work in a new dwelling that affects the heating, hot water, air-conditioning or ventilation provides suitable energy performance certificates to the owner and local authority;
(7) that consideration is made of high-efficiency alternative systems for new buildings;
(8) that building services are pressure-tested in an appropriate manner;
(9) that building services are commissioned in an appropriate manner;
(10) that the owner is provided with sufficient information about the building, the fixed building services and their maintenance requirements so that the building can be operated in an energy-efficient manner.

L1B

There are five main aspects to this part, covering the renovation of thermal elements, consequential improvements to thermal elements in buildings over $1000\,m^2$, energy performance certificates and the conservation of fuel and power in existing dwellings. This part requires:

(1)   that renovations to or replacement of thermal elements comply with Schedule 1;
(2)   that the person carrying out work in a new dwelling that affects the heating, hot water, air-conditioning or ventilation provides suitable energy performance certificates to the owner and local authority;
(3)   that fuel and power are conserved by limiting heat gains and losses;
(4)   that appropriately commissioned, energy-efficient fixed building services are provided with effective controls;
(5)   that the owner is provided with sufficient information about the building, the fixed building services and their maintenance requirements so that the building can be operated in an energy-efficient manner.

L2A

There are nine main aspects to this part, covering target CO_2 emissions, energy performance certificates and the conservation of fuel and power in new buildings other than dwellings. This part requires:

(1)   that fuel and power are conserved by limiting heat gains and losses;
(2)   that the Secretary of State approves the methodology of calculation of the energy performance;
(3)   that new buildings meet minimum energy performance requirements;
(4)   that the building does not exceed the CO_2 limits set for it;
(5)   that the person carrying out work in a new dwelling that affects the heating, hot water, air-conditioning or ventilation provides suitable energy performance certificates to the owner and local authority;
(6)   that consideration is made of high-efficiency alternative systems for new buildings;
(7)   that building services are pressure-tested in an appropriate manner;
(8)   that building services are commissioned in an appropriate manner;
(9)   that the owner is provided with sufficient information about the building, the fixed building services and their maintenance requirements so that the building can be operated in an energy-efficient manner.

L2B

There are five main aspects to this part, covering the renovation of thermal elements, consequential improvements to thermal elements in buildings over $1000\,m^2$, interpretation of the word 'building', the use of energy performance certificates and the conservation of fuel and power in new buildings other than dwellings. This part requires:

(1)   that renovations to or replacement of thermal elements comply with Schedule 1;
(2)   that the person carrying out work in a new dwelling that affects the heating, hot water, air-conditioning or ventilation provides suitable energy performance certificates to the owner and local authority;
(3)   that fuel and power are conserved by limiting heat gains and losses;
(4)   that appropriately commissioned, energy-efficient fixed building services are provided with effective controls;
(5)   that the owner is provided with sufficient information about the building, the fixed building services and their maintenance requirements so that the building can be operated in an energy-efficient manner.

2.5.12 Approved Document M Access to and use of buildings

There are four main aspects to this part:

(1)   that reasonable provision is made for people to gain access to and use a building and its facilities;

(2)   that suitable independent access is provided to an extension in a building other than a dwelling;

(3)   that, if sanitary conveniences are provided in any building that is to be extended, reasonable provision is made within the extension for sanitary conveniences;

(4)   that reasonable provision is made in the entrance storey for sanitary conveniences or, where this is not possible, reasonable provision is made for sanitary conveniences in the entrance storey or principal storey.

2.5.13 Approved Document P Electrical safety

This part requires that reasonable provision should be made in the design and installation of electrical installations in order to protect those who are operating, maintaining or altering the installations from fire or injury. The government document *New Rules for Electrical Safety in the Home* is available from the Planning Portal (http://www.planningportal.gov.uk/uploads/br/electrical_safety.pdf).

2.5.14 What happens if I do not comply with an Approved Document?

Just because you have not actually complied with an Approved Document (which is, after all, only meant as a guidance document), it does not mean that you are liable for any civil or criminal prosecution. If, however, you have contravened a Building Regulation, then not having complied with the recommendations contained in the Approved Documents may be held against you.

2.6 Health and safety responsibilities

Clients, contractors and designers may also have duties under health and safety legislation and may need to notify the Health and Safety Executive (HSE). Although a domestic client does not have duties under Construction (Design and Management) Regulations 2007 (CDM 2007), those who work for him or her on construction projects do (see CDM 2007 at http://www.hse.gov.uk/construction/cdm.htm).

2.7 Are there any exemptions?

The Building Regulations do not apply to:

*   a building included in the schedule of monuments maintained under Section 1 of the Ancient Monuments and Archaeological Areas Act 1979;

- a building on a site, being a building which is intended to be used only in connection with the disposal of buildings or building plots on that site;
- a building on the site of construction or civil engineering works which is intended to be used only during the course of those works and contains no sleeping accommodation;
- a building, other than a building containing a dwelling or used as an office or showroom, erected for use on the site of and in connection with a mine or quarry;
- a detached building designed and intended to shelter people from the effects of nuclear, chemical or conventional weapons, and not used for any other purpose;
- a detached building, having a floor area that does not exceed $15\,m^2$ and which contains no sleeping accommodation;
- a detached single-storey building, having a floor area that does not exceed $30\,m^2$ and which contains no sleeping accommodation;
- any building (other than a building containing a dwelling or a building used for office or canteen accommodation) erected on a site in respect of which a licence under the Nuclear Installations Act 1965 is, for the time being, in force;
- any building in which explosives are manufactured or stored under a licence granted under the Manufacture and Storage of Explosives Regulations 2005;
- certain detached buildings that are not frequented by people;
- greenhouses and agricultural buildings;
- temporary buildings that are not intended to remain where they are erected for more than 28 days;
- the extension of a building by the addition at ground level of a conservatory, porch, covered yard or covered way or a carport open on at least two sides where the floor area of that extension does not exceed $30\,m^2$, as long as, in the case of a glazed conservatory or porch, the glazing satisfies the requirements of Approved Document N of Schedule 1.

Purpose-built student living accommodation (including flats) should be treated as hotel/motel accommodation in respect of space requirements and internal facilities.

The Metropolitan Police Authority is also exempt from Building Regulations (except where Regulation 29 applies), and there are occasional deviations for other buildings referred to as 'Crown buildings', those belonging to 'statutory undertakers' (e.g. a water board), some buildings in inner London, those belonging to the United Kingdom Atomic Energy Authority (UKAEA), or buildings belonging to a 'public body' (i.e. local authority, county council and any other body 'that acts under an enactment for public purposes and not for its own profit').

 This can be rather a grey area and it is best to seek advice if you think that you come under this category.

2.8 Energy- and water-efficiency requirements

Building Regulations include a number of energy-efficiency requirements which apply equally to the erection of buildings and extensions and the carrying out of work within buildings and extensions. These regulations apply to all buildings with roofs and walls and that use energy to control the internal temperature, with the following exceptions:

- buildings within a Conservation Area;
- listed buildings;
- places of worship;
- scheduled monuments;
- stand-alone buildings with a floor area of less than $50\,m^2$;
- temporary buildings (if expected to be used for only two years or less).

Any changes to the energy status or the thermal elements within the building must comply with Approved Document L of Schedule 1. New buildings are required to conform to minimum energy-efficiency performance and CO_2 emission rates. Certificates must be produced by an appropriately qualified and registered energy assessor, and these must detail any changes to the energy-efficiency of an existing building and provide information on the energy-efficiency characteristics of a new building.

All new dwellings are subject to water-efficiency requirements, and this will include the need to carry out a wholesome water consumption calculation for the residents of the building and provide details of this calculation to the local authority within five days of work being complete.

2.9 Do I need Building Regulations approval?

If you are in any doubt about whether you need to apply for permission, you should contact your local building control body (BCB) before commencing any work (in all cases, you may require planning permission).

If you are considering carrying out building work to your property, then you may need to apply to your local authority for *Building Regulations approval*.

You will be required to submit a Building Regulations application prior to commencing any work for most types of building work (e.g. extensions, alterations, conversions and drainage works). Some extensions and small detached buildings are exempt from Building Regulations control – particularly if the increased area of the alteration or extension is less than $30\,m^2$. However, you may **still** be required to apply for planning permission.

Note: In May 2013 Parliament agreed to increase the size of single-storey rear extensions which can be built under permitted development and brought into force the associated neighbour consultation scheme. For a period of three years, between 30 May 2013 and 30 May 2016, householders will be able to build larger single-storey rear extensions under permitted development. In essence the permitted development size limits for extensions have doubled from $4\,m^2$ to $8\,m^2$ for detached houses, and from $3\,m^2$ to $6\,m^2$ for all other houses.

2.9.1 Building work needing formal approval

Building Regulations apply to any building work that involves:

- building a garage;
- electrical work;
- re-covering a roof;
- removing a chimney breast;
- removing internal walls;
- the 'material alteration' of a building (such as converting it into separate units);
- the 'material change of use' of a building.
- the erection of a new or re-erection of an existing building;
- the extension of a building (including loft conversions or adding a kitchen, bedroom or lounge);
- the insertion of insulation into a cavity wall;
- the installation or extension of a service or fitting which is controlled under the regulations, including:

  ○ creating new baths/shower rooms,
  ○ installing new electrical cabling,
  ○ replacing of a boiler, windows, doors, etc.,
  ○ making doors or windows larger;

- the underpinning of the foundations of a building;
- work affecting the thermal elements, energy status or energy performance of a building.

2.9.2 Other examples of work needing approval

These include:

- altered openings for new windows in roofs or walls;
- cellars (particularly in London);
- electrical installations;
- installation of new heating appliances;
- internal structural alterations, such as the removal of a load-bearing wall or partition;
- new chimneys or flues;
- replacing roof coverings (unless exactly like-for-like repairs).

2.9.3 Self-certification

There are certain types of work that are exempt from control as long as the work is carried out by a person authorized under a self-certification scheme. The installations listed in Table 2.2 are exempt from having to give building notice or deposit full plans, **provided** that the person carrying out the work is as indicated in the third column.

Table 2.2 Self-certification schemes providing exemption from giving building notice or depositing full plans

Area	Type of work	Person authorized to self-certify
Cavity walls	Insertion of insulating material into the cavity walls of an existing building	A person registered by Certsure, CIGA, Benchmark, NAPIT or Stroma in respect of that type of work
	Internal walls of a building	A person registered by BBA, Benchmark, CERTASS, Certsure, NAPIT or Stroma in respect of that type of work
	External walls of a building (not including insulation of demountable-clad buildings)	A person registered by BBA, Benchmark, CERTASS, Certsure, NAPIT or Stroma in respect of that type of work
	Both external and internal walls of a building ('hybrid insulation' – not including insulation of demountable-clad buildings)	A person registered by BBA, Benchmark, Certsure or NAPIT, in respect of that type of work
Electrical installations	In dwellings – installation of fixed low- or extra-low-voltage electrical installations	A person registered by APHC, Benchmark, BESCA, Certsure, NAPIT, OFTEC or Stroma in respect of that type of work
	In dwellings – fixed low- or extra-low-voltage electrical installations as part of other work being carried out by the registered person	A person registered by APHC, Benchmark, BESCA, Certsure, NAPIT, OFTEC or Stroma in respect of that type of work
	Buildings other than dwellings – installation of lighting or electrical heating systems	A person registered by Benchmark, BESCA, Certsure, NAPIT or Stroma in respect of that type of electrical work
Energy production	Installation in a building of a system to produce electricity, heat or cooling: (a) by microgeneration, or (b) from renewable sources (as defined in European Parliament and Council Directive 2009/28/EC of 23 April 2009 on the promotion of the use of energy from renewable sources)	A person registered by Ascertiva, Association of Plumbing and Heating Contractors, Benchmark, British Standards Institution, Building Engineering Services, ECA, HETAS, NAPIT, Oil Firing Technical Association or Stroma in respect of that type of work

Area	Type of work	Person authorized to self-certify
Gas	Installation of a heat-producing gas appliance	A person, or an employee of a person, who is a member of a class of persons approved in accordance with Regulation 3 of the Gas Safety (Installation and Use) Regulations 1998 who is on the Gas Safe Register
Hot water systems	Installation of a heating or hot water system connected to an oil-fired combustion appliance or its associated controls	A person registered by APHC, Benchmark, BESCA, Certsure, HETAS, NAPIT, OFTEC or Stroma in respect of that type of work
	Installation of a heating or hot water system connected to a heat-producing gas appliance, or associated controls	A person registered by APHC, Benchmark, BESCA, Capita, Certsure, HETAS, NAPIT, OFTEC or Stroma in respect of that type of work
	Installation of a heating or hot water system connected to a solid-fuel-burning combustion appliance or its associated controls	A person registered by APHC, BESCA, Certsure, HETAS, NAPIT, OFTEC or Stroma in respect of that type of work
	Installation of a heating or hot water system connected to an electric heat source or its associated controls	A person registered by APHC, Benchmark, BESCA, Certsure, HETAS, NAPIT, OFTEC or Stroma in respect of that type of work
Lighting	Installation of a lighting system or electric heating system, or associated electrical controls	A person registered by Ascertiva, Building Engineering Services, ECA, NAPIT or Stroma in respect of that type of work
Microgeneration and renewable technologies		A person registered by APHC, Benchmark, BSI, BESCA, BRE, Certsure, HETAS, NAPIT, OFTEC or Stroma
Oil-fired appliances	Installation of: (a) an oil-fired combustion appliance; or (b) oil storage tanks and the pipes connecting them to combustion appliances	A person registered by APHC, Benchmark, BESCA, Certsure, HETAS, NAPIT, OFTEC or Stroma in respect of that type of work
Plumbing and water supply systems	Installation of a sanitary convenience, sink, washbasin, bidet, fixed bath, shower or bathroom in a dwelling which does not involve work on shared or underground drainage	A person registered by APHC, Benchmark, BESCA, Certsure, HETAS or NAPIT in respect of that type of work
	Installation of a wholesome cold water supply or a softened wholesome cold water supply	A person registered by APHC, Benchmark, BESCA, Certsure, HETAS or NAPIT in respect of that type of work

Table 2.2 (Continued)

Area	Type of work	Person authorized to self-certify
	Installation of a supply of non-wholesome water to a sanitary convenience fitted with a flushing device which does not involve work on shared or underground drainage	A person registered by APHC, Benchmark, BESCA, Certsure, HETAS, NAPIT or Stroma in respect of that type of work
Roof coverings	Installation, as a replacement, of the covering of a pitched or flat roof and work carried out by the registered person as a necessary adjunct to that installation. This paragraph does not apply to the installation of solar panels	A person registered by NAPIT or the Competent Roofer Scheme in respect of that type of work
Solid-fuel appliances (including biomass)	Installation of a solid-fuel-burning combustion appliance	A person registered by APHC, Benchmark, BESCA, Certsure, HETAS or NAPIT in respect of that type of work
Ventilation systems	Installation of a mechanical ventilation or air-conditioning system or associated controls which does not involve work on a system shared with parts of the building occupied separately, in a building other than a dwelling	A person registered by BESCA, Certsure, NAPIT or Stroma in respect of that type of work
	Installation of an air-conditioning or ventilation system in a dwelling which does not involve work on systems shared with other dwellings	A person registered by BESCA, Certsure, NAPIT or Stroma in respect of that type of work
Replacement windows, doors, roof windows and rooflights	Installation, as a replacement, of a window, rooflight, roof window or door in an existing dwelling	A person registered by Fensa, BM Trada, Benchmark, BSI, CERTASS, Certsure, NAPIT, Network VEKA or Stroma in respect of that type of work
	Installation, as a replacement, of a window, rooflight, roof window or door in an existing building other than a dwelling	A person registered by BM Trada, CERTASS, Certsure, Stroma or Fensa in respect of that type of work

Where this regulation applies, the person authorized under the self-certification scheme must provide the local authority with a certificate stating what work has been carried out within 30 days of completion of the work. This does not apply for work carried out under Schedule 4 (descriptions of work where no building notice or deposit of full plans is required).

Any work may be carried out if it is necessary to ensure that the appliance (or service or fitting) complies with the requirements of Building Regulations as long as the person who installs the service or fitting is registered under the appropriate certification scheme.

2.9.4 Where can I obtain assistance in understanding the requirements?

Local councils can provide assistance with:

* advice about how to incorporate the most efficient energy safety measures into your scheme;
* advice about the use of materials;
* advice on electrical safety;
* advice on fire safety measures (including safe evacuation of buildings in the event of an emergency);
* deciding what type of application is most appropriate for your proposal;
* details of the stages at which local councils need to inspect your work;
* how to apply for Building Regulations approval;
* how to prepare your application (and what information is required);
* how to provide adequate access for disabled people;
* what your Building Regulation completion certificate means to you.

2.10 How do I obtain Building Regulations approval?

You, as the owner or builder, are required to get approval to carry out building work either from your local authority building control or an approved inspector building control service. It is your choice which route you take, and the responsibility for checking the Building Regulations have been met falls to the building control body (BCB). How to get approval depends on which service is being used.

An approved inspector is a company or individual who has been authorized under the Building Act to carry out building control work in England and Wales. There is a list of approved inspectors on the Construction Industry Council website, a link to which is provided at section 2.27.2. If you choose to use an approved inspector then you should jointly notify the local authority that the approved inspector is carrying out the building control function for the work. This notification is called an 'initial notice'.

If you choose to use a local authority, the procedures are set out in the Building Regulations and on the Planning Portal.

 Be sure that you understand which areas relate to pre-site procedures (those which must be completed before work commences) and those which relate to procedures that have to take place once work is under way on site.

There are three types of application for pre-site approval that you can make:

- Building notice – Plans are not required with this process, but you will need to be confident that the work will comply with Building Regulations, because if it does not you will be compelled to undertake any remedial work to make sure it does. Once you have given in your notice you will be advised by the authority if any of the work does not comply with the Building Regulations and you must provide the authority with any information it asks for before and during the project.

The authority does not consult with any other body in relation to a building notice and therefore you need to ensure the works comply with other regulations (particularly those relating to fire) prior to starting work.

- Full plans – A full plans application must be made well in advance of work starting. An application under full plans needs to contain detailed plans and other information showing all construction details.
- Regularization – This is a retrospective application which relates to works that have been carried out in the past without authorization. It applies to works carried out without Building Regulations consent which started on or after 11 November 1985. The purpose of the process is to regularize the unauthorized works and obtain a certificate of regularization.

Whatever method you adopt, it may save time and trouble if you make an appointment to discuss your scheme with the building control officer well before you intend carrying out any work.

The building control officer will be happy to discuss your intentions, including proposed structural details and dimensions together with any lists of the materials you intend to use, so that he or she can point out any obvious contravention of the Building Regulations before you make an official application for approval. At the same time the building control officer can suggest whether it is necessary to approach other authorities to discuss planning, sanitation, fire escapes and so on.

When the building is finished you must notify the local authority.

The building control officer will ask you to inform the office when crucial stages of the work are ready for inspection (by a surveyor) in order to make sure the work is carried out according to your original specification. Should the surveyor be dissatisfied with any aspect of the work, he or she may suggest ways to remedy the situation.

It would be to your advantage to ask for written confirmation that the work is satisfactory, as this will help to reassure a prospective buyer when you come to sell the property.

2.11 What are building control bodies?

Your local authority has a general duty to see that all building work complies with the Building Regulations. To ensure that your particular building work complies with the Building Regulations you must use one of the two services available to check and approve plans and to inspect your work as appropriate. The two services are the LABC service or the service provided by the private sector in the form of an approved inspector. Both building control bodies will charge for their services. Both may offer advice before work is started. Certain types of building work (e.g. electrical and plumbing work) can be self-certificated as compliant with Building Regulations by a member of a competent person scheme without the need to notify a BCB.

2.11.1 What will the local authority do?

This rather depends on the work you plan, which building control service you use and whether you are submitting a full plans application or a building notice application.

Where a local authority building control service is used, notice is given to the local authority of the commencement (at least two days before the work commenced) and completion of the work (not more than five days after the work is completed) and at certain other stages. The building control service will inspect the site as the work progresses to make sure that Building Regulations and other allied legislation are being complied with. You are required to give the local authority notice of when the work has reached a particular stage. Your local authority will explain about the notification procedures and tell you which stages of the work it needs to inspect. You should expect to pause work and give at least one whole day's notice for the local authority to carry out its inspection. It needs to inspect:

- excavation for a foundation (before covering up);
- the foundation itself (before covering up);
- any damp-proof course (before covering up);
- any concrete or material laid over a site (before covering up).

The local authority will advise you if the work does not comply with the Building Regulations. If a building or part of it is to be occupied before completion, the person carrying out the work is required to give the local authority at least five days' notice before the occupation.

 Don't forget that you will be required to pay a fee for the submission of an application. These fees are provided in section 4.19.

2.11.2 What is the difference between a full plans application and the building notice procedure?

With a full plans application, detailed plans need to be produced showing all constructional details and other relevant information. The plans need to be submitted well in advance of your intended commencement on site, and the local authority consults with any appropriate authorities (fire, etc.) before approving them and giving you a decision within five to eight weeks. At the end of the process the local authority may issue you with a completion certificate if you have requested it. A 'full plans' submission is required in the following circumstances:

- if the building use falls under Section 1 of the Fire Precautions Act 1971;
- if use falls under Part II of the Fire Precautions (Workplace) Regulations 1997;
- if work will be close to or over the top of drains shown on the 'map of sewers';
- where a new building will front on to a private street.

The building notice procedure requires much less detailed plans, and these can be submitted within two days of the commencement of work. However, you are required to ensure that your work complies with the regulations, as the authority does not consult with agencies such as the fire authority. At the end of the work you are not issued with a completion certificate.

In both cases, your application or notice should be submitted to the local authority and should be accompanied by any relevant calculations, to demonstrate compliance with safety requirements concerning the structure of the building. In both cases the work will be inspected on site and guidance will be given at each stage if the work fails to comply with Building Regulations.

 Approved plans are valid for at least three years.

2.11.3 What will the approved inspector do?

Approved inspectors are companies or individuals who have been authorized under the Building Act 1984 to carry out building control work in England and Wales. The Construction Industry Council (CIC) is responsible for deciding on all applications for approved inspector status, and a list of approved inspectors can be viewed at http://www.cic.org.uk/services/register.php.

If you use an approved inspector, the inspector will give you advice, check plans, issue a plans certificate, inspect the work, etc. as agreed between you. You and the inspector will jointly notify the local authority on what is termed an 'initial notice'. Once that has been accepted by the local authority, the approved inspector will then be responsible for the supervision of building work. Although the local authority will have no further involvement, you may still have to supply it with limited information to enable it to be satisfied about certain aspects linked to Building Regulations (e.g. about the point of connection to an existing sewer).

If the approved inspector is not satisfied with your proposals you may alter your plans according to his or her advice, or you may seek a ruling from the Secretary of State regarding any disagreement between you. The approved inspector might also suggest an alternative form of construction, and, provided that the work has not been started, you can apply to the local authority for a

relaxation of or a dispensation from one (or more) of the regulations' requirements and, in the event of a refusal by the local authority, you may appeal to
the Secretary of State.

If, however, you do not exercise these options and you do not do what
the approved inspector has advised to achieve compliance, the inspector will
not be able to issue a final certificate. The inspector will also be obliged to
notify the local authority so that it can consider whether to use its powers of
enforcement.

2.12 How do I apply for building control?

 Take advantage of the free advice that local authorities offer, and discuss your
ideas well in advance.

If your prospective work will involve any form of structure, you could need
building control approval. Some types of work may need both planning permission and building control approval; others may need only one or the other.
The process of assessing a proposed building project is carried out through an
evaluation of submitted information and plans and the inspection of work as
the building progresses.

2.12.1 What applications do not require submission plans?

The following building works do **not** require the submission of plans:

*   any work specified in an initial notice, an amendment notice or a public
    body's notice, which is in force;
*   electrical installation work which is completed by a competent firm regis-
    tered under the NICEIC Approved Contractor Scheme;
*   installation of a heating appliance by a person, or an employee of a person,
    approved in accordance with Regulation 3 of the Gas Safety (Installation
    and Use) Regulations 1998;
*   where Regulation 19 of the Building (Approved Inspectors, etc.) Regula-
    tions 2010 (local authority powers in relation to partly completed work)
    applies.

2.12.2 Other considerations

Depending on the type of work involved, you may need to get approval from
several sources before starting. The list below provides a few examples:

*   A solicitor might need to be consulted to see if any covenants or other
    forms of restriction are listed in the title deeds to your property and if
    any other person or party needs to be consulted before you carry out your
    work.

- If a building is listed or is within a Conservation Area or an Area of Outstanding Natural Beauty, special rules apply.
- There may be legal objections to alterations being made to your property.
- You may need planning permission for a particular type of development work.

2.13 Full plans application

This type of application can be used for any type of building work, but it **must** be used where the proposed premises are to be used as a factory, office, shop, hotel or boarding house or as railway premises. If you use the full plans procedure you must submit detailed drawings and other information showing all construction details, preferably well in advance of when work is to start on site, to the local authority, which will check your plans and consult any appropriate authorities (such as fire and water authorities). If your plans comply with the Building Regulations the local authority will then issue an approval notice before work starts. This process can take between five and eight weeks, depending on the project.

The full plans application may be accompanied by a request (from the person carrying out such building work) that, on completion of the work, he or she wishes the local authority to issue a completion certificate.

A full plans application requires the submission of fully detailed plans, specifications, calculations and other supporting details to enable the building control officer to ascertain compliance with the Building Regulations. The amount of detail depends on the size and type of building works proposed, but as a minimum will have to consist of:

- a description of the proposed building work or material change of use;
- a location plan showing where the building is located relative to neighbouring streets;
- a plan or plans showing what work will be completed.

Two copies of the full plans application need to be sent to the local authority, except in cases where the proposed building work relates to the erection, extension or material alteration of a building (other than a dwelling house or flat) and where fire safety imposes an additional requirement, in which case five copies are required. The LABC (an organization that coordinates local authority services regionally and nationally) has developed an online service for creating and submitting building control applications. You can apply to any local authority in England and Wales using the service (http://www.submit-a-plan.com) or make an online application on the Planning Portal (http://www.planningportal.gov.uk/england/public/planning/applications). Alternatively you may print out a copy of the application form and submit it by hand. The forms are available on the Planning Portal (http://www.planningportal.gov.uk/planning/applications/paperforms).

A full plans application will be thoroughly checked by the local authority, which is required to pass or reject your plans within a certain time limit (usually eight weeks), or it may add conditions to an approval (with your written agreement). If it is satisfied that the work shown on the plans complies with the regulations, you will be issued with an approval notice (within a period of five weeks to two months) showing that your plans were approved as complying with the Building Regulations. If the local authority is not satisfied with your plans, you may be asked to make amendments or provide more details. Alternatively, a conditional approval may be issued, which will either specify modifications that must be made to the plans or specify further plans that must be deposited.

If your plans are rejected, the reasons will be stated in the notice. If your plans are rejected, and you do not consider it is necessary to alter them, you will have two options available to you:

(1)   You may seek a 'determination' from the Secretary of State if you believe your work complies with the regulations (but you must apply before work starts).
(2)   If you acknowledge that your proposals do not necessarily comply with a particular requirement in the regulations and feel that it is too onerous in your particular circumstances, you may apply for a relaxation of or dispensation from that particular requirement from the local authority. You can make this sort of application at any time you like but it is obviously sensible to do so as soon as possible and preferably before work starts. If the local authority refuses your application, you may then appeal to the Secretary of State within a month of the date of receipt of the rejection notice.

Your local authority will inspect the building work while it is in progress and explain any notification procedures it requires you to follow at various stages of the work. It will issue you with a completion certificate once it is content that the completed work complies with the Building Regulations. You will need to request this certificate when you first make your application.

In the event of a disagreement with your local authority, the 'full plans' procedure enables you to ask for a determination from (in England) the Department for Communities and Local Government (DCLG) or (in Wales) the Welsh Assembly Government about whether your plans do or do not comply with the Building Regulations.

A full plans approval notice is valid for three years from the date of deposit of the plans, after which the local authority may send you a notice to declare the approval of no effect if the building work has not commenced.

2.13.1 Consultation with sewerage undertaker

Where applicable, the local authority shall consult the sewerage undertaker as soon as practicable after the plans have been deposited, and before issuing any completion certificate in relation to the building work.

2.13.2 Advantages of submitting a full plans application

The advantages of the full plans method are that:

*   you will get a (free) completion certificate, which will be issued on satis-
    factory completion of the work;
*   a formal notice of approval or rejection will be issued within five weeks
    (unless the applicant agrees to extend this to two months);
*   only when work starts on site (and the building control officer has com-
    pleted his or her initial visit) is the remaining part of the fee invoiced;
*   the plans can be examined and approved in advance (for an advance pay-
    ment of, typically, 25 per cent of the total fee).

2.14 Building notice procedure

If you use the building notice procedure, as with full plans applications, the
work will normally be inspected as it proceeds, but you will not receive any
notice indicating whether your proposal has been passed or rejected. Work
can then commence and regular site inspections will be made at agreed stages,
after which you will be advised where the work itself is found (by the building
control officer) not to comply with the regulations. A building notice cannot
be used for commercial developments, and there are specific exclusions in the
regulations as to when building notices cannot be used. These are:

*   for building work which is subject to Section 1 of the Fire Precautions Act
    1971;
*   for work which will be built close to or over the top of rainwater and foul
    drains shown on the 'map of sewers';
*   where Part II of the Fire Precautions (Workplace) Regulations 1997
    applies;
*   where a new building will front on to a private street.

You need to fill in a Building Notice application form and return it, along
with basic drawings and relevant information, to the building control office at
least two days before work commences. If you decide to use this procedure you
need to be confident that the work will comply with the Building Regulations,
as otherwise you have to carry out any remedial work that your local authority requests. You will be advised by the authority if the work does not comply
with the Building Regulations, and if at any stage your local authority requires
further information such as structural design calculations or plans you must
supply the details requested.

 The submission of a marked-up sketch showing the location of the building,
although not mandatory, is recommended.

Under the building notice procedure no approval notice is given. There is also
no procedure to seek a determination from the Secretary of State if there is a

disagreement between you and the local authority – unless plans are subsequently deposited. However, the advantage of the building notice procedure is that it will allow you to carry out **minor works** without the need to prepare full plans. You must, however, feel confident that the work will comply with the regulations or you risk having to correct any work you carry out at the request of the local authority.

A building notice is particularly suited to minor works (e.g. a householder wishing to install another WC). For such building work, detailed plans are unnecessary and most matters can be agreed when the building control officer visits your property. You do not need to have detailed plans prepared, but in some cases you may be asked to supply extra information.

As no formal approval is given, good liaison between the builder and the building control officer is essential to ensure that work does not have to be redone.

This type of application may be used for all types of building work, so long as no part of the premises is used for any of the purposes mentioned earlier under the full plans application.

A 'building notice' is valid for three years from the date the notice was given to the local authority, after which it will automatically lapse if the building work has not commenced. Under the building notice procedure a local authority is not required to issue a completion certificate and you are not able to approach DCLG for a determination if your local authority says your work does not comply with the Building Regulations.

2.14.1 What do I have to include in a building notice?

A building notice shall:

- state the name and address of the person intending to carry out the work;
- be signed by that person or on that person's behalf;
- contain, or be accompanied by:

  ○ a description of the proposed building work or material change of use,
  ○ particulars of the location of the building,
  ○ the use or intended use of that building.

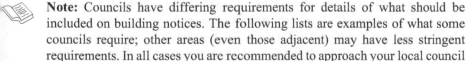

Note: Councils have differing requirements for details of what should be included on building notices. The following lists are examples of what some councils require; other areas (even those adjacent) may have less stringent requirements. In all cases you are recommended to approach your local council for further details.

New buildings and extensions

When planning a building extension, the building notice needs to be accompanied by:

- a plan to a scale of not less than 1:1250 showing the size and position of the building, or the building as extended, and its relationship to adjoining boundaries (commonly known as a block plan);
- particulars of the provisions to be made for the drainage of the building or extension.

Insertion of insulating material into the cavity walls of a building

For cavity wall insulations, the building notice needs to be accompanied by a statement which specifies:

- the name and type of insulating material to be used;
- the name of any European technical approval issuing body that has approved the insulating material;
- the requirements of Schedule 1 in relation to which the issuing body has approved the insulating material;
- any European Economic Area (EEA) national standard with which the insulating material conforms;
- the name of any body that has issued any current approval to the installer of the insulating material.

If the insulating material has been installed by an approved installer, it will usually submit a type approval notice on your behalf.

Provision of a hot water storage system

A building notice in respect of a proposed hot water system shall be accompanied by a statement which specifies:

- the name, make, model and type of hot water storage system to be installed;
- the name of the body (if any) that has approved or certified the system;
- the name of the body (if any) that has issued any current registered operative identity card to the installer or proposed installer of the system.

Electrical installations

All proposals to carry out electrical installation work **must** be notified to the local authority's BCB before work begins, **unless** the proposed installation work is undertaken by a competent person registered with an electrical self-certification scheme and does not include the provision of a new circuit.

Non-notifiable work (such as replacing a socket outlet or other fixed electrical equipment) can be completed by a DIY enthusiast (family member or friends), but needs to be installed in accordance with manufacturers' instructions and done in such a way that it does not present a safety hazard. This work does **not** need to be notified to a local authority BCB (unless it is installed in an area of high risk such as a kitchen or a bathroom, etc.), **but** all DIY electrical work (unless completed by a qualified professional – who is responsible for issuing a minor electrical installation certificate) will still need to be checked, certified and tested by a competent electrician.

Any work that involves adding a new circuit to a dwelling will need to be either notified to the BCB (which will then inspect the work) or carried out by a competent person who is registered under a government-approved Part P self-certification scheme.

Work involving any of the following will also have to be notified:

*   consumer unit replacements;
*   electric floor or ceiling heating systems;
*   extra-low-voltage lighting installations, other than pre-assembled, CE-marked lighting sets;
*   garden lighting or power installations;
*   installation of a socket outlet on an external wall;
*   installation of new central heating control wiring;
*   installation of outdoor lighting and/or power installations in the garden or that involves crossing the garden;
*   outdoor lighting and power installations;
*   small-scale generators such as micro-CHP units.
*   solar photovoltaic (PV) power supply systems.

Note: Where a person who is **not** registered to self-certify intends to carry out the electrical installation, then a Building Regulation (i.e. a building notice or full plans) application will need to be submitted together with the appropriate fee, based on the estimated cost of the electrical installation. The BCB will then arrange to have the electrical installation inspected at first fix stage and tested upon completion.

2.15 How long is a building notice valid?

The approved plans may be used (i.e. built to) for at least three years, **even if the Building Regulations change during this time**.

A building notice shall cease to have effect three years from the date when that notice was given to the local authority unless, before the expiry of that period:

*   the building work to which the notice related has commenced; or
*   the material change of use described in the notice was made.

2.16 What can I do if my plans are rejected?

In the first two cases below, the address to write to is the Department for Communities and Local Government (DCLG). In Wales, you should refer the matter to the Secretary of State for Wales.

If your plans were initially rejected, you can start work **provided** you give the necessary notice of commencement required under Regulation 14 of the Building Regulations and are satisfied that the building work itself now complies with the regulations. However, it would **not** be advisable to follow this course if you are in any doubt and have not taken professional advice. Instead:

*   you should resubmit your full plans application with amendments to ensure that the plans comply with Building Regulations; or
*   if you think your plans comply (and that the decision to reject is, therefore, unjustified), you can refer the matter to the Secretary of State for Environment, Food and Rural Affairs or the Secretary of State for Wales (as appropriate) for his or her determination, but usually only before the work has started; or
*   you could (in particular cases) ask the local authority to relax or dispense with its rejection. If the local authority refuses your application you could then appeal to the appropriate Secretary of State within one month of the refusal.

A fee is payable for determinations but not for appeals. The fee is half the plan fee (excluding VAT) subject to a minimum of £50 and a maximum of £500. The Department of the Environment, Transport and the Regions (DETR) or the Welsh Office will then seek comments from the local authority on your application (or appeal), which will be copied to you. You will then have a further opportunity to comment before a decision is issued by the Secretary of State.

2.16.1 Do my neighbours have the right to object to what is proposed in my Building Regulations application?

Every planning application submitted to a local authority must undergo a period of public consultation which varies in length between three and eight weeks. Indeed, in May 2013 Parliament agreed to increase the size of single-storey rear extensions which can be built under permitted development and brought into force the new neighbour consultation scheme.

All plans can be viewed on the council website, at the local town hall or at the offices of the local planning authority. If a neighbour wishes to comment on the application he or she will contact the local planning authority.

If a neighbour is affected by a planning application, the local planning authority will either notify the neighbour directly or display a site notice on or near the land to which the application relates. The neighbour is allowed to attend planning committee meetings to hear applications being considered and you are also allowed to have your say at these meetings but you must notify the council in advance of the meeting.

Unfortunately there is no comprehensive description of grounds for objection to planning applications available, because the government has given the main responsibility for dealing with planning applications to local authorities. Therefore the range of objections that can be made is wide and varies from council to council. Objections may be raised under other legislation, particularly if your proposal is subject to approval under the town and country planning legislation or the Party Wall Act.

While there is no requirement in the Building Regulations to consult neighbours (other than within the permitted development neighbour consultation) it would be prudent to do so. In any event, you should be careful that the work does not encroach on a neighbour's property, as this could well lead to bad feeling and possibly an application for an injunction for the removal of the work.

A free explanatory booklet on the Party Wall Act can be downloaded from the government website (https://www.gov.uk/government/uploads/system/uploads/attachment_data/file/207310/Party_Wall_etc._Act_1996_-_Explanatory_Booklet.pdf).

In a nutshell, this act says that if you intend to carry out building work which involves:

- building a new wall on or at the boundary of two properties;
- building on the boundary with a neighbouring property, especially if this involves knocking down and rebuilding a party wall;
- cutting into a party wall;
- excavating near an adjoining building, including digging below the foundation level of a neighbour's property;
- making a party wall taller, shorter or deeper;
- removing chimney breasts from a party wall;
- work on an existing wall shared with another property

then you **must** find out whether that work falls within the scope of the act. If it does, you must serve the statutory notice on all those defined by the act as 'adjoining owners'. You should, however, remember that reaching agreement with adjoining owners on a project that falls within the scope of the act does **not** remove the possible need for planning permission or Building Regulations approval.

Note: If you are not sure whether the act applies to the work that you are planning, you should seek professional advice (see 'Useful contact names and addresses' at the end of this book).

2.17 What happens if I wish to seek a determination but the work in question has started?

If you start work before you receive a decision on your full plans application, you will prejudice your ability to seek a determination from the Secretary of State if there is a dispute.

You will only need to seek a determination if you believe the proposals in your full plans application comply with the regulations but the local authority disagrees. You may apply for a determination either before or after the local authority has formally rejected your full plans application. The legal procedure is intended to deal with compliance of 'proposed' work only and, in general, applications relating to work which is substantially completed cannot be accepted. Exceptionally, however, applications for 'late' determinations may be accepted – but it is in your best interest always to ensure that you apply for a determination well before you start work.

2.18 When can I start work?

Again, it depends on whether you are using the local authority or the approved inspector and what work you are planning.

2.18.1 Using the local authority

Once you have given a building notice or submitted a full plans application, you can start work at any time. However, you must give the local authority a commencement notice at least two clear days (not including the day on which you give notice and any Saturday, Sunday or bank or public holiday) before you start.

2.18.2 Using an approved inspector

If you use an approved inspector you may, subject to any arrangements you may have agreed with the inspector, start work as soon as the initial notice is accepted by the local authority (or is deemed to have been accepted if nothing is heard from the local authority within five working days of the notice being given). Work may not start if the initial notice is rejected, however.

2.19 Planning officers

Before construction begins, planning officers determine whether the plans for the building or other structure comply with the Building Regulations and if they are suited to the engineering and environmental demands of the building site. Building inspectors are then responsible for inspecting the structural quality and general safety of buildings.

2.20 Building inspectors

Building inspectors examine the construction, alteration or repair of buildings, highways and streets, sewer and water systems, dams, bridges and other structures to ensure compliance with building codes and ordinances, zoning regulations and contract specifications.

Building codes and standards are the primary means by which building construction is regulated in the United Kingdom to ensure the health and safety of the general public. Inspectors make an initial inspection during the first phase of construction and then complete follow-up inspections throughout the construction project in order to monitor compliance with regulations.

The inspectors will visit the worksite before the foundation is poured to inspect the soil condition and positioning and depth of the footings. Later, they return to the site to inspect the foundation after it has been completed. The size and type of structure, as well as the rate of completion, determine the number of other site visits they must make. Upon completion of the project, they make a final comprehensive inspection.

2.21 Notice of commencement and completion of certain stages of work

A person who proposes carrying out building work shall not start work unless he or she has given the local authority notice that he or she intends to commence work and at least two days have elapsed since the end of the day on which he or she gave the notice.

2.21.1 Notice of completion of certain stages of work

The person responsible for completing the building work is also responsible for notifying the local authority a minimum of five days **prior** to commencing any work involving excavations for foundations, foundations themselves, any damp-proof course, any concrete or other material to be laid over a site, and drains or sewers.

Upon completion of this work (especially work that will eventually be covered up by later work) the person responsible for the building work shall give five days' notice of intention to backfill.

 This requirement does not apply in respect of any work specified in an initial notice, an amendment notice or a public body's notice that is in force.

A person who has laid, haunched or covered any drain or sewer shall (not more than five days after that work has been completed) give the local authority notice to that effect.

Where a building is being erected and that building (or any part of it) is to be occupied before completion, the person carrying out that work shall give the local authority at least five days' notice before the building, or any part of it, is occupied.

The person carrying out the building work shall **not**:

- cover up any foundation (or excavation for a foundation), any damp-proof course or any concrete or other material laid over a site; or
- cover up (in any way) any drains or sewers **unless** he or she has given the local authority notice that he or she intends to commence that work and at least one day has elapsed since the end of the day on which he or she gave the notice.

 Note: Where a person fails to comply with the above, then the local authority can insist that he or she shall cut into, lay open or pull down 'so much of the work as to enable the authority to ascertain whether these Regulations have been complied with, or not'. If the local authority then notifies the owner or builder that certain work contravenes the requirements in these regulations, then the owner or builder shall, after completing the remedial work, notify the local authority of its completion.

2.21.2 What kind of tests are local authorities likely to make?

To establish whether building work has been carried out in conformance with the Building Regulations, local authorities will test to ensure that all work has been carried out:

- in compliance with the requirements of Approved Document H of Schedule 1 (drainage and waste disposal);
- in a workmanlike manner;
- so as to enable them to ascertain whether the materials used comply with the provisions of these regulations;
- with adequate and proper materials which:
  - ○ are appropriate for the circumstances in which they are used,
  - ○ are adequately mixed or prepared, and
  - ○ are applied, used or fixed so as to adequately perform the functions for which they are designed.

2.22 What are the requirements relating to building work?

In all cases, building work shall be carried out so that:

(a) it complies with the applicable requirements contained in Schedule 1; and

(b) in complying with any such requirement there is no failure to comply with any other such requirement.

Building work shall be carried out so that, **after** it has been completed:

(a) any building which is extended or to which a material alteration is made; or
(b) any building in or in connection with which a controlled service or fitting is provided, extended or materially altered; or
(c) any controlled service or fitting

complies with the applicable requirements of Schedule 1 or, where it did not comply with any such requirement, is no more unsatisfactory in relation to that requirement than before the work was carried out.

2.23 Do I need to employ a professional builder?

Unless you have a reasonable working knowledge of building construction it would be advisable before you start work to get some professional advice (e.g. from an architect, a structural engineer or a building surveyor) and/or choose a recognized builder to carry out the work. It is also advisable to consult the LABC officer or an approved inspector in advance.

2.24 Unauthorized building work

If, for any reason, building work has been done without a building notice or full plans of the work being deposited with the local authority, or a notice of commencement of work being given, then the applicant may apply, in writing, to the local authority for a regularization certificate. This application will need to include:

* a description of the unauthorized work;
* a plan of the unauthorized work; and
* a plan showing any additional work that is required for compliance with the requirements relating to building work in the Building Regulations.

Local authorities may then 'require the applicant to take such reasonable steps, including laying open the unauthorized work for inspection by the authority, making tests and taking samples, as the authority think appropriate to ascertain what work, if any, is required to secure that the relevant requirements are met'.

When the applicant has taken any such steps required by the local authority, the local authority will notify the applicant:

* if no work is required to secure compliance with the relevant requirements;
* of the work that is required to comply with the relevant requirements;
* of the requirements that can be dispensed with or relaxed.

2.24.1 What happens if I do work without approval?

The local authority has a general duty to see that all building work complies with the regulations – except where it is formally under the control of an approved inspector. Where a local authority is controlling the work and finds after its completion that it does not comply, then the local authority may require you to alter or remove it. If you fail to do this, the local authority may serve a notice requiring you to do so and you will be liable for the costs.

2.24.2 What are the penalties for contravening the Building Regulations?

If you contravene the Building Regulations by building without notifying the local authority or by carrying out work which does not comply, the local authority can prosecute. If you are convicted, you are liable to a penalty not exceeding **£5000** (at the date of publication of this book) plus £50 for each day on which each individual contravention is not put right after you have been convicted. If you do not put the work right when asked to do so, the local authority has power to do the work itself and recover costs from you.

2.25 Why do I need a completion certificate?

A completion certificate certifies that the local authority is satisfied that the work complies with the relevant requirements of Schedule 1 of the Building Regulations, 'in so far as they have been able to ascertain after taking all reasonable steps'.

 A completion certificate is a valuable document that should be kept in a safe place!

2.26 How do I get a completion certificate when the work is finished?

The local authority shall give a completion certificate only when it has received the completion notice and has been able to ascertain that the relevant requirements of Schedule 1 (specified in the certificate) have been satisfied.

Where full plans are submitted for work that is also subject to the Fire Precautions Act 1971, the local authority must issue you with a completion certificate concerning compliance with the fire safety requirements of the Building Regulations once work has finished. In other circumstances, you may ask to be given one when the work is finished, but you must make your request **when you first submit your plans**.

If you use an approved inspector, the inspector must issue a final certificate to the local authority when the work is completed.

2.27 Where can I find out more?

If you do not understand the technical guidance, you can seek further help:

*   through the Planning Portal website: www.planningportal.gov.uk.
*   from your local authority building control service or from an approved inspector (if you are the person undertaking the building work);
*   from the scheme operator (if you are registered with a competent person scheme);
*   from a specialist or an industry technical body for the relevant subject (if your query is of a highly technical nature).

2.27.1 Local authority

Each local authority in England and Wales (i.e. unitary, district and London boroughs in England and county borough councils in Wales) has a building control section whose general duty is to see that work complies with the Building Regulations – except where it is formally under the control of an approved inspector. Most local authorities have their own website and these usually contain a wealth of useful information, the majority of which is downloadable as read-only pdf files.

Individual local authorities coordinate their services regionally and nationally (and provide a range of national approval schemes) via LABC Services. You can find out more about LABC Services through its website (http://www. labc.uk.com), but your LABC department will be pleased to give you information and advice. It may offer to let you see its copies of the Building Act 1984, the Building Regulations 2010 and the associated Approved Documents that provide additional guidance.

The *Fire and Building Regulations Procedural Guide*, which deals with procedures for building work to which the Fire Precautions Act 1971 applies, and the DETR leaflet on safety of garden walls are amongst the documentation and advice that are available, free of charge, from your local authority.

The DETR's (and the Welsh Office's) separate booklets on planning permission for small businesses and householders are also available free of charge from your local authority.

2.27.2 Approved inspectors

Approved inspectors are companies or individuals authorized under the Building Act 1984 to carry out building control work in England and Wales.

The Construction Industry Council (CIC) is responsible for deciding all applications for approved inspector status. You can find out more about the CIC's role (including how to apply to become an approved inspector) through its website. A list of approved inspectors can be viewed at http://www.cic.org. uk/services/register.php.

2.27.3 Planning Portal

Most of the documents relating to Building Regulations can be downloaded from the Planning Portal (http://www.planningportal.gov.uk/buildingregulations). They can also be purchased from The Stationery Office or from any main bookshop. Copies should also be available in public reference libraries.

2.27.4 Specialist advice

If you have a query of a highly technical nature you can seek specialist advice from an 'expert' in the field or from the relevant industry technical body. Advice on suitable technical bodies can be found on Constructionline, which is a register for pre-qualified contractors and consultants (http://www.constructionline.co.uk/static/suppliers/useful-links.html).

3

The requirements of the Building Regulations

3.1 Introduction

This chapter provides a breakdown of the key elements of each Approved Document, showing not only the regulation, but also an overview of the actual requirement 'in a nutshell'. This acts as a good starting point to understanding the regulations, and it is then expanded upon in much more detail by specific building elements in Chapter 6. The current editions of the Approved Documents are shown in Table 3.1. All statutory instruments can be accessed on the UK government legislation website (http://www.legislation.gov.uk) and the Planning Portal. They can also can be purchased from The Stationery Office or RIBA.

If you intend to carry out work that involves any of these areas you are recommended to view or download the appropriate instrument online or purchase a copy for reference.

Table 3.1 Current editions of Approved Documents

Section	Title	Edition	Latest amendment
A	Structure	2004	2010 and 2013
B	Fire safety	2006	2010 and 2013
C	Site preparation and resistance to moisture	2004	2010 and 2013
D	Toxic substances	1992	2002, 2010 and 2013
E	Resistance to the passage of sound	2003	2004, 2010 and 2013
F	Ventilation	2010	2013
G	Sanitation, hot water safety and water efficiency	2010	2010 and 2013
H	Drainage and waste disposal	2002	2010 and 2013
J	Combustion appliances and fuel storage systems	2010	2013
K	Protection from falling, collision and impact	2013	
L	Conservation of fuel and power (L1A and L2A)	2013	2013
	Conservation of fuel and power (L1B and L2B)	2010	
M	Access to and use of buildings	2004	2010 and 2013
P	Electrical safety	2013	
	Approved Document to support Regulation 7 – Materials and workmanship	2013	

 Note: The 2013 changes to Approved Document K, N and P currently only apply in England; the previous versions of these documents **still** apply in Wales.

Table 3.2 Part A – Structure

Number	Title	Regulation	Requirement (in a nutshell)
A1	Loading	(1) The building shall be constructed so that the combined dead, imposed and wind loads are sustained and transmitted by it to the ground: (a) safely; and (b) without causing such deflection or deformation of any part of the building, or such movement of the ground, as will impair the stability of any part of another building. (2) In assessing whether a building complies with sub-paragraph (1) regard shall be had to the imposed and wind loads to which it is likely to be subjected in the ordinary course of its use for the purpose for which it is intended.	The safety of a structure depends on: • the loading (see BS 6399: Parts 1 and 3); • properties of materials; • design analysis; • details of construction; • safety factors; • workmanship.
A2	Ground movement	The building shall be constructed so that ground movement caused by: (a) swelling, shrinkage or freezing of the subsoil; or (b) land-slip or subsidence (other than subsidence arising from shrinkage, in so far as the risk can be reasonably foreseen), will not impair the stability of any part of the building.	• Horizontal and vertical ties should be provided.
A3	Disproportionate collapse	The building shall be constructed so that in the event of an accident the building will not suffer collapse to an extent disproportionate to the cause.	

Table 3.3 Part B – Fire safety

Number	Title	Regulation	Requirement (in a nutshell)
B1	Means of warning and escape	*The building shall be designed and constructed so that there are appropriate provisions for the early warning of fire, and appropriate means of escape in case of fire from the building to a place of safety outside the building capable of being safely and effectively used at all material times.* B1 does not apply to any prison provided under Section 33 of the Prisons Act 1952 (1952 c.52; Section 33 was amended by Section 100 of the Criminal Justice and Public Order Act 1994 (c.33) and by S.I. 1963/597).	• There shall be an early warning fire alarm system for persons in the building. • There shall be sufficient escape routes that are suitably located to enable persons to evacuate the building in the event of a fire. • Safety routes shall be protected from the effects of fire. • In an emergency, the occupants of any part of the building shall be able to escape without any external assistance.

B2	Internal fire spread (linings)	(1) *To inhibit the spread of fire within the building, the internal linings shall:*
		
		  (a) *adequately resist the spread of flame over their surfaces; and*
		  (b) *have, if ignited, a rate of heat release or a rate of fire growth which is reasonable in the circumstances.*
		
		(2) *In this paragraph 'internal linings' means the materials or products used in lining any partition, wall, ceiling or other internal structure.*
			• The spread of flame over the internal linings of the building shall be restricted.
			• The heat released from the internal linings shall be restricted.

B3	Internal fire spread (structure)	(1) *The building shall be designed and constructed so that, in the event of fire, its stability will be maintained for a reasonable period.* (2) *A wall common to two or more buildings shall be designed and constructed so that it adequately resists the spread of fire between those buildings.*	Dependent on the use of the building, its size and the location of the element of construction: • load-bearing elements of a building structure shall be capable of withstanding the effects of fire for an appropriate period without loss of stability;

Table 3.3 Part B – Fire safety (Continued)

Number	Title	Regulation	Requirement (in a nutshell)
	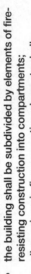	*For the purposes of this sub-paragraph a house in a terrace and a semi-detached house are each to be treated as a separate building.* *(3) Where reasonably necessary to inhibit the spread of fire within the building, measure shall be taken, to an extent appropriate to the size and intended use of the building, comprising either or both of the following:* *(a) sub-division of the building with fire-resisting construction;* *(b) installation of suitable automatic fire suppression systems.* *(4) The building shall be designed and constructed so that the unseen spread of fire and smoke within concealed spaces in its structure and fabric is inhibited.* B3(3) does not apply to material alterations to any prison provided under Section 33 of the Prisons Act 1952.	• the building shall be subdivided by elements of fire-resisting construction into compartments; • all openings in fire-separating elements shall be suitably protected in order to maintain the integrity of the element (i.e. the continuity of the fire separation); • any hidden voids in the construction shall be sealed and subdivided to inhibit the unseen spread of fire and products of combustion.

Table 3.3 Part B – Fire safety

Number	Title	Regulation	Requirement (in a nutshell)
B4	External fire spread	*(1) The external walls of the building shall adequately resist the spread of fire over the walls and from one building to another, having regard to the height, use and position of the building.* *(2) The roof of the building shall adequately resist the spread of fire over the roof and from one building to another, having regard to the use and position of the building.*	• External walls shall be constructed so as to have a low rate of heat release and thereby be capable of reducing the risk of ignition from an external source and the spread of fire over their surfaces. • The amount of unprotected area in the side of the building shall be restricted so as to limit the amount of thermal radiation that can pass through the wall. • The roof shall be constructed so that the risk of spread of flame and/or fire penetration from an external fire source is restricted.
B5	Access and facilities for the fire service	*(1) The building shall be designed and constructed so as to provide reasonable facilities to assist firefighters in the protection of life.* *(2) Reasonable provision shall be made within the site of the building to enable fire appliances to gain access to the building.*	• There is sufficient external access for fire appliances to be brought near to the building for use. • There is sufficient access into and within the building for firefighting personnel to effect search and rescue and fight fire. • The building has sufficient internal fire mains and other facilities to assist firefighters in their tasks. • The building has adequate means to vent heat and smoke from a fire in a basement. The extent to which these arrangements and facilities apply varies with the size and use of building.

Table 3.4 Part C – Site preparation and resistance to contaminants and moisture

Number	Title	Regulation	Requirement (in a nutshell)
C1	Site preparation and resistance to contaminants	(1) *The ground to be covered by the building shall be reasonably free from any material that might damage the building or affect its stability, including vegetable matter, topsoil and pre-existing foundations.* (2) *Reasonable precautions shall be taken to avoid danger to health and safety caused by contaminants on or in the ground covered, or to be covered by the building and any land associated with the building.* (3) *Adequate subsoil drainage shall be provided if it is needed to avoid:* (a) *the passage of the ground moisture to the interior of the building;* (b) *damage to the building, including damage through the transport of water-borne contaminants to the foundations of the building.* (4) *For the purpose of this requirement, 'contaminant' means: any substance that is or may become harmful to persons or buildings including substances which are corrosive, explosive, flammable, radioactive or toxic.*	Buildings should be safeguarded from the adverse effects of: • vegetable matter; • contaminants on or in the ground to be covered by the building; • groundwater.

Table 3.4 Part C – Site preparation and resistance to contaminants and moisture

Number	Title	Regulation	Requirement (in a nutshell)
C2	Resistance to moisture	The floors, walls and roof of the building shall adequately protect the building and people who use the building from harmful effects caused by: (a) ground moisture; (b) precipitation and wind-driven spray; (c) interstitial and surface condensation; and (d) spillage of water from or associated with sanitary fittings or fixed appliances.	• A solid or suspended floor shall be built next to the ground to prevent undue moisture from reaching the upper surface of the floor. • A wall shall be erected to prevent undue moisture from the ground reaching the inside of the building, and (if it is an outside wall) adequately resisting the penetration of rain and snow to the inside of the building. • The roof of the building shall be resistant to the penetration of moisture from rain or snow to the inside of the building. • All floors next to the ground, walls and roof shall not be damaged by moisture from the ground, rain or snow and shall not carry that moisture to any part of the building which it would damage.

Table 3.5 Part D – Toxic substances

Number	Title	Regulation	Requirement (in a nutshell)
D1	Cavity insulation	If insulating material is inserted into a cavity in a cavity wall, reasonable precautions shall be taken to prevent the subsequent permeation of any toxic fumes from that material into any part of the building occupied by people.	Fumes given off by insulating materials such as by urea formaldehyde (UF) foams should not be allowed to penetrate occupied parts of buildings to an extent where they could become a health risk to persons in the building by reaching an irritant concentration.

Table 3.6 Part E – Resistance to the passage of sound

Number	Title	Regulation	Requirement (in a nutshell)
E1	Protection against sound from other parts of the building and adjoining buildings	*Dwelling-houses, flats and rooms for residential purposes shall be designed and constructed in such a way that they provide reasonable resistance to sound from other parts of the same building and from adjoining buildings.*	Dwellings shall be designed so that the noise from domestic activity in an adjoining dwelling (or other parts of the building) is kept to a level that: • does not affect the health of the occupants of the dwelling; • will allow them to sleep, rest and engage in their normal activities in satisfactory conditions. Dwellings shall be designed so that any domestic noise that is generated internally does not interfere with the occupants' ability to sleep, rest and engage in their normal activities in satisfactory conditions.
E2	Protection against sound within a dwelling-house etc.	*Dwelling-houses, flats and rooms for residential purposes shall be designed and constructed in such a way that:* *(a) internal walls between a bedroom or a room containing a water closet, and other rooms; and* *(b) internal floors provide reasonable resistance to sound.* E2 does not apply to: (a) an internal wall which contains a door; (b) an internal wall which separates an en suite toilet from the associated bedroom; (c) existing walls and floors in a building which is subject to a material change of use.	Suitable sound-absorbing material shall be used in domestic buildings so as to restrict the transmission of noise from other rooms.

Table 3.6 Part E – Resistance to the passage of sound

Number	Title	Regulation	Requirement (in a nutshell)
E3	Reverberation in the common internal parts of buildings containing flats or rooms for residential purposes	The common internal parts of buildings which contain flats or rooms for residential purposes shall be designed and constructed in such a way as to prevent more reverberation around the common parts than is reasonable. 💡 E3 only applies to corridors, stairwells, hallways and entrance halls which give access to the flat or rooms for residential purposes.	Suitable sound-absorbing material shall be used in domestic buildings so as to restrict the transmission of echoes.
E4	Acoustic conditions in schools	(1) Each room or other space in a school building shall be designed and constructed in such a way that it has the acoustic conditions and the insulation against disturbance by noise appropriate to its intended use. (2) For the purposes of this Part – 'school' has the same meaning as in Section 4 of the Education Act 1996 [4]; and 'school building' means any building forming a school or part of a school.	Suitable sound-insulation materials shall be used within a school building so as to reduce the level of ambient noise (particularly echoing in corridors, etc.).

Table 3.7 Part F – Ventilation

Number	Title	Regulation	Requirement (in a nutshell)
F1	Means of ventilation	(1) *There shall be adequate means of ventilation provided for people in the building.* (2) *Fixed systems for mechanical ventilation and any associated controls must be commissioned by testing and adjusting as necessary to secure that the objective referred to in sub-paragraph (1) is met.* F1 does not apply to a building or space within a building: (a) into which people do not normally go; or (b) which is used solely for storage; or (c) which is a garage used solely in connection with a single dwelling.	Ventilation (mechanical and/or air-conditioning systems designed for domestic buildings) shall be capable of restricting the accumulation of moisture and pollutants originating within a building.
Requirements in the Building Regulations 2010			
39	Information about ventilation	(1) *This regulation applies where Part F(1) of Schedule 1 imposes a requirement in relation to building work.* (2) *The person carrying out the work shall not later than five days after the work has been completed give sufficient information to the owner about the building's ventilation system and its maintenance requirements so that the ventilation system can be operated in such a manner as to provide adequate means of ventilation.*	The installer shall provide details of the building's ventilation system to the owner.
42	Mechanical ventilation air flow rate testing	(1) *This regulation applies where paragraph F1(1) of Schedule 1 imposes a requirement in relation to the creation of a new dwelling by building work.* (2) *The person carrying out the work shall, for the purpose of ensuring compliance with paragraph F1(1) of Schedule 1* (a) *Ensure that testing of the mechanical ventilation air flow rate is carried out in accordance with a procedure approved by the Secretary of State; and*	Mechanical ventilation air flow rate testing shall be carried out.

Table 3.7 Part F – Ventilation

Number	Title	Regulation	Requirement (in a nutshell)
		(b) Give notice of the results of the testing to the local authority.	
		(3) The notice referred to in paragraph 2(b) shall	
		(a) Record the results and the date upon which they are based in a manner approved by the Secretary of State; and	
		(b) Be given to the local authority not later than five days after the final test is carried out.	
44	Commissioning	(1) This regulation applies to building work in relation to which paragraph F1(2) of Schedule 1 imposes a requirement, but does not apply to the provision or extension of any fixed system for mechanical ventilation or any associated controls where testing and adjustment is not possible.	The installer shall give the local authority a notice confirming that the fixed building services have been commissioned in accordance with procedures.
		(2) This regulation applies to building work in relation to which paragraph L1(b) of Schedule 1 imposes a requirement, but does not apply to the provision or extension of any fixed building service where testing and adjustment is not possible or would not affect the energy efficiency of that fixed building service.	
		(3) Where this regulation applies the person carrying out the work shall, for the purpose of complying with F1(2) or L1(b) of Schedule 1, give to the local authority a notice confirming that the fixed building services have been commissioned in accordance with a procedure approved by the Secretary of State.	
		(4) The notice shall be given to the local authority	
		(a) not later than the date on which the notice required by regulation 16(4) is required to be given; or	
		(b) where the regulation does not apply, not more than 30 days after the completion of the work.	

Table 3.7 Part F – Ventilation (Continued)

Number	Title	Regulation	Requirement (in a nutshell)
		Requirements in the Building (Approved Inspectors etc.) Regulations 2010 Application of provisions of the Principal Regulations	
20		*(1) Regulations 20 (provisions applicable to self-certification schemes), 27 CO_2 (emission rate calculations), 29 (energy performance certificates), 37 (wholesome water consumption calculation), 41 (sound insulation testing), 42 (mechanical ventilation air flow rate testing), 43 (pressure testing), and 44 (commissioning) of the Principal Regulations apply in relation to building work which is the subject of an initial notice as if references to the local authority were references to the approved inspector.*	
		(6) Regulation 44 of the Principal Regulations applies in relation to building work which is the subject of an initial notice as if for paragraph (4) there were substituted	
		(4) The initial notice shall be given to the approved inspector	
		(a) Subject to sub-paragraphs (b) and (c), not later than five days after the completion of the work to which the initial notice relates;	
		(b) Where regulation 17 of the Building (Approved Inspectors etc.) Regulations 2010 applies, not later than the date on which the initial notice ceases to be in force or, if earlier, the end of the period referred to in sub-paragraph (a);	
		(c) Where regulation 20 applies by virtue of regulation 20 of the Building (Approved Inspectors etc.) Regulations 2010, not later than the date on which the notice or certificate required by that regulation must be given.'	

Table 3.8 Part G – Hygiene

Number	Title	Regulation	Requirement (in a nutshell)
G1	Cold water supply	(1) *There must be a suitable installation for the provision of:* (a) *wholesome water to any place where drinking water is drawn off;* (b) *wholesome water or softened wholesome water to any washbasin or bidet provided in or adjacent to a room containing a sanitary convenience;* (c) *wholesome water or softened wholesome water to any washbasin, bidet, fixed bath or shower in a bathroom; and* (d) *wholesome water to any sink provided in any area where food is prepared.* (2) *There must be a suitable installation for the provision of water of suitable quality to any sanitary convenience fitted with a flushing device.*	• The cold water supply must be reliable. • The water supplied must be wholesome. • The pressure and flow rate must be sufficient for the operation of all appliances and locations planned in the building. • The installation must convey wholesome water or softened wholesome water without waste, misuse, undue consumption or contamination of water. For sanitary conveniences: • the water supplied may be wholesome, softened wholesome or of a suitable quality.
G2	Water efficiency	*Reasonable provision must be made by the installation of fittings and fixed appliances that use water efficiently for the prevention of undue consumption of water.* *G2 applies only when a dwelling is:* (a) *erected; or* (b) *formed by a material change of use of a building.*	For new dwellings: • white goods shall be installed in accordance with the requirements of this Approved Document; • a record of all installed sanitary appliances and relevant white goods (washing machines and dishwashers) shall be provided; • a record of the alternative sources of water supplied to the dwelling shall be provided.

Table 3.8 Part G – Hygiene (Continued)

Number	Title	Regulation	Requirement (in a nutshell)
Water efficiency of new dwellings			
36		*(1) The potential consumption of wholesome water by persons occupying a dwelling to which this regulation applies must not exceed 125 litres per person per day, calculated in accordance with the methodology set out in the document 'The Water Efficiency Calculator for New Dwellings', published in September 2009 by the DCLG.* *(2) This regulation applies to a dwelling which is:* *(a) erected; or* *(b) formed by a material change of use of a building within the meaning of regulation 5(a) or (b).*	• The estimated consumption of wholesome water for both cold and hot water systems shall not be greater than 125 litres per person per day of wholesome water
Wholesome water consumption calculation			
37		*(1) Where regulation 36 applies, the person carrying out the work must give the local authority a notice which specifies the potential consumption of wholesome water per person per day calculated in accordance with the methodology referred to in that regulation in relation to the completed dwelling.* *(2) The notice shall be given to the local authority not later than five days after the work has been completed.*	
Requirements in the Building (Approved Inspectors etc.) **Regulations 2010 Application of provisions of the Principal Regulations**			
20		*(1) Regulations 20 (provisions applicable to self-certification schemes), 27 CO$_2$ (emission rate calculations), 29 (energy performance certificates), 37 (wholesome water consumption calculation), 41 (sound insulation testing), 42 (mechanical ventilation air flow rate testing), 43 (pressure testing), and 44 (commissioning) of the Principal Regulations apply in relation to building work which is the subject of an initial notice as if references to the local authority were references to the approved inspector.*	

Table 3.8 Part G – Hygiene

Number	Title	Regulation	Requirement (in a nutshell)
		(4) Regulation 37(2) of the Principal Regulations applies in relation to building work which is the subject of an initial notice as if after 'work has been completed' there were inserted, 'or, if earlier the date on which in accordance with regulation 17 of the Building (Approved Inspectors etc.) Regulations 2010 the initial notice ceases to be in force'.	

G3	Hot water supply and systems	*(1) There must be a suitable installation for the provision of heated wholesome water or heated softened wholesome water to:*
		
		(a) any washbasin or bidet provided in or adjacent to a room containing a sanitary convenience;
		(b) any washbasin, bidet, fixed bath or shower in a bathroom; and
		(c) any sink provided in any area where food is prepared.
		
		(2) A hot water system, including any cistern or other vessel that supplies water to or receives expansion water from a hot water system, shall be designed, constructed and installed so as to resist the effects of temperature and pressure that may occur either in normal use, or in the event of such malfunctions as may reasonably be anticipated, and must be adequately supported.
		
		(3) A hot water system that has a hot water storage vessel shall incorporate precautions to:
		
		(a) prevent the temperature of the water stored in the vessel at any time exceeding 100° C; and
		(b) ensure that any discharge from safety devices is safely conveyed to where it is visible but will not cause a danger to persons in or about the building.
		
		G3(3) does not apply to a system which heats or stores water only for the purposes of an industrial process.
			• Water supplied shall be heated wholesome water or heated softened water.
			• The installation shall convey hot water without waste, misuse or undue consumption of water.
			• Hot water systems (including cisterns) shall safely contain the hot water:
			
			  ○ during normal operation of the hot water system;
			  ○ following failure of any thermostat used to control temperature; and
			  ○ during operation of any of the safety devices.
			
			• Storage vessels shall have a suitable vent pipe connecting the top to a point open to the atmosphere above the level of the water in the cold water storage cistern.
			• In addition to any thermostat, either the heat source or the storage vessel shall be fitted with a device to prevent the temperature of the stored water exceeding 100° C.

Table 3.8 Part G – Hygiene (Continued)

Number	Title	Regulation	Requirement (in a nutshell)
		(4) *The hot water supply to any fixed bath must be so designed and installed as to incorporate measures to ensure that the temperature of the water that can be delivered to that bath does not exceed 48° C.*	
		G3(4) applies only when a dwelling is: (a) erected; or (b) formed by a material change of use within the meaning of Regulation 5(a) or (b).	
G4	Sanitary conveniences and washing machines	(1) *Adequate and suitable sanitary conveniences must be provided in rooms provided to accommodate them or in bathrooms.* (2) *Adequate stand washing facilities must be provided in:* (a) *rooms containing sanitary conveniences; or* (b) *rooms or spaces adjacent to rooms containing sanitary conveniences.* (3) *Any room containing a sanitary convenience, a bidet, or any facility for washing hands provided in accordance with paragraph (2)(b) must be separated from any kitchen or any area where food is prepared.*	• The hot water system shall have pipework that allows hot water from the safety devices to discharge into a place open to the atmosphere where it will cause no danger to persons in or about the building. • Storage vessels shall have at least two independent safety devices that release pressure and, in doing so, prevent the temperature of the stored water at any time exceeding 100° C. Hot water outlet temperature devices used to limit the maximum temperature that is supplied at the outlet shall not be easily altered by building users.
G5	Bathrooms	*A bathroom must be provided containing a wash basin and either a fixed bath or a shower.* G5 applies only to dwellings and to buildings containing one or more rooms for residential purposes.	
G6	Food preparation areas	*A suitable sink must be provided in any area where food is prepared.*	

Table 3.9 Part H – Drainage and waste disposal

Number	Title	Regulation	Requirement (in a nutshell)

H1	Foul water drainage	(1) *An adequate system of drainage shall be provided to carry foul water from appliances within the building to one of the following, listed in order of priority:*
		
		  (a) *a public sewer; or, where that is not reasonably practicable,*
		  (b) *a private sewer communicating with a public sewer; or, where that is not reasonably practicable,*
		  (c) *either a septic tank which has an appropriate form of secondary treatment or another wastewater treatment system; or, where that is not reasonably practicable,*
		  (d) *a cesspool.*
		
		(2) *In this Part 'foul water' means wastewater which comprises or includes:*
		
		  (a) *waste from a sanitary convenience, bidet or appliance used for washing receptacles for foul waste; or*
		  (b) *water which has been used for food preparation, cooking or washing.*
		
		H1 does not apply to the diversion of water which has been used for personal washing or for the washing of clothes, linen or other articles to collection systems for reuse.
			The foul water drainage system shall:
			
			• convey the flow of foul water to a foul water outfall (i.e. sewer, cesspool, septic tank or settlement, i.e. holding, tank);
			• minimize the risk of blockage or leakage;
			• prevent foul air from the drainage system from entering the building under working conditions;
			• be ventilated;
			• be accessible for clearing blockages;
			• not increase the vulnerability of the building to flooding.

H2		(1) *Any septic tank and its form of secondary treatment system or cesspool shall be so sited and constructed that:*
		
		  (a) *it is not prejudicial to the health of any person;*
		  (b) *it will not contaminate any watercourse, underground water or water supply;*
		  (c) *there are adequate means of access for emptying and maintenance; and*
		  (d) *where relevant, it will function to a sufficient standard for the protection of health in the event of a power failure.*
			Wastewater treatment systems shall:
			
			• have sufficient capacity to enable breakdown and settlement of solid matter in the wastewater from the buildings;
			• be sited and constructed so as to prevent overloading of the receiving water.

Table 3.9 Part H – Drainage and waste disposal (Continued)

Number	Title	Regulation	Requirement (in a nutshell)
		(2) *Any septic tank, holding tank which is part of a wastewater treatment system or cesspool shall be:* (a) *of adequate capacity;* (b) *so constructed that it is impermeable to liquids; and* (c) *adequately ventilated.*	Cesspools shall have sufficient capacity to store the foul water from the building until they are emptied. Wastewater treatment systems and cesspools shall be sited and constructed so as not to: • be prejudicial to health or a nuisance; • adversely affect water sources or resources; • pollute controlled waters; and • be in an area where there is a risk of flooding.
		(3) *Where a foul water drainage system from a building discharges to a septic tank, wastewater treatment system or cesspool, a durable notice shall be affixed in a suitable place in the building containing information on any continuing maintenance requirement to avoid risks to health.*	Septic tanks and wastewater treatment systems and cesspools shall be constructed and sited so as to: • have adequate ventilation; • prevent leakage of the contents and ingress of subsoil water; and • have regard to water table levels at any time of the year and rising groundwater levels. Drainage fields shall be sited and constructed to: • avoid overloading of the soakage capacity; and • provide adequately for the availability of an aerated layer in the soil at all times.

Table 3.9 Part H – Drainage and waste disposal

Number	Title	Regulation	Requirement (in a nutshell)
H3	Rainwater drainage	(1) Adequate provision shall be made for rainwater to be carried from the roof of the building. (2) Paved areas around the building shall be so constructed as to be adequately drained. (3) Rainwater from a system provided pursuant to sub-paragraphs (1) or (2) shall discharge to one of the following, listed in order of priority: (a) an adequate soakaway or some other adequate infiltration system; or, where that is not reasonably practicable, (b) a watercourse; or, where that is not reasonably practicable, (c) a sewer. H3(2) applies only to paved areas: (a) which provide access to the building pursuant to requirement M1 (access and use), or requirement M2 (access to extension to buildings other than dwellings); (b) which provide access to or from a place of storage pursuant to paragraph H6(2) of Schedule 1 (solid waste storage); or (c) in any passage giving access to the building where this is intended to be used in common by the occupiers of one or more other buildings. H3(3) does not apply to the gathering of rainwater for reuse.	Rainwater drainage systems shall: • minimize the risk of blockage or leakage; • be accessible for clearing; • ensure that rainwater soaking into the ground is distributed sufficiently so that it does not damage foundations of the proposed building or any adjacent structure; • ensure that rainwater from roofs and paved areas is carried away from the surface either by a drainage system or by other means; • ensure that the rainwater drainage system carries the flow of rainwater from the roof to an outfall (e.g. a soakaway, a watercourse or a surface water or combined sewer).
H4	Building over sewers	(1) The erection or extension of a building or work involving the underpinning of a building shall be carried out in a way that is not detrimental to the building or building extension or to the continued maintenance of the drain, sewer or disposal main. (2) In this paragraph 'disposal main' means any pipe, tunnel or conduit used for the conveyance of effluent to or from a sewage disposal works, which is not a public sewer. (3) In this paragraph and paragraph H5 'map of sewers' means any records kept by a sewerage undertaker under Section 199 of the Water Industry Act 1991.	Building or extension or work involving underpinning shall: • be constructed or carried out in a manner which will not overload or otherwise cause damage to the drain, sewer or disposal main either during or after the construction; • not obstruct reasonable access to any manhole or inspection chamber on the drain, sewer or disposal main;

Table 3.9 Part H – Drainage and waste disposal (Continued)

Number	Title	Regulation	Requirement (in a nutshell)
		H4 applies only to work carried out: (a) over a drain, sewer or disposal main which is shown on any map of sewers; or (b) on any site or in such a manner as may result in interference with the use of, or obstruction of the access of any person to, any drain, sewer or disposal main which is shown on any map of sewers.	• in the event of the drain, sewer or disposal main requiring replacement, not unduly obstruct work to replace the drain, sewer or disposal main, on its present alignment; • reduce the risk of damage to the building as a result of the failure of the drain, sewer or disposal main.
H5	Separate systems of drainage	*Any system for discharging water to a sewer which is provided pursuant to paragraph H3 shall be separate from that provided for the conveyance of foul water from the building.* H5 applies only to a system provided in connection with the erection or extension of a building where it is reasonably practicable for the system to discharge directly or indirectly to a sewer for the separate conveyance of surface water which is: (a) shown on a map of sewers; or (b) under construction either by the sewerage undertaker or by some other person (where the sewer is the subject of an agreement to make a declaration of vesting pursuant to Section 104 of the Water Industry Act 1991; Section 104 was amended by Section 96 of Schedule 9 to the Water Act 2003, and is prospectively amended by Section 42 of the Flood and Water Management Act 2010 (c.29)).	Separate systems of drains and sewers shall be provided for foul water and rainwater where: • the rainwater is not contaminated; and • the drainage is to be connected either directly or indirectly to the public sewer system and either: ○ the public sewer system in the area comprises separate systems for foul water and surface water; or ○ a system of sewers which provides for the separate conveyance of surface water is under construction either by the sewerage undertaker or by some other person (where the sewer is the subject of an agreement to make a declaration of vesting pursuant to Section 104 of the Water Industry Act 1991).

Table 3.9 Part H – Drainage and waste disposal

Number	Title	Regulation	Requirement [in a nutshell]
H6	Solid waste storage	(1) *Adequate provision shall be made for storage of solid waste.* (2) *Adequate means of access shall be provided:* (a) *for people in the building to the place of storage; and* (b) *from the place of storage to a collection point (where one has been specified by the waste collection authority under Section 46 (household waste) or Section 47 (commercial waste) of the Environmental Protection Act 1990) (1990 c.43; Section 46 was amended by Section 19 of the London Local Authorities Act 2007 (2007 c. ii) and Section 47 was amended by Section 21 of the Act; Section 46 was also amended by Section 76 and Schedule 5 to the Climate Change Act 2008 (c..28)) or to a street (where no collection point has been specified).*	Solid waste storage shall be: • designed and sited so as not to be prejudicial to health; • of sufficient capacity having regard to the quantity of solid waste to be removed and the frequency of removal; • sited so as to be accessible for use by people in the building and of ready access from a street for emptying and removal.

Table 3.10 Part J – Combustion appliances and fuel storage systems

Number	Title	Regulation	Requirement (in a nutshell)

J1	Air supply	*Combustion appliances shall be so installed that there is an adequate supply of air to them for combustion, to prevent over-heating and for the efficient working of any flue.*
		
		J1 only applies to fixed combustion appliances (including incinerators).
			The building shall:
			
			• enable the admission of sufficient air for:
			
			  ○ the proper combustion of fuel and the operation of flues; and
			  ○ the cooling of appliances where necessary;
			
			• enable normal operation of appliances without the products of combustion becoming a hazard to health;
			• enable normal operation of appliances without their causing danger through damage by heat or fire to the fabric of the building;
			• have been inspected and tested to establish suitability for the purpose intended;
			• have been labelled to indicate performance capabilities.

J2	Discharge of products of combustion	*Combustion appliances shall have adequate provision for the discharge of products of combustion to the outside air.* J2 only applies to fixed combustion appliances (including incinerators).	
J3	Warning of release of carbon monoxide	*Where a fixed combustion appliance is provided, appropriate provision shall be made to detect and give warning of the release of carbon monoxide.* J3 only applies to fixed combustion appliances located in dwellings.	

Table 3.10 Part J – Combustion appliances and fuel storage systems

Number	Title	Regulation	Requirement (in a nutshell)
J4	Protection of building	*Combustion appliances and flue-pipes shall be so installed, and fireplaces and chimneys shall be so constructed and installed, as to reduce to a reasonable level the risk of people suffering burns or the building catching fire in consequence of their use.* J4 only applies to fixed combustion appliances located in dwellings.	Oil and liquid petroleum gas (LPG) fuel storage installations shall be located and constructed so that they are reasonably protected from fires that may occur in buildings or beyond boundaries. Oil storage tanks used wholly or mainly for private dwellings shall: • be reasonably resistant to physical damage and corrosion; • be designed and installed so as to minimize the risk of oil escaping during the filling or maintenance of the tank; • incorporate secondary containment when there is a significant risk of pollution; • be labelled with information on how to respond to a leak.
J5	Provision of information	*Where a hearth, fireplace, flue or chimney is provided or extended, a durable notice containing information on the performance capabilities of the hearth, fireplace, flue or chimney shall be affixed in a suitable place in the building for the purpose of enabling combustion appliances to be safely installed.*	
J6	Protection of liquid fuel storage systems	*Liquid fuel storage systems and the pipes connecting them to combustion appliances shall be so constructed and separated from buildings and the boundary of the premises as to reduce to a reasonable level the risk of the fuel igniting in the event of fire in adjacent buildings or premises.* J6 applies only to: (a) fixed oil storage tanks with capacities greater than 90 litres and connecting pipes; and (b) fixed liquefied petroleum gas storage installations with capacities greater than 150 litres and connecting pipes, which are located outside the building and which serve fixed combustion appliances (including incinerators) in the building.	

Table 3.10 Part J – Combustion appliances and fuel storage systems (Continued)

Number	Title	Regulation	Requirement (in a nutshell)
J7	Protection against pollution	*Oil storage tanks and the pipes connecting them to combustion appliances shall:*	
		(a) *be so constructed and protected as to reduce to a reasonable level the risk of the oil escaping and causing pollution; and*	
		(b) *have affixed in a prominent position a durable notice containing information on how to respond to an oil escape so as to reduce to a reasonable level the risk of pollution.*	
		J7 applies only to fixed oil storage tanks with capacities of 3500 litres or less, and connecting pipes, which:	
		(a) are located outside the building; and	
		(b) serve fixed combustion appliances (including incinerators) in a building used wholly or mainly as a private dwelling	
		but does not apply to buried systems.	

Table 3.11 Part K – Protection from falling, collision and impact

Number	Title	Regulation	Requirement (in a nutshell)
K1	Stairs, ladders and ramps	*Stairs, ladders and ramps shall be so designed, constructed and installed as to be safe for people moving between different levels in or about the building.* *K1 applies only to stairs, ladders and ramps which form part of the building.*	All stairs, steps and ladders shall provide reasonable safety between levels in a building. Stairs, etc. in public buildings may need to be of a higher standard than in a dwelling.
K2	Protection from falling	*(a) Any stairs, ramps, floors and balconies and any roof to which people have access, and* *(b) any light well, basement area or similar sunken area connected to a building,* *shall be provided with barriers where it is necessary to protect people in or about the building from falling.* *K2 (a) applies only to stairs and ramps which form part of the building.*	Pedestrian guarding should be provided for any part of a floor, gallery, balcony, roof or any other place to which people have access and any light well, basement area or similar sunken area next to a building.
K3	Vehicle barriers and loading bays	*(1) Vehicle ramps and any levels in a building to which vehicles have access shall be provided with barriers where it is necessary to protect people in or about the building.* *(2) Vehicle loading bays shall be constructed in such a way, or be provided with such features, as may be necessary to protect people in them from collision with vehicles.*	Vehicle barriers should be provided that are capable of resisting or deflecting the impact of vehicles. Loading bays shall be provided with an adequate number of exits (or refuges) to enable people to avoid being crushed by vehicles.
K4	Protection against impact with glazing	*Glazing with which people are likely to come into contact whilst moving in or about the building shall:* *(a) if broken on impact, break in a way which is unlikely to cause injury; or* *(b) resist impact without breaking; or* *(c) be shielded or protected from impact.*	All glazing installed in buildings shall be: • sufficiently robust to withstand impact from a falling or passing person; or • protected from a falling or passing person.

Table 3.11 Part K – Protection from falling, collision and impact (Continued)

Number	Title	Regulation	Requirement (in a nutshell)
K5.1	Protection from collision with open windows etc.	*Provision shall be made to prevent people moving in or about the building from colliding with open windows, skylights or ventilators.* 💡 K5.1 does not apply to dwellings.	All windows, skylights and ventilators shall be capable of being left open without danger of people colliding with them.
K5.2	Manifestation of glazing	*Transparent glazing with which people are likely to come into contact while moving in or about the building, shall incorporate features which make it apparent.* 💡 K5.2 does not apply to dwellings.	
K5.3	Safe opening and closing of windows etc.	*Windows, skylights and ventilators which can be opened by people in or about the building shall be so constructed or equipped that they may be opened, closed or adjusted safely.* 💡 K5.3 does not apply to dwellings.	
K5.4	Safe access for cleaning windows etc.	*Provision shall be made for any windows, skylights, or any transparent or translucent walls, ceilings or roofs to be safely accessible for cleaning.* 💡 K5.4 does not apply to: (a) dwellings; or (b) any transparent or translucent elements whose surfaces are not intended to be cleaned.	

Table 3.11 Part K – Protection from falling, collision and impact

Number	Title	Regulation	Requirement (in a nutshell)

K6	Protection against impact from and trapping by doors	(1) *Provision shall be made to prevent any door or gate:*
		
		  (a) *which slides or opens upwards, from falling onto any person; and*
		  (b) *which is powered, from trapping any person.*
		
		(2) *Provision shall be made for powered doors and gates to be opened in the event of a power failure.*
		
		(3) *Provision shall be made to ensure a clear view of the space on either side of a swing door or gate.*
		
		K6 does not apply to:
		
		  (a) dwellings; or
		  (b) any door or gate that is part of a lift.	

Table 3.12 Part L – Conservation of fuel and power

Number	Title	Regulation	Requirement (in a nutshell)

L1A *and* L2A	Conservation of fuel and power (new dwellings)	*Reasonable provision shall be made for the conservation of fuel and power in buildings by:*
		
		(a) limiting heat gains and losses:
		  (i) through thermal elements and other parts of the building fabric; and
		  (ii) from pipes, ducts and vessels used for space heating, space cooling and hot water services;
		
		(b) providing fixed building services which
		  (I) are energy efficient;
		  (ii) have effective controls; and are commissioned by testing and adjusting as necessary to ensure they use no more fuel and power than is reasonable in the circumstances.
			These are new regulations that introduce new energy-efficiency requirements and other relevant changes to the existing regulations for new dwellings.

24	Methodology of calculation of the energy performance of buildings	(1) The Secretary of State shall approve
		
		  (a) a methodology of calculation of the energy performance of buildings, including methods for calculating asset ratings and operational ratings of buildings; and
		  (b) ways in which the energy performance of buildings, as calculated in accordance with the methodology, shall be expressed.
		
		(2) In this regulation
		
		'asset rating' means a numerical indicator of the amount of energy estimated to meet the different needs associated with the standardised use of the building; and 'operational rating' means a numerical indicator of the amount of energy consumed during the occupation of the building over a period of time.
			The Secretary of State should approve the methods of calculation of energy performance.

Table 3.12 Part L – Conservation of fuel and power

Number	Title	Regulation	Requirement (in a nutshell)
25	Minimum energy performance requirements for buildings	Minimum energy performance requirements shall be set by the Secretary of State, in accordance with the methodology approved pursuant to regulation 24, for (a) new buildings (which shall include new dwellings), in the form of CO_2 emission rates; and (b) new dwellings, in the form of target fabric efficiency rates.	The Secretary of State should approve CO_2 emission rates.
26	Minimum energy performance requirements for new buildings	Where a building is erected, it shall not exceed the target CO_2 emission rate for the building that has been approved pursuant to regulation 25.	Buildings should not exceed their target CO_2 emission rates.
26A (L1A only)	Fabric energy efficiency rates for new dwellings	Where a dwelling is erected, it shall not exceed the target fabric energy efficiency rate for the dwelling that has been approved pursuant to regulation 25.	Buildings should not exceed their target fabric energy-efficiency rates.
27	CO_2 emission rate and fabric energy efficiency calculations	(1) This regulation applies where a building is erected and regulation 26 applies. (2) Not later than the day before the work starts, the person carrying out the work shall give the local authority a notice which specifies (a) the target CO_2 emission rate for the building. (b) the calculated CO_2 emission rate for the building as designed, and (c) a list of specifications to which the building is to be constructed. (3) Not later than five days after the work has been completed, the person carrying out the work shall give the local authority (a) a notice which specifies	The person carrying out the work should notify the local authority that he or she plans to start work and that he or she has completed work within the stipulated timescales (the day before work starts and no later than five days after completion).

Table 3.12 Part L – Conservation of fuel and power (Continued)

Number	Title	Regulation	Requirement (in a nutshell)
		(i) the target CO_2 emission rate for the building.	
		(ii) the calculated CO_2 emission rate for the building as constructed, and	
		(iii) whether the building has been constructed in accordance with the list of specifications referred to in paragraph (2)(c), and if not a list of any changes to those specifications; or	
		(b) a certificate of the sort referred to in paragraph (4) accompanied by the information referred to in subparagraph (a).	
		(4) A local authority is authorised to accept as evidence that the requirements of regulation 26 have been satisfied, a certificate to that effect by an energy assessor who is accredited to produce such certificates for that category of building.	
		(5) In this regulation, 'specifications' means specifications used for the calculation of the CO_2 emission rate.	
27A (L1A only)	Fabric energy efficiency rate calculations	(1) This regulation applies where a building is erected and regulation 26A applies.	The person carrying out the work should notify the local authority that he or she plans to start work and that he or she has completed work within the stipulated timescales (the day before work starts and no later than five days after completion).
		(2) Not later than the day before the work starts, the person carrying out the work shall give the local authority a notice which specifies	
		(a) the target fabric energy efficiency rate for the dwelling.	
		(b) the calculated fabric energy efficiency rate for the dwelling as designed, and	
		(c) a list of specifications to which the dwelling is to be constructed.	
		(3) Not later than five days after the work has been completed, the person carrying out the work shall give the local authority	

Table 3.12 Part L – Conservation of fuel and power

Number	Title	Regulation	Requirement (in a nutshell)
		(a) A notice which specifies	
		(i) the target fabric energy efficiency rate for the dwelling,	
		(ii) the calculated fabric energy efficiency rate for the dwelling as constructed, and	
		(iii) whether the dwelling has been constructed in accordance with the list of specifications referred to in paragraph (2)(c), and if not a list of any changes to those specifications; or	
		(b) A certificate of the sort referred to in paragraph (4) accompanied by the information referred to in subparagraph (a).	
		(4) A local authority is authorised to accept as evidence that the requirements of regulation 26A have been satisfied, a certificate to that effect by an energy assessor who is accredited to produce such certificates for that category of building.	
		(5) In this regulation, 'specifications' means specifications used for the calculation of the fabric energy efficiency rate.	
25A	Consideration of high-efficiency alternative systems for new buildings	*(1) Before construction of a new building starts, the person who is to carry out the work must analyse and take into account the technical, environmental and economic feasibility of using high-efficiency alternative systems (such as the following systems) in the construction, if available*	The person carrying out the work should consider the feasibility of using high-efficiency alternative systems in the building.
		(a) decentralised energy supply systems based on energy from renewable sources;	
		(b) cogeneration;	
		(c) district or block heating or cooling, particularly where it is based entirely or partially on energy from renewable sources; and	
		(d) heat pumps.	

Table 3.12 Part L – Conservation of fuel and power (Continued)

Number	Title	Regulation	Requirement (in a nutshell)
		(2) *The person carrying out the work must*	
		(a) *not later than the beginning of the day before the day on which the work starts, give the local authority a notice which states that the analysis referred to in paragraph (1)*	

		(i) *has been undertaken;*
		   (ii) *is documented; and*
		  (iii) *the documentation is available to the authority for verification purposes; and*	

		(b) *ensure that a copy of the analysis is available for inspection at all reasonable times upon request by an officer of the local authority.*	
		(3) *An authorised officer of the local authority may require production of the documentation in order to verify that this regulation has been complied with.*	
		(4) *The analysis referred to in paragraph (1)*	
		(a) *may be carried out for individual buildings or for groups of similar buildings or for common typologies of buildings in the same area; and*	
		(b) *in so far as it relates to collective heating and cooling systems, may be carried out for all buildings connected to the system in the same area.*	
		(5) *In this regulation*	
		(a) *'cogeneration' means simultaneous generation in one process of thermal energy and one or both of the following*	

Table 3.12 Part L – Conservation of fuel and power

Number	Title	Regulation	Requirement (in a nutshell)
		(i) electrical energy;	
		(ii) mechanical energy;	
		(b) 'district or block heating or cooling' means the distribution of thermal energy in the form of steam, hot water or chilled liquids, from a central source of production through a network of multiple buildings or sites, for the use of space or process heating or cooling;	
		(c) 'energy from renewable sources' means energy from renewable non-fossil sources, namely wind, solar, aerothermal, geothermal, hydrothermal and ocean energy, hydropower, biomass, landfill gas, sewage treatment plant gas and biogases; and	
		(d) 'heat pump' means a machine, a device or installation that transfers heat from natural surroundings such as air, water or ground to buildings or industrial applications by reversing the natural flow of heat such that it flows from a lower to a higher temperature. (For reversible heat pumps, it may also move heat from the building to the natural surroundings.)	
43	Pressure testing	*(1) This regulation applies to the erection of a building in relation to which paragraph L1(a)(i) of Schedule 1 imposes a requirement.*	Buildings should be pressure-tested.
		(2) Where this regulation applies, the person carrying out the work shall, for the purpose of ensuring compliance with regulation 26 and paragraph L1(a)(i) of Schedule 1	
		(a) ensure that:	
		(i) pressure testing is carried out in such circumstances as are approved by the Secretary of State; and	
		(ii) the testing is carried out in accordance with a procedure approved by the Secretary of State; and	

Table 3.12 Part L – Conservation of fuel and power (Continued)

Number	Title	Regulation	Requirement (in a nutshell)
		(b) subject to paragraph (5), give notice of the results of testing to the local authority.	
		(3) The notice referred to in paragraph (2)(b) shall:	
		(a) record the results and the data upon which they are based in a manner approved by the Secretary of State; and	
		(b) be given to the local authority no later than seven days after the final test is carried out.	
		(4) A local authority is authorised to accept, as evidence that the requirements of paragraph (2)(a)(ii) have been satisfied, a certificate to that effect by a person who is registered by the British Institute of Non-destructive Testing and Air Tightness Testing and Measuring Association in respect of pressure testing for the air tightness of buildings.	
		(5) Where such a certificate contains the information required by paragraph (3)(a), paragraph (2)(b) does not apply.	
44	Commissioning	(1) This regulation applies to building work in relation to which paragraph F1(2) of Schedule 1 imposes a requirement, but does not apply to the provision or extension of any fixed system for mechanical ventilation or any associated controls where testing and adjustment is not possible.	Buildings should be properly commissioned.
		(2) This regulation applies to building work in relation to which paragraph L1(b) of Schedule 1 imposes a requirement, but does not apply to the provision or extension of any fixed building service where testing and adjustment is not possible or would not affect the energy efficiency of that fixed building service.	
		(3) Where this regulation applies the person carrying out the work shall, for the purpose of ensuring compliance with paragraph	

Table 3.12 Part L – Conservation of fuel and power

Number	Title	Regulation	Requirement (in a nutshell)
		F1(2) or L1(b) of Schedule 1, give to the local authority a notice confirming that the fixed building services have been commissioned in accordance with a procedure approved by the Secretary of State. (4) The notice shall be given to the local authority (a) no later than the date on which the notice required by regulation 16(4) is required to be given; or (b) where that regulation does not apply, not more than 30 days after the completion of the work.	
40	Information about the use of fuel and power	(1) This regulation applies where paragraph L1 of Schedule 1 imposes a requirement relating to building work. (2) The person carrying out the building work shall not later than five days after the work has been completed provide the owner sufficient information about the building, the fixed building services and their maintenance requirements so that the building can be operated in such a manner as to use no more fuel and power than is reasonable in the circumstances.	The owner of the building should be provided with information about the use of fuel and power.
L1B and L2B	Conservation of fuel and power (existing dwellings)	Reasonable provision shall be made for the conservation of fuel and power in buildings by: (a) limiting heat gains and losses: (i) through thermal elements and other parts of the building fabric; and (ii) from pipes, ducts and vessels used for space heating, space cooling and hot water services; (b) providing fixed building services which (i) are energy efficient; (ii) have effective controls; and	These are new regulations that introduce new energy-efficiency requirements and other relevant changes to the existing regulations for existing dwellings.

Table 3.12 Part L – Conservation of fuel and power (Continued)

Number	Title	Regulation	Requirement (in a nutshell)
		(iii) are commissioned by testing and adjusting as necessary to ensure they use no more fuel and power than is reasonable in the circumstances; and Regulation 40 providing to the owner sufficient information about the building, the fixed building services and their maintenance requirements so that the building can be operated in such a manner as to use no more fuel and power than is reasonable in the circumstances.	
23	Requirements for the renovation or replacement of thermal elements	(1) Where the renovation of an internal thermal element (a) constitutes a major renovation; or (b) amounts to the renovation of more than 50% of the element's surface area; the renovation must be carried out so as to ensure that the whole of the element complies with paragraph L1(a)(i) of Schedule 1, in so far as that is technically, functionally and economically feasible. . (2) Where the whole or any part of an individual element is proposed to be replaced and the replacement (a) constitutes a major renovation; or (b) (in the case of part replacement) amounts to the replacement of more than 50% of the thermal element's surface area; the whole of the thermal element must be replaced so as to ensure that it complies with paragraph L1(a)(i) of Schedule 1, in so far as that is technically, functionally and economically feasible	

Table 3.12 Part L – Conservation of fuel and power

Number	Title	Regulation	Requirement (in a nutshell)
28	Consequential improvements to energy performance	(1) Paragraph (2) applies to an existing building with a total useful floor area over 1000 m² where the proposed building work consists of or includes (a) an extension; (b) the initial provision of any fixed building services; or (c) an increase to the installed capacity of any fixed building services. (2) Subject to paragraph (3), where this paragraph applies, such work, if any, shall be carried out as is necessary to ensure that the building complies with the requirements of Part L of Schedule 1. (3) Nothing in paragraph (2) requires work to be carried out if it is not technically, functionally or economically feasible.	
29	Energy performance certificates	(1) This regulation applies where (a) a building is erected; or (b) a building is modified so that it has a greater or fewer number of parts designed or altered for separate use than it previously had, where the modification includes the provision or extension of any of the fixed services for heating, hot water, air conditioning or mechanical ventilation. (2) The person carrying out the work shall (a) give an energy performance certificate for the building to the owner of the building; and (b) give to the local authority a notice to that effect, including the reference number under which the energy performance certificate has been registered in accordance with regulation 30(4). (3) The energy performance certificate and notice shall be given no later than five days after the work has been completed.	

Table 3.12 Part L – Conservation of fuel and power (Continued)

Number	Title	Regulation	Requirement (in a nutshell)
		(4) The energy performance certificate must	
		(a) express the asset rating of the building in a way approved by the Secretary of State under regulation 24;	
		(b) include a reference value such as a current legal standard or benchmark;	
		(c) be issued by an energy assessor who is accredited to produce energy performance certificates for that category of building; and	
		(d) include the following information	
		(i) the reference number under which the certificate has been registered in accordance with regulation 30(4);	
		(ii) the address of the building;	
		(iii) an estimate of the total useful floor area of the building;	
		(iv) the name of the energy assessor who issued it;	
		(v) the name and address of the energy assessor's employer, or, if he is self-employed, the name under which he trades and his address;	
		(vi) the date on which it was issued; and	
		(vii) the name of the approved accreditation scheme of which the energy assessor is a member.	
		(5) …	
		(6) Certification for apartments or units designed or altered for separate use in blocks may be based	
		(a) except in the case of a dwelling, on a common certification of the whole building for blocks with a common heating system; or	
		(b) on the assessment of another representative apartment or unit in the same block.	

Table 3.12 Part L – Conservation of fuel and power

Number	Title	Regulation	Requirement (in a nutshell)
		(7) Where	
		(a) a block with a common heating system is divided into parts designed or altered for separate use; and	
		(b) one or more, but not all, of the parts are dwellings,	
		certification for those parts which are not dwellings may be based on a common certification of all the parts which are not dwellings.	
		(8) Certification for a building which consists of a single dwelling may be based on the assessment of another representative building of similar design and size with a similar actual energy performance quality, provided such correspondence is guaranteed by the energy assessor issuing the energy performance certificate.	
		(9) An energy performance certificate is only valid if	
		(a) it was entered on the register no more than 10 years before the date on which it is made available; and	
		(b) no other energy performance certificate for the building has since been entered on the register.	
		(10) An energy performance certificate must not contain any information or data (except for the address of the building) from which a living individual (other than the energy assessor or his employer) can be identified.	
29A	Recommendation reports	(1) In these Regulations a 'recommendation report' means recommendations made by an energy assessor for the cost-effective improvement of the energy performance of a building.	
		(2) A recommendation report must include	

Table 3.12 Part L – Conservation of fuel and power (Continued)

Number	Title	Regulation	Requirement (in a nutshell)
		(a) recommended cost-effective measures that could be carried out in connection with a major renovation of the building envelope or fixed building services;	
		(b) recommended cost-effective measures for individual building elements that could be carried out without the necessity for a major renovation of the building envelope or fixed building services;	
		(c) an indication as to how the owner or tenant can obtain more detailed information about improving the energy efficiency of the building, including more detailed information about the cost-effectiveness of the recommendations; and	
		(d) information on the steps to be taken to implement the recommendations.	
		(3) Any cost-effective measure which the energy assessor recommends must be technically feasible for the building to which the recommendation report relates.	
		(4) In this regulation 'building element' means a controlled service or fitting or a thermal element of the building envelope.	

Table 3.13 Part M – Access to and use of buildings

Number	Title	Regulation	Requirement (in a nutshell)
M1	Access and use	*Reasonable provision shall be made for people to:* (a) *gain access to; and* (b) *use* *the building and its facilities.*	In addition to the requirements of the Disability Discrimination Act 1995, Approved Document M also requires that precautions are taken to ensure that: • new non-domestic buildings and/or dwellings (e.g. houses and flats used for student living accommodation, etc.);

Table 3.13 Part M – Access to and use of buildings

Number	Title	Regulation	Requirement (in a nutshell)
		M1 does not apply to: (a) an extension of, or material alteration of, a dwelling; or (b) any part of a building which is used solely to enable the building to be inspected, repaired or maintained.	• extensions to existing non-domestic buildings; • non-domestic buildings that have been subject to a material change of use (e.g. so that they become a hotel, boarding house, institution, public building or shop) are capable of allowing people, regardless of their disability, age or gender, to: • gain access to buildings; • gain access within buildings; • use the facilities of the buildings (both as visitors and as people who live or work in them); and • use sanitary conveniences in the principal storey of any new dwelling.
M2	Access to extensions of buildings other than dwellings	*Suitable independent access shall be provided to the extension where reasonably practical.* M2 does not apply where suitable access to the extension is provided through the building that is extended.	
M3	Sanitary conveniences in extensions to buildings other than dwellings	*If sanitary conveniences are provided in any building that is to be extended, reasonable provision shall be made within the extension for sanitary conveniences.* M3 does not apply where there is reasonable provision for sanitary conveniences elsewhere in the building, such that people occupied, or otherwise having occasion to enter the extension, can gain access to and use those sanitary conveniences.	

Table 3.13 Part M – Access to and use of buildings (Continued)

Number	Title	Regulation	Requirement (in a nutshell)
M4	Sanitary conveniences in dwellings	(1) Reasonable provision shall be made in the entrance storey for sanitary conveniences, or where the entrance contains no habitable rooms, reasonable provision for sanitary conveniences shall be made in either the entrance storey or principal storey. (2) In this paragraph 'entrance storey' means the storey which contains the principal entrance and 'principal storey' means the storey nearest to the entrance storey which contains a habitable room, or if there are two storeys equally near, either such storey.	

Table 3.14 Part P – Electrical safety: design and installation of electrical installations

Number	Title	Regulation	Requirement (in a nutshell)
P1	Design and installation	Reasonable provision shall be made in the design and installation of electrical installations in order to protect persons operating, maintaining or altering the installations from fire or injury.	These requirements only apply to electrical installations that are intended to operate at low or extra-low voltage: • in (or attached to) a dwelling; • in common parts of a building serving one or more dwellings (but excluding power supplies to lifts); • in a building that receives its electricity from a source located within (or sharesd with) a dwelling; • in a garden or in or on land associated with a building where the electricity is from a source located within or shared with a dwelling.

All electrical work shall be carried out in accordance with Amendment No. 1 (2011) to BS 7671:2008 (*Requirements for Electrical Installations* – commonly referred to as the *IEE Wiring Regulations* (17th edition).

4

Planning permission

 It is your responsibility for seeking, or not seeking, planning permission. If permission is required, it should be granted before any work begins.

In the UK you are required to ask for permission to undertake building work including most types of new buildings, major changes to existing buildings or significant changes to the use of a building or piece of land. This consent is known as planning permission. The planning system is there to manage the development of buildings and land and was put in place to ensure that buildings are constructed and land used in a way that does not adversely affect other people who live and work in the area. The responsibility for planning is delegated to local planning authorities; they are the people who can decide whether or not a development may go ahead. It is essential that you obtain the approval of the local planning authorities before undertaking a building project.

There have been major wholesale changes to the planning regulations because of the Coalition Government's initiative to reduce red tape and as a consequence of the Localism Act 2011. The previous technical guidance relating to planning permission has been substantially reduced, and these changes are covered in this chapter. It is anticipated that there will be further changes to the Planning Act 2008 in due course.

A new streamlined planning practice is being introduced which the government hopes will provide support for growth and create jobs and homes in the UK. In addition the new practice will provide protection for the natural and historic environment and give communities power to make local decisions. The planning website (the Planning Portal) has been updated, and the new pages set out guidance on a range of issues, including:

- a new affordability test for determining how many homes should be built;
- discouragement of any plans to introduce a new parking tax on people's driveways and parking spaces;
- encouragement of more town centre parking spaces and an end of aggressive 'anti-car' traffic-calming measures like speed bumps;
- introduction of new neighbourhood planning guidance to help communities with their own plans;
- new local green space guidance to help councils and local communities to plan for open spaces and to protect local green spaces which are special to local people,

- new neighbourhood planning guidance to help more communities start their own plans;
- provision of housing for older people – to encourage councils to build more bungalows and plan positively for an ageing population;
- the opening up of planning appeal hearings to filming.

As discussed in Chapter 2, there is a good deal of confusion between the terms 'planning permission' and 'Building Regulation approval'. While both are associated with gaining permission to complete building work, the process of receiving planning permission does not automatically confer Building Regulation approval, and vice versa. Additionally, you **may** require **both** sets of approval before you can start your building project. As the government advice on what is permitted is subject to change you may find that the information in this chapter becomes out of date. It should therefore be considered as a guide and not an authoritative statement of the law. If you are in doubt over a particular area you should check on the Planning Portal or with your local planning officer.

The difference between Building Regulations and planning permission is that:

- Building Regulations set standards for the design and construction of buildings to ensure the safety and health of people in or about those buildings, and also include requirements to ensure that fuel and power are conserved and facilities are provided for people, including those with disabilities, to access and move around inside buildings; while
- planning seeks to guide the way our towns, cities and countryside develop. This includes the use of land and buildings, the appearance of buildings, landscaping considerations, highway access and the impact that the development will have on the general environment.

Controversial or very large proposals, especially if they are of national significance, are usually decided on by the Secretary of State. Some other areas are given special protection against certain types of development. This includes:

- controlling the spread of towns and villages into open countryside, e.g. Green Belts;
- protecting attractive landscape, e.g. National Parks;
- protecting interesting plants and/or wildlife;
- protecting monuments or buildings of historic or architectural interest.

In addition you might require consent or be required to follow certain procedures if you want to carry out work to trees in a Conservation Area or work on a listed building or if you plan work that requires the pruning or felling of a tree that is protected by a Tree Preservation Order. In all these cases you should contact your local authority planning department before carrying out any work.

You are allowed to make certain changes to your home without having to apply to the local council for permission. This includes minor building works such as building a boundary wall below a certain height. These changes are known as **permitted development** and apply where the effect of the development on neighbours or the surrounding environment is likely to be small or if the change of use is within the same use class.

In May 2013 Parliament agreed to increase the size of single-storey rear extensions which can be built under permitted development and brought into force the associated neighbour consultation scheme. For a period of three years, between 30 May 2013 and 30 May 2016, householders will be able to build larger single-storey rear extensions under permitted development. In essence the permitted development size limits for extensions have doubled from 4 m^2 to 8 m^2 for detached houses, and from 3 m^2 to 6 m^2 for all other houses. There is no fee involved, but in order for you to get permission to build one of these larger extensions the planning authority still require a written description of the proposal, a plan of the site, the addresses of any adjoining properties (including those at the rear) and a contact address for the developer. During the 42-day consultation period the local authority will serve a notice on adjoining owners or occupiers. The neighbour has a 21-day period in which to raise an objection with the local authority. The authority will take this objection into account when it makes its decision and will consider if the impact on the amenity of all adjoining properties is acceptable; no other issues will be considered. The development can go ahead if the local authority gives the developer permission in writing within the 42-day period. If the local authority fails to notify the developer of its decision within the 42-day determination period, the development may go ahead.

Many people are initially reluctant to approach local authorities because they think the authorities are likely to be obstructive. In fact the reality is quite the reverse. The planning team's purpose is to protect all of us from irresponsible builders and developers. They are normally most sympathetic and helpful to a builder or householder who has asked for their advice in order to comply with the statutory requirements.

4.1 What are planning controls?

The process of managing the development of buildings and land (anything from work on a house to building a new school) is known as planning control. Councils use planning controls to protect the character of an area and to provide for the needs of future communities. They do this by regulating the use, siting and appearance of buildings and other constructions within the area. The local authority might refuse permission on the grounds that the planned scheme would not blend sympathetically with its surroundings. What might seem to be a minor development in itself could have far-reaching implications that you had not previously considered. Your property could also be affected by legal restrictions such as a right of way, which could prejudice planning permission.

The actual details of planning requirements are complex, but with a domestic development they are mainly concerned with the construction work that is being carried out. Structures such as walls and fences are also considered because their height or siting could infringe the rights of neighbours and other members of the community. The planning authority will also need to approve any change of use, such as converting a house into flats or running a business from premises previously occupied as a dwelling only.

The new planning system is 'plan-led' and requires each local planning authority to prepare a folder of documents outlining how planning will be managed for that area (called the Local Development Framework). Planning applications must be considered with regard to this framework.

4.1.1 Why are planning controls needed?

The purpose of the planning system is to protect the environment, public amenities and public facilities. The system **will not** protect the interests of one person over another. Within the framework of legislation approved by Parliament, councils should ensure that development is permitted where it is needed. However, they must also ensure that the character and amenity of the area are not adversely affected by new buildings or changes in the use of existing buildings or land.

There is mixed opinion on the subject of the planning system, with some considering the system should be used to prevent any change in their local environment, while others think that planning controls are an unnecessary interference in their individual rights. The present position is that **all** major works need planning permission from the council but many minor works do not. Parliament thinks this strikes the right balance because it enables councils to protect the character and amenity of their area, while allowing individuals to have a reasonable degree of freedom to alter their property.

If you live in a listed building of historic or architectural interest or your house is in a Conservation Area, you should seek advice before considering any alterations.

4.2 Who needs planning permission?

Anyone who is planning a development which involves building work must make sure that they comply with planning rules and building regulations. Planning regulations have to cover many different situations, and so the provisions that affect the average householder are quite detailed. It is therefore important that you seek advice before you start work, and the local planning authority or building control service will be happy to discuss your plans with you. The rules and requirements will vary according to whether you actually own land, a house, or a flat or maisonette. Generally speaking the principles and procedures for making planning applications are the same for owners of houses and for freeholders (or leaseholders) of flats and maisonettes.

The penalty for failure to comply with the regulations can be severe and could result in the owner being required to make good any infringements as well as facing a hefty fine. 'Making good' can involve restoration work or the complete demolition of all the work you have carried out and reinstatement of the building to how it was before.

 If you are a leaseholder, you may first need to get permission from your landlord or management company.

4.2.1 What is meant by 'building works'?

In the context of the planning permission, 'building works' include:

*   major alteration to an existing building;
*   new buildings;
*   significant changes to the use of a building or piece of land.

Planning permission is not required for minor building works which are known as permitted development.

4.3 Who controls planning permission?

The government's planning policy for England is set out in the National Planning Policy Framework. The UK has a plan-led planning system which requires each local planning authority to prepare a folder of documents outlining how planning will be managed for that area. At a local level planning policies are set out in your Local Development Framework; planning applications must be considered with regard to this framework.

4.3.1 What is the National Planning Policy Framework?

The National Planning Policy Framework was published on 27 March 2012 and is a major part of the reforms to simplify the planning system. It sets out the planning policies for England and states how the policies are to be applied. The Framework provides guidance for local planning authorities and decision takers, both in drawing up plans and in making decisions about planning applications. There are no specific policies for nationally significant infrastructure projects in the Framework, as these are determined by the Secretary of State on a case-by-case basis. The Framework is the result of extensive consultation and is based upon the presumption in favour of sustainable development. It aims to:

*   ensure the local plan is the keystone of the planning system;
*   establish a presumption in favour of sustainable development (which means that development is not held up unless to approve it would be against our collective interest);

- guarantee strong protections for the natural and historic environment, and to require improvements to put right some of the neglect that has taken place;
- make planning much simpler and more accessible;
- raise design standards so that the requirements for design are the most exacting yet.

The National Waste Management Plan for England will create a national waste planning policy; therefore the Framework does not contain specific waste policies. Similarly the Framework should be read in conjunction with the government's planning policy for traveller sites.

4.3.1.1 Are there any initiatives to speed up the planning process?

A range of initiatives were introduced in March 2012 in an effort to speed up the planning process. These are:

- a review of the supporting guidance on planning (aiming to simplify it);
- a trial period of three years (effective July 2013) during which time businesses and home owners will be permitted to extend their premises by up to 6–8 m^2 without having to get planning permission;
- instructions to the Planning Inspectorate to respond quickly to all major economic and housing-related appeals, to ensure applicants receive a response in the quickest possible time;
- new powers for the Planning Inspectorate to take over the role of making planning decisions in an area if the local authority has a record of consistently slow or poor-quality decisions;
- plans to give developers extra time to get their sites up and running before planning permission expires;
- plans to permit the reuse of existing agricultural, retail and commercial buildings without the need to submit a planning application, thereby supporting small-business growth;
- work on proposals to speed up the process for determining planning appeals.

4.3.2 What is the Local Development Framework?

The way that planning is managed in your area is set out in the Local Development Framework. The framework is a collection of documents prepared by district councils that outline the planning strategy for the local area. These plans are examined by independent planning inspectors to ensure that they are positively prepared, justified, effective and consistent with national policy. In preparing the Local Development Framework the council is responsible for consulting local people and for ensuring that their views are taken into account, thereby giving them a chance to influence the way in which their area is affected. The published framework is used as a guide to the location of development over a ten-year period, and it is these local plans that are taken into account when deciding planning applications.

You can find details about a particular planning case through your local planning department or on its website.

4.4 What is planning permission?

In the UK you are required to ask for permission to undertake building work. This permission is known as planning permission and is required for most types of development, including:

- building work;
- engineering work;
- materially (significantly) changing the use of a building or piece of land;
- mining work.

Some minor building work and changes in the way you use land or buildings or some minor building works are automatically allowed. These categories of work are known as permitted development.

Other areas are given special protection against certain types of development. These apply to National Parks and Conservation Areas or if the property is a listed building. In order that your local planning authority can advise you of any reason why the development may not be permitted or if you need planning permission it is recommended that you contact it to discuss your proposal before any work begins.

You can also ask for planning permission after development work has been carried out. This is called retrospective planning permission.

4.4.1 What are permitted development rights?

Certain types of minor changes can be carried out to your house without you having to apply for planning permission. These changes are known as 'permitted development rights' and come from a general planning permission which is granted by Parliament. Permitted development rights, which apply to many common projects for houses, **do not** apply to flats, maisonettes or other buildings. Permitted development rights are also more restricted in certain areas of the country; these areas are known as 'designated areas' and include Conservation Areas, National Parks, Areas of Outstanding Natural Beauty and the Norfolk or Suffolk Broads.

4.4.1.1 Can permitted development rights be withdrawn?

In some areas the local planning authority may remove some of your permitted development rights by issuing an Article 4 direction. This is most common in Conservation Areas, where the character of an area is acknowledged to be important and could be threatened by certain types of development. In most cases you should be aware if your property is affected. If a property is affected you will have to submit a planning application for work which does not normally require permission.

4.4.2 What happens if I fail to obtain or comply with planning permission?

If a person planning some building work fails to obtain planning permission or fails to comply with the details of the permission then he or she will be considered to have committed a planning breach. There are two types of circumstances when a planning breach occurs: either where a development takes place without permission being granted (it could have been refused or never applied for) or when a development has been given permission subject to certain conditions but those conditions are not followed.

A planning breach in itself is not illegal, but as a result it is possible that your local planning authority will serve an enforcement notice which requires you to put things back as they were if it considers you have broken planning control rules. This is particularly likely if the breach involves a previously rejected development or if it considers what you are doing, or have done, is harmful to your neighbourhood. The decisive issue for the local planning authority when considering an enforcement notice is whether the breach would unacceptably affect public amenity or the existing use of land and buildings meriting protection in the public interest. You can appeal against both refusals of permission and enforcement notices, but if the verdict comes out against you and you still refuse to comply you may be prosecuted.

It is illegal to disobey an enforcement notice unless it has been successfully appealed against.

4.4.3 What is the most recent legislation relating to town planning?

The principal order which applies to planning permission is the Town and Country Planning (General Permitted Development) Order 1995. In this order you will find the classes of development for which a grant of planning permission is automatically given. There have been a number of amendments to this order, and those that apply as at November 2013 are in Table 4.1.

All applications for planning permission will also have to take into account the following acts and regulations.

4.4.3.1 The Planning (Listed Buildings and Conservation Areas) Act 1990

Under the terms of the Planning (Listed Buildings and Conservation Areas) Act 1990, local councils must maintain a list of buildings within their boroughs

Table 4.1 Amendments to Town and Country Planning (General Permitted Development) Order 1995

1996	The Town and Country Planning (General Permitted Development) (Amendment) Order 1996
1998	The Town and Country Planning (General Permitted Development) (Amendment) Order 1998
1999	The Town and Country Planning (Environmental Impact Assessment) (England and Wales) Regulations 1999
	The Town and Country Planning (General Permitted Development) (Amendment) Order 1999
2001	The Town and Country Planning (General Permitted Development) (Amendment) (England) Order 2001
2003	The Thurrock Development Corporation (Area and Constitution) Order 2003
2004	The Milton Keynes (Urban Area and Planning Functions) Order 2004
2005	The Town and Country Planning (General Permitted Development) (Amendment) (England) Order 2005
	The Town and Country Planning (General Permitted Development) (England) (Amendment) (No. 2) Order 2005
2006	The Town and Country Planning (General Permitted Development) (Amendment) (England) Order 2006
	The Town and Country Planning (Application of Subordinate Legislation to the Crown) Order 2006
2007	The Town and Country Planning (General Permitted Development) (Amendment) (England) Order 2007
2008	The Town and Country Planning (General Permitted Development) (Amendment) (England) Order 2008 (SI 2008 675)
	The Town and Country Planning (General Permitted Development) (Amendment) (No. 2) (England) Order 2008 (SI 2008 2362)
2010	The Town and Country Planning (General Permitted Development) (Amendment) (No.2) (England) Order 2010
	The Town and Country Planning (General Permitted Development) (Amendment) (England) Order 2010
2011	The Town and Country Planning (General Permitted Development) (Amendment) (England) Order 2011
2012	The Town and Country Planning (General Permitted Development) (Amendment) (No. 2) (England) Order 2012
	The Town and Country Planning (General Permitted Development) (Amendment) (England) Order 2012

that have been classified as being of special architectural or historic interest. Councils are also required to keep maps showing which properties are within Conservation Areas.

4.4.3.2 Town and Country Planning (Control of Advertisements) (England) Regulations 2007

In accordance with the Town and Country Planning (Control of Advertisements) (England) Regulations 2007, councils need to maintain a publicly available register of applications and decisions for consent to display advertisements.

4.4.3.3 The Local Government (Access to Information) (Variation) Order 2006

The Local Government (Access to Information) (Variation) Order 2006 ensures that information relating to proposed development by councils cannot be treated as exempt when the planning decision is made.

4.4.3.4 Town and Country Planning (General Permitted Development) (Amendment) (England) Order 2012

Every council must keep the following registers available for public inspection in accordance with the Town and Country Planning (General Permitted Development) (Amendment) (England) Order 2012:

• applications for a certificate of lawfulness of existing or proposed use or development;
• enforcement notices and any related stop notices;
• planning applications, including accompanying plans and drawings.

All applications for planning permission must receive publicity.

4.4.4 Latest policy consultations and guidance

The Planning Portal contains details of the latest policy, consultations and guidance in its 'Planning Policy and Legislation' pages. The most recent at the time of writing were the changes to permitted development rights for change of use from commercial to residential which were published on 24 January 2013. This guidance has been included in this edition.

4.5 What types of planning permission are available?

There are a number of different types of planning permission available. In addition to the traditional consents of outline, reserved and full planning, you can apply for other (multiple) consents if your work is covered by a number of areas; for example, you can apply for full planning permission, consent under a Tree Preservation Order and Conservation Area consent.

It is recommended that you apply for planning permission online via the Planning Portal, as the system will automatically adjust to ensure you are presented with all the forms that you require. If you submit a paper application you have to locate all the correct forms, ensure they contain no errors and submit them with the correct fee. If you fail to do this your application will not be valid. It is now possible to apply for the following consents online:

- advertisement consent;
- approval of conditions;
- consent under Tree Preservation Orders;
- Conservation Area consent;
- full planning consent;
- householder planning consent;
- lawful development certificate (LDC);
- listed building consent;
- notification of proposed works to trees in Conservation Areas;
- outline planning consent;
- prior notification;
- removal/variation of conditions;
- reserved matters.

At the time of writing it is not possible to apply for the following online:

- application for non-material amendments;
- extending the time limits of existing planning permissions.

4.5.1 Advertisement consent

If your proposal is to display an advertisement or sign you will need to make a separate application. No certificate of ownership is needed, but it is illegal to display signs on the property without the consent of the owner. The term 'advertisements and signs' includes the following:

- advance signs and directional signs;
- captive balloon advertising (not balloons in flight);
- estate agents' boards;
- fascia signs and projecting signs;
- flag advertisements;
- models and devices;
- placards and boards;
- pole signs and canopy signs;
- posters and notices;
- price markers and price displays;
- town and village name signs;
- traffic signs.

When considering this type of application the local authorities are required to take account of amenity and public safety. Therefore in some areas certain advertisements are not regulated by the planning authority, and others have 'deemed consent'. In most cases whether or not you need to apply for consent will depend on the size, position and illumination of the advert.

In most cases if you are planning any of the following types of advertising you will need planning permission:

- advertisements using specialized structures for their display, such as poster hoardings and most non-highway authority roadside advance warning or directional signs;
- signs positioned above 4.6 m in relation to buildings above the level of the bottom part of first-floor windows or on gable ends;
- the majority of illuminated signs.

You must take particular care if a sign is to be displayed on or close to a listed building. The sign must not detract from the character and appearance of the building. This may mean that you may require separate listed building consent if the sign is in the vicinity of a listed building even if the type of sign you pro-pose is normally permitted.

4.5.2 Approval of conditions

If you failed to provide sufficient details in your original application then the planning authority may have approved your plans on the condition that you provide further details before work commences. An application in this category is not an application for planning permission or listed building consent; it is in connection with a previously permitted application.

You can apply online for the approval of conditions or submit a standard paper form to your authority. On the form (which will give details of the permitted application) you can provide information on how you intend to meet the condition or conditions that were placed on you; you should also provide contact information so that the authority can respond to your appli-cation. You will also need to attach a site location plan and plans of the proposed work (usually the same ones that you submitted originally). Once your application has been submitted the authority has eight weeks in which to consider it.

There are no national requirements for this type of application. Each author-ity may have local-level requirements requiring certain information or support-ing documents (such as plans or photographs) to be submitted with an appli-cation. Any local requirements will be available to view on the authority's website or can be confirmed by contacting the authority directly.

4.5.3 Consent under Tree Preservation Orders

You will need to seek approval to carry out **any** work on a protected tree or trees; this includes pruning overhanging branches even in your own garden. Protected trees include **all trees** growing in a Conservation Area or trees in other areas which are the subject of a Tree Preservation Order (TPO). Consent under Tree Preservation Orders is required in addition to other forms of plan-ning permission and applies even if you are carrying out development where full planning permission is not needed or where full planning permission has already been granted.

In all cases you **must** give your local planning authority six weeks' notice before carrying out work on trees located in a Conservation Area, even if the trees are not yet subject to a TPO. This gives the authority an opportunity to consider whether an order should be made to protect the trees. More information about TPO procedures can be found in section 4.5.3.1. You must apply for permission to work on a protected tree where you are carrying out development even if full planning permission has been granted or it is not needed. Your local planning authority should be able to provide you with details of any Tree Preservation Orders in your area.

Where the tree or trees are in a National Park or the Broads you should consult the relevant authority rather than the local planning authority.

If the tree is dead or dangerous you should speak to both the local planning authority and a competent tree surgeon about your proposed work to get their advice. Work on trees is potentially dangerous, and it therefore should only be carried out by a trained, competent and appropriately insured tree surgeon, and your local planning authority will want any work to be carried out to a good standard (usually BS 3998). If you carry out the work without this you may be required to prove that the work was exempt. In some cases the local planning authority or tree surgeon may charge for giving this advice.

A number of changes were made to the Tree Preservation Order system in England in April 2012. Further guidance on TPO procedures, including exemptions and felling licences, can be found in the following documents:

*   *Protected Trees: A Guide to Tree Preservation Procedures*;
*   *Tree Preservation Orders: A Guide to the Law and Good Practice.*

Both are available on the GOV.UK website, along with other information, or from DCLG Publications (https://www.gov.uk/government/policies/improving-the-energy-efficiency-of-buildings-and-using-planning-to-protect-the-environment/supporting-pages/tree-preservation-orders).

You may be prosecuted if work is carried out on a preserved tree without local planning authority consent or if the required notice has not been given or an exemption is misused.

4.5.3.1 Approval of Tree Preservation Order applications

Once the local planning authority has received your application under a TPO you can expect it to visit the site to assess the proposed work. During this visit alternative works will be discussed. If you do not receive a decision within eight weeks or two months (depending on the age of your TPO) of your application you will have a right of appeal to the Planning Inspectorate.

4.5.4 Conservation Area consent

Conservation Areas are areas of special architectural or historic interest where it is considered that the character or appearance of the area should be preserved or enhanced. Special characteristics are not confined solely to buildings and can include other features which contribute to particular views and the familiar local scene, for example:

* areas where there are public and private spaces, such as gardens, parks and greens;
* building and paving materials which are considered a characteristic of the area;
* buildings that are used in a particular way;
* roads, paths and boundaries that are laid out in a particular way;
* trees and street furniture that can be considered characteristic of the area.

The planning authority in a Conservation Area has the duty to protect or improve the character or appearance of the area and is given extra powers to control works and the demolition of buildings. Conservation Areas are able to give protection across a broader area of land than the individual listing of build-ing. In this way all the features within the area, whether or not they are listed, can be recognized as part of its character.

The demolition of an unlisted building in a Conservation Area, without the consent of the local planning authority, is a criminal offence.

If you live in a Conservation Area, you will need Conservation Area consent to carry out the following work:

* the substantial demolition of any unlisted structure or building within a Conservation Area;
* the demolition of a building bigger than 115 m³;
* the demolition of a gate, fence, wall or railing if it is next to a highway or public open space and is over 1 m high'
* the demolition of a gate, fence, wall or railing if it is over 2 m high elsewhere.

Note: Conservation Area consent does not apply to an ecclesiastical build-ing in ecclesiastical use, listed buildings, or the demolition of a scheduled monument or a building in some other categories. If you are unsure what the situation is in your area then you should seek advice from your local planning authority.

4.5.5 Full planning consent

Full planning consent is a detailed planning application for development. The category of 'development' includes making a material change in the use of land or a building; it also covers engineering, building or other works which are proposed in, on, over or under land. Full planning permission is required for structural alterations or additions to buildings, including:

- anything outside the garden of the property (e.g. stables in a separate paddock);
- any works relating to a flat;
- applications to change the number of dwellings (e.g. flat conversions or building a separate house in the garden);
- changes to use of part or all of the property to non-residential uses;
- demolition of buildings;
- other work normally undertaken by a builder;
- rebuilding.

Building works such as maintenance, improvements or other alterations inside the building, or works which do not 'materially affect' the look of the outside of the building do not require planning permission. Examples of this type of work are:

- certain uses for agriculture or forestry;
- changes of use if the intended use will be incidental to the existing use(s);
- internal building works;
- putting up boundary walls and fences below a certain height;
- small alterations to the outside such as installing alarm boxes.

 Note: Building Regulation approval may still however be required, and you are advised to contact your local authority building control section.

Details of how to apply for full planning consent are in Section 4.6.1 of this chapter and are available online via the Planning Portal. It is recommended that you speak to the planning department of your local authority before commencing any work, especially if you are unsure if your proposal requires planning permission.

4.5.6 Householder planning consent

Anyone who proposes to alter or enlarge a single house, including works within the boundary of the property, should submit a 'householder application for works' to his or her local planning authority. They types of project that these applications are for include:

- alterations;
- conservatories;
- dormer windows;
- extensions;
- fences;
- garages, car ports or outbuildings;
- loft conversions;
- porches;

- satellite dishes;
- swimming pools;
- vehicular access including footway crossovers;
- walls.

There are certain building projects that householders can carry out under permitted development rules. In May 2013 Parliament agreed to increase the size of single-storey rear extensions which can be built under permitted development and brought into force the associated neighbour consultation scheme. For a period of three years, between 30 May 2013 and 30 May 2016, householders will be able to build larger single-storey rear extensions under permitted development. In essence the permitted development size limits for extensions have doubled from $4\,m^2$ to $8\,m^2$ for detached houses, and from $3\,m^2$ to $6\,m^2$ for all other houses. There is no fee involved, but in order for you to get permission to build one of these larger extensions the planning authority still require a written description of the proposal, a plan of the site, the addresses of any adjoining properties (including those at the rear) and a contact address for the developer. During the 42-day consultation period the local authority will serve a notice on adjoining owners or occupiers. The neighbour has a 21-day period in which to raise an objection with the local authority. The authority will take this objection into account when it makes its decision and will consider if the impact on the amenity of all adjoining properties is acceptable; no other issues will be considered. The development can go ahead if the local authority gives the developer permission in writing within the 42-day period. If the local authority fails to notify the developer of its decision within the 42-day determination period, the development may go ahead.

4.5.7 Lawful development certificate

A lawful development certificate (LDC) offers peace of mind that an existing use or proposed use of a building is lawful or that the proposal doesn't require planning permission. It is used to establish whether an existing use of land, or some operational development, or some activity in breach of a planning condition, is lawful or if a proposed use of buildings or other land, or some operations proposed to be carried out in, on, over or under land, would be lawful. Examples when an application for a lawful development certificate should be made include:

- when planning enforcement action is taken by the local planning authority and the owner believes it is immune from action because the time limit for taking enforcement action has passed;
- when an owner discovers, in the course of a sale of the land, that planning permission has never been granted, and needs to show a prospective purchaser that no enforcement action can be taken by the local planning authority.

It is not compulsory to have an LDC but there may be times when you need one to confirm that the use, operation or activity is lawful for planning control purposes. In certain circumstances where no previous planning permission has been applied for, and if no action has been taken following the development of a property, an application can be submitted for a lawful development certificate.

An application for a lawful development certificate is also sometimes used in cases involving intensification of use or where the precise nature of the existing use is difficult to describe, such as:

* intensification;
* mixed uses;
* secondary uses;
* subdivision of the planning unit.

There are time limits within which local planning authorities are able to take planning enforcement action against breaches. These were introduced under the Planning and Compensation Act 1991 and are:

* four years for building, engineering, mining or other operations in, on, over or under land, without planning permission (this development becomes immune from enforcement action four years after the operations are substantially completed);
* four years for the change of use of a building, or part of a building, to use as a single dwelling house (enforcement action can no longer be taken once the unauthorized use has continued for four years without any enforcement action being taken);
* ten years for all other developments (the period runs from the date the breach of planning control was committed).

 If your application is partly or wholly refused or is granted differently from what you asked for, or is not determined within the time limit of eight weeks, you can appeal. Appeals are made to the Planning Inspectorate.

4.5.8 Listed building consent

 It is a criminal offence to carry out work which needs listed building consent without obtaining it beforehand.

Buildings which are of special architectural or historic interest are protected by listed building control. Each building is listed in its entirety and both the outside and inside are protected, as is anything that is fixed to the building or which was within the grounds before 1948. In order to be 'listed' a building, object or structure has to be assessed to be of national importance. All such buildings are included on the List of Buildings of Special Architectural or Historic Interest. This list includes a wide variety of structures, from castles and cathedrals to milestones and village pumps, and the majority of listings are a result of survey work conducted by English Heritage. However, anyone is permitted to ask the Department for Culture, Media and Sport to list a particular building if he or she feels it is of importance. The criteria used to assess whether a building should have listed status are:

- architectural interest: buildings of importance because of their design, decoration and craftsmanship;
- group value: buildings that form part of an architectural ensemble, such as squares, terraces or model villages;
- historic association: buildings that demonstrate close historic association with nationally important people or events;
- historic interest: buildings which illustrate an aspect of the nation's social, economic, cultural or military history.

4.5.8.1 Types of buildings that are eligible for listed status

The following types of building are eligible for listed status:

- a limited number of outstanding buildings built after 1939 (which are at least ten years old and usually more than 30 years old);
- all buildings built before 1700 that survive in anything like their original condition;
- buildings of definite quality and character from the period between 1840 and 1914 – the selection is designed to include the major works of principal architects;
- most buildings from the period 1700–1840, although selection is necessary;
- selected buildings of high quality or historic interest from the period between 1914 and 1939.

4.5.8.2 Grades of listed buildings

- Grade I – buildings of exceptional interest;
- Grade II* – particularly important and more than special interest;
- Grade II – buildings of special interest, warranting every effort being made to preserve them.

A listed building may require planning permission even if planning permission would not usually be required for that type of building or when the proposed activity does not constitute development. You will need to apply for listed building consent in either of the following circumstances:

- you want to alter or extend a listed building in a manner which would affect its character;
- you want to demolish a listed building.

 You may also need listed building consent for any works to separate buildings within the grounds of a listed building. Check the position carefully with the council.

When listed building consent is granted it is often subject to conditions which:

- preserve a particular feature or features of the building, either as part of it or after it is removed;
- make good any damage caused to the building by the works after work is completed;
- reconstruct the building or any parts of it following the proposed works, using the original materials as far as possible, and any alterations within the building as laid down in the conditions.

When a listed building is to be demolished the permission is only granted on the condition that no demolition occurs before agreement is reached on how the site will be redeveloped and planning permission has been granted for the redevelopment.

 Note: At the time of writing you cannot make an application for 'outline' listed building consent. This is so that the impact of the works on the building can be properly assessed by reference to detailed plans and other drawings. For works involving the external alteration or extension of a listed building, where works affect the character of the building a full application or householder application **and** listed building consent should be submitted.

4.5.9 Notification of proposed works to trees in Conservation Areas

Protected trees include all trees growing in a Conservation Area as well as any tree that is covered by a Tree Preservation Order (TPO). In a Conservation Area you must give notice before you commence any work on trees that have a trunk diameter of more than 75 mm when measured at 1.5 m from ground level (or more than 100 mm if reducing the number of trees to benefit the growth of other trees). The protection applies even if you only wish to prune overhanging branches from the protected tree even if it is in your garden. You should give your local authority six weeks' notice for work on trees in a Conservation Area even if the trees are not yet the subject of a Tree Preservation Order. This gives the authority an opportunity to consider whether an order should be made to protect the trees.

You must still apply for permission to work on a protected tree where you are carrying out development even if full planning permission has been granted or it is not needed.

If the tree is dead or dangerous you should speak to the local planning authority and a competent tree surgeon about your proposed work to get their advice. If you carry out the work without this you may be required to prove that the work was exempt. In some cases the local planning authority or tree surgeon may charge for giving this advice. Work on trees is potentially dangerous and it therefore should only be carried out by a trained, competent and appropriately insured tree surgeon. The local planning authority will want the work carried out to a good standard (usually BS 3998) and may be able to offer guidance on contractors in your area.

More information about TPO procedures can be found in section 4.5.9.1.

You may be prosecuted if work is carried out on a preserved tree without local planning authority consent or if the required notice has not been given or an exemption is misused.

In order to carry out work on trees in Conservation Areas you will need to fill out the form 'Notification of Proposed Works to Trees in Conservation Areas'. This form may be submitted electronically via the Planning Portal, by email or by fax. Whichever method you choose, the local planning authority use the same method to communicate with you. You will need to have the following information to hand when you complete the form:

- Additional information: You should ensure that you provide sufficient evidence to support your case.
- Applicant's and agent's names and addresses: Do not include the tree surgeon's details unless he or she has agreed to be your agent for this application.
- Employee of the authority or council member: If the applicant or the applicant's agent is a member of the council's staff, an elected member of the council or related to a member of staff or elected member of the council, he or she must declare it, as anyone in these categories should play no part in the determination of the planning application.
- Condition of tree(s): Any pests, diseases or fungi that are present should be described and any evidence from an appropriate expert should be provided.
- Describe the works and reasons for it: You must give reasons why you plan these works and what type of work needs to be done to each tree. Use of the terms 'cut back some branches', 'lop some branches' or 'trim some branches' is inadequate. Where possible you should describe the operation in the following terms:
    - crown lifting – the removal of lower branches to increase the clearance between the ground and the crown of the tree (this is used to prevent low branches obstructing paths or to permit more light to reach the garden);
    - crown reduction – to change the shape of the tree by shortening branches over the whole tree (used to prevent branches from damaging gutters and roofs);
    - crown thinning – to reduce the density of the tree without changing its shape (used to let more light into windows or gardens).
- Identification of tree(s) and description of work: You will need to know the species of tree and, if applicable, the numbering of trees from the TPO. Make sure any other trees are identified by using a different numbering sequence.
- Make sure you know what you are applying for.
- Other structural damage: You need to provide evidence in respect of structural damage (e.g. garden walls, drains, paving and drive surfaces) and this should be provided by an appropriate expert.

- Poor tree surgery: Any proposal that endangers the health or condition of a tree or greatly reduces its amenity is unlikely to be allowed.
- Replacement trees: It is usually a condition of approval that a replacement tree be planted. You should indicated your preference of tree should your application be successful.
- Sketch plan identifying the trees: You should provide a sketch plan which shows the names of roads, boundaries and the number or name of any adjoining properties. It may be useful to show the position of the tree(s) in relation to any building and give approximate distances between the tree(s) and other features. It is also sensible to show any other trees on the property and make it clear which tree(s) you want to work on. You may use photographs, particularly where you plan to work on a specific part of the tree.
- Subsidence: You should submit any reports you have received from a structural engineer or chartered surveyor. These should be supported by technical analysis from other experts, e.g. soil analysis.
- Tree location: You need to include the details of all the affected addresses here, especially if the trees grow on another property or straddle a number of properties.
- Tree ownership: The tree or trees don't have to be yours but you will need permission from the owner before you carry out any work. The owner of the tree is usually the owner of the land on which it grows. You will need to make the ownership details clear, particularly if the owner lives at a different address to that of the site where the tree or trees stand or if there is joint responsibility or you wish to carry out work on trees on a number of properties.
- Tree Preservation Order details.

4.5.9.1 Approval of Tree Preservation Order applications

In a Conservation Area the local planning authority has six weeks within which it can object to your proposals. During this time the local planning authority may take the opportunity to protect trees of amenity value, and it will usually carry out a site visit and may discuss alternative work with you. If the local planning authority objects, it will make a TPO on the tree(s) concerned. You may carry out the work if you have not received a decision within six weeks of your notice being given. However, if a TPO has been made you will need to apply for consent.

4.5.10 Outline planning consent

If you want to find out if a proposed development would be acceptable to the local planning authority before submitting a detailed proposal you can apply for outline planning consent. This allows you to provide fewer details about the proposal. However, you need to be in a position to ask for approval of the details (known as 'reserved matters') once permission has been granted. The level of detail required differs for each application but your local authority will tell you how much detail it requires for your project. You can also get 'pre-application advice' from your local authority before you send in your application.

 If you have been granted outline planning permission you must make a 'reserved matters' application within three years of the consent or your permission will be denied.

4.5.11 Prior notification

Prior notification relates to a proposal for development which involves agriculture, forestry or telecommunication projects or demolition work. In essence the prior approval procedure means that the principle of development has already been agreed but that the local planning authority should still be informed of your intention to carry out the work. More information on prior notification can be found in pdf guides on the Planning Portal and are discussed briefly below.

 You should not use an application for prior notification if a specific planning application is required. In this case, a separate planning application form must also be completed.

4.5.11.1 Agricultural and forestry developments

When an application is submitted for agricultural or forestry developments you should not commence work before the application is made. A prior notification application form should be submitted for agricultural units of 5 ha or larger for the following types of development:

- assembly or placing of a tank or cage for use in fish farming in any waters;
- building a fish tank;
- building, extending or significantly altering a building or altering a private way on forestry land;
- building a significant extension or significant alteration of a building on agricultural land;
- certain excavations and waste disposal;
- formation or alteration of a private way;
- proposed excavation or deposit of waste;
- proposed roads.

Once an application has been received the local planning authority has 28 days to decide whether to allow or refuse the development or to insist that a full application is made. At the end of this time the local planning authority must inform the applicant of its decision. The following are examples of work which requires planning permission:

- agricultural buildings with a floor area greater than 465 m^2;
- agricultural developments higher than 12 m;
- any development on land which is not currently used in connection with an agricultural business (e.g. hobby farming);
- any development that is not necessary for the purposes of agriculture or forestry on the holding (agriculture does not usually include keeping or rearing horses);
- buildings or excavations for livestock or slurry storage within 400 m of a protected building;
- development on an agricultural holding of less than 0.4 ha;
- development on separate parcels of land of less than: 1 ha on units of 5 ha or more; or 0.4 ha on units of less than 5 ha;
- development which is not for agricultural purposes;
- development within 25 m of the hardened part of a trunk or classified road;
- excavations or engineering operations connected with fish farming.

4.5.11.2 Demolition

Although planning permission is not usually required to demolish a building you will need approval from the local planning authority to demolish certain buildings, such as:

- any building except a dwelling house or one next to a dwelling house;
- buildings in Conservation Areas if they are under 50 m^3 capacity;
- fences or enclosures outside Conservation Areas;
- listed buildings.

You should give the local planning authority six weeks' notice of your intention to demolish a building in order to give it the opportunity to make sure that the demolition has a minimal impact on the local community. It will also determine if prior approval will be required for both the demolition and any proposed restoration of the site afterwards.

You do not need to apply for permission if the demolition is:

- on land which is the subject of planning permission for its redevelopment, granted on an application, or deemed to be granted, under Part III of the act;
- required or permitted to be carried out by or under any enactment;
- required to be carried out by virtue of a relevant obligation.

 Note: Further information on the demolition of buildings can be found in Circular 10/95: Planning Controls over Demolition.

4.5.11.3 Telecommunications developments

Some types of telecommunication development fall into the category of 'permitted development' (e.g. mobile telephone masts). The prior approval procedure means that the principle of development has already been agreed but that the local planning authority has 56 days to consider the appearance and siting of the proposal and to let the mast operator know of its decision. The 56-day period cannot be extended.

The prior approval procedure applies to the construction, installation, alteration or replacement of:

* an antenna (including any supporting structure) which exceeds the height of the building or structure (other than a mast) by 4 m or more at the point where it is installed or to be installed;
* a public call box;
* a ground-based mast of up to and including 15 m in height;
* a mast of up to and including 15 m in height installed on a building or structure;
* development ancillary to radio equipment housing (for example, fences or access roads);
* radio equipment housing with a volume of 2.5 m^3.

 Note: More information is available in the Mobile Phone Network Development Code of Best Practice and the National Planning Policy Framework.

4.5.12 Removal or variation of conditions

The local planning authority may make planning permission subject to conditions that control or limit the way in which the planning permission is implemented. The types of conditions that can be applied are to:

* carry out any works required for the reinstatement of land at the end of a specified period;
* discontinue a use of the land which had been authorized;
* end a specified period;
* regulate development or use of any land under the control of the applicant;
* remove any buildings or works authorized by the permission;
* require the applicant to carry out work on his or her land.

It is possible to apply to remove or change a condition that has been imposed or to apply for permission to develop without complying with a condition that had previously been imposed on a planning permission. The local planning authority will consider the application and decide whether to give permission subject to different conditions or unconditionally to approve the work. However, it can also refuse the application and insist that the original conditions still apply.

Listed building consent is often issued subject to conditions which required the work to be carried out in a particular way or which will safeguard the treatment of the building. If it becomes apparent that a condition is no longer appropriate the applicant can ask for these conditions to be changed. This allows for better solutions for the treatment of the building to be applied or for structural problems that arise to be dealt with. The authority may also take this opportunity to add consequential new conditions to its consent.

 Note: More details regarding the use of conditions are contained in Circular 11/95: *The Use of Conditions in Planning Permissions*.

4.5.13 Reserved matters

An application under reserved matters can be made within three years of outline permission being granted. It permits the applicant to submit information that was not included in the outline application, such as details of proposed:

- appearance;
- landscaping;
- layout;
- scale – including the size of the development (including height, width and length of proposed buildings).

The application does not necessarily deal with all of the outstanding details but must be made in line with the outline approval. This includes any conditions which were attached to the permission.

Any permission under reserved matters usually lasts for two years from the date of approval (or three years from the date of outline planning permission). It is not possible for work to begin until all of the reserved matters have been approved.

 If your proposals have changed you may need to reapply for outline or full planning permission.

4.6 Applying for planning permission

You don't have to own the land to make a planning application, but you will need to disclose your interest in the property. This means that you can apply for planning permission if you plan to buy land, with the intention of developing it, subject to planning approval. If you are in the planning stages for your work and you know planning permission will be required, it is wise to get the plans passed before you go to any expense or make any decisions that you may find hard to reverse – such as signing a contract for work. An architect (surveyor or general contractor) can be asked to prepare and submit your plans on your behalf if you like, but it is the name of the owner and person requiring the development that goes on the application, even if all the correspondence goes between the architect and the planning department.

It can take up to eight weeks, or longer, to get planning permission, so apply early. You should prepare a plan showing the position of the site in question (a site plan) so that the authority can determine exactly where the building is located. You must also submit another, larger-scale, plan to show the relationship of the building to other premises and highways (a block plan). In addition, it would help the council if you supplied drawings to give a clear idea of what the new proposal will look like, together with details of both the colour and the kind of materials you intend using. You may prepare the drawings yourself, provided you are able to make them accurate. If your plans are rejected, you will still have to pay your architect or whoever prepared your plans for submission, but you will not have to pay any penalty clauses to the building contractor. There are three types of plans (namely site, block and building) that can accompany your application, and the choice of which to use will depend on the work proposed.

- **Site plan:** A site plan indicates the development location and relationship to neighbouring property and roads, etc. The minimum scale is 1:2500 (or 1:1250 in a built-up area). The land to which the application refers is outlined in red ink. Adjacent land, if owned by the applicant, is outlined in blue ink.
- **Block plan:** A block plan is a detailed plan of a construction or structural alteration that shows the existing and proposed building, all trees, waterways, ways of access, pipes and drainage and any other important features. The minimum scale is 1:1500.
- **Building plan:** Building plans are the detailed drawings of the proposed building works and show plans, elevations and cross-sections that accurately describe every feature of the proposal. These plans are normally very thorough and include types of material, colour and texture, the layers of foundations, floor constructions and roof constructions, etc.

 If you require a location plan or a site plan these can now be purchased from one of the Planning Portal's accredited suppliers. If your local planning authority requires any further supporting documents these can also be uploaded.

Under normal circumstances you will have to pay a fee in order to seek planning permission, but there are exceptions. The planning department will advise you. A list of fees at the time of writing is at section 4.19.

 If you do work without getting planning permission and it was required, you can be forced to undo all the changes you have made.

4.6.1 The application process

The easiest way to apply for planning permission is via the online service on the Planning Portal but it is still possible to submit paper applications and you can get blank pdf-format forms from the Planning Portal 'forms cabinet'.

The online system is considered to be 'best practice' and the majority of applications for England and Wales are now made online. The advantage of the online system is that it automatically guides you through the application process, making sure you complete all the appropriate forms. You can also get help at each stage and there is access to a support team. The online application process was simplified in August 2013 with the launch of the new electronic application system.

4.6.1.1 Application forms and plans

It is important to make sure that you make your planning application correctly. The following checklist may help:

- Apply online using the Planning Portal or download the correct forms.
- Fill in and sign the relevant certificate relating to land ownership.
- Fill in the relevant parts of the forms and remember to sign and date them.
- If submitting a paper copy make sure you submit the correct number and type of supporting plans. Each application should be accompanied by a site plan of not less than 1:2500 scale, and detailed plans, sections and elevations, where relevant.
- Read the 'Notes for Applicants' carefully – again available from the planning department or the local council's website.

It is in your own interest to provide plans of good quality and clarity, and so it is probably advisable to get help from an architect, surveyor or similarly qualified person to prepare the plans and carry out the necessary technical work for you. You can ask a builder or architect to make the application on your behalf, particularly if the development you are planning is in any way complicated; you will have to include measured drawings with the application form.

4.6.1.2 Making an online application

The online application process is simple to follow and to begin you need to access the Planning Portal home page and click on 'Start an online planning application'. The process is then completed in three steps, which can be saved as you go along. You can view and access your submitted application for two years from the date of submission. If you wish to alter or add to a submission this can be done online and linked to the initial application (but only if you made the initial application online).

STEP 1: NAME AND LOCATION

You will need to have the site location details (with either postcode or grid reference) to hand and to choose a reference for your application so you can find it once it has been started. If your application affects neighbouring properties or you do not have a postcode then you should enter a grid reference.

STEP 2: APPLICATION TYPE

The next stage of the application requires you to choose which type of application you require. The number of consents available via the online service have been increased and you can now apply for the following consents online:

- advertisement consent;
- approval of conditions;
- consent under Tree Preservation Orders;
- Conservation Area consent;
- full planning consent;
- householder planning consent;
- lawful development certificate (LDC);
- listed building consent;
- notification of proposed works to trees in Conservation Areas;
- outline planning consent;
- prior notification;
- removal/variation of conditions;
- reserved matters.

This stage has combined two of the old steps and provides a drop-down list of all application types. You are therefore able to apply for multiple permissions and there is a link that will help you select those that apply to your case. You can contact your local planning authority if you are still unsure.

Note: Currently some forms are not available to complete online; at the time of writing these forms are 'Extending the time limits of existing planning permissions' and 'Application for non-material amendments'.

STEP 3: CONFIRM DETAILS

At this stage you are able to save your application. You will need to register yourself with the portal and thereafter you will have access to the overview screen for the application you have just created. Here you will see:

- **Any forms you should fill out:** The forms contain a series of questions that must all be answered to allow the local planning authority to determine your application. You should save your answers as you go along and answer the questions in order. You will find that the system will automatically remove any questions that are not required. Once you have filled in the form you should find that all questions show as 'complete'. If there are any questions that you feel are irrelevant you should enter 'not applicable'.
- **Any supporting documents:** You should attach all supporting documents, as the local planning authority cannot process your application without them. To avoid delays, it is best to submit everything online wherever possible.

  ○ Mandatory documents. There are two levels of mandatory documents:

    - National: The application service will tell you what mandatory documentation you need to provide in support of your application.
    - Local: The local planning authority will have produced a list of any specific documentation that is required to accompany the application in addition to the national requirements.

- ○ Optional documents. You can also attach any documentation that you think will help the local planning authority decide on the application. All plans and drawings must have a scale bar, key dimensions, the direction of north, original paper size and scale (e.g. 1:200 at A3) clearly marked on them.

- **A fee calculator:**The fee calculator will help you work out what fee is required for your application. If there are any changes the fee will be checked by the local planning authority and it will confirm whether the fee is correct.
- **Details of how to make payment and submit the application:**You will not be able to submit the application until you have completed the 'Check' stage. Once you have completed all the stages the system will check the entire application and identify any **incomplete sections**. **Once all the stages are complete** you can pay for the application and submit it to the local planning authority. The payment options displayed are those offered by the local planning authority processing your application.

 Note: Further guidance on electronic document submission 'best practice' is available on the Planning Portal (http://www.planningportal.gov.uk/uploads/1app/1app guidance note england en.pdf). You should note that individual files should not be larger than 5 MB and the system is only capable of accepting the following file types:

- images/plans: pdf, bmp, gif, jpg/jpeg, png, tif/tiff;
- documents: pdf, doc, rtf, txt, xls, plt;
- video: avi, mpg/mpeg, wmv.

4.7 What does the local planning authority do with my application?

Once your local planning authority has received your application you will get a confirmation email with a unique reference number (this email is not a formal acceptance of your electronic submission by your local planning authority). Once the local planning authority has received your application, supporting documents and payment, it will register and validate it within its normal work-flow processes and timescales. If the local planning authority needs more infor-mation or has any queries it will contact you directly.

4.8 How long will the process take?

 Discuss the proposal with a representative of the planning department before you submit your application. The representative will do his or her best to help you meet the requirements and give you an idea of local timescales.

You can expect to receive a decision from the planning department within eight weeks and, **once granted, planning permission is valid for three years**. If the work is not begun within that time, you can apply to extend the period for a further two years. Thereafter you will have to apply for planning permission again.

If the council cannot make a decision within eight weeks then it must obtain your written consent to extend the period. If it has not done so, you can appeal to the Secretary of State for Environment, Food and Rural Affairs or, in Wales, the National Assembly for Wales. Appeals can take several months to decide and it may be quicker to reach agreement with the council.

4.9 What factors affect planning permission?

A number of factors can affect whether or not you need to apply for planning permission or how successful your application will be:

- bats;
- consultations;
- crime prevention;
- design;
- environmental health;
- nature and wildlife;
- neighbours;
- other considerations – such as the Community Infrastructure Levy – that might apply;
- other permissions that you may require – such as an alcohol or badger licence, listed buildings consent or permission under the Party Wall Act;
- roads and highways.

4.9.1 Bats

There are two main types of roost that affect humans: buildings (houses, churches, farms, ancient monuments and industrial buildings) are popular summer roosts, while underground caves (mines, cellars and tunnels) are popular hibernation roosts. Bats make up nearly one-quarter of the mammal species throughout the world, and they are given special protection under the Wildlife and Countryside Act 1981. Bats form social colonies; female bats gather together in maternity colonies for a few weeks during the summer to give birth to and rear their young. During the winter, clusters of hibernating bats gather together in sheltered places for long periods.

Natural England or the Countryside Council for Wales (CCW) must be notified of any proposed action (e.g. remedial timber treatment, renovation, demolition and extensions) that is likely to disturb bats or their roosts. Natural England or CCW must then be allowed time to advise on how best to prevent inconvenience both to bats and to householders. Information on bats and the law is included in the Natural England booklet *Focus on Bats* (ISBN 978-1-84754-019-5), which can be obtained free of charge from your local Natural England office (http://www.naturalengland.org.uk/ourwork/regulation/wildlife/species/bats.aspx). If any timber treatment is carried out, only chemicals safe for use in bat roosts should be used. A list of suitable chemicals is available from Natural England on request. Any pre-treated timber used should have been treated using the CCA (copper–chrome–arsenic) method, which is safe for bats.

The pipistrelle, the smallest of the European bats, has been found lurking in many strange places, including vases, under floorboards and between the panes of double glass.

The types of stone barns and traditional buildings found in the UK have lots of potential bat-roosting sites, the most likely places being gaps in stone rubble walls, under slates or within beam joints. These sites can be used throughout the year by varying numbers of bats, but could be particularly important for winter hibernation. As a result, the following points should be followed when considering or undertaking any work on a stone barn or similar building, particularly where bats are known to be in the area. A survey for the presence of bats should be carried out by a member of the local bat group (contact via Natural England) before any work is done to a suitable barn during the summer bat-breeding period. Restrictions include the following:

* Pointing of walls should not be carried out between mid-November and mid-April or during maternity roosting periods of the summer, to avoid potentially entombing any bats.

    ○ In maternity roosts there are much higher numbers of bats (typically between 30 and 300), but they will be highly visible at feeding times.

* If walls are to be pointed, areas of wall high up on all sides of the building should be left unpointed to preserve some potential roosting sites.

    ○ If bats are found whilst work is in progress, work should be stopped and Natural England contacted for advice on how to proceed.

* Work should not commence during the winter hibernation period (mid-November to mid-April).

    ○ Any bats present during the winter are likely to be torpid (i.e. unable to wake up and fly away), and are therefore particularly vulnerable.

If you kill a bat, whether accidentally or not, there will be a £5000 fine per bat.

4.9.2 Consultations

Before you make your application it is advisable to consult anyone who might be affected by your proposal, including local councils and neighbours. In more complicated cases you may need to consult other bodies such as:

- the Environment Agency or local water and sewage company – sewerage, water or flooding problems;
- the Highways Authority – road safety and traffic issues;
- the Health and Safety Executive – use of any dangerous chemicals in your proposed development.

4.9.3 Crime prevention

Alterations and additions to your house may make you more vulnerable to crime than you realize. Even if you consider your home is secure against burglary you should consider what additional security precautions should be taken in light of the changes you plan. The crime prevention officer at your local police station can provide helpful advice on ways of reducing the risk. Alternatively, information on crime prevention issues is available in 'Secured by Design – An Official UK Police Initiative' and on the Home Office crime reduction website. Areas to consider include:

- extending your alarm to cover any extra rooms or a new garage;
- security locks for windows;
- the location of drainpipes and climbing plants.

4.9.4 Design

Personal taste can vary and different types of property will suit different styles of development work. You should consider how your property will look after the work is finished; for example, if you use the same materials and similar style in an extension it will look better. If your extension or building is well designed it will probably be more attractive and acceptable to your neighbours and you may find it adds more value to your house when you come to sell it. Although it is impossible to give a single definition of good design in this context, there may be many ways of producing a good result. You could look for guidance on designs from the council's planning department or you may wish to use a suitably qualified, skilled and experienced designer.

4.9.5 Environmental health

The safety of people living or working in an area is protected under environmental health legislation. The local environmental health officers work alongside the planning departments and are responsible for making sure developments do not cause air pollution, provide unfit housing or provide unhygienic food preparation premises. If a development is likely to infringe environmental health legislation it may be necessary for an 'environmental impact assessment' to be submitted.

4.9.6 Nature and wildlife

Some properties may also be the home for a range of protected species of animals, plants and habitats that are protected by the Wildlife and Countryside Act 1981 or under European legislation. These could be affected by work you wish to carry out. Natural England can give advice on what species are protected by legislation, and what course of action should be taken. Your local planning authority should also be able to advise on any species or habitats that may be affected by your proposals. Tree Preservation Orders may control the extent to which you can fell or even prune a tree, even if it is on your property. Trees in Conservation Areas are particularly protected, and you will need to supply at least six weeks' notice before working on them.

Barn owls occupy barns and similar buildings as roosting sites in some areas. The owls are more obvious than bats and, therefore, perhaps easier to take into account. Barn owls are also fully protected by law and should not be disturbed during their breeding season. Special owl boxes can be incorporated into walls during building work, details of which can be obtained from Natural England.

4.9.7 Neighbours

Many of us live in close proximity to others, and neighbours should be the first individuals you talk to when you are planning work on your property. Your alteration could infringe their access to light or a view. Disputes are notorious for causing bad feeling, but with a little consideration at an early stage you can avoid a good deal of unpleasantness later. It is always best to submit an application in the early stages of your planning – if you try to be clever by submitting plans at the last minute (in the hope that neighbours will not have time to react) then you could be in for an expensive mistake! It is much better to do things properly and up front.

Your neighbours will probably be as concerned about how the work will affect them as you would be if you were in their position. If you carry out work that seriously overshadows a neighbour's window and that window has been there for 20 years or more, you may be affecting their 'right to light' and you could be open to legal action. You also should consider if you need to notify your neighbour under the Party Wall Act. Any new building on or at the boundary of two properties, work to an existing party wall or party structure or excavation near to and below the foundation level of neighbouring buildings is affected and this may include:

- building a new wall on or at the boundary of two properties;
- cutting into a party wall;
- digging below the foundation level of a neighbour's property;
- knocking down and rebuilding a party wall;
- making a party wall taller, shorter or deeper;
- removing chimney breasts from a party wall.

It is best to consult a lawyer if you think you need advice. By simply modifying your proposals you could address some of your neighbour's worries. But even if you decide not to change what you want to do it is usually better to have told your neighbours what you are proposing before you apply for planning permission and before any building work starts.

Plans for the local area can normally be viewed at the local town hall, but most planning applications will involve consultation with neighbours and statutory consultees such as the highway authority and the drainage authorities. The extent of consultation will, quite naturally, reflect on the nature and scale of the proposed development – together with its location. Applications to make an alteration to your property can also be refused because you live in an Area of Outstanding Natural Beauty, a National Park or a Conservation Area or if your property is listed. Any alterations to public utilities such as drains or sewers, or changes to public access such as footpaths will require consultation with the local council. The council will have to approve your plans; even a sign on or above your property may need to be of a certain size or shape. The council also needs to approve new street names and house numbers and names.

 If you do need to make a planning application the council will still ask your neighbours for their views. If you or any of the people you are employing to do the work need to go on to a neighbour's property, you will, of course, need to obtain the neighbour's consent before doing so.

4.9.8 Other considerations

There are other issues in addition to the planning and building control regime that you should consider before starting work, in particular whether or not the Community Infrastructure Levy applies.

4.9.8.1 Community Infrastructure Levy

The Community Infrastructure Levy is a planning charge which came into force on 6 April 2010. Certain developments are liable for a charge under the Community Infrastructure Levy and you should check if your local planning authority has chosen to set a charge in its area. Further advice can be obtained from your local planning authority, Planning Aid, a local councillor or Member of Parliament or the Local Government Ombudsman.

4.9.9 Other permissions you may require

You may need additional permissions as well as obtaining planning permission before a development can proceed. The most commonly required consents include:

•   alcohol licences;
•   ancient monuments;

- badger licences;
- coal authority permits;
- common land;
- Conservation Areas;
- covenants and private rights;
- environmental permits;
- European protected species;
- flood defence;
- footpaths, bridleways or restricted byways;
- hazardous substances;
- listed buildings;
- stopping up or diverting a highway;
- the Party Wall Act.

4.9.10 Roads and highways

The local highways authority should be consulted where highways are affected during or after construction work. If your local council is not the highways authority for the road involved it will be able to help you contact the right person.

4.10 What involvement can the community have in planning matters?

Under the Local Government Act 1972 the public have the right to inspect and copy the following:

- reports for the public part of the meeting;
- the agenda for a council committee or subcommittee meeting;
- the minutes of such meetings and any background papers, including planning applications, used in preparing reports.

There is no charge to inspect a document but councils will charge for making photocopies. All documents are available from three clear days before a meeting.

The public can also play an active role in the planning system by making comment on decisions that affect them or their community. The main ways individuals can get involved are:

- commenting on planning-related appeals;
- commenting on planning applications which affect their community;
- reporting planning control breaches;
- taking part in public consultations for Local Development Frameworks.

If you want to be involved in the planning process you will need to be able to access planning information and be willing to contribute an opinion; you may need to seek expert help to represent a personal position effectively. You should make your views known in good time so that you do not hinder the planning timetable.

4.10.1 Can I comment on a proposed development?

Your local planning authority will invite comments on a formal planning application via public notices, on its website or through letters to adjacent homes and businesses. Larger developments are also advertised in local newspapers and sometimes via a developer's own public consultation process. Your local planning authority will publish the details of a proposal, including architectural drawings, at the local planning authority offices and on its website. If you wish to comment on a proposal you must make your comments to the local planning authority within its set time period; otherwise your views will not be taken into account. In order to reach its verdict the local planning authority often consults a large number of organizations, including parish councils and organizations with special expertise, e.g. the Environment Agency or English Heritage.

4.10.2 How can I find out about a development?

Once a planning application has been made, the local planning authority will invite comment on the proposed development by posting notices near the site, writing letters to those closest to the proposed development or taking an advertisement in a local newspaper. It may also keep local civic and environment societies informed of all applications in the area. All the details of the proposal are available for inspection at the local council offices or the council's website.

4.10.3 How can I make my views known?

There will be a limited amount of time in which you can comment by sending comments to the local planning office. It is very important to meet any deadline or your submission may not be taken into account. It is also possible to attend any committee meetings dealing with planning applications. In many cases members of the public can speak briefly to ensure that the committee is aware of their views. However, only elected members of the local planning authority can vote on the application decision itself.

4.10.4 Can I appeal after a planning decision?

It is not possible for a third party to appeal against a local planning authority's decision once it has been made. You should therefore make sure that your views are made known during the consultation period.

4.10.5 Can I complain about an application?

If you disagree with the way a local planning authority has handled a planning application you may be able to complain to the Local Government Ombudsman. You should note that the Ombudsman has no power to alter the decision and cannot investigate a complaint just because you do not agree with the decision. If the local authority administration has not been entirely correct the Ombudsman may recommend that the authority take action to mitigate the effect on you, and pay you compensation, if appropriate.

Note: If you are the planning applicant you should appeal to the Secretary of State, through the Planning Inspectorate; the Ombudsman will not usually look at your complaint.

4.10.6 How do I complain about a planning breach?

A planning breach in itself is not illegal and it is possible for the council to permit a retrospective application. A planning breach may occur when a development that has been given permission subject to conditions breaks one or more of those conditions or when a development takes place without permission being granted. If you think this is the case you can report it to the local planning authority. You can contact your local planning authority to report a development that has breached a planning control. The authority will organize its own administrative process for enforcing alleged breaches of planning control and will record and investigate complaints about alleged breaches of planning controls. If it decides to exercise discretion and not take formal enforcement action following a complaint, the local planning authority should be prepared to explain its reasons to any organization or person who has asked for an alleged breach to be investigated.

If the breach involves a previously rejected development (or a retrospective application fails) the council can issue an enforcement notice. The main grounds for this will be whether the existing use of the land and buildings merits protection or if the work would unacceptably affect public amenity.

It is illegal to disobey an enforcement notice unless it is successfully appealed against.

4.11 What can I do if my planning application is refused?

4.11.1 Can I amend my application?

You can amend an application you have submitted on the Planning Portal. It is free to change the supporting documentation but charges may apply if you want to make major changes to the application.

A second application is normally exempt from a fee.

If the council refuses permission or imposes conditions, it must give reasons. If you are unhappy or unclear about the reasons for refusal or the conditions imposed on your application you should talk to staff at the planning department. Ask them if changing your plans might make a difference. If your application has been refused, you may be able to submit another application under a new Planning Portal reference number (this option is only available if the original application was made via the portal) or by submitting another paper application.

Remember, the planning department will always grant planning permission unless there are very sound reasons for refusal, in which case the department must explain the decision to you so that you can amend your plans accordingly and resubmit them for further consideration. The following are some of the main objection areas that your application may meet.

4.11.1.1 The property is a listed building

Listed buildings are protected for their special architectural or historic value. A listed building consent may be needed for alterations, but grants could be available towards repair and restoration! If a building is listed it probably has some historic importance and will have been listed by the Department for Environment, Food and Rural Affairs. This could apply to houses, factories, warehouses and even walls or gateways. Most alterations which affect the external appearance or design will require listed building consent in addition to other planning consents.

4.11.1.2 The property is in a Conservation Area

This is an area defined by the local authority which is subject to special restrictions in order to maintain the character and appearance of that area. Again, other planning consents may be needed for an area designated as Green Belt, an Area of Outstanding Natural Beauty, a National Park or a Site of Special Scientific Interest.

4.11.1.3 The application does not comply with the Local Development Plan

Local authorities often publish a development plan, which sets out policies and aims for future development in certain areas. These are to maintain specific environmental standards, and can include very detailed requirements, such as minimum or maximum dimensions of plot sizes, number of dwellings per acre, height and style of dwellings, etc. It is important to check if a plan exists for your area, as proposals can meet with some fierce objections from residents protecting their environment.

4.11.1.4 The property is subject to a covenant

This is an agreement between the original owners of the land and the persons who acquired it for development. Covenants were implemented to safeguard residential standards and can include things such as the size of outbuildings, banning the use of front gardens for parking cars, or even just the colours of exterior paintwork.

4.11.1.5 There is existing planning permission

A previous resident or owner may have applied for planning permission, which may not have expired yet. This could save time and expense if a new application can be avoided.

4.11.1.6 The proposal infringes a right of way

If your proposed development would obstruct a public path that crosses your property, you should discuss the proposals with the council at an early stage. The granting of planning permission will **not** give you the right to interfere with, obstruct or move the path. A path cannot be legally diverted or closed unless the council has made an order to divert or close it to allow the development to go ahead. The order must be advertised and anyone may object. You must not obstruct the path until any objections have been considered and the order has been confirmed. You should bear in mind that confirmation is not automatic; for example, an alternative line for the path may be proposed, but not accepted.

If you are considering a planning application, you should consider the above questions. Normally your retained expert – architect, surveyor or builder – can advise and help you to get an application passed. Most information can be collected from your local planning department; if you need to find out about covenants, look for the appropriate Land Registry entry.

4.11.2 Can I appeal if my application is refused?

If you think the council's decision is unreasonable, you can appeal to the Secretary of State or (in Wales) to the National Assembly for Wales. Appeals must be made within six months of the date of the council's notice of decision. You can also appeal if the council does not issue a decision within eight weeks. There are three processes that an appeal can follow: written representations, a hearing or a local inquiry. The Planning Inspectorate provides an online system for submitting, tracking and commenting on appeals electronically. The planning authority will supply you with the necessary appeal forms.

Appeals are intended as a last resort, and they can take several months to decide. Recent statistics indicate that on average only one appeal in three is successful. It is often quicker to discuss with the council whether changes to your proposal would make it more acceptable.

Appeals are available for most planning controls. Table 4.2 provides the time limits for different appeals in the UK.

Table 4.2 Time limits for planning appeals

Type of appeal	When to appeal	Further comments
Householder planning application	• 12 weeks from the date on the decision notice. Note: If the local planning authority has failed to determine your householder planning application or you are appealing against the grant of permission subject to conditions to which you object, please follow the time limits under 'Planning application' below.	If an enforcement notice has been served for the same or a very similar development the time limit is: • 28 days from the date of the local planning authority decision if the enforcement notice was served before the decision was made yet not longer than two years before the application was made; • 28 days from the date the enforcement notice was served on or after the date the decision was made (unless this extends the appeal period beyond 12 weeks).
Planning application	• Six months from the date on the decision notice; or • six months from the expiry of the period which the local planning authority had to determine the application. Note: The local planning authority determination period is usually eight weeks (13 weeks for major developments and 28 days for non-material amendment applications). If you have agreed a longer period with the local planning authority, the time limit runs from that date.	If an enforcement notice has been served for the same or a very similar development within the previous two years, the time limit is: • 28 days from the date of the local planning authority decision if the enforcement notice was served before the decision was made yet not longer than two years before the application was made; • 28 days from the date the enforcement notice was served on or after the date the decision was made (unless this extends the appeal period beyond six months).
Listed building consent or Conservation Area consent application	• Within six months of receipt of the decision; or • within six months of the expiry of the period which the local planning authority had to determine the application. Note: The local planning authority determination period is usually eight weeks. If you have agreed a longer period with the local planning authority, the time limit runs from that date.	

Table 4.2 (Continued)

Type of appeal	When to appeal	Further comments
Advertisement consent	• Within eight weeks of the date you receive the decision; or • within eight weeks of the expiry of the period which the local planning authority had to determine the application. **Note:** The local planning authority determination period is usually eight weeks. If you have agreed a longer period with the local planning authority, the time limit runs from that date.	
Advertisement discontinuance notice	Before the effective date of the notice (the effective date will be shown on the notice).	
Enforcement notice	Before the effective date of the notice (the effective date will be shown on the notice).	
Listed building enforcement notice	Before the effective date of the notice (the effective date will be shown on the notice).	
Conservation Area enforcement notice	Before the effective date of the notice (the effective date will be shown on the notice).	
Certificate of lawful use or development	No time limit. **Note:** The local planning authority determination period is usually eight weeks, although you may agree a longer period with the local planning authority. An appeal on the grounds of non-determination can only be made following the expiry of the determination period.	

Table 4.2 (Continued)

Type of appeal	When to appeal	Further comments
Community Infrastructure Levy appeals	• Chargeable amount appeals (Regulation 114). Before the end of the period of 60 days beginning with the day on which the liability notice stating the original chargeable amount was issued. • Apportionment of liability appeals (Regulation 115). Before the end of the period of 28 days beginning with the day on which the demand notice stating the amount payable was issued. • Charitable relief appeals (Regulation 116). Before the end of the period of 28 days beginning with the date of the decision of the collecting authority on the claim for charitable relief. • Surcharge appeals (Regulation 117). Before the end of the period of 28 days beginning with the day on which the surcharge is imposed. • Deemed commencement appeals (Regulation 118). Before the end of the period of 28 days beginning with the day on which the demand notice is issued. • CIL stop notice appeals (Regulation 119). Before the end of the period of 60 days beginning with the day on which the CIL stop notice takes effect.	

The publication *Procedural Guide: Planning Appeals and Called-In Planning Applications – England*, published on 28 August 2013, is available at http://planningguidance.planningportal.gov.uk. The new *Procedural Guide* provides in advance the information that, for appeals received on or after 1 October 2013, and called-in planning applications where the date of the call-in letter is 1 October 2013 or later, the Inspector may initiate an award of costs. It also contains information about reporting, recording and filming at hearings and inquiries, including the use of digital and social media.

 Be careful not to proceed without approval, as you might find yourself obliged to restore the property to its original condition.

4.12 What are the most common stumbling blocks?

In no particular order of priority, the main stumbling blocks to achieving planning permission are:

* adequacy of parking;
* archaeology;
* design, appearance and materials;
* effect on a listed building or Conservation Area;
* effect on the street or area (but not loss of private view);
* government advice;
* ground contamination;
* hazardous materials;
* landscaping;
* light pollution;
* local planning policies;
* nature conservation;
* noise and disturbance from the use (but not from construction work);
* overlooking and loss of privacy;
* previous appeal decisions;
* previous planning decisions;
* road access;
* size, layout and density of buildings;
* traffic generation and overall highway safety.

4.13 What matters cannot be taken into account?

Matters that cannot be taken into account are:

* competition;
* disturbance from construction work;

- identity or personal characteristics of the applicant;
- loss of property value;
- loss of view;
- matters controlled under other legislation, such as Building Regulations (e.g. structural stability, drainage, fire precautions, etc.);
- moral issues;
- need for development;
- private issues between neighbours (e.g. land and boundary disputes, damage to property, private rights of way, deeds, covenants, etc.);
- Sunday trading.

4.14 Before you start work

There are many kinds of alterations and additions to houses and other buildings that do not require planning permission. Whether or not you need to apply, you should think about the following before you start work.

4.14.1 Lighting

If your plan includes putting in external lighting for security or other purposes, you need to make sure that others are not disturbed by the intensity and direction of the light. Excessive exposure to poorly designed lighting causes disturbance to many people, so, in particular, make sure that beams are **not** pointed directly at windows of other houses. Passive infrared detectors (PIRs) and/or timing devices on security lights should be adjusted so that they minimize nuisance to neighbours and are set so that they are not triggered by traffic or pedestrians passing outside your property.

4.14.2 Covenants

Even if you do not need to apply for planning permission you may be required to get someone else's agreement before carrying out some kinds of work to your property because of a covenant or another restriction in the title to your property or condition in the lease. You can check this yourself or consult a lawyer. You will probably need to use the professional services of an architect or surveyor when planning a loft conversion. Such services should include considerations of planning control rules and any covenants.

4.14.3 Listed buildings

Buildings are listed because they are considered to be of special architectural or historic interest, and as a result require special protection. Listing protects the **whole** building, both inside and out, and possibly also adjacent buildings if they were erected before 1 July 1948. The prime purpose of having a building listed is to protect the building and its surroundings from changes that will materially alter the special historic or architectural importance of the building or its setting. Listed buildings are covered in section 4.5.8.

4.14.3.1 Listed building owner's responsibilities

If you are the owner of a listed building or come into possession of one, you are tasked with ensuring that the property is maintained in a reasonable state of repair. The council may take legal action against you if it has cause to believe that you are deliberately neglecting the property, or have carried out works without consent. Enforcement action may be instigated.

There is no statutory duty to effect improvements, but you must not cause the building to fall into any worse state than it was in when you became its owner. This may necessitate some works, even if they are just to keep the building wind- and watertight. However, you may need listed building consent in order to carry these works out!

If you are selling a listed building you may wish to indemnify yourself against future claims: speak to your solicitor.

A photographic record of the property when it came into your possession may be useful, although you may also have inherited incomplete or unimplemented works from your predecessor, which you will become liable for.

4.14.4 Conservation Area

Conservation Areas are 'areas of special architectural or historic interest the character and appearance of which it is desirable to preserve or enhance'. Tighter regulations apply to developments in Conservation Areas and to developments affecting listed buildings. Separate Conservation Area consent and/ or listed building consent may be needed in addition to planning consent and Building Regulation consent. As the title indicates, these designations cover more than just a building or property curtilage and most local authorities have designated Conservation Areas within their boundary. Although councils are not required to keep any statutory lists, you can usually identify Conservation Areas from a local plan's 'proposals maps' and appendices. Some councils may keep separate records or even produce leaflets for individual areas. Conservation Areas are discussed in detail in section 4.5.4.

If you live or work in a Conservation Area, grants may be available towards repairing and restoring your home or business premises.

Local planning authorities may ask for more information to accompany your normal planning application concerning proposals within (or adjoining) a Conservation Area. This may include:

* a description of the works and the effect (if any) you think they may have on the character and appearance of the Conservation Area;

- a set of scale drawings showing the present and proposed situation, including building elevations, internal floor plans and other details as necessary;
- a site plan to 1:1250 or 1:2500 scale showing the property in relation to the Conservation Area.

For major works you may need to involve an architect with experience of works affecting Conservation Areas.

4.14.4.1 Trees in Conservation Areas

Nearly all trees in Conservation Areas are automatically protected, and it is necessary to obtain the council's approval for works to trees in Conservation Areas before they are carried out. Trees in Conservation Areas are generally treated in the same way as if they were protected by a Tree Preservation Order (TPO). There are certain exceptions (where a tree is dead or in a dangerous condition), but it is always advisable to seek the opinion of your council's tree officer to ensure your proposed works are acceptable. Even if you are certain that you do not need permission, notifying the council may save the embarrassment of an official visit if a neighbour contacts the council to tell it what you are doing. If you wish to lop, top or fell a tree within a Conservation Area you must give six weeks' notice, in writing, to the local authority. This is required in order that it can check to see if the tree is already covered by a TPO, or consider whether it is necessary to issue a TPO to control future works on that tree.

4.14.4.2 Tree Preservation Orders

Trees are possibly the biggest cause of upset in town and country planning, and many neighbours fall out over tree-related issues. They may be too tall, block out natural light, have overhanging branches or shed leaves on to other property, or the roots may cause damage to property. When purchasing a property the official searches carried out by your solicitor should reveal the presence of a TPO on the property or whether your property is within a Conservation Area within which trees are automatically protected. However, not all trees are protected by the planning regulations system – but trees that have protection orders on them must not be touched unless specific approval is granted. Do not overlook the fact that a TPO could have been put on a tree on your land before you bought it, and is still enforceable. TPOs are also covered in section 4.5.3.

 It is an offence to carry out work on a tree with a TPO on it without prior permission.

If you have a tree on your property that is particularly desirable – either an uncommon species or a mature specimen – then you can request a TPO for it. However, this will mean that in years to come you, and others, will be unable to lop it, remove branches or fell it unless you apply for permission.

Trees covered by TPOs remain the responsibility of the landowner, both in terms of any maintenance that may be required from time to time and for any damage they may cause. The council must formally approve any works to a tree covered by a TPO. If you cut down, uproot or wilfully damage a protected tree or carry out works such as lopping or topping which could be likely to seriously damage or destroy the tree, there are fines on summary conviction of up to £20,000, or, on indictment, the fines are unlimited. Other offences concerning protected trees could incur fines of up to £2500.

Although there are certain circumstances in which permission to carry out works to a protected tree are not required, it is generally safe to say that you should always write to your council to seek its permission before undertaking any works. You should provide details of the trees on which you intend to do work, the nature of that work – such as lopping or topping – and the reasons why you think this is necessary. The advice of a qualified tree surgeon may also be helpful.

 You may be required to plant a replacement tree if the protected tree is to be removed.

4.14.5 What about nature conservation issues?

Many traditional buildings, particularly farm buildings, provide valuable wild-life habitats for protected species such as barn owls and bats.

Planning permission will not normally be granted for conversion and reuse of buildings if protected species would be harmed. However, in many cases, careful attention to the timing and detail of building work can safeguard or re-create the habitat value of a particular building. Guidance notes prepared by Natural England (formerly English Nature) are available from councils or from Natural England's website (http://www.naturalengland.org.uk).

4.15 What happens if I don't get planning permission?

If you build something which needs planning permission without obtaining permission first, you may be forced to put things right later, which could prove troublesome and extremely costly. You might even have to remove an unau-thorized building.

4.16 Greener homes

Home energy generation through renewable energy technologies is increas-ingly popular and many domestic microgeneration projects fall under permit-ted development. Grants may be available to offset the cost of some of the installations, and the government encourages the public to cut their energy use – and save money. Further details of finance and incentives can be found on the Microgeneration Certification Scheme website (http://www.microgenera-tioncertification.org). As an alternative to fossil fuels they help home owners to meet their own energy requirements and reduce carbon dioxide emissions. The most common home energy generation systems are listed below.

4.16.1 Biomass

Biomass is produced from organic materials, either directly from plants or indirectly from industrial, commercial, domestic or agricultural products.

4.16.2 Heat pumps

Ground source heat pumps transfer heat from the ground into a building to provide space heating and, in some cases, to pre-heat domestic hot water.

4.16.3 Hydroelectricity

Hydroelectrical power systems use running water turning a turbine to produce electricity. A micro hydro plant is one that generates less than 100 kW.

4.16.4 Micro combined heat and power

A combined heat and power device simultaneously generates both heat and power and is a mature technology widely used in industry. Recovering the heat from a power-generating process leads to high overall efficiencies.

4.16.5 Solar electricity

Solar photovoltaic (PV) uses energy from the sun to create electricity to run appliances and lighting.

4.16.6 Solar thermal

Solar water heating uses energy from the sun to work alongside your conventional water heater.

4.16.7 Wind turbines

Small-scale wind turbines use the wind's lift forces to turn aerodynamic blades that turn a rotor which creates electricity.

4.17 The Planning Inspectorate

The Planning Inspectorate is responsible for national infrastructure planning, processing planning and enforcement appeals, and holding examinations into local plans and Community Infrastructure Levy charging schedules. The Inspectorate also deals with a wide variety of other planning-related casework, including listed building consent appeals, advertisement appeals, and reporting on planning applications called in for decision by the Department for Communities and Local Government (DCLG) or, in Wales, the Welsh Government. More information is available from the Inspectorate's Media Centre. You can also find out more about the national planning process, as it applies in England and Wales, and view details of anticipated and live project applications for development consent on its website.

4.18 Planning Portal specialist pages

The Planning Portal contains specialist guidance within its pages on the following areas.

4.18.1 Specialist guides

The Planning Portal specialist guides provide guidance to the procedures of the planning system for specialist areas and sectors of the economy. The guides provide practical advice that will help you prepare a planning application for your project, including:

* guidance for sports clubs;
* guidance for farmers;
* guidance on mobile phone masts.

4.18.2 Working with professionals

These pages offer guidance for professionals considering online submission of a planning application for the first time as well as advice for more experienced online submitters. The pages include details of the Smarter Planning Partnership between planning agents, local planning authorities and the Planning Portal, which supports and encourages best practice in the online submission of planning applications.

4.18.3 Working with local planning authorities to improve planning services

The pages offer details of the Planning Portal's partnerships with local planning authorities to change the culture of planning and provide accessible, open and transparent online planning services to citizens and professionals. This section offers practical information and resources to support local planning authorities, including:

* Community Infrastructure Levy: information for local authorities;
* information on plans as a common cause of invalid applications;
* Local planning authority online application statistics and website referrals;

- National Planning Casework Unit;
- National Planning Policy Framework – local planning authority helpline;
- online appeal service help;
- smarter planning, supporting development managers.

4.19 How much does it cost?

Most applications will require a fee to be paid. However, for some consents, e.g. listed building and Conservation Area consent, no fee is required. The council cannot deal with your application until the correct fee is paid, and the fee is not refundable if your application is withdrawn or refused. You can use the online fee calculator tool as a step in the online application process, as well as using it as a stand-alone tool. In most cases you will also be required to pay a fee when the work is commenced. These fees are dependent on the type of work that you intend to carry out. The fees outlined in Sections 14.19.1 to 14.19.12 are typical of the charges made by local authorities during 2012 when applications were submitted, and are extracted from the Town and Country Planning (Fees for Applications and Deemed Applications, Requests and Site Visits) (England) Regulations 2012.

If you are unsure about the appropriate fee for your application, you are strongly advised to contact your local planning authority in advance of submitting your application, as an incorrect fee will delay the processing of your application.

Note: A change introduced by the new regulations in 2012 directs that fees in respect of deemed planning applications be paid in full to the local planning authority rather than half to the local planning authority and half to the Secretary of State. As a result of this change the Planning Inspectorate has reviewed its procedures for dealing with deemed planning application fees received in association with enforcement appeals. These changes apply to appeals against enforcement notices issued on or after 22 November 2012. The main changes are that:

The Planning Inspectorate will:

- continue to handle requests to extend the fee payment deadline (so long as such requests are made before expiry of the deadline);
- continue to set the deadline for paying the fee for the deemed planning application (this remains a requirement within the regulations);
- no longer issue invoices setting out the amount payable;
- no longer routinely check or calculate the amount payable for the deemed planning application;
- return any cheques sent to them in error to the sender.

Local planning authorities will need to:

- action any refunds as required by the regulations;
- confirm to the Planning Inspectorate case officer in writing (to the enforcement team email address which can be found at the top of any letters from the case officer) if the fee has not been received by the deadline set by the Planning Inspectorate;
- confirm to the Planning Inspectorate case officer in writing (to the enforcement team email address which can be found at the top of any letters from the case officer) once the fee has been received;
- ensure that the explanatory note which accompanies the enforcement notices states:
  - the correct fee for the deemed planning application,
  - that the fee payable is double the amount payable for a normal planning application,
  - that the full amount is payable to the local planning authority;
- inform the Planning Inspectorate case officer immediately if a fee payment fails (e.g. if a cheque bounces), as a new fee payment deadline may need to be set, or the deemed planning application may be deemed to have lapsed.

4.19.1 Scale of fees – all outline applications

All outline applications — Where the site does not exceed 2.5 ha £385 per 0.1 ha. Where the site exceeds 2.5 ha, £9527 and an additional £115 for each additional 0.1 ha, maximum £125,000.

4.19.2 Scale of fees – householder applications

Householder applications — Alterations and extensions to a single dwelling including works within the boundary, £172.

4.19.3 Scale of fees – full applications

Full applications (and first submissions of reserved matters) — Alterations/extensions to two or more dwellings (or flats), including works within boundaries, £339. New dwellings up to and including 50, £385per dwelling. New dwelling in developments over 50 dwellings, £19,049 and an additional £115 for each additional unit, maximum £250,000.

Full applications (and first submissions of reserved matters) for the erection of buildings (not dwellings, agricultural buildings, glasshouses, plant or machinery) — Where there is no increase in gross floor space or it is no more than 40 m^2, £195. Where the area of gross floor space to be created by the development is greater than 40 m^2 but less than 75 m^2, £385. Where the area of gross floor space to be created by the development is greater than 75 m^2 but does not exceed 3750 m^2, £385for each 75 m^2 or part thereof. Where the area of gross floor space to be created by the development does not exceed 3750 m^2, £19,049 and an additional £115 for each 75 m^2 in excess of 3750 m^2, subject to a maximum in total of £250,000.

The erection of buildings (on land used for agriculture or agricultural purposes)	Where the area of gross floor space to be created by the development does not exceed 465 m^2, £80. Where the area of gross floor space to be created by the development exceeds 465 m^2 but does not exceed 540 m^2, £385. Where the area of the gross floor space to be created by the development exceeds 540 m^2 but does not exceed 4215 m^2, £385 for the first 540 m^2, and an additional £385 for each 75 m^2 in excess of 540 m^2. Where the area of gross floor space to be created by the development exceeds 4215 m^2, £19,049; and an additional £115 for each 75 m^2 in excess of 4215 m^2, subject to a maximum in total of £250,000.
Erection of glasshouses on land used for agriculture)	Works not creating more than 465 m^2 £80. Works creating more than 465 m^2 £2150.
Erection, alteration or replacement of plant and machinery	On sites not exceeding 5 ha, £385 per 0.1 ha (or part thereof). On sites over 5 ha, £19,049 and an additional £115 for each additional 0.1 ha up to maximum fee £250,000.

4.19.4 Scale of fees – applications other than building works

Car parks, service roads or other accesses	For existing uses £195.
Waste (use of land for disposal of refuse or waste materials or deposit of material remaining after the extraction or storage of minerals)	On sites not exceeding 15 ha, £195 per 0.1 ha (or part thereof). On sites over 15 ha, £29,112 and an additional £115 for each additional 0.1 ha up to maximum fee £65,000.
Operations connected with exploratory drilling for oil and natural gas	On sites not exceeding 7.5 ha, £385per 0.1 ha (or part thereof). On sites over 7.5 ha, £28,750 and an additional £115 for each additional 0.1 ha up to maximum fee £250,000.
Other operations (winning and working of minerals)	On sites not exceeding 15 ha, £195 per 0.1 ha (or part thereof). On sites over 15 ha, £29,112 and an additional £115 for each additional 0.1 ha up to maximum fee £65,000.
Other operations (not in any of the above categories)	£195 per 0.1 ha (or part thereof) up to maximum fee £1690.

4.19.5 Scale of fees – lawful development certificate

LDC for existing use which is in breach of a planning condition	Same as full.
LDC for existing use where it is lawful not to comply with a particular condition	£195.
LDC for proposed use	Half the normal planning fee.

4.19.6 Scale of fees – prior approval

Approvals for agricultural or forestry buildings and operations and demolition of buildings £80.

Telecommunications code systems operators £385.

Prior approval of a proposed change of use to a state-funded school £80.

Prior approval of proposed change of use of agricultural building to a flexible use within shops, financial and professional services, restaurants and cafés, business, storage or distribution, hotels, or assembly or leisure £80.

Notification of a proposed change of use to dwelling(s) £80.

4.19.7 Scale of fees – reserved matters

Application for approval of reserved matters following outline approval — Full fee due or, if full fee already paid, £385.

4.19.8 Scale of fees – approval/variation/ discharge of condition

Application for removal or variation of a condition following grant of planning permission £195.

Request for confirmation that one or more planning conditions have been complied with — £28 per request for householder; otherwise £97 per request.

4.19.9 Scale of fees – change of use

Change of use of a building to use as one or more separate dwellings — Developments of not more than 50 dwellings £385 for each.

Change of use of a building to use as one or more separate dwellings — Developments of more than 50 dwellings £19,049 and an additional £115 for each additional unit, maximum £250,000.

Other changes of use of a building or land £385.

4.19.10 Scale of fees – advertising

Adverts relating to the business on the premises £110.

Advance signs directing the public to a business (unless business can be seen from the sign's position) £110.

Other advertisements (e.g. hoardings) £385.

4.19.11 Scale of fees – application for new planning permission to replace an extant planning permission

Applications for major developments	£575.
Applications in respect of householder developments	£57.
Applications in respect of other developments	£195.

4.19.1 Scale of fees – application for a non-material amendment following a grant of planning permission

Applications in respect of householder developments	£28.
Applications in respect of other developments	£195.

 For the online fee calculator, visit http://www.planningportal.gov.uk/planning/usefultools.

4.20 Sustainable homes

The Code for Sustainable Homes aims to create sustainable homes and reduce carbon emissions. The code was launched in 2006, and since then any developer of any new home in England can choose to be assessed against the code. The Code for Sustainable Homes is still operational and remains the government's national sustainability standard for new homes.

4.20.1 How does the code work?

The code measures the sustainability of a home against design categories (see below) that rate the whole home as a complete package using stars (one star for entry level, i.e. above Building Regulations, and six stars for the highest level), depending on the extent to which it has achieved the required standards. The code sets minimum standards for energy and water use at each level and, within England, replaces the EcoHomes scheme, developed by the Building Research Establishment (BRE).

Notwithstanding the removal of the requirement for a Home Information Pack (HIP), the Code for Sustainable Homes is still operational and remains the UK government's national sustainability standard for new homes.

4.20.2 Is there a connection between 'sustainable homes' and 'sustainable locations'?

It should be noted that the code principally deals with sustainable homes as opposed to the *sustainability* of locations. This role will still be covered by the existing planning system, which is recognized as a means of ensuring that developments are located on sustainable sites and that developments are such that they assist in the reduction of the need to travel.

4.20.3 How many sustainable homes have been certified?

The latest official statistics on the Code for Sustainable Homes and average energy efficiency (SAP ratings) were released on 24 May 2013 in the publication *Code for Sustainable Homes and Energy Performance of Buildings: March 2013*, showing how many homes have been certified to the standards set out in the Code for Sustainable Homes and at which level.

4.20.4 Do I still need to provide a sustainability certificate?

A sustainability certificate is no longer required. In May 2010 the UK government announced the suspension of the requirement for sellers to give a sustainability certificate to buyers of newly constructed homes. However, the Code for Sustainable Homes is still operational and remains the government's national sustainability standard for new homes.

4.21 Home Information Pack (HIP)

The UK government suspended Home Information Packs (HIPs). Homes marketed for sale on or after 21 May 2010 do not require a HIP.

5

Requirements for planning permission and Building Regulations approval

You can make some changes to your home without having to get permission as long as the work does not affect the external appearance of the building. This is called 'permitted development'. Outside these rights you will find that the majority of building work requires you to have planning permission and/or Building Regulations approval prior to work commencing.

This chapter shows whether planning permission and/or Building Regulations apply to particular projects (which are listed in alphabetical order). Table 5.3 at the end of the chapter contains a useful summary of all the projects and what permissions are required. Be advised that there may be variations in the planning requirements, and to some extent the Building Regulations, from one area of the country to another, so you should consult your local authority for specific guidance which relates to your area.

In this chapter:

- all measurements are taken externally;
- designated areas are:

  ○ Areas of Outstanding Natural Beauty,
  ○ Conservation Areas,
  ○ National Parks,
  ○ the Norfolk or Suffolk Broads,
  ○ World Heritage Sites;

- the regulations depend on previous works on the site (which can date back to 1948); where the word 'original' is used in planning regulations this means the house as it was first built or as it was on 1 July 1948:

  ○ any extensions added since that date count towards the allowance; you should always check with planning control staff if you are unsure.

 Following devolution this guidance currently applies to England. Guidance specific to Scotland and Wales is available by following the links on the Planning Portal.

5.1 Advertising

Planning permission	Building Regulations approval
No If the advertisement is less than 0.3 m² and not illuminated.	Possibly Domestic adverts and signs are not usually subject to planning control, but they must be kept in a safe condition.

Advertisement signs on buildings and on land often need planning consent. Some smaller signs and those that are not illuminated may not need consent, but it is always advisable to check with local development control staff.

You can display certain small signs at the front of residential premises, such as election posters, notices of meetings, jumble sales and car for sale, etc. and information such as the house name or number and information signs such as 'Beware of the dog'. Temporary notices relating to local events, such as fêtes and concerts, may be displayed for a short period as long as they are under 0.6 m². Estate agents' boards are covered by different rules, but in general these should not be bigger than 0.5 m² on each side.

Professional and business types of display and permanent signs may come under the category of 'advertising control', for which planning consent is required. Signs for businesses are a complex area, and in addition to any planning permissions they must:

- be kept clean and tidy;
- be kept in a safe condition;
- be removed carefully where so required by the planning authority;
- not obscure, or hinder the interpretation of, official road, rail, waterway or aircraft signs, or otherwise make hazardous the use of these types of transport;
- only be displayed with the permission of the owner of the site on which they are displayed.

Illuminated signs and all advertising signs outside commercial premises need to be approved. Most local authorities can give advice, by way of booklets or leaflets, on what signs are allowed, are not allowed or need approval. If you are in any doubt you can download a free booklet, *Outdoor Advertisements and Signs: A Guide for Advertisers*. This can also be downloaded from the Planning Portal (https://www.gov.uk/government/publications/outdoor-advertisements-and-signs-a-guide-for-advertisers).

 It is illegal to post notices on empty shops' windows, doors and buildings, and also on trees. This is commonly known as 'fly posting' and can carry heavy fines under the Town and Country Planning Act.

5.2 Aerials, satellite dishes and flagpoles

Planning permission	Building Regulations approval

Satellite dishes, aerials and antennas on buildings up to 15m high

No Unless: • there are more than two antennas on the property overall; • a single antenna is more than 100 cm in any linear dimension; • if installing two antennas, one is more than 100 cm and the other is more than 60 cm in any linear dimension; • the cubic capacity of each antenna is more than 30 litres; • the antenna is fitted to a chimney stack and is greater than 60 cm in any linear dimension; • the antenna sticks out more than 60 cm above the roof line or chimney stack, whichever is the lower.	No But make sure that the fixing point is stable and the installation is safe.
Possibly If the building is in a designated area and the capacity of each antenna is greater than 35 litres or it is intended to install the antenna on a chimney, wall or roof slope visible from a road or Broads waterway.	No As long as the installation is safe and the fixing point stable.

Satellite dishes, aerials and antennas on buildings over 15m high

No Unless: • there are more than four antennas on the property overall; • the size of any antenna is more than 130 cm in any linear dimension; • the cubic capacity of each antenna is more than 35 litres; • the antenna is fitted to a chimney stack and is greater than 60 cm in any linear dimension; • the antenna sticks out more than 300 cm above the highest part of the roof line; • the building is in a designated area and it is intended to install the antenna on a chimney, wall or roof slope visible from a road or Broads waterway.	No But make sure that the fixing point is stable and the installation is safe.

Planning permission		Building Regulations approval	
Flagpole			
No	Flagpoles, etc. erected in your garden are treated under the same rules as outbuildings, and cannot exceed 3 m in height.	No	As long as the installation is safe and the fixing point stable.

Note: These limits refer to buildings as a whole and not flats within them. If you wish to install any of the following on a flat you should contact the local authority.

You should get specific advice if you plan to install a large satellite dish or aerial, such as a short wave mast, as the rules differ between authorities. Unless it is a stand-alone antenna, flagpole or mast greater than 3 m in height you will not need to apply for permission.

Normally there is no need for planning permission to attach an aerial or satellite dish to your house or its chimneys. However, if it rises significantly higher than the roof's highest point then it may contravene local regulations or covenants. In certain circumstances, you will need to apply for planning permission to install a satellite dish on your house (see the Department for Communities and Local Government's free booklet *A Householder's Planning Guide for the Installation of Satellite Television Dishes*, which can be obtained from your local council. Conservation Areas have specific local rules on aerials and satellite dishes, so you need to approach your local planning department to find out the particular rules for your area. Certainly, if your house is a listed building, you may need listed building consent to install a satellite dish on your house.

Before you go to any expense by purchasing or renting an antenna, check whether you need planning permission, listed building consent, or permission from the landlord or owner. You are responsible for placing antennas in the appropriate position.

The Planning Portal interactive house has a satellite dish locator. This provides a user-friendly way to check which parts of your house offer a suitable location for a satellite dish. Further guidance on the installation of antennas can be found in:

*   Town and Country Planning (General Permitted Development) Order 1995, SI 1995/0418;
*   Town and Country Planning (General Permitted Development) (Amendment) Order 1998, SI 1998/0462;
*   Town and Country Planning (General Permitted Development) (Amendment) (No. 2) (England) Order 2008, SI 2008/2362.

Remember, if you are a leaseholder, you may need to obtain permission from the landlord.

5.3 Basements

Planning permission	Building Regulations approval
No Provided that: • a light well is not added; • it is not a listed building; • it is not providing a separate unit of accommodation; • the external appearance of the building will not change; • the usage is not significantly changed; • you are not creating a new basement.	Yes Covering: • ceiling height; • damp-proofing; • fire escapes; • foundation work; • Party Wall Act; • underpinning; • ventilation wiring; • water supplies.

The creation of living space in basements, and its planning regime, is evolving and continues to be under review at the time of writing. In all circumstances before you carry out any work you are advised to contact your local planning authority to discuss any local policy changes.

5.4 Biomass-fuelled appliances

Planning permission	Building Regulations approval
No Provided any outside flue on the rear or side elevation is no more than 1 m above the highest part of the roof.	Yes Particularly in relation to electrical and plumbing work, ventilation, noise and general safety.
Yes If the building is listed or in a designated area, Conservation Area or World Heritage Site.	Yes Particularly in relation to electrical and plumbing work, ventilation, noise and general safety.

If the project also requires an outside building to store fuel or related equipment the same rules apply to that building as for other extensions and garden outbuildings. It is recommended you use an installer who belongs to either the Microgeneration Certification Scheme or the relevant competent person scheme.

5.5 Ceilings and floors

Planning Permission	Building Regulations approval
No If you are replacing a floor or ceiling.	Possibly If more than 25 per cent of a ceiling below a cold loft space or flat roof is being replaced then Building Regulations will apply and the thermal insulation of that ceiling should be improved.
Possibly If you live in a listed building you should contact your local planning authority.	Possibly If more than 25 per cent of a ceiling below a cold loft space or flat roof is being replaced then Building Regulations will apply and the thermal insulation of that ceiling should be improved.

The condensation risk in the roof space should be assessed and any provision in line with the requirements of Approved Document C should be made. If a dwelling has a pitched roof you will need to provide new or additional loft insulation unless the loft is already boarded out and the boarding is not to be removed as part of the work. Under a flat roof in a dwelling the insulation should be placed between and over joists as required to achieve the target U-value set out in the Approved Document. Where the ceiling height would be adversely affected, a lower performance target may be appropriate.

Note: Technical guidance can be found in Chapter 6, sections 6.5 (floors) and 6.7 (ceilings) and Approved Documents C and L.

5.6 Central heating

Planning permission		Building Regulations approval	
No	Unless it is a listed building or in a Conservation Area.	No	If electric and it is installed by an approved person and complies with Part P
No	If an external flue is required as long as:	No	If gas, solid fuel or oil fired and it is installed by an approved person and complies with Part J
	• any flues on the rear or side elevation are no more than 1m above the highest part of the roof;		
	• no flue is fitted the principal or side elevation if the building fronts onto a highway in Conservation Area, World Heritage site, national park, area of outstanding natural beauty or the Broads;		
	• you seek consent for internal and external work if the building is listed or in a designated area (even if you enjoy Permitted Development rights it is advisable to check with your local planners.)		

If the project requires an outside building to store fuel or related equipment the same rules apply to that building as for other extensions and garden outbuildings. Planning permission is not normally required for installation or replacement of a boiler or heating system if all the work is internal.

5.6.1 Repairing a heating system

Where permission is required it may not be necessary to apply in advance of carrying out the work, so if an emergency occurs you can carry out repairs straight away as long as they comply with the requirements, and after the event you should apply for retrospective approval and a completion certificate. If an existing system has been altered or replaced then the person who last worked on the system is responsible for the safe running of that system and should issue a certificate to show that the necessary checks have been carried out.

Guidance on what is reasonable provision for compliance with the energy-efficiency requirements is given in the Department for Communities and Local Government (DCLG) publication *Domestic Heating Compliance Guide*.

5.6.2 Replacing boilers

If a new system is to be installed then the installer should proceed as if the work is being carried out in a new building. If you want to change your existing hot water central heating boiler or if you decide to change to one of these boilers from another form of heating system you will need Building Regulations approval because of the safety issues and the need for energy-efficiency. The installer must follow the guidelines set out in Approved Document J, particularly with regard to air supplies, hearths, flue linings and chimney labelling where the flue outlet can be positioned. You can fulfil this requirement if you employ an installer who is registered under an approved scheme such as:

- gas boiler – the installer should be Gas Safe registered from 1 April 2009;
- oil-fired boiler – the installer should be registered with one of the relevant competent person schemes;
- solid-fuel-fired boiler – the installer should be registered with one of the relevant competent person schemes.

Each boiler must have a minimum efficiency of 86 per cent for gas and 85 per cent for oil, and any replacement gas boiler will probably have to be a condensing boiler unless there is sufficient reason why one cannot be installed. This assessment should be carried out by a registered installer on that type of boiler. This type of assessment is described in the Department for Communities and Local Government publication *Domestic Heating Compliance Guide*, and more advice can be found in the information leaflet *Gas and Oil Central Heating Boilers: Advice to Householders*.

If an oil-fired heating system is to have a new storage tank then there are guidelines set out in Approved Document J which should be followed for fire safety reasons and to limit risks of oil pollution. Following the guidance in Approved Document J may mean that replacement oil storage tanks cannot be sited where the old tank was. It is however possible for local authorities to set aside the guidance on tank location if it considers that this is unreasonable in all the circumstances.

In some cases householders may decide to engage a firm that is not a member of one of the approved competent person schemes. In this case the installer will not be able to self-certify that the work is compliant, and the firm or the householder will need to give notice to the local authority of the intention to carry out the boiler work in advance, and pay a notification fee. Local authority building control may choose to check the works have been carried out to the necessary standards and may employ a registered installer to do this.

When works have been completed the installer should then produce for you a commissioning certificate such as a Benchmark certificate and notify the local authority building control department either directly or, if a member of a competent person scheme, via the scheme operator. In due course the local authority should supply you with a Building Regulations completion certificate that indicates compliance.

Guidance on installing boiler and other combustion appliances, and the building provisions that are necessary to safely accommodate them (air supplies, hearths, fireplaces, flues and chimneys) can be found in Approved Document J.

5.7 Change of use

Planning Permission	Building Regulations approval
No Provided the present and proposed use fall within the same 'class' or the Town and Country Planning (Use Classes) Order says that a change of class is permitted to another specified class which is listed in paragraph 5.7.2.	Yes To certain changes of use, even though you may think that the work involved is not 'building work'
Material changes of use	
Yes	Yes
Working from home	
No Provided that: • the activities you undertake are not considered unusual in a residential area; • the business does not disturb your neighbours at unreasonable hours or create other forms of nuisance such as noise or smells; • there will be no increase in traffic or people calling; • you still intend to use the building mainly as a private residence.	No Unless building work is carried out
Associated building work	
Yes	Yes

The use of buildings or land for a different purpose may need consent even if no building or engineering works are proposed. Again, it is always advisable to check with development control staff. When you are assessing use of your property, whatever business you carry out the key test is to identify if it is still mainly a home or if it has become business premises. If you are in doubt you may apply to your council for a certificate of lawful use for the proposed activity, to confirm it is not a change of use and still the lawful use.

 Note: There are additional changes of use permitted development rights which apply from 30 May 2013; these are covered in this section.

5.7.1 What is meant by material change of use?

A material change of use is where there is a change in the purposes for which or the circumstances in which a building is used, so that after that change:

- a building no longer comes within the exemptions in Schedule 2 to the Building Regulations, where previously it did;
- the building contains a flat, where previously it did not;
- the building contains a room for residential purposes, where previously it did not;
- the building is used as a dwelling, where previously it was not;
- the building is used as a hotel or a boarding house, where previously it was not;
- the building is used as a public building, where previously it was not:
  - ○ a theatre, public library, hall or other place of public resort,
  - ○ a school or other educational establishment,
  - ○ a place of public worship;
- the building is used as a shop, where previously it was not;
- the building is used as an institution, where previously it was not;
- the building, which contains at least one dwelling, contains a greater or lesser number of dwellings than it did previously;
- the building, which contains at least one room for residential purposes, contains a greater or lesser number of such rooms than it did previously.

Whenever such changes occur, the building must be brought up to the standards required by all existing Approved Documents. This may involve Building Regulations approval even though you do not consider the works to be 'building works'.

5.7.2 Use classes

Table 5.1 shows the types of use which may fall within each class of use.

Table 5.1 Use classes and types of use

Class	Description of types of use
A1 Shops	Shops, retail warehouses, hairdressers, undertakers, travel and ticket agencies, post offices (but not sorting offices), pet shops, sandwich bars, showrooms, domestic hire shops, dry cleaners, funeral directors, internet cafés.
A2 Financial and professional services	Financial services such as banks and building societies, professional services (other than health and medical services), including estate and employment agencies and betting offices.
A3 Restaurants and cafés	For the sale of food and drink for consumption on the premises – restaurants, snack bars and cafés.
A4 Drinking establishments	Public houses, wine bars or other drinking establishments (but not night clubs).
A5 Hot food takeaways	For the sale of hot food for consumption off the premises.
B1 Business	Offices (other than those that fall within A2), research and development of products and processes, light industry appropriate in a residential area.
B2 General industrial	Use for an industrial process other than one falling within class B1 (excluding incineration purposes, chemical treatment or landfill or hazardous waste).
B8 Storage or distribution	This class includes open-air storage.
C1 Hotels	Hotels, boarding and guest houses where no significant element of care is provided (excludes hostels).
C2 Residential institutions	Residential care homes, hospitals, nursing homes, boarding schools, residential colleges and training centres.
C2A Secure residential institutions	Use for a provision of secure residential accommodation, including use as a prison, young offenders institution, detention centre, secure training centre, custody centre, short-term holding centre, secure hospital, secure local authority accommodation, military barracks.
C3 Dwelling houses	This class is formed of three parts:
C3(a)	Use by a single person or a family (a couple whether married or not, people related to one another, with members of the family of one of the couple to be treated as members of the family of the other), an employer and certain domestic employees (such as an au pair, nanny, nurse, governess, servant, chauffeur, gardener, secretary and personal assistant), a carer and the person receiving the care, and a foster parent and foster child.
C3(b)	Use by up to six people living together as a single household and receiving care (e.g. supported housing schemes, such as those for people with learning disabilities or mental health problems).

Table 5.1 Use classes and types of use (Continued)

Class	Description of types of use
C3(c)	Groups of people (up to six) living together as a single household. This allows for those groupings that do not fall within the C4 definition of houses in multiple occupation, but which fell within the previous C3 use class, to be provided for (e.g. a small religious community may fall into this section, as could a home owner who is living with a lodger).
C4 Houses in multiple occupation	Small, shared dwelling houses occupied by between three and six unrelated individuals, as their only or main residence, who share basic amenities such as a kitchen or bathroom.
D1 Non-residential institutions	Clinics, health centres, crèches, day nurseries, day centres, schools, art galleries (other than for sale or hire), museums, libraries, halls, places of worship, church halls, law courts, non-residential education and training centres.
D2 Assembly and leisure	Cinemas, music and concert halls, bingo and dance halls (but not night clubs), swimming baths, skating rinks, gymnasiums or areas for indoor or outdoor sports and recreation (except for motor sports, or where firearms are used).
Sui generis	Certain uses do not fall within any use class and are considered 'sui generis'. Such uses include: theatres, houses in multiple occupation, hostels providing no significant element of care, scrap yards, petrol filling stations and shops selling and/or displaying motor vehicles, retail warehouse clubs, night clubs, launderettes, taxi businesses, amusement centres and casinos.

 Note: The fee for prior approval for change of use is £80.

5.7.3 Changes to permitted development rights which apply from 30 May 2013

A number of changes to permitted development rights took effect on 30 May 2013. These include:

- agricultural buildings:
  - ○ agricultural buildings between $150\,m^2$ and $500\,m^2$ require prior approval (to cover flooding, highways and transport impacts, and noise),
  - ○ agricultural buildings under $500\,m^2$ can change to other uses (A1, A2, A3, B1, B8, C1 and D2);
- buildings with A1, A2, A3, A4, A5, B1, D1 and D2 uses will be permitted to change use for a single period of up two years to A1, A2, A3 and B1 use;
- premises in B1(a) office use can change to C3 residential use:
  - ○ subject to prior approval for flooding, highways and transport issues and contamination;
- premises in B1, C1, C2, C2A and D2 use classes can change use permanently to a state-funded school:

- ○   subject to prior approval for highways and transport impacts and noise;
- thresholds for business change of use from B1 or B2 to B8 and from B2 or B8 to B1 increased in May 2013 from $235\,m^2$ to $500\,m^2$.

5.7.4 Community Right to Build

A new type of neighbourhood development order was introduced under the Localism Act. The Community Right to Build order allows certain community organizations to put forward smaller-scale developments on a specific site without the need for planning permission, giving the community the opportunity to put forward facilities that they want to see and which will benefit them as a group, and can be used for projects such as:

- to save a former school;
- to save a local football ground;
- to save a local library;
- to save a local pub;
- to save or create a village shop;
- to save or create a park;
- to save or create sports centres.

You can use the Community Right to Bid to 'pause' the sale of buildings or land. Community Right to Build orders are subject to a limited number of exclusions, such as proposals needing to fall below certain thresholds so that an environmental impact assessment is not required. Proposals are subject to testing by an independent person and a community referendum. More information is available on the Community Right to Bid website (http://mycommunityrights.org.uk/community-right-to-bid).

5.7.5 Material changes of use

Where there is a material change of use of a **whole** building to a hotel, boarding house, institution, public building or shop (restaurant, bar or public house) the building must be upgraded so that it complies with Approved Document M1 (access and use).

If an existing building undergoes a change of use so that part of it can be used as a hotel, boarding house, institution, public building or shop, the work being carried out must ensure that:

- people can gain access from the site boundary and any on-site car parking space;
- sanitary conveniences are provided in that part of the building or it is possible for people (no matter what their disability) to use sanitary conveniences elsewhere in the building.

When a building is subject to a material change of use, then:

- any thermal element that is being retained should be upgraded;

- any existing window (including roof window or rooflight) or door which separates a conditioned space from an unconditioned space (or the external environment) and which has a U-value that is worse than 3.3 W/m^2 K should be replaced.

5.7.6 Material alterations

Where work results in a building or controlled service or fitting that no longer complies with a relevant requirement where previously it did, or if a previous compliance is made unsatisfactory, the material alteration will need to comply with the requirements for conservation of heat and energy.

5.7.6.1 Material alterations (domestic buildings)

If a building is subject to a material alteration by:

- making an existing element part of the thermal envelope of the building (where previously it was not);
- providing (or extending) a controlled service;
- providing a controlled fitting;
- renovating a thermal element;
- substantially replacing a thermal element,

then, in addition to the requirements of Part L, all applicable requirements from the following Approved Documents must be taken into account:

- Paragraph B1 (means of warning and escape);
- Paragraph B3 (internal fire spread – structure);
- Paragraph B4 (external fire spread);
- Paragraph B5 (access and facilities for the fire service);
- Part A (structure);
- Part B2 (internal fire spread – linings);
- Part M (access to and use of buildings).

5.7.6.2 Material alterations (buildings other than dwellings)

When an existing element becomes part of the thermal element of a building (where previously it was not) and it has a U-value worse than 3.3 W/m^2 K it should be replaced (unless it is a display window or high-usage door).

5.7.7 Electrical work in extensions, material alterations or a material change of use

Where any electrical installation work is classified as an extension, a material alteration or a material change of use, the work must consider and include:

- confirmation that the mains supply equipment is suitable and can carry the additional loads envisaged;

- the amount of additions and alterations that will be required to the existing fixed electrical installation in the building;
- that the earthing and bonding systems are satisfactory and meet the requirements;
- the necessary additions and alterations to the circuits which feed them;
- the protective measures required to meet the requirements;
- that the rating and the condition of existing equipment (belonging to both the consumer and the electricity distributor) are sufficient.

 Note: BS 7671 and this book's partner publication *Wiring Regulations in Brief* (3rd edition) offer guidance on the types of installations that might be encountered during alteration work.

5.7.8 What are the requirements relating to material change of use?

Where there is a material change of use of the whole of a building, any work carried out shall ensure that the building complies with the following documents:

- Access and facilities for the fire service (B5);
- Bathrooms (G5);
- Combustion appliances (J1, J2 and J3);
- Conservation of fuel and power – buildings other than dwellings (L2);
- Conservation of fuel and power – dwellings (L1);
- Dwelling houses and flats formed by material change of use (E4);
- Electrical safety (P1 and P2);
- External fire spread – roofs (B4)(2);
- Foul water drainage (H1);
- Internal fire spread – linings (B2);
- Internal fire spread – structure (B3);
- Means of warning and escape (B1);
- Resistance to moisture (C1)(2);
- Sanitary conveniences and washing facilities (G4);
- Solid waste storage (H6);
- Ventilation (F1).

In the case of a building exceeding 15 m in height it should also comply with external fire spread – walls (B4 1).

Material change of use	Requirement	Approved Document
The building is used as a dwelling where previously it was not.	Resistance to moisture	C2, E1, E2, E3
The public building consists of a new school.	Acoustic conditions in schools	E4
The building contains a flat where previously it did not.	Resistance to the passage of sound	E1, E2, E3

Material change of use	Requirement	Approved Document
The building is used as a hotel or a boarding house where previously it was not.	Structure	A1, A2, A3, E1, E2, E3
The building is used as an institution where previously it was not.	Structure	A1, A2, A3
The building is used as a public building, where previously it was not.	Structure	A1, A2, A3, E1, E2, E3
The building is not a building described in Classes I to VI in Schedule 2 where previously it was.	Structure	A1, A2, A3
The building, which contains at least one room for residential purposes, contains a greater or lesser number of dwellings than it did previously.	Structure	A1, A2, A3, E1, E2, E3
The building, which contains at least one dwelling, contains a greater or lesser number of dwellings than it did previously.	Resistance to the passage of sound	E1, E2, E3

In some circumstances (particularly when a historic building is undergoing a material change of use and where the special characteristics of the building need to be recognized) it may **not** be practical to improve sound insulation to the standards set out in Part E1 or resistance to contaminants and water as set out in Part C. In these cases, the aim should be to improve the insulation and resistance where it is practically possible – always provided that the work does not prejudice the character of the historic building, or increase the risk of long-term deterioration to the building fabric and/or fittings.

 Note: BS 7913:1998, *The Principles of the Conservation of Historic Buildings*, provides guidance on the principles that should be applied when proposing work on historic buildings.

5.7.9 Mixed use development

In mixed use developments the requirements of the regulations may differ depending on whether the area concerned is part of a building used as a dwelling or part of a building which has a non-domestic use. In these cases the requirements for non-domestic use shall apply in any shared parts of the building.

5.8 Conservatories

Planning permission	Building Regulations approval
No Provided that:	No Provided that:

Planning permission

No Provided that:

- extensions of more than one storey do not extend beyond the rear wall of the original house by more than 3 m and are not within 7 m of any boundary opposite the rear wall of the house;
- maximum eaves and ridge height of the extension are no higher than the existing house;
- maximum eaves height of an extension within 2 m of the boundary is 3 m;
- maximum height of a single-storey rear extension is 4 m;
- no extension is forward of the principal elevation or side elevation fronting a highway;
- no extension is higher than the highest part of the roof;
- no more than half the area of land around the 'original house' would be covered by additions or other buildings;
- there are no verandas, balconies or raised platforms;
- on designated land:

  - there is no permitted development for rear extensions of more than one storey;
  - there is no cladding of the exterior;
  - there are no side extensions;

- roof pitch of extensions higher than one storey should match the existing house;
- side extensions are single-storey with a maximum height of 4 m and width no more than half that of the original house;
- single-storey rear extensions must not extend beyond the rear wall of the original house by more than 3 m if an attached house or 4 m if a detached house. Outside Article 1(5) designated land and Sites of Special Scientific Interest the limit is increased to 6 m if an attached house and 8 m if a detached house until 30 May 2016.

  - These increased limits are subject to the neighbour consultation scheme.

Building Regulations approval

No Provided that:

- glazing and any fixed electrical installations comply with the applicable Building Regulations requirements;
- it is built at ground level and has a floor area less than 30 m^2;
- it is separated from the house by external-quality walls, doors or windows;
- there is an independent heating system with separate temperature and on/off controls.

Note: Any new structural opening between the conservatory and the existing house will require Building Regulations approval, even if the conservatory itself is an exempt structure.

A conservatory has to be separated from the rest of the house to be exempt (e.g. with patio doors between it and the main house).

Conservatories and sun lounges attached to a house are classed as permitted development, subject to the conditions above and provided that the glazing complies with the safety glazing requirements of the Building Regulations (Part K). Your local authority building control department or an approved inspector can supply further information on safety glazing. It is advisable to ensure that a conservatory is not constructed so that it restricts ladder access to windows serving a room in the roof or a loft conversion, particularly if that window is needed as an emergency means of escape in the case of fire.

As part of the new larger single-storey rear extensions scheme (for three years between 30 May 2013 and 30 May 2016) the government has introduced a mandatory neighbour consultation scheme. It is always good to keep your neighbours informed of what is happening, and be prepared to alter the plans you had for locating the building if they object – it is better in the long run, and can avoid lengthy delays.

Whilst it will provide useful space, a badly designed conservatory can be a huge energy drain for the house as a whole. If you want a conservatory or sun lounge separated from the house, this will need planning consent under similar rules for outbuildings. More guidance that is applicable to the windows, doors and electrics in a conservatory is contained in the relevant sections of this book.

The regulations described here apply only to houses, and not to other buildings, maisonettes or flats. There may be a charge under the Community Infrastructure Levy if the development is over $100\,\text{m}^2$.

 Note: The Government Technical Guidance document has not at the time of writing been updated to reflect the permitted development changes for larger single-storey rear extensions which came into force on 30 May 2013. It will be updated and republished in due course.

 For the most up-to-date guidance use the conservatories mini-guide on the Planning Portal.

5.9 Conversions

Planning permission	Building Regulations approval
Yes For flats – even where construction works may not be intended.	Yes Unless you are not proposing any building work to make the change.
No For loft conversions, as long as you do not alter or extend the roof space, and observe the following requirements: • a volume allowance of 40 m³ additional roof space for terraced houses; • a volume allowance of 50 m³ additional roof space for detached and semi-detached houses; • materials to be similar in appearance to the existing house; • no extension beyond the plane of the existing roof slope of the principal elevation that fronts the highway; • no extension to be higher than the highest part of the roof; • no verandas, balconies or raised platforms; • roof extensions not to be permitted development in designated areas; • roof extensions, apart from hip to gable ones, to be set back, as far as practicable, at least 20 cm from the original eaves; • side-facing windows to be obscure-glazed; any opening to be 1.7 m above the floor.	Yes To ensure: • reasonable sound insulation between the conversion and the rooms below; • safe escape from fire; • safely designed stairs to the new floor; • the stability of the structure (including the existing roof) is not endangered; • the structural strength of the new floor is sufficient. For further advice contact building control to discuss your proposal. You must also find out whether work you intend to carry out falls within the Party Wall etc. Act 1996.
Yes For conversion to shops and offices.	Yes

You will probably need planning permission whether or not building work is proposed. More information can be found in the relevant conversions mini-guides on the Planning Portal. The areas you should consider include:

• ensuring that the stability of the structure (including the existing roof) is not endangered;
• ensuring there is reasonable sound insulation between the conversion and adjacent rooms;
• including safely designed stairs to the new floor;
• making sure the structural strength of the new floor is sufficient;
• providing a safe escape from fire.

Note: Work on a loft conversion or a roof may affect bats. You need to consider protected species when planning work of this type. A survey may be needed, and if bats are using the building a licence may be needed (http://www.naturalengland.org.uk/ourwork/regulation/wildlife/species/bats.aspx).   (See Chapter 4 for more information on bats and other wildlife.)

You may wish to make these alterations to enhance the storage facilities available or to increase the living space of the home. If you plan to make the loft space more accessible or more habitable by, for example, installing a stair to it and improving it by boarding it out and lining the walls, rafters, etc., more extensive work is likely to be required and the Building Regulations are likely to apply. Common projects include:

- Boarding-out for storage: In most homes, the existing timber joists that form the 'floor' of the loft space are not designed to support a significant weight. An excessive additional load, for example from storage, may mean that the joists are loaded beyond their design capacity. You may require a Building Regulations application to building control.
- Creating a liveable space: A 'liveable space' is where you intend to use the room as a normal part of your house (this includes spare bedrooms which may be used infrequently). In an existing loft space of a home it is likely to require a range of alterations. Many of these could have an adverse impact on the building and its occupants if they are not properly thought out, planned and undertaken in accordance with the requirements of the legislation.

When considering a loft conversion you should take into account the following areas:

- doors and windows;
- drainage;
- electrics;
- existing walls and foundations;
- external walls;
- fire safety;
- internal walls;
- kitchens and bathrooms;
- new dormers;
- new internal elements;
- roofs;
- stairs.

It is recommended that you contact building control to discuss your proposal and for further advice and you must also find out whether work you intend to carry out falls within the Party Wall etc. Act 1996.

5.9.1 Do I need approval to convert my buildings?

The changes of use shown in Table 5.2 do not require planning permission.

Table 5.2 Changes of use not requiring planning permission

From	To
A2 (professional and financial services) when premises have a display window at ground level	A1 (shop)
A3 (restaurants and cafés)	A1 or A2
A4 (drinking establishments)	A1, A2 or A3
A5 (hot food takeaways)	A1, A2 or A3
B1 (business) (permission limited to change of use relating to not more than 235 m^2 of floor space)	B8 (storage and distribution)
B2 (general industrial)	B1 (business)
B2 (general industrial) (permission limited to change of use relating to not more than 235 m^2 of floor space)	B8 (storage and distribution)
B8 (storage and distribution) (permission limited to change of use relating to not more than 235 m^2 of floor space)	B1 (business)
C3 (dwelling houses)	C4 (houses in multiple occupation)
C4 (houses in multiple occupation)	C3 (dwelling houses)
Casinos (sui generis)	D2 (assembly and leisure)

In addition the changes of use shown in Table 5.2, planning applications are not required for change of use in the following circumstances:

- from A1 or A2 to A1 plus up to two flats above;
- from A2 to A2 plus up to two flats above.

These changes are reversible without an application only if the part that is now a flat was, respectively, in either A1 or A2 use immediately before it became a flat.

5.9.1.1 Converting an old building

Planning permission	Building Regulation approval
Yes	Yes

Throughout the UK there are many underused or redundant buildings, particularly farm buildings which may no longer be required, or suitable, for agricultural use. Such buildings, which are usually made of weathered stone and slate, contribute substantially to the character and appearance of the landscape and the built environment of an area. Their layout and proportions and the type of building materials used usually display local building methods and skills.

The best way to safeguard traditional buildings is to maintain their original. However, with changing patterns of land use and farming methods, change to buildings' use or conversion may have to be considered. The conversion or reuse of traditional buildings may, in the right locations, assist in providing employment opportunities, housing for local people or holiday accommodation. Applicants and developers are encouraged to refer to the local plan for comprehensive guidance and to seek advice from a planning officer if further assistance is necessary.

All councils place the highest priority on good design and proposals. Those that fail to respect the character and appearance of traditional buildings will not be permitted. Sensitive conversion proposals should ensure that: existing ridge and eave lines are preserved; new openings are avoided as far as possible; traditional matching materials are used; and the impact of parking and garden areas is minimized. Buildings that are listed as being of 'special architectural or historic interest' require skilled treatment to conserve internal and external features.

In many instances, traditional buildings that are of simple, robust form with few openings may only be suitable for use as storage or workshops. Other uses, such as residential, may be inappropriate. Most local development plans stipulate that conversion proposals 'should relate to buildings of traditional design and construction which enhance the natural beauty of the landscape' as opposed to 'non-traditional buildings, buildings of inappropriate design, or buildings constructed of materials which are of a temporary nature'.

Applicants are strongly advised to employ qualified architects or designers in preparing conversion proposals. Informal discussions with a planning officer at an early stage in considering design solutions are also encouraged.

5.9.1.2 Isolated buildings

The subject of planning permission for the conversion or reuse of isolated buildings is currently under review and is the subject of a government consultation. Permission may be given for such buildings to be used for living accommodation, small-scale storage or workshop uses, or camping purposes. You should check with your local planning authority for the most up-to-date information. An isolated building is normally:

• a building, or part of a building, which stands alone in the open countryside; or
• a building, or part of a building, in a group which occupies a remote location in relation to other buildings in the area by dint of the character of the surroundings, and the nature and availability of access and essential services.

Assessing whether or not a particular building should be regarded as isolated may not always be straightforward, and in such instances early discussion with a planning officer at the National Park authority is advised. Under the National Planning Policy Framework new homes in isolated parts of the countryside should be avoided unless there are special circumstances such as the exceptional quality or innovative nature of the design of the dwelling.

5.9.1.3 Structural condition

Planning permission will not normally be granted for reconstruction if substantial collapse occurs during work on the conversion of a building.

Buildings that are put forward for conversion should be large enough to accommodate the proposed use without the necessity for major alterations, extension or reconstruction. In cases of doubt regarding the structural condition of any particular building, the authority will require the submission of a full structural survey to accompany a planning application. The authority can advise on this requirement and, if necessary, on persons who are suitably qualified to undertake such work and who practise locally.

A list of local consulting engineers can be found via your search engine on the internet.

5.9.1.4 Workshop conversions

Redundant farm buildings and buildings of historic interest are often well suited to workshop use, and this type of conversion normally requires minimal alterations. Potential problems of traffic generation and unneighbourliness can usually be addressed by the imposition of appropriate conditions. The local authority will generally favourably consider proposals that make good use of traditional buildings by promoting local employment opportunities. In some instances grants may be available from other agencies to assist the conversion of buildings to workshop use.

5.9.1.5 Residential conversions

When reviewing proposals for converting a traditional building, the local authority will pay particular attention to the overall objectives of the housing policies of the local plan. If land that can be used for a new housing development is limited, residential conversions can make a valuable contribution to the local housing stock and support the social and economic well-being of rural communities. The local plan will require that residential conversions should, in most instances, contribute to the housing needs of the locality. Permission for such conversions is, in some districts, only granted subject to a condition restricting occupancy to local persons.

5.9.1.6 Renovation

Grants may be available towards the cost of renovating historical and important local buildings that have fallen into disrepair, or towards the cost of renovation

works to retain agricultural buildings in farming use, so as to retain their importance as landscape features. Further advice on the workings of the scheme may
be obtained from the Environmentally Sensitive Area project officers or the
local authority's building conservation officer.

'Local persons' are normally defined as persons working, about to work or
having last worked in the locality or who have resided for a period of three
years within the locality.

5.10 Decoration and repairs inside and outside a building

Generally speaking, you do not need to apply for planning permission:

*   for internal alterations;
*   for minor improvements, such as painting your house or replacing
    windows;
*   for repairs or maintenance;
*   for the insertion of windows, skylights or rooflights (but, if you want
    to create a new bay window, this will be treated as an extension of the
    house);
*   for the installation of solar panels which do not project significantly beyond
    the roof slope (rules for listed buildings and houses in Conservation Areas
    are different, however);
*   to re-roof your house (but additions to the roof are treated as extensions to
    the house).

Occasionally, you may need to apply for planning permission for some of these
works because your council has made an Article 4 direction withdrawing permitted development rights.

Planning permission		Building Regulations approval	
Repairs to a house, shop or office			
No	As long as the repairs are of a minor nature (e.g. replacing the felt to a flat roof, repointing brickwork, or replacing floorboards).	No	
Yes	For major works such as removing a substantial part of a wall and rebuilding it, or underpinning a building.	Yes	
Internal decoration, repair and maintenance			
No	For internal decoration, repair and maintenance.	Yes	For some types of repair work.

Planning permission	Building Regulations approval

External decoration, repair and maintenance

| No   Provided it does not make the building any larger or affect the thermal element of the building. | No   Provided it does not make the building any larger or affect the thermal element of the building. |

To alter the position of a WC, bath, etc. within a house, shop or flat

| No   If you are refitting a kitchen or bathroom with new units and fittings. | No |
| No | Yes   If you are fitting a bathroom or kitchen where there was not one before, or drainage and electrical work relating to a refit. |

To alter the construction of fireplaces, hearths or flues within a house, shop or flat

| No   To fit or replace an external flue, chimney or soil and vent pipe. | Yes   If you are installing a flue. |

If the property is a listed building or in a Conservation Area

| Yes   • For any external work, especially if it will alter the visual appearance, or use alternative materials.
• To alter, repair or maintain a gate, fence, wall or other means of enclosure. | Yes |

To insert cavity wall insulation

| No   As long as there is no change in external appearance or the building is not listed or in a Conservation Area. | Yes   All insulation has to comply with the relevant Building Regulations both when installed during construction and when fitted retrospectively. If such an upgrade is not technically or functionally feasible, the element should be upgraded to the best standard which can be achieved within a simple payback of no greater than 15 years. If you are installing loft insulation as part of a roof renovation project, where more than 25 per cent of the roof is being renewed, then the level of insulation should meet the standards required by Building Regulations Approved Documents. Care should be taken not to block any ventilation at the edges. |

Planning permission		Building Regulations approval
To apply cladding		
No	As long as it is not a listed building or in a Conservation Area.	Yes To re-render or replace timber cladding to external walls, depending on the extent of the work. Where 25 per cent or more of the wall is re-clad, re-rendered or re-plastered internally, or 25 per cent or more of the external wall is rebuilt, the thermal insulation will need to be improved and Building Regulations will apply.

5.11 Demolition

Planning permission		Building Regulations approval
No	Unless the council has made an Article 4 direction restricting permitted development rights in respect of demolition.	Yes Six weeks' prior notice must be given to the local authority building control.
Possibly	The council may wish to agree with you how you propose to carry out the demolition. This is called a 'prior approval application'.	Yes For a partial demolition to ensure that the remaining part of the house (or adjoining buildings/extensions) is structurally sound.

You must have good reasons and you will require a 'prior approval application' before knocking a building down and should get formal confirmation of whether the council wishes to agree with how you propose to carry out the demolition. Reasons such as making way for rebuilding or improvement would be incorporated in the same planning application. There are severe penalties for demolishing something illegally.

You do not need to make a planning application to demolish a listed building or to demolish a building in a Conservation Area. However, you may need listed building or Conservation Area consent.

The regulations covering the demolition of internal walls are covered in section 5.29.

You will not automatically get planning permission to build any replacement structure or to change the use of the site if you demolish a building (even if it has been damaged).

5.11.1 What about dangerous buildings?

The local authority can make an order restricting a building's use until such time as a magistrates' court is satisfied that all necessary works have been completed. These works are controllable by the local authority under Sections 77 and 78 of the Building Act 1984. In inner London the legislation is under the London Building (Amendment) Act 1939. This involves responding to all reported instances of dangerous walls, structures and buildings within each local authority's area on a 24-hour, 365-days-a-year basis.

If a building or structure poses a potential danger to the safety of people, the local authority will take the appropriate action to remove the danger. It has power to require the owners to remedy the defects, or it can direct its own contractors to carry out works to make the building or structure safe. This includes demolishing the structure, or any dangerous part of it, and the removal of any rubbish resulting from the demolition. In addition, the local authority may provide advice on the structural condition of buildings to the fire brigade during firefighting.

If you are concerned that a building or other structure may be in a dangerous condition, then you should report it to the local council.

5.11.2 Can I be made to demolish a dangerous building?

If the local authority considers that a building is so dangerous that it should be demolished, it is entitled to issue a notice to the owner requiring the owner or occupier:

- to shore up any building adjacent to the building to which the notice relates;
- to weatherproof any surfaces of an adjacent building that are exposed by the demolition;
- to repair and make good any damage to an adjacent building caused by the demolition or by the negligent act or omission of any person engaged in it;
- to remove material or rubbish resulting from the demolition and clear the site;
- to disconnect, seal and remove any sewer or drain in or under the building;
- to make good the surface of the ground that has been disturbed in connection with this removal of drains, etc.;
- in accordance with the Water Act 1945 (interference with valves and other apparatus) and the Gas Act 1972 (public safety), arrange with the relevant statutory undertakers (e.g. water board, British Gas or electricity supplier) for the disconnection of gas, electricity and water supplies to the building;
- to leave the site in a satisfactory condition following completion of all demolition work.

In certain circumstances, the owner of an adjacent building may be liable to assist in the cost of shoring up his or her part of the building and waterproofing the surfaces. It could be worth checking this point with the local authority! As detailed in section 5.11, anyone carrying out demolition work is required to give the local authority six weeks' notice of his or her plans.

Before complying with this notice, the owner must give the local authority 48 hours' notice of commencement.

5.12 Doors and windows

Planning permission	Building Regulations approval

Repairs, maintenance and minor improvements

No Unless it is a listed building or in a Conservation Area or your council has made an Article 4 direction withdrawing permitted development rights.	Yes If you use an unregistered installer or do it yourself you will need to get approval from the local authority building control.
	No If the work is carried out by an approved person and complies with the requirements of Part F.

New windows in an upper-floor side elevation

No   As long as they are:

- obscure-glazed; and
- non-opening or fitted more than 1.7 m above the floor level.

New rooflights or skylights

No   As long as:

- they do not protrude more than 150 mm beyond the plane of the roof slope;
- they are no higher than the highest part of the roof.

Windows on the side elevation roof slope must be obscure-glazed and either non-opening or more than 1.7 m above the floor level.

Shop windows

Yes To replace shop windows. Further information is available from your local building control or from the Glass and Glazing Federation (GGF) website (http://www.ggf.org.uk).	Yes To replace shop windows.

If you are a leaseholder, you may first need to get permission from your landlord or management company.

External windows and doors are known as 'controlled fittings' and require certain standards to be met when they are replaced. To ensure you comply with the requirements you should use an installer registered with the relevant competent person scheme, who will be able to carry out the work without the need to involve the local authority building control and provide you with a certificate once the work has been carried out. Alternatively, you can do it yourself or use an unregistered installer; however, in this case you will need approval from your local authority or an approved inspector, who will check the replacement fittings comply and, if satisfied, issue a certificate of compliance. More information about competent person schemes can be found on the GOV.UK website and section 2.9.3 of this book. The following areas should be considered when replacing external doors and windows:

- Access to the building: If the main entrance doors are replaced in a dwelling that has been constructed since 1999 the threshold must remain level to enable people, including those with disabilities, to have continued access to the dwelling.
- Fire safety: There are two aspects to be considered:

  ○ Fire spread – external doors and windows may need to resist a fire, be self-closing or be fixed shut.
  ○ Means of escape – a replacement window should have an opening at least the same size as the window it replaces. The means of escape should be considered for any new window installed to an extension or existing dwelling. If an escape window is required then the criteria set out below should be followed:

    - width and height – either of these is not to be any less than 450 mm;
    - clear openable area – no less than 0.33 m^2;
    - sill height – no less than 800 mm and no more than 1100 mm from floor level;
    - only one window per room is generally required.

- Safety glazing: Safety glass is required for the following areas:

  ○ glazed area within windows below 800 mm from floor level;
  ○ glazed area within windows that are 300 mm or less from a door and up to 1500 mm from floor level;
  ○ glazed doors up to 1500 mm from floor level.

- Thermal heat loss: In order to make the building more energy-efficient, replacement windows and doors need to comply with the requirements of the Building Regulations, particularly their U-values (the amount of heat that can to pass through the glass and framework). This is covered in Approved Document L1B, and Chapter 6.
- Ventilation: Windows and doors provide ventilation to rooms, and the type and extent of ventilation that are required will depend on the use and size of the room.

You should note that a new bay window will be treated as an extension and may require permission.

Note: As FENSA does not apply to commercial premises or new-build properties, **all** replacements of windows in offices and other commercial premises (including the replacement of shop fronts) will require local authority building control approval.

5.13 Drains and sewers

Planning permission		Building Regulations approval	
No	For repairs, maintenance and very minor works.	Yes	Work on drains and sewers should be carried out in compliance with Approved Document H of the Building Regulations.
Possibly	If permitted development rights in the area extend beyond repair, maintenance and very minor works.		
Yes	Internal or external works to listed buildings are likely to require listed building consent.		

If you fail to confirm who owns the drain or sewer or to comply with relevant legislation you could be faced with the bill for legal and remedial action.

Sewers can be publicly or privately owned. Public sewers are owned by the sewerage undertaker (you will find its address on your water bill). Private sewers are owned by the properties they serve. Building work on and around a sewer needs permission of the sewer owner. You may find that a drain or sewer on your property starts as your responsibility as it exits the building but becomes the responsibility of the sewerage undertaker as it joins with pipes from neighbours (e.g. along the back of a row of terraced houses). You should therefore check who is responsible for each part that will be affected by your work and obtain permission before undertaking any work.

You are strongly advised to seek advice from a builder, architect or drainage engineer or from local authority building control before committing to or commencing work.

There are two systems of drainage: 'foul' and 'surface water'. Both have above-ground and underground elements. The two systems should be kept separate. In simple terms a drain serves a single property, whilst a sewer serves more than one property.

- Foul drainage – takes used water from toilets, sinks, basins, baths, showers, bidets, dishwashers and washing machines. Any pipework that is above ground is referred to as sanitary pipework; the underground pipework is referred to as foul drains and foul sewers.

- Surface water drainage – takes rainwater, snow or ice from hard surfaces. Any pipes and gutters that are above ground are referred to as roof drainage; the underground pipework is referred to as surface water drains and surface water sewers.

5.13.1 Underground drainage

Building over an existing drain or sewer can damage pipes, which could cause leaks or blockages and lead to odour nuisance, health problems and environmental damage. Drains should be kept clear of obstruction and away from any foundations. Any drain that is below, or close to, a proposed extension may need to be moved or protected, and any work that takes place over or close to a public sewer will require written agreement from your sewerage undertaker before work commences.

In order to carry the flow and to avoid blockages, the drain or sewer that you intend to connect to generally needs to be at least 0.8 m lower than the ground-floor level. If it is less than this, you should seek advice from a builder, architect or drainage engineer.

If you want to connect to an underground drain or sewer you will need to find the depth and location and then adapt your plans accordingly. The first indication that there is a drain in the vicinity is often the presence of a drain cover. If you lift the cover you can see the direction, size and depth of pipes but you should not enter the chamber and you should make sure that the cover is replaced securely. Maps of public sewers may be inspected free of charge at the offices of the sewerage undertaker or local authority. Private sewers and drains are not normally mapped, and if their location needs to be discovered there are firms that carry out CCTV surveys and are able to advise on the condition of the drains.

If you increase the size of your gutters and rainwater pipes, or add new rainwater pipes or lay a new patio or driveway and these discharge on to the ground you need to make sure the water will not damage foundations or flow on to neighbouring property. You should consider the use of a soakaway or some other way of allowing the water to soak into the ground or store it for toilet flushing or garden watering. Approved Document H gives advice on soakaways. Where it is impractical to use infiltration (e.g. because of nearby foundations, impermeable or contaminated ground, or high groundwater), it is preferable to discharge it to a watercourse or, failing this, to a surface water sewer or, as a last resort, to a combined sewer. Surface water must not be discharged into a foul drain or sewer.

5.13.2 Pipes from kitchen and bathroom appliances

Pipes should be sized for the flow of water, to minimize the risk of blockage and to allow air movement. Advice on pipe sizes is given in Approved Document H. Sanitary pipework should be designed with access hatches, or be capable of being dismantled, in order to deal with blockages. Rodding eyes and access chambers should be used to enable all parts of the underground drainage to be cleared and to allow removal of blockages.

5.13.3 Ventilating pipes

Sanitary pipework needs to be ventilated to avoid air from the pipework and drains from escaping into the building. The normal way of doing this is to extend the pipework (known as the ventilating pipe) to outside the building, leaving the end open (but protected with a mesh to prevent birds getting in). To stop smells entering a building, the open end of the ventilating pipe should be at least 3 m to the side of, or extended to 0.9 m above, any opening into a building. If the drainage is already ventilated, additional ground floor appliances (e.g. a WC and washbasin) may be connected directly to the drain without a ventilating pipe. It may be possible to use proprietary valves to avoid the need for ventilating pipes.

5.14 Electrical work in the home or garden

Planning permission		Building Regulations approval	
No	Unless it is in a listed building or a Conservation Area.	No	As long as: • you comply with Part P (and other relevant Building Regulations Approved Documents); • all electrical work follows the safety standards in BS 7671 (the 'wiring regulations'). Your contract with the electricity supply company will have conditions about electrical safety which must not be broken. In particular, you should not interfere with the company's equipment, which includes the cables to your consumer unit up to and including the separate isolator switch, if provided.

These rules were introduced to reduce the number of deaths, injuries and fires caused by faulty installations.

 Note: The Building Regulations only set standards for electrical installation work in dwellings in Part P. BS 7671:2008, incorporating Amendment No. 1:2011 *Requirements for Electrical Installations* applies to the design, erection and verification of all electrical installations in residential, industrial, commercial, professional, public, agricultural and horticultural premises. If you have any queries about work in these buildings you should contact the Health and Safety Executive.

Repairs, replacements and maintenance work or extra power points or lighting points or other alterations to existing circuits are classed as non-notifiable work and can be completed by a DIY enthusiast (family member or friend). The installation will **still** need to be in accordance with the manufacturer's instructions and done in such a way that it does not present a safety hazard. Even minor electrical work can be a risk. If qualified electricians carry out the work they should give you a minor works certificate, which means that they have tested the work to make sure it is safe. If you do the work yourself you may wish to engage a qualified electrician to check it for you. This category of minor works includes:

- adding a fused spur (which is a socket that has a fuse and a switch that is connected to an appliance, e.g. heater) to an existing circuit (but not in a kitchen, bathroom or outdoors);
- any repair or maintenance work;
- installing cabling at extra-low voltage for signalling, cabling or communication purposes (e.g. telephone cabling, cabling for fire alarm or burglar alarm systems, or cabling for heating control systems).
- installing or upgrading main or supplementary equipotential bonding;
- replacing any electrical fitting (e.g. socket outlets, light fittings, control switches).

This work **does** need to be notified to the local authority building control body if it is installed in an area of high risk such as a kitchen, a bathroom or outdoors. If you are not sure whether the work you want to undertake is notifiable, you should contact your local authority building control department for advice.

The competent person scheme in relation to electrical safety ensures that an electrician who is registered by an organization authorized by the Secretary of State is able to certify that the work carried out is safe, without you having to notify building control. Once works are complete the electrician will arrange for you to receive a Building Regulations compliance certificate within 30 days of completion of the work and provide you with a completed electrical installation certificate. Your local authority will then also be notified about the work by the electrician. You should check which scheme the electrician belongs to and obtain his or her membership number. You will then be able to check with the Competent Person Register to make sure the electrician is registered. The GOV.UK website has a list of organizations which run the competent person schemes, and this is shown in section 2.9.3 of this book.

If you or the electrician you employ to carry out the works is not registered as a competent person under one of the relevant competent person schemes you will need to make a Building Regulations application to building control. The building control body will then arrange to have the electrical installation inspected at first fix stage and tested upon completion. Alternatively you could use an approved inspector.

More information can be found in the document *Rules for Electrical Safety in the Home*, which is available as a pdf on the Planning Portal.

5.15 Extensions

 This guidance reflects increases to the size limits for single-storey rear extensions which apply between 30 May 2013 and 30 May 2016, and the associated neighbour consultation scheme.

Planning permission	Building Regulations approval
No Provided that: • extensions of more than one storey do not extend beyond the rear wall of the original house by more than 3m; • materials are similar in appearance to the existing house; • maximum eaves and ridge height of the extension is no higher than the existing house; • maximum eaves height of an extension within 2m of the boundary is 3m; • maximum height of a single-storey rear extension is 4m; • no extension is forward of the principal elevation or side elevation fronting a highway; • no extension is higher than the highest part of the roof; • no more than half the area of land around the 'original house' would be covered by additions or other buildings; • there are no verandas, balconies or raised platforms; • the roof pitch of extensions higher than one storey matches the existing house; • side extensions are single-storey with a maximum height of 4m and width no more than half that of the original house; • single-storey rear extensions must not extend beyond the rear wall of the original house by more than 3m if an attached house or 4m if a detached house. For extensions outside of Article 1(5) designated land and Sites of Special Scientific Interest the limit is increased to 6m if an attached house and 8m if a detached house until 30 May 2016: ◦ these increased limits (between 3m and 6m and between 4m and 8m respectively) are subject to the neighbour consultation scheme; • two-storey extensions are no closer than 7m to the rear boundary; • upper-floor, side-facing windows are obscure-glazed and any opening is 1.7m above the floor.	Yes Most extensions of properties require approval under the Building Regulations.

Planning permission	Building Regulations approval
On designated land	
No Provided that in addition to the above: • there is no cladding of the exterior; • there are no side extensions; • there are no rear extensions of more than one storey.	Yes Most extensions of properties require approval under the Building Regulations.

As part of the new larger single-storey rear extensions scheme (for three years between 30 May 2013 and 30 May 2016) the government has introduced a mandatory neighbour consultation scheme. You can download the notification of a proposed larger home extension and its relevant guidance notes from the Planning Portal.

You need approval to build an extension if it would 'materially alter the appearance of the building'. Major alterations and extensions nearly always need approval. However, some extensions such as porches, garages and conservatories may be 'permitted development' and, therefore, do not need planning consent.

Check with your local authority planning department if you are not sure.

If a building is extended, or undergoes a material alteration, the completed building must comply with the relevant Building Regulations. Should this not be feasible, the alterations should be 'no more unsatisfactory than before'. In order to satisfy the requirements of the regulations when building an extension you should find out if the work you intend to carry out falls within the Party Wall Act and ensure that you comply with the Approved Documents in the following areas:

• doors and windows;
• drainage;
• electrics;
• energy-efficiency;
• external walls;
• flooring;
• foundations;
• internal walls;
• kitchens and bathrooms;
• roofs;
• structural opening;
• ventilation;
• walls below ground level.

These items are discussed separately in this chapter and more detail is contained in Chapter 6 of this book. In addition the Planning Portal has a useful interactive guide, and more detail is in the mini-guide on extensions, which can be downloaded as a pdf file.

The area of windows, roof windows and doors in extensions should **not** be greater than 25 per cent of the floor area of the extension **plus** the area of any windows or doors that, as a result of the extension works, no longer exist or are no longer exposed.

You may also require planning permission if your house has previously been added to or extended, or if the original planning permission for your house imposed restrictions on future development (i.e. permitted development rights may have been removed by an 'Article 4 direction', which is often the case with more recently constructed houses).

 Any building that has been added to your property and which is more than $10 \, m^3$ in volume and that is within 5 m of your house is treated as an extension of the house and so reduces the allowance for further extensions without planning permission.

If your development is over $100 \, m^2$ it may be liable for a charge under the Community Infrastructure Levy.

You **will** also require planning permission if you want to make additions or extensions to a flat or maisonette. You will need to apply for planning permission for the following:

- All additional buildings which are more than $10 \, m^3$ in volume, if you live in a Conservation Area, a National Park, an Area of Outstanding Natural Beauty or the Norfolk Broads. Wherever they are in relation to the house, these buildings will be treated as extensions of the house and reduce the allowance for further extensions.
- Alterations to the roof, including dormer windows (but permission is not normally required for skylights).
- An addition which would be nearer to any highway than the nearest part of the original house, unless there would be at least 20 m between your house (as extended) and the highway. The term 'highway' includes all public roads, footpaths, bridleways and byways.
- An extension that is higher than the highest part of the roof of the original house, or any part of the extension that is more than 4 m high and is within 2 m of the boundary of your property.
- An extension to your house that comes within 5 m of another building belonging to your house (e.g. a garage or shed). The volume of that building counts against the allowance given above.
- A terraced house, end-of-terrace house, or any house in a Conservation Area, a National Park, an Area of Outstanding Natural Beauty or the Norfolk Broads where the volume of the original house would be increased by more than 10 per cent or $50 \, m^3$ (whichever is the greater).
- Changes that create a separate dwelling, such as self-contained living accommodation or a granny flat.

- Detached or semi-detached houses where the volume of the original house would be increased by more than 15 per cent or 70 m³ (whichever is the greater). In any case the volume of the original house would be increased by more than 115 m³.
- Extensions nearer to a highway than the nearest part of the original house (unless the house, as extended, would be at least 20 m away from the highway).
- Extensions or additions that exceed the limits on height and volume.
- Where more than half the area of land around the original house would be covered by additions or other buildings – although you may not have built an extension to the house, a previous owner may have done so.

 You should measure the height of buildings from the ground level immediately next to it.

There are a number of classes of new buildings or extensions of existing buildings that are exempt (some partially and others completely) from the regulations. They may however require planning permission.

Class of work	Approved Documents A–K and M	Approved Document L	Approved Document P
Class 1 (Buildings controlled under other legislation)	Exempt	May apply	Exempt
Class 2 (Buildings not frequented by people)	Exempt	May apply	Exempt
Class 3 (Greenhouses)	Exempt	May apply	
Class 3 (Agricultural buildings)	Exempt	May apply	Exempt
Class 4 (Temporary buildings)	Exempt	May apply	Exempt
Class 5 (Ancillary buildings)	Exempt	May apply	Exempt
Class 6 (Small detached buildings)	Exempt	May apply	
Class 7 (Extensions)	Exempt	Exempt	

For further information you should refer to Regulation 9 of the Building Regulations.

5.16 External walls

Planning permission		Building Regulations approval	
Repairs, maintenance or minor improvements			
No	As long as it is for repairs, maintenance or minor improvements, such as painting your house and it is not a listed building or in a Conservation Area or designated area.	Yes	If 25 per cent or more of an external wall is re-rendered, re-clad, re-plastered or re-lined internally or where 25 per cent or more of the external leaf of a wall is rebuilt.
		Yes	If you want to insert insulation into a cavity wall.
Cladding the outside of a house with stone, artificial stone, pebble dash, render, timber, plastic or tiles			
Yes	If you live in a Conservation Area, a National Park, an Area of Outstanding Natural Beauty or the Broads.	Possibly	Depending on the extent of the work involved in re-cladding the building.
No	If you live outside the areas above and any cladding is of a similar appearance to material used in the construction of the house.	Possibly	Depending on the extent of the work involved in re-cladding the building.

There are two types of external wall. A cavity wall has two skins of masonry; the outer skin can be of brickwork or blockwork and the inner skin is generally of blockwork. The gap between the two skins is called the cavity and its depth will vary. A solid wall has one skin of masonry, which can consist of brickwork or blockwork. It is difficult to meet the high standards of thermal insulation needed in buildings with a solid wall construction unless a surface finish is applied. Redecorating the surface of an existing wall would not normally require approval.

If you are converting an existing external wall in a conversion project then the existing walls will need to be checked for their adequacy in terms of:

*   if they need to be upgraded, whether this will involve the addition of a new internal skin – possibly constructed of lightweight studwork, and the detail at the foot of the new skin will need careful planning to ensure that damp-proofing arrangements are sound and that any new timbers are protected from damp;
*   thermal resistance and changes to 'thermal elements';
*   weather resistance;
*   weight and structural stability.

You should also consider the following requirements of Building Regulations that will need to be satisfied with respect to external walls:

*   decoration and renovation;
*   fire protection;

- thermal resistance and changes to 'thermal elements';
- weather resistance;
- weight (loading).

If an opening is formed in a wall, the structure above the opening, even if it is relatively small, will need to be supported. Lintels are normally used to provide such support, in one of two ways. All lintels should have a suitable bearing on to the wall at each side of the opening. It is best to consult with the manufacturer or an engineer for the correct-size lintel and bearing.

5.17 Fascias

Planning permission	Building Regulations approval
No Unless you live in a listed building or designated area (Conservation Area, National Park, Area of Outstanding Natural Beauty).	No

You are advised to check that the replacement work does not reduce the ventilation provided to the roof void, as this could cause condensation to occur within the roof void, which can then lead to damp on the timbers. If the existing system has vents installed, a similar system will need to be maintained.

Fascia boards are boards attached to the end of the rafters/truss at the eaves, where the guttering is attached for the roof rainwater drainage. These are also placed at the ends of gable roofs to cover the rafters/truss.

Soffits boards are placed on the underside of the eaves where the roof overhangs the walls. This is where the ventilation holes are generally provided.

5.18 Fences, gates and garden walls

Planning permission	Building Regulations approval
Yes • If it would be over 1 m high and next to a highway used by vehicles (or the footpath of such a highway), or over 2 m high elsewhere; or • if your right to put up or alter fences, walls and gates is removed by an Article 4 direction or a planning condition; • if your house is a listed building or in the curtilage of a listed building; • if it forms a boundary with a neighbouring listed building or its curtilage.	No

5.18.1 Do I need approval to take down or alter a garden wall or boundary wall?

You will not need to apply for planning permission to take down a fence, wall or gate, or to alter, maintain or improve an existing fence, wall or gate (no matter how high) if you don't increase its height. Similarly planning permission is not required for hedges as such. However, there may be a planning condition or a covenant that restricts planting (for example, on 'open plan' estates), in which case you may need planning permission and/or other consent.

If the fence, wall or gate is classed as a 'party fence wall' then you must notify the adjoining owner of the work planned. In normal circumstances, the only restriction on walls and fences is the height allowed. This is in case its height might obscure a driver's view of other traffic, pedestrians or road users. If there is a valid reason for a wall or fence higher than the prescribed dimensions, then it is possible to get planning consent. There may be security issues that would support an application for a high fence. If it has no effect on other people's valid interests and does not impair any amenity qualities in an area, there is no reason why a request should be refused. You will need to apply for planning permission if you put an ornament on your gate post and together with the post it stands higher than 1 m tall.

 Note: Although Building Regulations do not apply, the structures must be structurally sound and maintained.

Some walls have historic value and they, as well as arches and gateways, can be listed. Modifications, extensions and removal of these must have planning consent. You might need Conservation Area consent to take down a fence, wall or gate in a Conservation Area.

One of the most common types of masonry that suffers from collapse is garden walls, and it is therefore a good idea to inspect garden and boundary walls occasionally to check if any repairs are necessary. You may find they need rebuilding. In particular you should check:

* for climbing plants, like ivy, which may cause damage;
* for crumbling surfaces on the brickwork;
* for any trees near the wall;
* for any cracks in the wall;
* if the wall is thick enough for its height;
* that the top of the wall is firmly attached;
* that the wall is upright;
* that there is sufficient mortar and that it is in good condition.

5.19 Flats and maisonettes

There is a different planning regime for flats and maisonettes, and the permitted development rights which apply to many common projects for houses may not apply to flats. In particular, the following apply.

Planning permission		Building Regulations approval	

To extend a flat

Yes	You must apply for planning permission.	Yes	You may need to consult the fire service regarding issues relating to fire escapes. The Housing Act 2004 requires that subdivided buildings meet standards and houses in multiple occupation are licensed.

To subdivide a house or single flat

Yes		Yes	You may also need to consult the fire service, and the property will need to be licensed under the Housing Act 2004.

Loft conversion in a top-floor flat

No	Provided that it is internal works. Local requirements differ, so it is recommended that you check with the local planning authority. Permission is required where you extend or alter the roof space. You should also check whether you own the roof space which you wish to convert. If you are a leaseholder, you may need to get permission from your landlord, freeholder or management company.	Yes	
Yes	If you intend to extend or alter the roof space.	Yes	

Painting the exterior

No	Unless you live in an area where an Article 4 direction applies.	No	

Satellite dishes

Possibly	Planning permission may be required in certain circumstances. Check with the local council.	No	

Windows

Possibly	To fit new double-glazed windows.	Yes	In relation to thermal performance and other areas such as safety, air supply, means of escape and ventilation.

Planning permission		Building Regulations approval	
No	If you are replacing like with like or adding internal secondary glazing.	Yes	In relation to thermal performance and other areas such as safety, air supply, means of escape and ventilation.

Note: Consult the local planning authority for advice before starting work on windows, because local policy and interpretation of the rules vary from council to council.

In addition to planning permission you may also require listed building or Conservation Area consent, as work on a building that affects its special historic character without consent is a criminal offence. Local policy and interpretation of the rules covering windows in flats vary from council to council, and you are advised to contact your local planning authority for advice before starting work.

Don't forget to get permission from your landlord, freeholder or management company if you are a leaseholder.

You may be able to convert space over a shop (and certain space over premises with a display window), or over a ground-floor office, into up to two flats without putting in a planning application if you comply with the following requirements:

- The space is in the same class of use as the shop or office to start with (either Class A1 or Class A2).
- The space is not in a separate planning unit from the shop.
- You will not change the outside appearance of the building.
- If there is a display window at ground-floor level, you will not incorporate any of the ground floor into the flat.

The regulations define this as a 'material change of use' and specify the requirements with which, as a result of that change of use, the whole or part of the building must comply (e.g. those concerned with escape and other fire precautions, hygiene, sound insulation and energy conservation).The whole, or at least part, of the building may therefore need to be upgraded to make it comply with the specified requirements.

5.20 Flues, chimneys or soil and vent pipes

Planning permission	Building Regulations approval
Domestic flues	
No As long as: • flues are not fitted on the principal or side elevation that fronts a highway in a designated area; • flues on the rear or side elevation of the building are no more than 1 m above the highest part of the roof; • the building is not listed or in a designated area.	Yes Particularly with regard to ventilation and general safety. Installation should be carried out by a suitably qualified installer.
Flues for biomass and combined heat and power systems (non-domestic)	
No As long as: • it is the first installation of a flue as part of either a biomass heating system or a combined heat and power system, but further installations will require planning permission; • the building is not listed or a scheduled monument; • the capacity of the system is no more than 45 kilowatts thermal; • the flue is no more than 1 m higher than the highest part of the roof or the height of an existing flue which is being replaced, whichever is the higher.	Yes Particularly with regard to ventilation and general safety. Installation should be carried out by a suitably competent person.

5.21 Fuel tanks

Planning permission	Building Regulations approval
No As long as: • in National Parks, the Broads, Areas of Outstanding Natural Beauty and World Heritage Sites the maximum area to be covered by buildings, enclosures, containers and pools more than 20 m from the house is limited to 10 m³; • maximum height is 2.5 m within 2 m of a boundary; • maximum overall height is 3 m;	No If you are installing an oil tank and/or connecting pipework and you employ an installer registered with one of the related competent person schemes, you will not need to involve a building control service.

Planning permission	Building Regulations approval
• it is not at the side of properties on designated land; • it is not forward of the principal elevation fronting a highway; • it is not more than 3500 litres in capacity; • not more than half the area of land around the 'original house' would be covered by additions or other buildings.	
Yes Any container within the curtilage of a listed building will require planning permission.	

Storage of oil, or any other liquids, especially petrol, diesel and chemicals, is strictly controlled and would not be allowed on residential premises. If you are considering installing an external oil storage tank for central heating, the storage tanks, and the pipes connecting them to combustion appliances, should be constructed and protected so as to reduce the risk of the oil escaping and causing pollution. If the installation is above ground the requirements will be applied to achieve adequate shielding of the tank from any surrounding fire and, in the case of an oil tank, to contain any oil leakages so that groundwater is not contaminated. Where new oil connecting pipework is proposed, a fire valve will be needed at the point where the pipe enters the building.

 Note: This includes liquid petroleum gas tanks as well as oil storage tanks.

5.22 Garages and car ports

Planning permission	Building Regulations approval
No As long as: • there is a maximum height of 2.5 m in the case of a building, enclosure or container within 2 m of a boundary of the curtilage of the dwelling house; • no more than half the area of land around the 'original house' would be covered by additions or other buildings; • there is no outbuilding on land forward of a wall forming the principal elevation; • there are no verandas, balconies or raised platforms;	**No** If the floor area of the building is between 15 m^2 and 30 m^2 you will not normally be required to apply for Building Regulations approval providing that the building contains no sleeping accommodation and is either at least 1 m from any boundary or is constructed of substantially non-combustible materials.

Planning permission	Building Regulations approval
• outbuildings and garages are single-storey with maximum eaves height of 2.5 m and maximum overall height of 4 m with a dual pitched roof or 3 m for any other roof; • in National Parks, the Broads, Areas of Outstanding Natural Beauty and World Heritage Sites the maximum area to be covered by buildings, enclosures, containers and pools more than 20 m from the house is limited to 10 m².	
Yes • On designated land buildings, enclosures, containers and pools at the side of properties will require planning permission. • Within the curtilage of listed buildings any outbuilding will require planning permission.	

A typical floor for a garage generally consists of hardcore, sand blinding, damp-proof membrane (DPM) and concrete. It is considered to be good practice to place reinforced mesh within the concrete, as this can reduce the chances of cracking from the load (weight) of a vehicle. The floor will not normally need to be insulated.

5.23 Hardstanding for a car, caravan or boat

Planning permission		Building Regulations approval	
No	Provided it is within your boundary, is at or near ground level and does not require significant works of embanking or terracing.	No	Unless you introduce steps where none existed before.

You do not need to apply for planning permission to build a hardstanding for a car provided that it is within your boundary and is not used for a commercial vehicle. You should check local council rules, as there are different rules depending on what you use a hardstanding for. Provided there are no covenants limiting the installation of a hardstanding for the parking of cars, caravans or boats, you do not need permission for a hardstanding on your own land or to gain access to it within the confines of your land. You would need permission for a hardstanding leading on to a public highway. However, there are still rules for commercial parking (e.g. taxis or commercial delivery vans), and a 'change of use' to trade premises would probably need to be granted for this to be allowed.

There are no laws to prevent you, or your family, from making use of a parked caravan while it is on your land or drive, but you cannot actually live in it, as this would be classed as an additional dwelling. In addition, you cannot use a parked caravan for business use, as this would constitute a change of use of the property. If you want to put a caravan on your land to lease out as holiday accommodation or for friends or family to stay in while they visit you, then this would require planning permission. Rules on siting of static caravans or mobile homes are quite stringent.

Access from a new hardstanding to an unclassified roadway does not require planning permission. However, the busier the road, the less likely a new drive-way or footway will be allowed to meet it. If the access crosses a pedestrian thoroughfare, pavement or roadside verge, then the planning department will gain approval from the highways department. If this is the case, highways approval is required in addition to planning consent. The basic principle is to maintain safety and eliminate hazards. Your local authority highways department will be able to tell you if a road is classified or unclassified.

You should check if there are any local covenants limiting changes in access to your premises or for a hardstanding and parking of vehicles on it.

5.24 Heat pumps

Planning permission	Building Regulations approval
Domestic ground source or water source heat pumps	
No Unless the building is listed or in a Conservation Area.	Yes It is advisable to contact an installer who can provide the necessary advice, preferably one who belongs to either the Microgeneration Certification Scheme or the relevant competent person scheme.
Domestic air source heat pumps	
Yes As long as all the limits and conditions listed below are met: • All parts of the air source heat pump must be at least 1 m from the property boundary. • Development is permitted only if the air source heat pump installation complies with the Microgeneration Certification Scheme Planning Standards or equivalent standards. • It is not installed on a wall if that wall fronts a highway and any part of that wall is above the level of the ground storey if it is within a Conservation Area or World Heritage Site.	Yes It is advisable to contact an installer who can provide the necessary advice, preferably one who belongs to either the Microgeneration Certification Scheme or the relevant competent person scheme.

Planning permission	Building Regulations approval

- On a flat roof all parts of the pump are at least 1 m from the external edge of that roof.
- Only the first installation of an air source heat pump will be permitted development, and only if there is no existing wind turbine on a building or within the curtilage of that property. Additional wind turbines or air source heat pumps at the same property require an application for planning permission.
- The installation is not on a pitched roof.
- The installation is not within the curtilage of a listed building or within a site designated as a scheduled monument.
- The pump is not on a wall or roof which fronts a highway or nearer to any highway which bounds the property than any part of the building if it is in a Conservation Area or World Heritage Site.
- The volume of the air source heat pump's outdoor compressor unit (including housing) must not exceed $0.6\,m^3$.

In addition, the following conditions must be met. The air source heat pump must be:

- removed as soon as reasonably practicable when it is no longer needed for microgeneration;
- sited, so far as is practicable, to minimize its effect on the external appearance of the building and its effect on the amenity of the area;
- used solely for heating purposes.

Non-domestic ground source or water source heat pumps

No   As long as all the following conditions are observed:

- Only the first stand-alone installation will be permitted development. Further installations will require planning permission from the local authority.
- The total area of excavation must not exceed 0.5 ha.
- When no longer needed for microgeneration, pumps should be removed as soon as reasonably practicable and the land should, as far as reasonably practicable, be restored to its condition before the development took place, or to the condition agreed in writing between the local planning authority and the developer.

It is advisable to contact an installer who can provide the necessary advice, preferably one who belongs to either the Microgeneration Certification Scheme or the relevant competent person scheme.

Planning permission	Building Regulations approval
• The total surface area covered by the water source heat pump (including any pipes) must not exceed 0.5 ha. **Non-domestic air source heat pumps**	It is advisable to contact an installer who can provide the necessary advice, preferably one who belongs to either the Microgeneration Certification Scheme or the relevant competent person scheme.

 Note: You may wish to discuss with the local planning authority for your area whether all of these limits and conditions will be met.

You should make reasonable provision to limit heat losses from pipes as set out in DCLG's *Domestic Building Services Compliance Guide*.

 Note: The new Approved Documents L1A and L2A contain new requirements for pressure-testing and commissioning of heat pumps.

5.25 Hedges

Planning permission	Building Regulations approval
No Unless the hedge obscures the view of traffic at a junction or access to a main road.	No

You do not need planning permission for hedges but there may be a condition attached to the planning permission for your property which restricts the planting of hedges or trees, or the use and nature of hedges can be controlled through planning legal covenants (e.g. on an 'open plan' estate or where a sight line might be blocked). You will need to obtain the council's consent to relax or remove the condition before planting a hedge or tree screen. If you are unsure about this, you can check with the planning department of your council.

Hedges should not be allowed to block out natural light, and the positioning of fast-growing hedges should be checked with your local authority. Recent incidents regarding hedging of the fast-growing leylandii trees have led to changes in the planning rules, where hedges previously had no restrictive laws.

Foundations can be affected by roots, as can soil moisture. This should be considered when planting or removing hedges or building new structures, as certain plant species can affect foundations over 20 m away.

Note: High hedges are dealt with under Part 8 of the Anti-Social Behaviour Act 2003, which came into operation in England on 1 June 2005.

5.26 Hydroelectricity

Planning permission		Building Regulations approval	
Yes	Some form of environmental assessment will be essential for this type of project.	Yes	Particularly for electrical installations.

Note: This is a complex area and the Environment Agency must also be consulted about water extraction licences. More details can be found at http://www. environment-agency.gov.uk/business/topics/water/32020.aspx.

5.27 Infilling

Planning permission		Building Regulations approval	
Possibly	Consult your local planning officer.	Yes	If it is a new development.

If you are considering using an unused, but adjoining, piece of land to build a house (e.g. build a new house on land that used to be a large garden) you may require planning permission. Often there may be no official grounds for denying consent, but residents and individuals can impose quite some delay. It is worth testing the likelihood of a successful application by talking to the neighbours and judging opinions. Planning consent is often quite difficult to obtain in these cases, as this sort of development normally causes a lot of opposition when it is in a settled residential area, and people do not like change. New developments will undoubtedly also need to follow Building Regulations. This, and all site visits from inspectors, is normally arranged by your building contractor.

5.28 Installing a swimming pool

Planning permission		Building Regulations approval	
Possibly	Consult your local planning officer.	Yes	For an indoor pool.

Swimming pools and saunas are subject to special requirements specified in Sections 702 and 703 of BS 7671:2008 (incorporating Amendment No. 1:2011). Installing a covered swimming pool will be covered by the rules that apply to sheds and outbuildings.

5.29 Internal walls

Planning permission	Building Regulations approval
No	Yes In order to satisfy the requirements of the regulations for internal walls you should consider the following areas: • fire safety; • load-bearing walls; • sound insulation; • structural support; • ventilation.

Care should be taken before removing any internal wall. These walls can have a number of functions that could affect the building and the safety of the occupants within the building. When erecting a new internal wall you should ensure that there is adequate separation – in terms of fire resistance and thermal insulation – between the new habitable space and the remaining space. Any door provided in such a wall should have adequate fire resistance and be self-closing. Depending on the use of the new habitable room, the new separating wall may also need to provide sound insulation.

5.30 Kitchens and bathrooms

Planning permission	Building Regulations approval
No Not unless it is part of a house extension or the property is a listed building.	No Unless drainage or electrical works form part of the refit.
No	Yes If a bathroom or kitchen is to be provided in a room where there wasn't one before, particularly in relation to ventilation, drainage, structural stability, and electrical and fire safety.

Under the Greener Homes initiative home owners are encouraged to pay attention to the appliances and fittings by looking for A-rated kitchen appliances and using aerated taps to reduce water use. The use of mains gas for cooking as well as heating or the installation of an electric induction hob in preference to a conventional electric or halogen hob is also recommended. In bathrooms you are asked to reduce water consumption by buying low-flush toilets, low-flow showers and basin taps, and a smaller-capacity bath.

5.31 Laying a patio, decking or driveway

Planning permission	Building Regulations approval
No As long as: • no significant embanking or terracing works are required; • it is not a listed building.	No As long as alterations do not make access to the dwelling any less satisfactory than it was before.
Decking	
No As long as: • the decking is no more than 30 cm above the ground; • together with extensions, outbuildings, etc. the decking covers no more than 50 per cent of the garden.	No As long as the structure does not require planning permission.

Although you do not usually need to apply for planning permission to install a patio or driveway, you may need approval from the local council if the pathway crosses a pavement. There are separate rules if you intend to pave over your front garden. Surface water from hardstandings must not be allowed to run on to the highway, where it could lead to accidents or cause a nuisance.

 You are strongly advised to seek advice from a builder, architect or drainage engineer or from local authority building control before committing to or commencing work.

In order to avoid flooding, it is preferable for these to be sloped towards permeable ground or to be made of pervious materials. Pervious materials include both porous materials (e.g. as reinforced grass or gravel, porous concrete or porous asphalt) and permeable materials (e.g. clay bricks or concrete blocks, designed to allow water to flow through joints or voids). As well as minimizing environmental impact, this avoids the cost of drainage.

Where it is impractical to drain on to pervious ground or use pervious paving, it is preferable to keep the extra surface water on site, in order to avoid increasing flood risk elsewhere. This can be achieved by using a soakaway or some other way of allowing the water to soak into the ground (referred to as 'infiltration'). Where it is impractical to use infiltration (e.g. because of nearby foundations, impermeable or contaminated ground, or high groundwater), it is preferable to discharge water to a watercourse or, failing this, to a surface water sewer or, as a last resort, to a combined sewer. Surface water must not be discharged into a foul drain or sewer.

5.32 Lighting

Planning permission		Building Regulations approval	
No	Not unless the property is a listed building, but you should make sure that the intensity and direction of light do not disturb others.	Yes	You should either use an installer who is registered with the competent person scheme or make an application to your local authority's building control department or approved inspectors.

Beams from your security lights should **not** point directly at windows of other houses, and security lights fitted with passive infrared detectors (PIRs) and/or timing devices should be adjusted so that they minimize nuisance to neighbours and are set so that they are not triggered by traffic or pedestrians passing outside your property. In accordance with Approved Document L1B, you are required to install efficient electric lighting to your house in specific circumstances, including:

• when your dwelling has been extended;
• when your existing lighting system is being replaced as part of rewiring works.

If you are installing an external light which is supplied from your electrical system and fixed to the exterior surface of your house then you should ensure that reasonable provisions are made to enable effective control and/or use of energy-efficient lamps.

 Note: For a new building (other than a dwelling) the target emissions rate (TER) will need to be calculated in accordance with Approved Document L2A.

5.33 Micro combined heat and power

Planning permission		Building Regulations approval	
No	• As long as flues on the rear or side elevation of the building are a maximum of 1 m above the highest part of the roof. • If the building is listed or in a designated area even if you enjoy permitted development rights it is advisable to check with your local planning authority before a flue is fitted. Consent is also likely to be needed for internal alterations. • In a Conservation Area or World Heritage Site the flue should not be fitted on the principal or side elevation if it would be visible from a highway.	Yes	The *Low or Zero Carbon Energy Sources: Strategic Guide* (LZC) supports the inclusion of low or zero carbon energy sources in Part L of the Building Regulations and Approved Documents L1A, L1B, L2A and L2B. Chapter 4 of the guide deals with micro combined heat and power.

You should make reasonable provision to limit heat losses from pipes as set out in DCLG's *Domestic Building Services Compliance Guide*.

 Note: The new Approved Documents L1A and L2A contain new requirements for pressure-testing and commissioning of heat pumps.

5.34 New homes

Planning permission	Building Regulations approval
Yes	Yes

All new houses or premises of any kind require planning permission. Private individuals will normally only encounter this if they intend to buy a plot of land to build on, or buy land with existing buildings that they want to demolish to make way for a new property to be built.

Research has shown that more than half the population would like to build their own home at some stage in their lives. A website has been designed as the result of a joint initiative between the government and the custom-build housing industry to provide encouragement and impartial advice to people who want to build their own home to suit their family's needs. It forms part of the government's housing strategy to bring about a custom-build housing revolution. The site includes an interactive guide to self-build, where users can key in details of their own situation (how much money they have, where they want to build, the size of house they want and various other things) and the guide will automatically calculate whether it's feasible. If not, users can adjust their circumstances until they can realistically get their self-build project under way. The site has practical information about how to find a plot of land, where to get a self-build mortgage, the different types of construction methods you can use, and a host of other issues.

 If you are hiring a professional, be sure to find out exactly who does what and that approval is obtained before going to too much expense, should a refusal arise.

In all cases, unless you are an architect or a builder, you **must** seek professional advice. If you are using a solicitor to act on your behalf in purchasing a plot on which to build, he or she will include the planning questions within all the other legal work, as well as investigating the presence of covenants or existing planning consent, together with other constraints or conditions. The architect, surveyor or contractor you hire will then need to take into account the planning requirements as part of his or her planning and design procedures. He or she will normally handle planning applications for any type of new development.

5.35 Outbuildings

Planning permission		Building Regulations approval	
No	As long as: <ul><li>there is a maximum height of 2.5m in the case of a building, enclosure or container within 2m of a boundary of the curtilage of the dwelling house;</li><li>no more than half the area of land around the 'original house' would be covered by additions or other buildings;</li><li>there is no outbuilding on land forward of a wall forming the principal elevation;</li><li>there are no verandas, balconies or raised platforms;</li><li>outbuildings and garages are single-storey with maximum eaves height of 2.5m and maximum overall height of 4m with a dual pitched roof or 3m for any other roof;</li><li>in National Parks, the Broads, Areas of Outstanding Natural Beauty and World Heritage Sites the maximum area to be covered by buildings, enclosures, containers and pools more than 20m from the house is limited to 10m^2.</li></ul>	No	If the floor area of the building is between 15m^2 and 30m^2 you will not normally be required to apply for Building Regulations approval providing that the building contains no sleeping accommodation and is either at least 1m from any boundary or is constructed of substantially non-combustible materials.
Yes	<ul><li>On designated land buildings, enclosures, containers and pools at the side of properties will require planning permission.</li><li>Within the curtilage of listed buildings any outbuilding will require planning permission.</li></ul>		
Polytunnels			
Yes	If it is a listed building or in a Conservation Area, National Park or Area of Outstanding Natural Beauty.	Possibly	Depending on the work involved.
No	As long as the polytunnel is no nearer to the road than the nearest part of the house (unless there is more than 20m between the tunnel and the road) and is less than 3m high.	Possibly	Depending on the work involved.

Many kinds of buildings and structures can be built in your garden or on the land around your house without the need to apply for planning permission. These can include sheds, garages, greenhouses, accommodation for pets and domestic animals (e.g. chicken houses), summer houses, ancillary garden buildings such as swimming pools, ponds and sauna cabins, enclosures (including tennis courts), and many other kinds of structure which could be considered incidental to the enjoyment of the dwelling house.

Outbuildings intended to go in the garden of a house do not normally require any planning permission, as long as they are associated with the residential amenities of the house and a few requirements are adhered to, such as position and size. If your new building exceeds $10\,m^3$ (and/or comes within $5\,m$ of the house) it would be treated as an extension and would count against your overall volume entitlement.

When building an outbuilding you should consider where the foundations are to go, any walls that will be built below ground and the type of flooring you plan to use.

Floors in storage buildings, annexes and summer houses can take one of the three general types outlined in Chapter 6, section 6.5. The exact specification for the floor will depend on how the building is to be used and whether it is to be heated. More information on each floor type can be found in section 5.15.

More information can be found in the outbuildings mini-guide on the Planning Portal. However, to summarize, permission is required for:

- a building or structure which would result in more than half of the grounds of your house being covered by buildings or structures;
- a propane gas (LPG) tank;
- a storage tank holding more than 3500 litres;
- any building or structure nearer to a highway than the nearest part of the original house, unless more than $20\,m$ away from a highway;
- structures not required for domestic use;
- structures over $3\,m$ high (or $4\,m$ if the structure has a ridged roof).

You will also need to apply for planning permission if any of the following cases apply:

- your house is a listed building and you want to put up a building or structure with a volume of more than $10\,m^3$;
- more than half the area of land around the original house would be covered by additions or other buildings;
- the building or structure is not to be used for domestic purposes and is to be used instead, for example, for parking a commercial vehicle, running a business or storing goods in connection with a business;
- you want to put up a building or structure which is more than $3\,m$ high, or more than $4\,m$ high if it has a ridged roof (measured from the highest ground next to it);
- you want to put up a building or structure which would be nearer to any highway than the nearest part of the original house, unless there would be at least $20\,m$ between the new building and any highway (the term 'highway' includes public roads, footpaths, bridleways and byways).

 Erecting any type of outbuilding can be a potential minefield, and it is best to consult the local planning officer before commencing work.

5.35.1 External water storage tanks

An application is required under Building Regulations to check any new drainage which runs to a tank that recycles rainwater collected from the roof and ground. If you are considering installing an external water tank you should seek guidance from your local authority, especially if the tank is to be mounted on a roof.

5.36 Paving your front garden

Planning permission		Building Regulations approval	
No	If the replacement driveway of any size uses permeable surfacing or if the rainwater is directed to a lawn or border to drain naturally.	No	As long as alterations do not make access to the dwelling any less satisfactory than it was before.
Yes	If the surface to be covered is more than 5 m² and you plan to lay a traditional impermeable driveway.	No	As long as alterations do not make access to the dwelling any less satisfactory than it was before.

Climate change has increased the frequency of heavy rainfall and flooding in the UK. In 2007, the UK experienced serious flooding, which was in many cases caused by drains being unable to cope with the amount of rainwater flowing into them. It is predicted that this kind of heavy rainfall event and flooding may occur more often in the future. As the drains in most urban areas were built many years ago they were not designed to cope with increased rainfall, and as more water enters the system from new developments and paved front gardens the problem gets worse.

It may not seem that paving over one or two gardens is a problem, but the combined effect of lots of people in a street or area doing this can increase the risk of flooding. Hard surfaces such as concrete and asphalt also collect pollution (oil, petrol, brake dust, etc.) that is washed off into the drains. The drains carry rainwater directly to streams or rivers where the pollution can damage wildlife and the wider environment. In older areas the rainwater may go into the foul water sewer, which normally takes household waste from bathrooms and kitchens to the sewage treatment works. These overflow into streams and rivers in heavy rainfall, putting untreated sewage into watercourses.

In order to avoid flooding, it is preferable for any paved area to be sloped towards permeable ground or to be made of pervious materials. Pervious materials include both porous materials (e.g. as reinforced grass or gravel, porous concrete or porous asphalt) and permeable materials (e.g. clay bricks or concrete blocks, designed to allow water to flow through joints or voids). As well as minimizing environmental impact, this avoids the cost of drainage. Surface water from hardstandings must not be allowed to run on to the highway, where it could lead to accidents or cause a nuisance.

Where it is impractical to drain on to pervious ground or use pervious paving, it is preferable to keep the extra surface water on site, in order to avoid increasing flood risk elsewhere. This can be achieved by using a soakaway or some other way of allowing the water to soak into the ground (referred to as 'infiltration'). Where it is impractical to use infiltration (e.g. because of nearby foundations, impermeable or contaminated ground, or high groundwater), it is preferable to discharge the water to a watercourse or, failing this, to a surface water sewer or, as a last resort, to a combined sewer. Surface water must not be discharged into a foul drain or sewer.

Further details on permeable surfaces (such as loose gravel), hard permeable and porous surfaces, rain gardens and soakaways, and wheel tracks can be found on the Planning Portal.

- **Loose gravel:** This is the simplest type of construction. The driveway sub-base is covered by a surface layer of gravel or shingle. Gravel with different shapes and colours is available to make the surface more decorative. A strip of block paving or asphalt at the entrance can limit the loss and spread of gravel from the drive.
- **Hard permeable and porous surfaces:** Hard surfacing which allows water to soak into it can be built with porous asphalt, porous concrete blocks, concrete or clay block permeable paving. The material has open voids across the surface of the material or around the edges of blocks that allow water to soak through. To work effectively permeable surfaces should be laid over a sub-base which differs from traditional hardcore, which has a lot of fine material in it (sand and silt) that stops water passing through it easily. For permeable and porous driveways different sub-base materials are required that allow water to pass through and also store the water for a while if it cannot soak into the ground as fast as the rain falls. Various materials are available, and two examples are known as 4/20 and Type 3 sub-base. Materials for a permeable sub-base are described as open graded and consist only of larger pieces of stone that have spaces between to store water.
- **Rain gardens:** An area of garden can be formed into a rain garden – a depression to collect and store rainwater running from conventional impermeable surfaces (asphalt, concrete and block paving), before slowly allowing it to soak into the ground or to flow to the drains. Rain gardens are widely used in the USA and elsewhere but are a relatively new concept in the UK. The depressions can be located along the edge of the drive or as a larger area in the garden at a low point. The depression can be planted with suitable plants to help slow run-off, or gravel or cobbles can be used as decorative features. There may be a gravel-filled trench below it to increase the storage capacity and allow water to soak into the ground more easily.
- **Soakaways:** These are a similar idea except that water is piped into a gravel-filled trench or special container and allowed to soak into the ground. In some areas many houses have the roof downpipes connected to soakaways. They are more suitable for houses with larger front gardens, as they require space and need to be located a suitable distance from buildings.
- **Wheel tracks:** To keep hard surfaces to a minimum a driveway can be created that has just two paved tracks where the wheels go. These can be surfaced with blocks, asphalt or concrete, but to provide a durable construction

they should have a sub-base below. The area between and around the tracks can be surfaced in gravel or planted with grass or suitable low-growing plants. Water must drain from the tracks into the surrounding permeable area. A typical width is between 300 mm and 600 mm for each track.

If you are making a new access into the garden across the footpath you will need to obtain permission from the local council to drop the kerbs, and the pavement may need strengthening. This is to protect any services buried in the ground such as water pipes.

5.37 Plumbing

Planning permission		Building Regulations approval	
No	Unless it is a listed building.	No	If it is installed by an approved person and complies with Part J.
		Yes	If you use an unregistered installer or do it yourself you will need to get approval from local authority building control.

5.38 Porch

Planning permission		Building Regulations approval	
No	Unless: • any part is less than 2 m from a boundary adjoining a highway; • any part is more than 3 m high; • the floor area exceeds 3 m².	No	As long as glazing and electrical installations comply with Building Regulations and: • the front entrance door between the existing house and the new porch must remain in place; • if the house has ramped or level access for disabled people, the porch must not adversely affect access.

More information can be found in the porches mini-guide on the Planning Portal. Your local authority building control department or an approved inspector can supply further information on safety glazing. It is advisable to ensure that a porch is not constructed so that it restricts ladder access to windows serving a room in the roof or a loft conversion, particularly if that window is needed as an emergency means of escape in the case of fire.

The permitted development allowances described here apply to houses, not flats, maisonettes or other buildings.

The regulations depend on previous works on the site (which can date back to 1948). You should always check with planning control staff if you are unsure.

5.39 Roof

Planning permission	Building Regulations approval
No Provided that: • any side-facing windows are obscure-glazed and any opening is 1.7 m above the floor; • the alterations do not project more than 150 mm from the existing roof plane; • the alterations are no higher than the highest part of the roof.	Yes If you want to carry out repairs on or re-cover less than 25 per cent of the area of a pitched or flat roof, you will not normally need to submit a Building Regulations application. You will need approval, however, if: • the performance of the new covering will be significantly different to that of the existing covering in the event of a fire; • you are replacing or repairing more than 25 per cent of the roof area, in which case the roof thermal insulation will normally have to be improved; • you carry out structural alterations.

Work on a roof may affect bats. You need to consider protected species when planning work of this type. A survey may be needed, and if bats are using the building a licence may be needed. When performing work on any roof, care should be taken to ensure the roof will continue to perform effectively and without any movement.

After a period of time the roof on existing buildings will need to be replaced. In most situations, this work will need Building Regulations approval. Some repairs to flats roofs will not require an application for approval under the Building Regulations. However, if a roof with integral insulation is to be replaced then you may be required to upgrade this 'thermal element' of the structure and reduce the amount of heat that was originally lost, by upgrading the insulation.

If the existing roof covering is to be replaced with a different material to its original, for example slate to tiles, then approval under the Building Regulations is likely to be needed to ensure the roof will be adequate in terms of structural stability (applicable where the replacement tile will be significantly heavier or lighter than the existing) and also meets requirements in respect of fire safety and energy-efficiency. If the new roof covering is significantly heavier or lighter than the existing one, the roof structure may need modifying and/or strengthening, and you are advised to check with a structural engineer or surveyor before commencing works. With a warm deck, the insulation is placed over the rafters and then a felt is placed on top. The battening and tiling are then fixed down over it. The thickness of insulation will vary depending on the manufacturer's specification. With a cold deck, the insulation can be placed between the rafters or it can be placed between the ceiling joists. The thickness of insulation in both cases will vary depending on the material you use and the manufacturer's specification. The roof should have vents installed along the eaves to both front and rear or from side to side. In the case where the insulation is placed between the rafters then vents should also be placed along the ridge.

As a roof is defined as a thermal element, the work to re-cover a roof should also include improving the thermal insulation properties of the roof.

A rooflight is a window that is installed within a pitched roof or flat roof normally to give more light to rooms or spaces within the home. Approval under the Building Regulations will generally be needed for the installation of a new rooflight for the following reasons:

- Any rooflight that is installed will need to prove that it has sufficient insulation against heat loss, i.e. is energy-efficient.
- If the rooflight is in close proximity to a boundary, the fire performance of the rooflight will need to be considered.
- The roof will have to be able to carry the load (weight) of the new rooflight. If the roof cannot do this then it will need to be strengthened.
- To install a rooflight, the roof structure will generally need to be altered to create the opening.

5.40 Shops

Planning permission		Building Regulations approval	
To change residential use to a shop			
Yes	• This is classed as a material change of use, as the proposed new use is a different use class within the planning system. • The majority of local authorities have designated shopping or commercial areas within their planning policies where they may consider such a change of use suitable. It is likely to be very difficult to gain permission outside these areas, such as in a wholly residential area. • While permission is not required before work starts, if permission is refused the work will have to be undone and the authority may take formal enforcement action over a change of use without planning permission.	Yes	The regulations define converting a home into a shop as a 'material change of use' and specify that the building must comply with the Building Regulations in respect of escape and other fire precautions, hygiene, energy conservation, and access to and use of buildings. • The building may therefore need to be upgraded to make it comply with the specified requirements. • You should also check with the local fire authority, usually the county council, to see what 'ongoing' fire precautions legislation will apply when the building is in use.
To convert a shop to a café (use class A3) or public house (A4) or takeaway (A5)			
No	Premises in shop use (A1) are able to change to café use (A3) without planning permission for a single period of up to two years.	Possibly	You may wish to contact your local building control body for further advice.

Planning permission	Building Regulations approval
• The majority of local authorities will have designated shopping or commercial areas within their planning policies (which often contain policies on striking a balance of uses in an area). • Some authorities may have specific policies to retain a level of retail uses. • Permission will be required for commercial extractor flues if cooking is intended to be carried out on the premises. • The placing of tables and chairs on the highway is very likely to require a licence from the highway authority to allow it to assess factors such as pedestrian movements, sight lines and road safety. • Changes in signage are likely to require express advertisement consent. • While permission is not required for the change of use before work starts, if permission is refused the work will have to be undone and the authority may take formal enforcement action over a change of use without planning permission.	
Yes For other material changes of use, such as to a public house (A4) or takeaway (A5), because the proposed new use is a different use class within the planning system. • The majority of local authorities will have designated shopping or commercial areas within their planning policies (which often contain policies on striking a balance of uses in an area). • Some authorities may have specific policies to retain a level of retail uses. • Permission will also be required for commercial extractor flues if cooking is intended to be carried out on the premises. • The placing of tables and chairs on the highway is very likely to require a licence from the highway authority to allow it to assess factors such as pedestrian movements, sight lines and road safety.	**Possibly** You may wish to contact your local building control body for further advice.

Planning permission	Building Regulations approval
• Changes in signage are likely to require express advertisement consent. • While permission is not required for the change of use before work starts, if permission is refused the work will have to be undone and the authority may take formal enforcement action over a change of use without planning permission.	

To convert a shop to an office (use class B1) or storage (B8) or other uses

No	Premises in shop use (A1) are able to change to office use (B1) without planning permission for a single period of up to two years.	Possibly	You may wish to contact your local building control body for further advice.

    • The majority of local authorities will have designated shopping or commercial areas within their planning policies (which often contain policies on striking a balance of uses in an area).
    • Some authorities may have specific policies to retain a level of retail uses.
    • Changes in signage are likely to require express advertisement consent.
    • While permission is not required for the change of use before work starts, if permission is refused the work will have to be undone and the authority may take formal enforcement action over a change of use without planning permission.
    • You may also require the consent of the landlord or landowner.

Yes	For material changes of use, such as to storage (B8), because the proposed new use is a different use class within the planning system.	You may wish to contact your local building control body for further advice.

    • The majority of local authorities will have designated shopping or commercial areas within their planning policies (which often contain policies on striking a balance of uses in an area).
    • Some authorities may have specific policies to retain a level of retail uses.

Planning permission	Building Regulations approval
• Changes in signage are likely to require express advertisement consent. • While permission is not required for the change of use before work starts, if permission is refused the work will have to be undone and the authority may take formal enforcement action over a change of use without planning permission. • You may also require the consent of the landlord or landowner.	

To erect, change or alter a shop's adverts, fascia or projecting signs

Yes All advertisements require consent either from the local authority or from the legislation – the display of an advert without consent is a criminal offence. Some adverts relating to the business at the site may have deemed consent from the legislation governing adverts. For example, many traditional, non-illuminated fascia signs and hanging signs are likely to have deemed consent subject to a number of conditions and limitations such as size. If an advert meets all the criteria of deemed consent you will not require express consent. Because inappropriate adverts can harm an area visually (and even economically) many authorities have produced supplementary guidance on the design of suitable adverts and commercial signage. Other adverts without deemed consent will require express advertisement consent from the local authority. If the building is listed you will also require listed building consent. You must obtain consent, as it is required before an advertisement is displayed, and the local planning authority may take action (including prosecution) if no consent is in place. You may also require the consent of the landlord or landowner.	

5.41 Solar panels

Planning permission	Building Regulations approval

Solar panels mounted on a house or on a building within the grounds of a house

No   As long as all the following conditions are observed:

- Panels on a building should be sited, so far as is practicable, to minimize the effect on the external appearance of the building and the amenity of the area.
- When no longer needed for microgeneration, panels should be removed as soon as reasonably practicable.

All the following limits must be met:

- Panels should not be installed above the highest part of the roof (excluding the chimney) and should project no more than 200 mm from the roof slope or wall surface.
- The panels must not be installed on a building that is within the grounds of a listed building.
- The panels must not be installed on a site designated as a scheduled monument.
- Wall mounted only – if your property is in a Conservation Area or World Heritage Site, panels must not be fitted to a wall which fronts a highway.

Yes   The ability of the roof to carry the weight of the panels needs to be checked and electrical work needs to be carried out by an approved contractor.

Stand-alone solar panel installations (panels not on a building)

No   All the following conditions must be observed:

- Panels on a building should be sited, so far as is practicable, to minimize the effect on the amenity of the area.
- When no longer needed for microgeneration, panels should be removed as soon as reasonably practicable.

All the following limits must be met:

- If your property is in a Conservation Area or World Heritage Site, no part of the solar installation should be nearer to any highway bounding the house than the part of the house that is nearest to that highway.

Yes   The ability of the existing roof to carry the load (weight) of the panels will need to be checked and proven. Some strengthening work may be needed.
  Building Regulations also apply to other aspects of the work such as electrical installation. It is advisable to contact an installer who can provide the necessary advice, preferably one who belongs to the competent person scheme.

Planning permission	Building Regulations approval

Planning permission

- No part of the installation should be higher than 4 m.
- Only the first stand-alone solar installation will be permitted development. Further installations will require planning permission.
- Panels should not be installed within the boundary of a listed building or a scheduled monument.
- The installation should be at least 5 m from the boundary of the property.
- The size of the array should be no more than 9 m² or 3 m wide by 3 m deep.

Solar panels (non-domestic)

No   All the following conditions must be observed:

- Panels should be sited, so far as is practicable, to minimize the effect on the external appearance of the building and the amenity of the area.
- When no longer needed for microgeneration, panels should be removed as soon as reasonably practicable.

All the following limits must be met:

- Equipment mounted on a roof must not be within 1 m of the external edge of the roof.
- Equipment mounted on a wall must not be within 1 m of a junction of that wall with another wall or with the roof of the building.
- If the building is on designated land the equipment must not be installed on a wall or a roof slope which fronts a highway.
- Solar panels installed on a wall or a pitched roof should project no more than 200 mm from the wall surface or roof slope.
- The panels must not be installed on a listed building or on a building that is within the grounds of a listed building.
- The panels must not be installed on a site designated as a scheduled monument.
- Where panels are installed on a flat roof the highest part of the equipment should not be more than 1 m above the highest part of the roof (excluding the chimney).

Building Regulations approval

The ability of the existing roof to carry the load (weight) of the panels will need to be checked and proven. Some strengthening work may be needed.

Building Regulations also apply to other aspects of the work such as electrical installation. It is advisable to contact an installer who can provide the necessary advice, preferably one who belongs to the competent person scheme.

Planning permission		Building Regulations approval	

Solar panels mounted on a non-domestic building

No	All the following conditions must be observed:	Possibly	The ability of the existing roof to carry the load (weight) of the panel will need to be checked and proven. Some strengthening work may be needed.

All the following conditions must be observed:

- Panels should be sited, so far as is practicable, to minimize the effect on the amenity of the area.
- When no longer needed for microgeneration, panels should be removed as soon as reasonably practicable.

All the following limits must be met:

- Only the first stand-alone solar installation will be permitted development. Further installations will require planning permission from the local authority.
- If the property is in a designated area no part of the solar installation should be nearer to any highway bounding the grounds of the property than the part of the building that is nearest to that highway.
- No part of the installation should be higher than 4 m.
- Panels should not be installed within the boundary of a listed building or a scheduled monument.
- The installation should be at least 5 m from the boundary of the property.
- The size of the array should be no more than 9 m² or 3 m wide by 3 m deep.

Building Regulations also apply to other aspects of the work such as electrical installation. It is advisable to contact an installer who can provide the necessary advice, preferably one who belongs to the competent person scheme.

In many cases installing solar panels on domestic land is likely to be considered 'permitted development', with no need to apply to the council for planning permission. You may wish to discuss with the local planning authority for your area whether all of the limits and conditions will be met.

 Note: Permitted development rights for solar panels are available for both single houses and buildings which consist wholly of flats. If you are a leaseholder you may need to get permission from your landlord, freeholder or management company.

In many cases installing solar panels on non-domestic land is likely to be considered 'permitted development', with no need to apply to the council for planning permission. There are, however, important limits and conditions which must be met to benefit from the permitted development rights.

Non-domestic land for the purposes of these permitted development rights is broad and can include businesses and community buildings. Permitted development rights are also available for domestic properties. You may wish to discuss with the local planning authority for your area whether all of the limits and conditions will be met.

5.41.1 Stand-alone solar panel installations in the grounds of a non-domestic building

The permitted development regime for solar panels has different limits on projections and in relation to protected areas.

5.42 Structural alterations inside

Planning permission		Building Regulations approval	
No	Unless it is a listed building or within a Conservation Area.	Possibly	If it is a listed building or within a Conservation Area.
Yes	If the alterations are major, such as removing or part-removing a load-bearing wall or altering the drainage system.	Yes	If you wish to build or remove an internal wall or make openings in an internal wall.

You will need approval if your alterations are to the structure, such as the removal or part-removal of a load-bearing wall, joist, beam or chimney breast, or would affect fire precautions of a structural nature either inside or outside your house. You also need approval if, in altering a house, work is necessary to the drainage system or to maintain the means of escape in case of fire.

 You need approval to make internal alterations to a shop or office, and you may also require the consent of the landlord or landowner.

5.43 Trees

Planning permission		Building Regulations approval
No	You can fell or lop trees on your property unless the trees are protected by a Tree Preservation Order or you live in a Conservation Area.	No

The use and nature of trees can be controlled through planning conditions and legal covenants. Foundations can be affected by tree roots and soil moisture. Many trees are protected by Tree Preservation Orders (TPOs), which mean that, in general, you need the council's consent to prune or fell them. Nearly all trees in Conservation Areas are automatically protected. This should be considered when planting or removing trees or building new structures, as certain tree species can affect foundations over 20 m away.

Ask the council for a copy of the free leaflet *Protected Trees: A Guide to Tree Preservation Procedures*.

5.44 Underpinning

Planning permission	Building Regulations approval
No But if you live in a listed building or designated area (Conservation Area, National Park or Area of Outstanding Natural Beauty) you should check with your local planning authority before carrying out any work.	Yes The regulations specifically define underpinning as 'building work', and appropriate measures must be applied to ensure the underpinning stabilizes the movement of the building. Particular attention will need to be given to any sewers and drains near the work.

Underpinning is a method of construction that sees the depth of the foundations to a building being increased. The soil beneath the existing foundation is excavated and is replaced with foundation material, normally concrete, in phases. The reasons for underpinning are generally:

•   that the existing foundations of the building have moved – this is caused by poor soil or changes to the soil conditions (e.g. subsidence has occurred);
•   that there has been a decision to add another storey to the building, either above or below ground level, and the depth of the existing foundations is inadequate to support the modified building or load (weight) of it.

Underpinning requires close attention to design, methodology and safety procedures. **If not carried out properly, this kind of work poses very real risks and could see damage to or collapse of the existing home.** Gaining such approval will usually involve the preparation of a structural design of the underpinning, including the process to be carried out during construction. An initial step, before substantial commencement of the work, will generally be for a trial hole to be dug next to the existing footings for a structural engineer or surveyor to make an assessment of the circumstances of the case.

The exact method to be employed for underpinning will depend on the many circumstances of the case. To avoid excessively undermining the existing foundations, causing further damage to the structure above, the excavations for the underpinning should be carried out to the engineer's instructions and details.

You are therefore advised to employ experienced people for the design (for example, an experienced designer and structural engineer) and construction (for example, somebody with experience of underpinning and general building work) of the project.

A typical method is for short sections of underpinning to be carried out one at a time. The excavation for each section of underpinning will normally be inspected by a design engineer and a building control surveyor before it is concreted.

5.45 Warehouses and industrial buildings

Planning permission	Building Regulations approval
No As long as:	Yes

As long as:

- no new building is higher than 5m, if within 10m of the boundary, and in other cases no new building is higher than the highest building within the boundary or 15m, whichever is lower;
- no new building exceeds a gross floor space of 100m²;
- until 30 May 2016 outside designated land and Sites of Special Scientific Interest no new building exceeds a gross floor space of 200m².

No extension or alteration should make the building higher than 5m, if within 10m of the boundary. In all other cases any extension or alteration must not be higher than the building being extended or altered. You must obey all the following conditions:

- Any new, extended or altered industrial building must relate to the current use of the building or the provision of staff facilities or be for research and development of products or processes.
- Any new, extended or altered warehouse must relate to the current use of the building or the provision of staff facilities.
- Development must be within the curtilage of an existing industrial building or warehouse.
- Developments that would reduce space available for parking or turning vehicles are not permitted development.
- No development should come within 5m of the curtilage boundary.
- No development should be carried out within the cartilage of a listed building.
- No new, extended or altered building should provide staff facilities:

  ○ between 7 p.m. and 6.30 a.m. for employees other than those present at the premises of the undertaking for the purposes of their employment;
  ○ at all, if a notifiable quantity of hazardous substance is present.

- On designated land any new, extended or altered buildings should use materials similar in external appearance to those used for the existing industrial building or warehouse.
- On designated land the cumulative limit for new buildings and extensions is 10 per cent of the original building or 500m², whichever is the lesser.
- Until 30 May 2016 in Sites of Special Scientific Interest the cumulative limit for new buildings and extensions is 25 per cent of the original building or 1000m², whichever is the lesser.
- Until 30 May 2016 the cumulative limit for new buildings and extensions must not exceed the gross floor space of the original building by more than 50 per cent or 1000m², whichever is the lesser.

You are strongly advised to read all the interim guidance on the new permitted development rules.

Where there is any doubt as to whether a development would be permitted development, advice from the local planning authority should be sought. To be certain that a proposed development is lawful and does not require an application for planning permission, it is possible to apply for a 'lawful development certificate' from the local authority.

The local planning authority may also have removed some permitted development rights by issuing what is known as an Article 4 direction or may have removed those rights on the original, or any subsequent, planning permission for the site. This will mean a planning application will be needed for development which normally does not need one. Before undertaking any development, checks should be undertaken with the local planning authority to determine whether any restrictions on permitted development have been made.

5.46 Wind turbines

Planning permission	Building Regulations approval
Wind turbine: building mounted	
No As long as:	Yes With regard to size, weight and force exerted on fixed points. Building Regulations also apply to other aspects of the work such as electrical installation. It is advisable to contact an installer who can provide the necessary advice. You could use an installer registered with the relevant competent person scheme (as listed in row 17 of Schedule 3 of the Building Regulations). A registered installer may be authorized to self-certify the work to comply with this aspect of the Building Regulations without involving local authority building control.

No   As long as:

- it is removed as soon as reasonably practicable when no longer needed for microgeneration.
- it is sited, so far as practicable, to minimize its effect on the external appearance of the building and its effect on the amenity of the area to a dwelling (consult your local planning officer);
- the building-mounted wind turbine installation complies with the Microgeneration Certification Scheme Planning Standards or equivalent standards (read more about the scheme);
- in Conservation Areas, the building-mounted wind turbine would not be on a wall or roof slope which fronts a highway;
- no part (including blades) of the building-mounted wind turbine protrudes more than 3m above the highest part of the roof (excluding the chimney) or exceeds an overall height (including building, hub and blade) of 15m, whichever is the lesser;
- no part of the building-mounted wind turbine (including blades) is within 5m of any boundary.

Planning permission	Building Regulations approval
• This only applies to installations on detached houses (not blocks of flats) and other detached buildings within the boundaries of a house or block of flats. A block of flats must consist wholly of flats (e.g. should not also contain commercial premises). • Only the first installation of any wind turbine will be permitted development, and only if there is no existing air source heat pump at the property. Additional wind turbines or air source heat pumps at the same property require an application for planning permission. • Permitted development rights do not apply to a turbine within the curtilage of a listed building or within a site designated as a scheduled monument or on designated land other than Conservation Areas. • The distance between ground level and the lowest part of any wind turbine blade must not be less than 5 m. • The installation must not be sited on safeguarded land. An Aviation Safeguarding Tool can be used to check whether the installation will be on safeguarded land. • The swept area of any building-mounted wind turbine blade must be no more than 3.8 m². • Use non-reflective materials on blades.	

Wind turbine: stand-alone

No	As long as: • no part of the stand-alone wind turbine (including blades) would be in a position which is less than a distance equivalent to the overall height of the turbine (including blades) plus 10 per cent of its height when measured from any point along the property boundary; • the stand-alone wind turbine installation complies with the Microgeneration Certification Scheme Planning Standards or equivalent standards; • in Conservation Areas, the stand-alone wind turbine is not installed so that it is nearer to any highway which bounds the curtilage (garden or grounds) of the house or block of flats than the part of the house or block of flats which is nearest to that highway.	Yes	With regard to size, weight and force exerted on fixed points. Building Regulations also apply to other aspects of the work such as electrical installation. It is advisable to contact an installer who can provide the necessary advice. You could use an installer registered with the relevant competent person scheme (as listed in row 17 of Schedule 3 of the Building Regulations). A registered installer may be authorized to self-certify the work to comply with this aspect of the Building Regulations without involving local authority building control.

Planning permission	Building Regulations approval
• Only the first installation of any wind turbine will be permitted development, and only if there is no existing air source heat pump at the property. Additional wind turbines or air source heat pumps at the same property require an application for planning permission. • Permitted development rights do not apply to a turbine within the curtilage of a listed building or within a site designated as a scheduled monument or on designated land other than Conservation Areas. • The distance between ground level and the lowest part of any wind turbine blade must not be less than 5m. • The highest part of the stand-alone wind turbine must not exceed 11.1m. • The installation must not be sited on safeguarded land. The Aviation Safeguarding Tool can be used to check whether the installation will be on safeguarded land. • The swept area of any stand-alone wind turbine blade must be no more than 3.8m². In addition, the following conditions must be met. The wind turbine must: • use non-reflective materials on blades; • be removed as soon as reasonably practicable when no longer needed for microgeneration; • be sited, so far as is practicable, to minimize its effect on the external appearance of the building and its effect on the amenity of the area.	

 Note: These limits refer to buildings as a whole and not flats within them. If you wish to install wind turbines on a flat you should contact the local authority.

Planning permission may be required to install a wind turbine, either free-standing or attached to a dwelling. It is recommended that you seek the advice of your local planning authority in relation to wind turbines, as permissions vary depending on the region of the UK.

 Remember, if you are a leaseholder, you may need to obtain permission from the landlord.

To support the development of the microgeneration industry and to drive the quality and reliability of installations a Microgeneration Certification Scheme has been developed in partnership with the industry and other organizations representing consumer interests. The Microgeneration Certification Scheme includes clear standards to support the installation of wind turbines and air source heat pumps. The main purpose of the scheme is to build consumer confidence in microgeneration technologies and to help move the industry to a sustainable position.

It includes certification for products and installer companies, and a code of practice based on the Office of Fair Trading Consumer Code. Permitted development rights for wind turbines and air source heat pumps will only be accorded for equipment installed by an installer who has been certificated through the scheme using a certificated product. The installer is therefore responsible for ensuring that the installation meets permitted development noise standards at the time of installation. For further details, see the Microgeneration Certification Scheme's website at: http://www.microgeneration-certification.org/.

5.47 Working from home

Planning permission	Building Regulations approval	
No	No	Unless your proposed project involves work which will affect the structure of the building, the means of escape and other fire precautions, or affect the access to and use of buildings, the regulations will regard your work as a 'material alteration' (and therefore 'building work'), which must then comply with the regulations.

Table 5.3 is a synopsis of the detailed sections on basic requirements for planning permission and Building Regulations approval which is contained in Chapter 5. In all circumstances it is recommended that you talk to your local planning officer before contemplating any work. The cost of a local phone call could save you a lot of money (and stress) in the long term!

Table 5.3 Basic requirements for planning permission and Building Regulations approval

Planning permission		Building Regulations approval	
Advertising			
No	If the advertisement is less than 0.3 m² and not illuminated.	Possibly	Domestic adverts and signs are not usually subject to planning control, but they must be kept in a safe condition.
Aerials, satellite dishes and flagpoles			
Satellite dishes, aerials and antennas on buildings up to 15 m high			
No	Unless: • there are more than two antennas on the property overall; • a single antenna is more than 100 cm in any linear dimension; • if installing two antennas, one is more than 100 cm and the other is more than 60 cm in any linear dimension; • the cubic capacity of each antenna is more than 30 litres; • the antenna is fitted to a chimney stack and is greater than 60 cm in any linear dimension; • the antenna sticks out more than 60 cm above the roof line or chimney stack, whichever is the lower.	No	But make sure that the fixing point is stable and the installation is safe.
Possibly	If the building is in a designated area and the capacity of each antenna is greater than 35 litres or it is intended to install the antenna on a chimney, wall or roof slope visible from a road or Broads waterway.	No	As long as the installation is safe and the fixing point stable.
Satellite dishes, aerials and antennas on buildings over 15 m high			
No	Unless: • there are more than four antennas on the property overall; • the size of any antenna is more than 130 cm in any linear dimension; • the cubic capacity of each antenna is more than 35 litres;	No	But make sure that the fixing point is stable and the installation is safe.

Table 5.3 Basic requirements for planning permission and Building Regulations approval

Planning permission		Building Regulations approval
No	• the antenna is fitted to a chimney stack and is greater than 60 cm in any linear dimension; • the antenna sticks out more than 300 cm above the highest part of the roof line; • the building is in a designated area and it is intended to install the antenna on a chimney, wall or roof slope visible from a road or Broads waterway.	As long as the installation is safe and the fixing point stable.
Flagpole		
No	Flagpoles, etc. erected in your garden are treated under the same rules as outbuildings, and cannot exceed 3 m in height.	No
Basements		
No	Provided that: • a light well is not added; • it is not a listed building; • it is not providing a separate unit of accommodation; • the external appearance of the building will not change; • the usage is not significantly changed; • you are not creating a new basement.	Yes Covering: • ceiling height; • damp-proofing; • fire escapes; • foundation work; • Party Wall Act; • underpinning; • ventilation wiring; • water supplies.
Biomass-fuelled appliances		
No	Provided any outside flue on the rear or side elevation is no more than 1 m above the highest part of the roof.	Yes Particularly in relation to electrical and plumbing work, ventilation, noise and general safety.

Table 5.3 Basic requirements for planning permission and Building Regulations approval (Continued)

Planning permission		Building Regulations approval	
Yes	If the building is listed or in a designated area, Conservation Area or in a World Heritage Site.	Yes	Particularly in relation to electrical and plumbing work, ventilation, noise and general safety.
Ceilings and floors			
No	If you are replacing a floor or ceiling.	Possibly	If more than 25 per cent of a ceiling below a cold loft space or flat roof is being replaced then Building Regulations will apply, and the thermal insulation of that ceiling should be improved.
Possibly	If you live in a listed building you should contact your local planning authority.	Possibly	If more than 25 per cent of a ceiling below a cold loft space or flat roof is being replaced then Building Regulations will apply, and the thermal insulation of that ceiling should be improved.
Central heating			
No	Unless it is a listed building or in a Conservation Area.	No	If electric and it is installed by an approved person and complies with Part P.
No	If an external flue is required as long as: • any flues on the rear or side elevation are no more than 1 m above the highest part of the roof; • no flue is fitted on the principal or side elevation if the building fronts on to a highway in a Conservation Area, World Heritage Site, National Park, Area of Outstanding Natural Beauty or the Broads;	No	If gas, solid fuel or oil fired and it is installed by an approved person and complies with Part J.

Table 5.3 Basic requirements for planning permission and Building Regulations approval

Planning permission		Building Regulations approval	
	• you seek consent for internal and external work if the building is listed or in a designated area (even if you enjoy permitted development rights it is advisable to check with your local planners).		
Change of use			
No	Provided the present and proposed use fall within the same 'class' or the Town and Country Planning (Use Classes) Order says that a change of class is permitted to another specified class (as listed in section 5.7.2).	Yes	To certain changes of use, even though you may think that the work involved is not 'building work'.
Material changes of use			
Yes		Yes	
Working from home			
No	Provided that: • the activities you undertake are not considered unusual in a residential area; • the business does not disturb your neighbours at unreasonable hours or create other forms of nuisance such as noise or smells; • there will be no increase in traffic or people calling; • you still intend to use the building mainly as a private residence.	No	Unless building work is carried out.
Associated building work			
Yes		Yes	
Conservatories			
No	Provided that: • extensions of more than one storey do not extend beyond the rear wall of the original house by more than 3m and are not within 7m of any boundary opposite the rear wall of the house;	No	Provided that: • glazing and any fixed electrical installations comply with the applicable Building Regulations requirements;

Table 5.3 Basic requirements for planning permission and Building Regulations approval (Continued)

Planning permission	Building Regulations approval

• maximum eaves and ridge height of the extension are no higher than the existing house;
• maximum eaves height of an extension within 2 m of the boundary is 3 m;
• maximum height of a single-storey rear extension is 4 m;
• no extension is forward of the principal elevation or side elevation fronting a highway;
• no extension is higher than the highest part of the roof;
• no more than half the area of land around the 'original house' would be covered by additions or other buildings;
• there are no verandas, balconies or raised platforms;
  ○ on designated land:
  ○ there is no permitted development for rear extensions of more than one storey;
  ○ there is no cladding of the exterior;
  ○ there are no side extensions;
• roof pitch of extensions higher than one storey should match the existing house;
• side extensions are single-storey with a maximum height of 4 m and width no more than half that of the original house;
  ○ single-storey rear extensions must not extend beyond the rear wall of the original house by more than 3 m if an attached house or by 4 m if a detached house. Outside Article 1(5) designated land and Sites of Special Scientific Interest the limit is increased to 6 m if an attached house and 8 m if a detached house until 30 May 2016.
  ○ These increased limits are subject to the neighbour consultation scheme.
	• it is built at ground level and has a floor area less than 30 m²;
	• it is separated from the house by external-quality walls, doors or windows;
	• there is an independent heating system with separate temperature and on/off controls.
	
	Note: Any new structural opening between the conservatory and the existing house will require Building Regulations approval, even if the conservatory itself is an exempt structure.

Conversions

Yes For flats – even where construction works may not be intended.	Yes Unless you are not proposing any building work to make the change.

Table 5.3 Basic requirements for planning permission and Building Regulations approval

Planning permission	Building Regulations approval
No For loft conversions, as long as you do not alter or extend the roof space, and observe the following requirements: • a volume allowance of 40 m³ additional roof space for terraced houses; • a volume allowance of 50 m³ additional roof space for detached and semi-detached houses; • materials to be similar in appearance to the existing house; • no extension beyond the plane of the existing roof slope of the principal elevation that fronts the highway; • no extension to be higher than the highest part of the roof; • no verandas, balconies or raised platforms; • roof extensions not to be permitted development in designated areas; • roof extensions, apart from hip to gable ones, to be set back, as far as practicable, at least 20 cm from the original eaves; • side-facing windows to be obscure-glazed; any opening to be 1.7 m above the floor.	Yes To ensure: • reasonable sound insulation between the conversion and the rooms below; • safe escape from fire; • safely designed stairs to the new floor; • the stability of the structure (including the existing roof) is not endangered; • the structural strength of the new floor is sufficient. For further advice contact building control to discuss your proposal. You must also find out whether work you intend to carry out falls within The Party Wall etc. Act 1996.
Yes For conversion to shops and offices.	Yes
To convert an old building	
Yes	
Decoration and repairs inside and outside a building	
Repairs to a house, shop or office	
No As long as the repairs are of a minor nature (e.g. replacing the felt to a flat roof, repointing brickwork, or replacing floorboards).	No
Yes For major works such as removing a substantial part of a wall and rebuilding it, or underpinning a building.	Yes

Table 5.3 Basic requirements for planning permission and Building Regulations approval (Continued)

Planning permission		Building Regulations approval	
Internal decoration, repair and maintenance			
No	For internal decoration, repair and maintenance.	Yes	For some types of repair work.
External decoration, repair and maintenance			
No	Provided it does not make the building any larger or affect the thermal element of the building.	No	Provided it does not make the building any larger or affect the thermal element of the building.
To alter the position of a WC, bath, etc. within a house, shop or flat			
No	If you are refitting a kitchen or bathroom with new units and fittings.	No	
No		Yes	If you are fitting a bathroom or kitchen where there was not one before, or drainage and electrical work relating to a refit.
To alter the construction of fireplaces, hearths or flues within a house, shop or flat			
No	To fit or replace an external flue, chimney or soil and vent pipe.	Yes	If you are installing a flue.
If the property is a listed building or in a Conservation Area			
Yes	• For any external work especially if it will alter the visual appearance, or use alternative materials. • To alter, repair or maintain a gate, fence, wall or other means of enclosure.	Yes	

Table 5.3 Basic requirements for planning permission and Building Regulations approval

Planning permission		Building Regulations approval
To insert cavity wall insulation		
No	As long as there is no change in external appearance or the building is not listed or in a Conservation Area.	Yes — All insulation has to comply with the relevant Building Regulations both when installed during construction and when fitted retrospectively. If such an upgrade is not technically or functionally feasible, the element should be upgraded to the best standard which can be achieved within a simple payback of no greater than 15 years. If you are installing loft insulation as part of a roof renovation project, where more than 25 per cent of the roof is being renewed, then the level of insulation should meet the standards required by Building Regulations Approved Documents. Care should be taken not to block any ventilation at the edges.
To apply cladding		
No	As long as it is not a listed building or in a Conservation Area.	Yes — To re-render or replace timber cladding to external walls, depending on the extent of the work. Where 25 per cent or more of the wall is re-clad, re-rendered or re-plastered internally, or 25 per cent or more of the external wall is rebuilt, the thermal insulation will need to be improved and Building Regulations will apply.
Demolition		
No	Unless the council has made an Article 4 direction restricting permitted development rights in respect of demolition.	Yes — Six weeks' prior notice must be given to the local authority building control.

Table 5.3 Basic requirements for planning permission and Building Regulations approval (Continued)

Planning permission		Building Regulations approval	
Possibly	The council may wish to agree with you how you propose to carry out the demolition. This is called a 'prior approval application'.	Yes	For a partial demolition to ensure that the remaining part of the house (or adjoining buildings/extensions) is structurally sound.
Doors and windows			
Repairs, maintenance, and minor improvements			
No	Unless it is a listed building or in a Conservation Area or your council has made an Article 4 direction withdrawing permitted development rights.	Yes	If you use an unregistered installer or do it yourself you will need to get approval from the local authority building control.
		No	If the work is carried out by an approved person and complies with the requirements of Part F.
New windows in an upper-floor side elevation			
No	As long as they are: • obscure-glazed; and • non-opening or fitted more than 1.7 m above the floor level.		
New rooflights or skylights			
No	As long as: • they do not protrude more than 150 mm beyond the plane of the roof slope; • they are no higher than the highest part of the roof. Windows on the side elevation roof slope must be obscure-glazed and either non-opening or more than 1.7 m above the floor level.		
Shop windows			
Yes	To replace shop windows. Further information is available from your local building control or from the Glass and Glazing Federation (GGF) website (http://www.ggf.org.uk).	Yes	To replace shop windows.

Table 5.3 Basic requirements for planning permission and Building Regulations approval

Planning permission	Building Regulations approval
Drains and sewers	
No For repairs, maintenance and very minor works.	Yes Work on drains and sewers should be carried out in compliance with Approved Document H of the Building Regulations.
Possibly If permitted development rights in the area extend beyond repair, maintenance and very minor works.	
Yes Internal or external works to listed buildings are likely to require listed building consent.	
Electrical work in the home or garden	
No Unless it is in a listed building or a Conservation Area.	No As long as: • you comply with Part P (and other relevant Building Regulations Approved Documents); • all electrical work follows the safety standards in BS 7671 (the 'wiring regulations'). Your contract with the electricity supply company will have conditions about electrical safety which must not be broken. In particular, you should not interfere with the company's equipment, which includes the cables to your consumer unit up to and including the separate isolator switch, if provided.

Table 5.3 Basic requirements for planning permission and Building Regulations approval (Continued)

Planning permission		Building Regulations approval
Extensions		
No	Provided that: • extensions of more than one storey do not extend beyond the rear wall of the original house by more than 3 m; • materials are similar in appearance to the existing house; • maximum eaves and ridge height of the extension is no higher than the existing house; • maximum eaves height of an extension within 2 m of the boundary is 3 m; • maximum height of a single-storey rear extension is 4 m; • no extension is forward of the principal elevation or side elevation fronting a highway; • no extension is higher than the highest part of the roof; • no more than half the area of land around the 'original house' would be covered by additions or other buildings; • there are no verandas, balconies or raised platforms; • the roof pitch of extensions higher than one storey matches the existing house; • side extensions are single-storey with a maximum height of 4 m and width no more than half that of the original house; • single-storey rear extensions must not extend beyond the rear wall of the original house by more than 3 m if an attached house or 4 m if a detached house. For extensions outside of Article 1(5) designated land and Sites of Special Scientific Interest the limit is increased to 6 m if an attached house and 8 m if a detached house until 30 May 2016: ◦ these increased limits (between 3 m and 6 m and between 4 m and 8 m respectively) are subject to the neighbour consultation scheme; • two-storey extensions are no closer than 7 m to the rear boundary; • upper-floor, side-facing windows are obscure-glazed and any opening is 1.7 m above the floor.	Yes	Most extensions of properties require approval under the Building Regulations.

Table 5.3 Basic requirements for planning permission and Building Regulations approval

Planning permission		Building Regulations approval
On designated land		
No	Provided that in addition to the above: • there is no cladding of the exterior; • there are no side extensions; • there are no rear extensions of more than one storey.	Yes	Most extensions of properties require approval under the Building Regulations.
External walls		
Repairs, maintenance or minor improvements		
No	As long as it is for repairs, maintenance or minor improvements, such as painting your house and it is not a listed building or in a Conservation Area or designated area.	Yes	If 25 per cent or more of an external wall is re-rendered, re-clad, re-plastered or re-lined internally or where 25 per cent or more of the external leaf of a wall is rebuilt.
		Yes	If you want to insert insulation into a cavity wall.
Cladding the outside of a house with stone, artificial stone, pebble dash, render, timber, plastic or tiles			
Yes	If you live in a Conservation Area, a National Park, an Area of Outstanding Natural Beauty or the Broads.	Possibly	Depending on the extent of the work involved in re-cladding the building.
No	If you live outside the areas above and any cladding is of a similar appearance to material used in the construction of the house.	Possibly	Depending on the extent of the work involved in re-cladding the building.
Fascias		
No	Unless you live in a listed building or designated area (Conservation Area, National Park, Area of Outstanding Natural Beauty).	No	

Table 5.3 Basic requirements for planning permission and Building Regulations approval (Continued)

Planning permission		Building Regulations approval
Fences, gates and garden walls		
Yes	• If it would be over 1 m high and next to a highway used by vehicles (or the footpath of such a highway), or over 2 m high elsewhere; or • if your right to put up or alter fences, walls and gates is removed by an Article 4 direction or a planning condition; • if your house is a listed building or in the curtilage of a listed building; • if it forms a boundary with a neighbouring listed building or its curtilage.	No
Flats and maisonettes		
To extend a flat		
Yes	You must apply for planning permission.	Yes
		You may need to consult the fire service regarding issues relating to fire escapes. The Housing Act 2004 requires that subdivided buildings meet standards and houses in multiple occupation are licensed.
To subdivide a house or single flat		
Yes		Yes
		You may also need to consult the fire service, and the property will need to be licensed under the Housing Act 2004.
Loft conversion in a top-floor flat		
No	Provided that it is internal works. Local requirements differ, so it is recommended that you check with the local planning authority. Permission is required where you extend or alter the roof space. You should also check whether you own the roof space which you wish to convert. If you are a leaseholder, you may need to get permission from your landlord, freeholder or management company.	Yes
Yes	If you intend to extend or alter the roof space.	Yes

Table 5.3 Basic requirements for planning permission and Building Regulations approval

Planning permission		Building Regulations approval	
Painting the exterior			
No	Unless you live in an area where an Article 4 direction applies.	No	
Satellite dishes			
Possibly	Planning permission may be required in certain circumstances. Check with the local council.	No	
Windows			
Possibly	To fit new double-glazed windows.	Yes	In relation to thermal performance and other areas such as safety, air supply, means of escape and ventilation.
No	If you are replacing like with like or adding internal secondary glazing.	Yes	In relation to thermal performance and other areas such as safety, air supply, means of escape and ventilation.
Flues, chimneys or soil and vent pipes			
Domestic flues			
No	As long as: • flues are not fitted on the principal or side elevation that fronts a highway in a designated area; • flues on the rear or side elevation of the building are no more than 1 m above the highest part of the roof; • the building is not listed or in a designated area.	Yes	Particularly with regard to ventilation and general safety. Installation should be carried out by a suitably qualified installer.

Table 5.3 Basic requirements for planning permission and Building Regulations approval (Continued)

Planning permission	Building Regulations approval
Flues for biomass and combined heat and power systems (non-domestic)	
No As long as: • it is the first installation of a flue as part of either a biomass heating system or a combined heat and power system, but further installations will require planning permission; • the building is not listed or a scheduled monument; • the capacity of the system is no more than 45 kilowatts thermal; • the flue is no more than 1 m higher than the highest part of the roof or the height of an existing flue which is being replaced, whichever is the higher.	Yes Particularly with regard to ventilation and general safety. Installation should be carried out by a suitably competent person.
Fuel tanks	
No As long as: • in National Parks, the Broads, Areas of Outstanding Natural Beauty and World Heritage Sites the maximum area to be covered by buildings, enclosures, containers and pools more than 20m from the house is limited to 10m³; • maximum height is 2.5m within 2m of a boundary; • maximum overall height is 3m; • it is not at the side of properties on designated land; • it is not forward of the principal elevation fronting a highway; • it is not more than 3500litres in capacity; • not more than half the area of land around the 'original house' would be covered by additions or other buildings.	No If you are installing an oil tank and/or connecting pipework and you employ an installer registered with one of the related competent person schemes, you will not need to involve a building control service.
Yes Any container within the curtilage of a listed building will require planning permission.	

Table 5.3 Basic requirements for planning permission and Building Regulations approval

Planning permission		Building Regulations approval
Garages and car ports		
No	As long as: • there is a maximum height of 2.5 m in the case of a building, enclosure or container within 2 m of a boundary of the curtilage of the dwelling house; • no more than half the area of land around the 'original house' would be covered by additions or other buildings; • there is no outbuilding on land forward of a wall forming the principal elevation; • there are no verandas, balconies or raised platforms; • outbuildings and garages are single-storey with maximum eaves height of 2.5 m and maximum overall height of 4 m with a dual pitched roof or 3 m for any other roof; • in National Parks, the Broads, Areas of Outstanding Natural Beauty and World Heritage Sites the maximum area to be covered by buildings, enclosures, containers and pools more than 20 m from the house is limited to 10 m^2.	If the floor area of the building is between 15 m^2 and 30 m^2 you will not normally be required to apply for Building Regulations approval providing that the building contains no sleeping accommodation and is either at least 1 m from any boundary or is constructed of substantially non-combustible materials.
Yes	• On designated land buildings, enclosures, containers and pools at the side of properties will require planning permission. • Within the curtilage of listed buildings any outbuilding will require planning permission.	
Hardstanding for a car, caravan or boat		
No	Provided it is within your boundary, is at or near ground level and does not require significant works of embanking or terracing.	No	Unless you introduce steps where none existed before.
Heat pumps		
Domestic ground source or water source heat pumps		
No	Unless the building is listed or in a Conservation Area.	Yes	It is advisable to contact an installer who can provide the necessary advice, preferably

Table 5.3 Basic requirements for planning permission and Building Regulations approval (Continued)

Planning permission	Building Regulations approval
	one who belongs to either the Microgeneration Certification Scheme or the relevant competent person scheme.
Domestic air source heat pumps	It is advisable to contact an installer who can provide the necessary advice, preferably one who belongs to either the Microgeneration Certification Scheme or the relevant competent person scheme.
Yes As long as all the limits and conditions listed below are met:	Yes
• All parts of the air source heat pump must be at least 1 m from the property boundary.	
• Development is permitted only if the air source heat pump installation complies with the Microgeneration Certification Scheme Planning Standards or equivalent standards.	
• It is not installed on a wall if that wall fronts a highway and any part of that wall is above the level of the ground storey if it is within a Conservation Area or World Heritage Site.	
• On a flat roof all parts of the pump are at least 1 m from the external edge of that roof.	
• Only the first installation of an air source heat pump will be permitted development, and only if there is no existing wind turbine on a building or within the curtilage of that property. Additional wind turbines or air source heat pumps at the same property require an application for planning permission.	
• The installation is not on a pitched roof.	
• The installation is not within the curtilage of a listed building or within a site designated as a scheduled monument.	
• The pump is not on a wall or roof which fronts a highway or nearer to any highway which bounds the property than any part of the building if it is in a Conservation Area or World Heritage Site.	
• The volume of the air source heat pump's outdoor compressor unit (including housing) must not exceed 0.6m³.	

Table 5.3 Basic requirements for planning permission and Building Regulations approval

Planning permission	Building Regulations approval
In addition, the following conditions must be met. The air source heat pump must be: • removed as soon as reasonably practicable when it is no longer needed for microgeneration; • sited, so far as is practicable, to minimize its effect on the external appearance of the building and its effect on the amenity of the area; • used solely for heating purposes. ***Non-domestic ground source or water source heat pumps*** No As long as all the following conditions are observed: • Only the first stand-alone installation will be permitted development. Further installations will require planning permission from the local authority. • The total area of excavation must not exceed 0.5 ha. • When no longer needed for microgeneration, pumps should be removed as soon as reasonably practicable and the land should, as far as reasonably practicable, be restored to its condition before the development took place, or to the condition agreed in writing between the local planning authority and the developer. • The total surface area covered by the water source heat pump (including any pipes) must not exceed 0.5 ha. ***Non-domestic air source heat pumps***	 It is advisable to contact an installer who can provide the necessary advice, preferably one who belongs to either the Microgeneration Certification Scheme or the relevant competent person scheme. It is advisable to contact an installer who can provide the necessary advice, preferably one who belongs to either the Microgeneration Certification Scheme or the relevant competent person scheme.

Table 5.3 Basic requirements for planning permission and Building Regulations approval (Continued)

Planning permission		Building Regulations approval	
Hedges			
No	Unless the hedge obscures the view of traffic at a junction or access to a main road.	No	
Hydroelectricity			
Yes	Some form of environmental assessment will be essential for this type of project.	Yes	Particularly for electrical installations.
Infilling			
Possibly	Consult your local planning officer.	Yes	If a new development.
Installing a swimming pool			
Possibly	Consult your local planning officer.	Yes	For an indoor pool.
Internal walls			
No		Yes	In order to satisfy the requirements of the regulations for internal walls you should consider the following areas: • fire safety; • load-bearing walls; • sound insulation; • structural support; • ventilation.
Kitchens and bathrooms			
No	No unless it is part of a house extension or the property is a listed building.	No	Unless drainage or electrical works form part of the refit.

Table 5.3 Basic requirements for planning permission and Building Regulations approval

Planning permission	Building Regulations approval
No	Yes · If a bathroom or kitchen is to be provided in a room where there wasn't one before, particularly in relation to ventilation, drainage, structural stability, and electrical and fire safety.
Laying a patio, decking or driveway	
No · As long as: • no significant embanking or terracing works are required; • it is not a listed building.	No · As long as alterations do not make access to the dwelling any less satisfactory than it was before.
Decking	
No · As long as: • the decking is no more than 30 cm above the ground; • together with extensions, outbuildings, etc. the decking covers no more than 50 per cent of the garden.	No · As long as the structure does not require planning permission.
Lighting	
No · Not unless the property is a listed building, but you should make sure that the intensity and direction of light do not disturb others.	Yes · You should either use an installer who is registered with the competent person scheme or make an application to your local authority's building control department or approved inspectors.
Micro combined heat and power	
No • As long as flues on the rear or side elevation of the building are a maximum of 1 m above the highest part of the roof.	Yes · The *Low or Zero Carbon Energy Sources: Strategic Guide* (LZC) supports

Table 5.3 Basic requirements for planning permission and Building Regulations approval (Continued)

Planning permission	Building Regulations approval
• If the building is listed or in a designated area even if you enjoy permitted development rights it is advisable to check with your local planning authority before a flue is fitted. Consent is also likely to be needed for internal alterations. • In a Conservation Area or World Heritage Site the flue should not be fitted on the principal or side elevation if it would be visible from a highway.	the inclusion of low or zero carbon energy sources in Part L of the Building Regulations and Approved Documents L1A, L1B, L2A and L2B. Chapter 4 of the guide deals with micro combined heat and power.
New homes	
Yes	Yes
Outbuildings	
No As long as: • there is a maximum height of 2.5 m in the case of a building, enclosure or container within 2 m of a boundary of the curtilage of the dwelling house; • no more than half the area of land around the 'original house' would be covered by additions or other buildings; • there is no outbuilding on land forward of a wall forming the principal elevation; • there are no verandas, balconies or raised platforms; • outbuildings and garages are single-storey with maximum eaves height of 2.5 m and maximum overall height of 4 m with a dual pitched roof or 3 m for any other roof; • in National Parks, the Broads, Areas of Outstanding Natural Beauty and World Heritage Sites the maximum area to be covered by buildings, enclosures, containers and pools more than 20m from the house is limited to 10m².	No If the floor area of the building is between 15m² and 30m² you will not normally be required to apply for Building Regulations approval providing that the building contains no sleeping accommodation and is either at least 1 m from any boundary or is constructed of substantially non-combustible materials.
Yes • On designated land buildings, enclosures, containers and pools at the side of properties will require planning permission. • Within the curtilage of listed buildings any outbuilding will require planning permission.	

Table 5.3 Basic requirements for planning permission and Building Regulations approval

Planning permission		Building Regulations approval	

Polytunnels

| Yes | If it is a listed building or in a Conservation Area, National Park or Area of Outstanding Natural Beauty. | Possibly | Depending on the work involved. |
| No | As long as the polytunnel is no nearer to the road than the nearest part of the house (unless there is more than 20m between the tunnel and the road) and is less than 3m high. | Possibly | Depending on the work involved. |

Paving your front garden

| No | If the replacement driveway of any size uses permeable surfacing or if the rainwater is directed to a lawn or border to drain naturally. | No | As long as alterations do not make access to the dwelling any less satisfactory than it was before. |
| Yes | If the surface to be covered is more than 5m² and you plan to lay a traditional impermeable driveway. | No | As long as alterations do not make access to the dwelling any less satisfactory than it was before. |

Plumbing

| No | Unless it is a listed building. | No | If it is installed by an approved person and complies with Part J. |
| | | Yes | If you use an unregistered installer or do it yourself you will need to get approval from the local authority building control. |

Porch

| No | Unless:
• any part is less than 2m from a boundary adjoining a highway;
• any part is more than 3m high;
• the floor area exceeds 3m². | No | As long as glazing and electrical installations comply with Building Regulations and: |

Table 5.3 Basic requirements for planning permission and Building Regulations approval (Continued)

Planning permission	Building Regulations approval
	• the front entrance door between the existing house and the new porch must remain in place; • if the house has ramped or level access for disabled people, the porch must not adversely affect access.
Roof	
No Provided that: • any side-facing windows are obscure-glazed and any opening is 1.7 m above the floor; • the alterations do not project more than 150 mm from the existing roof plane; • the alterations are no higher than the highest part of the roof.	Yes If you want to carry out repairs on or re-cover less than 25 per cent of the area of a pitched or flat roof, you will not normally need to submit a Building Regulations application. You will need approval, however, if: • the performance of the new covering will be significantly different to that of the existing covering in the event of a fire; • you are replacing or repairing more than 25 per cent of the roof area, in which case the roof thermal insulation would normally have to be improved; • you carry out structural alterations.
Shops	
To change residential use to a shop	
Yes • This is classed as a material change of use, as the proposed new use is a different use class within the planning system.	Yes The regulations define converting a home into a shop as a 'material

Table 5.3 Basic requirements for planning permission and Building Regulations approval

Planning permission	Building Regulations approval
	change of use' and specify that the building must comply with the Building Regulations in respect of escape and other fire precautions, hygiene, energy conservation, and access to and use of buildings.
• The majority of local authorities have designated shopping or commercial areas within their planning policies where they may consider such a change of use suitable. It is likely to be very difficult to gain permission outside these areas, such as in a wholly residential area. • While permission is not required before work starts, if permission is refused the work will have to be undone and the authority may take formal enforcement action over a change of use without planning permission.	• The building may therefore need to be upgraded to make it comply with the specified requirements. • You should also check with the local fire authority, usually the county council, to see what 'ongoing' fire precautions legislation will apply when the building is in use.

To convert a shop to a café (use class A3) or public house (A4) or takeaway (A5)

Planning permission	Building Regulations approval
No	Possibly
Premises in shop use (A1) are able to change to café use (A3) without planning permission for a single period of up to two years.	You may wish to contact your local building control body for further advice.
• The majority of local authorities will have designated shopping or commercial areas within their planning policies (which often contain policies on striking a balance of uses in an area). • Some authorities may have specific policies to retain a level of retail uses. • Permission will be required for commercial extractor flues if cooking is intended to be carried out on the premises. • The placing of tables and chairs on the highway is very likely to require a licence from the highway authority to allow it to assess factors such as pedestrian movements, sight lines and road safety. • Changes in signage are likely to require express advertisement consent.	

Table 5.3 Basic requirements for planning permission and Building Regulations approval (Continued)

Planning permission		Building Regulations approval	
	• While permission is not required for the change of use before work starts, if permission is refused the work will have to be undone and the authority may take formal enforcement action over a change of use without planning permission.		
Yes	For other material changes of use, such as to a public house (A4) or takeaway (A5), because the proposed new use is a different use class within the planning system.	Possibly	You may wish to contact your local building control body for further advice.
	• The majority of local authorities will have designated shopping or commercial areas within their planning policies (which often contain policies on striking a balance of uses in an area).		
	• Some authorities may have specific policies to retain a level of retail uses.		
	• Permission will also be required for commercial extractor flues if cooking is intended to be carried out on the premises.		
	• The placing of tables and chairs on the highway is very likely to require a licence from the highway authority to allow it to assess factors such as pedestrian movements, sight lines and road safety.		
	• Changes in signage are likely to require express advertisement consent.		
	• While permission is not required for the change of use before work starts, if permission is refused the work will have to be undone and the authority may take formal enforcement action over a change of use without planning permission.		

To convert a shop to an office (use class B1) or storage (B8) or other uses

No	Premises in shop use (A1) are able to change to office use (B1) without planning permission for a single period of up to two years.	Possibly	You may wish to contact your local building control body for further advice.
	• The majority of local authorities will have designated shopping or commercial areas within their planning policies (which often contain policies on striking a balance of uses in an area).		
	• Some authorities may have specific policies to retain a level of retail uses.		
	• Changes in signage are likely to require express advertisement consent.		

Table 5.3 Basic requirements for planning permission and Building Regulations approval

Planning permission	Building Regulations approval
• While permission is not required for the change of use before work starts, if permission is refused the work will have to be undone and the authority may take formal enforcement action over a change of use without planning permission. • You may also require the consent of the landlord or landowner.	You may wish to contact your local building control body for further advice.
Yes For material changes of use, such as to storage (B8), because the proposed new use is a different use class within the planning system.	
• The majority of local authorities will have designated shopping or commercial areas within their planning policies (which often contain policies on striking a balance of uses in an area). • Some authorities may have specific policies to retain a level of retail uses. • Changes in signage are likely to require express advertisement consent. • While permission is not required for the change of use before work starts, if permission is refused the work will have to be undone and the authority may take formal enforcement action over a change of use without planning permission. • You may also require the consent of the landlord or landowner.	

To erect, change or alter a shop's adverts, fascia or projecting signs

Yes All advertisements require consent either from the local authority or from the legislation – the display of an advert without consent is a criminal offence. Some adverts relating to the business at the site may have deemed consent from the legislation governing adverts. For example, many traditional, non-illuminated fascia signs and hanging signs are likely to have deemed consent subject to a number of conditions and limitations such as size. If an advert meets all the criteria of deemed consent you will not require express consent. Because inappropriate adverts can harm an area visually (and even economically) many authorities have produced supplementary guidance on the design of suitable adverts and commercial signage. Other adverts without deemed consent will require express advertisement consent from the local authority. If the building is listed you will also require listed building consent. You must obtain consent, as it is required before an advertisement is displayed, and the local planning authority may take action (including prosecution) if no consent is in place. You may also require the consent of the landlord or landowner.	

Table 5.3 Basic requirements for planning permission and Building Regulations approval (Continued)

Planning permission	Building Regulations approval
Solar panels	
Solar panels mounted on a house or on a building within the grounds of a house	
No As long as all the following conditions are observed:	Yes The ability of the roof to carry the weight of the panels needs to be checked and electrical work needs to be carried out by an approved contractor.
• Panels on a building should be sited, so far as is practicable, to minimize the effect on the external appearance of the building and the amenity of the area.	
• When no longer needed for microgeneration, panels should be removed as soon as reasonably practicable.	
All the following limits must be met:	
• Panels should not be installed above the highest part of the roof (excluding the chimney) and should project no more than 200 mm from the roof slope or wall surface.	
• The panels must not be installed on a building that is within the grounds of a listed building.	
• The panels must not be installed on a site designated as a scheduled monument.	
• Wall mounted only – if your property is in a Conservation Area or World Heritage Site, panels must not be fitted to a wall which fronts a highway.	
Stand-alone solar panel installations (panels not on a building)	
No All the following conditions must be observed:	Yes The ability of the existing roof to carry the load (weight) of the panel will need to be checked and proven. Some strengthening work may be needed. Building Regulations also apply to other aspects of the work such as electrical installation. It is advisable to contact an installer who can provide the necessary advice, preferably one who belongs to the competent person scheme.
• Panels on a building should be sited, so far as is practicable, to minimize the effect on the amenity of the area.	
• When no longer needed for microgeneration, panels should be removed as soon as reasonably practicable.	
All the following limits must be met:	
• If your property is in a Conservation Area or World Heritage Site, no part of the solar installation should be nearer to any highway bounding the house than the part of the house that is nearest to that highway.	

Table 5.3 Basic requirements for planning permission and Building Regulations approval

Planning permission	Building Regulations approval

• No part of the installation should be higher than 4 m.
• Only the first-stand alone solar installation will be permitted development. Further installations will require planning permission.
• Panels should not be installed within the boundary of a listed building or a scheduled monument.
• The installation should be at least 5 m from the boundary of the property.
• The size of the array should be no more than 9 m^2 or 3 m wide by 3 m deep.
 	
Solar panels (non-domestic)
 	
No   All the following conditions must be observed:
  • Panels should be sited, so far as is practicable, to minimize the effect on the external appearance of the building and the amenity of the area.
  • When no longer needed for microgeneration, panels should be removed as soon as reasonably practicable.
 	
All the following limits must be met:
  • Equipment mounted on a roof must not be within 1 m of the external edge of the roof.
  • Equipment mounted on a wall must not be within 1 m of a junction of that wall with another wall or with the roof of the building.
  • The panels must not be installed on a listed building or on a building that is within the grounds of a listed building.
  • If the building is on designated land the equipment must not be installed on a wall or a roof slope which fronts a highway.
  • Solar panels installed on a wall or a pitched roof should project no more than 200 mm from the wall surface or roof slope.
  • The panels must not be installed on a listed building or on a building that is within the grounds of a listed building.
  • The panels must not be installed on a site designated as a scheduled monument.
  • Where panels are installed on a flat roof the highest part of the equipment should not be more than 1 m above the highest part of the roof (excluding the chimney).
	The ability of the existing roof to carry the load (weight) of the panel will need to be checked and proven. Some strengthening work may be needed. Building Regulations also apply to other aspects of the work such as electrical installation. It is advisable to contact an installer who can provide the necessary advice, preferably one who belongs to the competent person scheme.

Table 5.3 Basic requirements for planning permission and Building Regulations approval (Continued)

Planning permission		Building Regulations approval	

Solar panels mounted on a non-domestic building

No	All the following conditions must be observed: • Panels should be sited, so far as is practicable, to minimize the effect on the amenity of the area. • When no longer needed for microgeneration, panels should be removed as soon as reasonably practicable. All the following limits must be met: • Only the first stand-alone solar installation will be permitted development. Further installations will require planning permission from the local authority. • If the property is in a designated area no part of the solar installation should be nearer to any highway bounding the grounds of the property than the part of the building that is nearest to that highway. • No part of the installation should be higher than 4 m. • Panels should not be installed within the boundary of a listed building or a scheduled monument. • The installation should be at least 5 m from the boundary of the property. • The size of the array should be no more than 9 m² or 3 m wide by 3 m deep.	Possibly	The ability of the existing roof to carry the load (weight) of the panel will need to be checked and proven. Some strengthening work may be needed. Building Regulations also apply to other aspects of the work such as electrical installation. It is advisable to contact an installer who can provide the necessary advice, preferably one who belongs to the competent person scheme.

Structural alterations inside

No	Unless it is a listed building or within a Conservation Area.	Possibly	If it is a listed building or within a Conservation Area.
Yes	If the alterations are major, such as removing or part-removing a load-bearing wall or altering the drainage system.	Yes	If you wish to build or remove an internal wall or make openings in an internal wall.

Trees

No	You can fell or lop trees on your property unless the trees are protected by a Tree Preservation Order or you live in a Conservation Area.	No	

Table 5.3 Basic requirements for planning permission and Building Regulations approval

Planning permission		Building Regulations approval
Underpinning		
No	But if you live in a listed building or designated area (Conservation Area, National Park or Area of Outstanding Natural Beauty) you should check with your local planning authority before carrying out any work.	Yes	The regulations specifically define underpinning as 'building work', and appropriate measures must be applied to ensure the underpinning stabilizes the movement of the building. Particular attention will need to be given to any sewers and drains near the work.
Warehouses and industrial buildings		
No	As long as: • no new building is higher than 5m, if within 10m of the boundary, and in other cases no new building is higher than the highest building within the boundary or 15m, whichever is lower; • no new building exceeds a gross floor space of 100m^2; • until 30 May 2016 outside designated land and Sites of Special Scientific Interest no new building exceeds a gross floor space of 200m^2.	Yes	
Wind turbines		
Wind turbine: building mounted		
No	As long as: • it is removed as soon as reasonably practicable when no longer needed for microgeneration. • it is sited, so far as practicable, to minimize its effect on the external appearance of the building and its effect on the amenity of the area to a dwelling (consult your local planning officer);	Yes	With regard to size, weight and force exerted on fixed points. Building Regulations also apply to other aspects of the work such as electrical installation. It is advisable to contact an installer who can provide the necessary advice. You could use an installer registered with

Table 5.3 Basic requirements for planning permission and Building Regulations approval (Continued)

Planning permission	Building Regulations approval
• the building-mounted wind turbine installation complies with the Microgeneration Certification Scheme Planning Standards or equivalent standards (read more about the scheme); • in Conservation Areas, the building-mounted wind turbine would not be on a wall or roof slope which fronts a highway; • no part (including blades) of the building-mounted wind turbine protrudes more than 3 m above the highest part of the roof (excluding the chimney) or exceeds an overall height (including building, hub and blade) of 15 m, whichever is the lesser; • no part of the building-mounted wind turbine (including blades) is within 5 m of any boundary. • This only applies to installations on detached houses (not blocks of flats) and other detached buildings within the boundaries of a house or block of flats. A block of flats must consist wholly of flats (e.g. should not also contain commercial premises). • Only the first installation of any wind turbine will be permitted development, and only if there is no existing air source heat pump at the property. Additional wind turbines or air source heat pumps at the same property require an application for planning permission. • Permitted development rights do not apply to a turbine within the curtilage of a listed building or within a site designated as a scheduled monument or on designated land other than Conservation Areas. • The distance between ground level and the lowest part of any wind turbine blade must not be less than 5 m. • The installation must not be sited on safeguarded land. An Aviation Safeguarding Tool can be used to check whether the installation will be on safeguarded land. • The swept area of any building-mounted wind turbine blade must be no more than 3.8 m^2. • Use non-reflective materials on blades.	the relevant competent person scheme (as listed in row 17 of Schedule 3 of the Building Regulations). A registered installer may be authorized to self-certify the work to comply with this aspect of the Building Regulations without involving local authority building control.

Table 5.3 Basic requirements for planning permission and Building Regulations approval

Planning permission		Building Regulations approval
Wind turbine: stand-alone		
No	As long as:	Yes
	• no part of the stand-alone wind turbine (including blades) would be in a position which is less than a distance equivalent to the overall height of the turbine (including blades) plus 10 per cent of its height when measured from any point along the property boundary;	With regard to size, weight and force exerted on fixed points. Building Regulations also apply to other aspects of the work such as electrical installation. It is advisable to contact an installer who can provide the necessary advice. You could use an installer registered with the relevant competent person scheme (as listed in row 17 of Schedule 3 of the Building Regulations). A registered installer may be authorized to self-certify the work to comply with this aspect of the Building Regulations without involving local authority building control.
	• the stand-alone wind turbine installation complies with the Microgeneration Certification Scheme Planning Standards or equivalent standards;	
	• in Conservation Areas, the stand-alone wind turbine is not installed so that it is nearer to any highway which bounds the curtilage (garden or grounds) of the house or block of flats than the part of the house or block of flats which is nearest to that highway.	
	• Only the first installation of any wind turbine will be permitted development, and only if there is no existing air source heat pump at the property. Additional wind turbines or air source heat pumps at the same property require an application for planning permission.	
	• Permitted development rights do not apply to a turbine within the curtilage of a listed building or within a site designated as a scheduled monument or on designated land other than Conservation Areas.	
	• The distance between ground level and the lowest part of any wind turbine blade must not be less than 5 m.	
	• The highest part of the stand-alone wind turbine must not exceed 11.1 m.	
	• The installation must not be sited on safeguarded land. The Aviation Safeguarding Tool can be used to check whether the installation will be on safeguarded land.	
	• The swept area of any stand-alone wind turbine blade must be no more than 3.8 m².	

Table 5.3 Basic requirements for planning permission and Building Regulations approval (Continued)

Planning permission	Building Regulations approval
In addition, the following conditions must be met. The wind turbine must: • use non-reflective materials on blades; • be removed as soon as reasonably practicable when no longer needed for microgeneration; • be sited, so far as is practicable, to minimize its effect on the external appearance of the building and its effect on the amenity of the area.	
Working from home	
No	No
	Unless your proposed project involves work which will affect the structure of the building, the means of escape and other fire precautions, or affect the access to and use of buildings, the regulations will regard your work as a 'material alteration' (and therefore 'building work'), which must then comply with the regulations.

6

Meeting the requirements of the Building Regulations

6.1 Foundations

To support the weight of the structure, most brick-built buildings are supported on a solid concrete base which is called the 'foundation'. Timber-framed houses are usually built on a concrete foundation with a 'strip' or 'raft' construction to spread the weight.

6.1.1 Requirements

Structure
  The building shall be constructed so that:

1. *the combined dead, imposed and wind loads are sustained and transmitted by it to the ground, safely and without causing any building deflection/ deformation or ground movement that will affect the stability of any part of the building;*
2. *ground movement caused by swelling, shrinkage or freezing of the sub- soil, land-slip or subsidence will not affect the stability of any part of the building.*

(Approved Document A)

 Amendments to the 2010 version of Approved Document A came into force in 2013.

The walls and floors of the building shall adequately protect the building and people who use the building from harmful effects caused by ground moisture, precipitation and wind-driven spray, interstitial and surface condensation, and spillage of water from or associated with sanitary fittings or fixed appliances.
  All floors next to the ground, walls and roof shall not be damaged by mois- ture from the ground, rain or snow and shall not carry that moisture to any part of the building that it would damage.

(Approved Document C2)

Amendments to the 2010 version of Approved Document C came into force in 2013.

Site preparation and resistance to contaminants and water

1. *The ground to be covered by the building shall be reasonably free from any material that might damage the building or affect its stability, including vegetable matter, topsoil and pre-existing foundations.*
2. *Reasonable precautions shall be taken to avoid danger to health and safety caused by contaminants on or in the ground covered, or to be covered, by the building and any land associated with the building.*
3. *Adequate subsoil drainage shall be provided if it is needed to avoid:*

    (a) *the passage of ground moisture to the interior of the building;*
    (b) *damage to the building, including damage through the transport of water-borne contaminants to the foundations of the building.*

(Approved Document C1)

Rainwater drainage
Rainwater drainage systems shall ensure that rainwater soaking into the ground is distributed sufficiently so that it does not damage foundations of the proposed building or any adjacent structure.

(Approved Document H3)

Note: For the purpose of this requirement, *contaminant* means any substance which is or may become harmful to persons or buildings, including substances that are corrosive, explosive, flammable, radioactive or toxic.

6.1.1.1 Potential problems

There may be known and/or recorded conditions of ground instability, such as geological faults, landslides, disused mines, and unstable strata of similar nature, which affect or may potentially affect a building site or its environs.

There may also be:

- unsuitable material, including vegetable matter, topsoil and pre-existing foundations;
- contaminants on or in the ground covered, or to be covered, by the building and any land associated with the building;

Note: The Contaminated Land (England) Regulations 2006 (as amended 2012) empower local authorities to oversee the management of cause and effects of land contamination and this process is subject to controls under the Town and Country Planning Acts, and guidance in the National Planning Policy Framework.

- groundwater.

These conditions should be taken into account before proceeding with the design of a building or its foundations.

6.1.1.2 What about hazards?

Hazards associated with the ground may include:

- chemical and biological contaminants;
- gas generation from biodegradation of organic matter;
- naturally occurring radioactive radon gas and gases produced by some soils and minerals;
- physical, chemical or biological hazards;
- underground storage tanks or foundations;
- unstable fill or unsuitable hardcore containing sulphate;
- the effects of vegetable matter, including tree roots.

In the most hazardous conditions the only complete remedy is the total removal of contaminants from the ground to be covered by the building. In other cases, remedial measures can reduce the risks to acceptable levels. These measures should only be undertaken with the benefit of expert advice. Where the removal would involve handling large quantities of contaminated materials you are advised to seek expert advice.

Even when these actions have been completed successfully, the ground to be covered by the building will **still** need to have at least 100 mm of concrete laid over it!

6.1.1.3 What about contaminated ground?

Potential building sites which are likely to contain contaminants can be identified at an early stage from planning records or from local knowledge (e.g. previous uses). In addition to solid and liquid contaminants, problems can also arise from natural contamination such as methane and the radioactive radon gas (and its decay product).

The following list contains examples of sites that are most likely to contain contaminants:

- asbestos works;
- ceramics, cement and asphalt manufacturing works;
- chemical works;
- dockyards and dockland;
- engineering works (including aircraft manufacturing, railway engineering works, shipyards, and electrical and electronic equipment manufacturing works);
- gas works, coal carbonization plants and ancillary by-product works;
- industries making or using wood preservatives;
- landfill and other waste disposal sites;
- metal mines, smelters, foundries, steelworks and metal-finishing works;
- munitions production and testing sites;

- oil storage and distribution sites;
- paper and printing works;
- power stations;
- railway land, especially larger sidings and depots;
- road vehicle fuelling, service and repair: garages and filling stations;
- scrap yards;
- sewage works, sewage farms and sludge disposal sites;
- tanneries;
- textile works and dye works.

If any signs of possible contaminants are present, the local authority's environmental health officer should be told at once. If the officer confirms the presence of any of these contaminants (see Table 6.1) then he or she will require either the removal of the contaminants or some other prescribed action to be completed before any planning permission for building work can be sought.

6.1.1.4 What about gaseous contaminants?

Radon (measured in becquerels per cubic metre of air, Bq/m^3) is a colourless, odourless, naturally occurring radioactive gas formed by decaying uranium that occurs naturally in all rocks and soils. Normally the gas that escapes from rock or soil is immediately diluted by the atmosphere and thus poses no harm to humans. However, when radon is trapped in an enclosed space, it can seep out of the ground and build up in houses, buildings and indoor workplaces, and studies have established that exposure to radon is the second largest cause of lung cancer in the UK after smoking. Some parts of the country (in particular the West Country) have higher natural levels than elsewhere, and precautions against radon may be necessary.

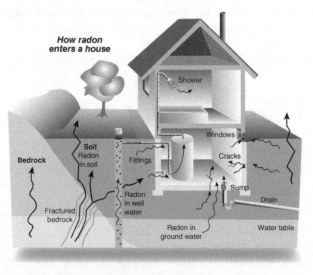

Figure 6.1 Common entry points of radon gas and household airflow.

Essentially a house acts as a chimney. Air moves up through the house due to warm air rising through the house and due to suction effects on the roof caused by wind action. The warm air eventually finds its way out via gaps at the top of the house. This air is then replaced by more air entering lower down in the house. The highest levels of radon are generally found in the small hours of the morning and in the middle of winter – in other words, the coldest times, when buildings are tightly closed.

The average level in UK homes is 20 Bq/m^3 and, for levels below 100 Bq/m^3, individual risk remains relatively low and not a cause for concern. However, the risk increases as the radon level increases.

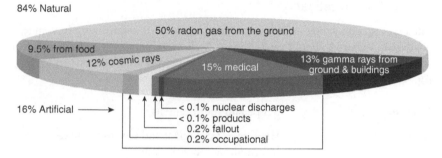

Figure 6.2 Average annual dose of radon to UK population.

Guidance on whether an area is susceptible to radon (and appropriate protective measures) can be obtained from BRE Report BR 211, *Radon: Guidance on Protective Measures for New Buildings (Including Supplementary Advice for Extensions, Conversions and Refurbishment)* (2007), and the report also contains a series of maps showing areas at risk.

As an alternative approach to using these maps, radon risk reports may be used. These reports are available from:

* UK Radon, www.UKradon.org, for small domestic and workplace buildings (and extensions) that have an existing postal address;
* BGS Georeports, www.shop.bgs.ac.uk/Georeports, for other development sites;
* Public Health England (formerly the Health Protection Agency), radon@phe.gov.uk, for large workplaces.

 Use of the alternative radon risk reports approach will provide a more accurate assessment of whether radon protective measures are necessary and, if needed, the level of protection that is appropriate.

For workplaces, see the Ionising Radiation (Medical Exposure) Regulations 2000 (SI 2000 No. 1059) and/or the BRE guide *Radon in the Workplace*, which includes a national reference level for radon in workplaces. The Health and

Safety Executive also provides guidance on protection from radon in the work-place (www.hse.gov.uk/radiation/ionising/radon.htm).

Landfill gas is generated by the action of anaerobic micro-organisms on biodegradable material in landfill sites, and generally consists of methane and carbon dioxide together with small quantities of volatile organic compounds (VOCs), which give the gas its characteristic odour. Landfill gas can migrate under pressure through the subsoil and through cracks and fissures into buildings.

Methane and carbon dioxide can also be produced by organically rich soils and sediments such as peat and river silts, and a wide range of VOCs can be present as a result of petrol, oil and solvent spillages.

6.1.1.5 Site investigation

Site investigation is now the recommended method for determining how much unsuitable material should be removed before commencing building work. This investigation will normally consist of a number of well-defined stages, for example:

Planning stage	Scope and requirements.
Desktop study	Historical, geological and environmental information about the site.
Site reconnaissance or walkover survey	Identification of actual and potential physical hazards and the design of the main investigation.
Main investigation and reporting	Intrusive and non-intrusive sampling and testing to provide soil parameters.

Note: BS EN 1997-2:2007 supported by BS 5930:1999 provides comprehensive guidance on site investigation.

6.1.1.5.1 Risk assessment

The site investigation may identify certain risks, which will require a risk assessment. There are three types of assessment:

* preliminary (once the need for a risk assessment has been identified, and depending on the situation and the outcome);
* generic quantitative risk assessment (GQRA);
* detailed quantitative risk assessment (DQRA).

Each risk assessment should include:

Hazard identification	Developing the conceptual model to establish contaminant sources, pathways and receptors. This preliminary site assessment consists of a desk study and a site walkover in order to gather sufficient

information to obtain an initial understanding of the potential.

Hazard assessment Identifying what pollutant linkages may be present and analysing the potential for unacceptable risks.

Risk estimation Establishing the scale of the possible consequences by considering the degree of harm that may result and to which receptors.

Risk evaluation Deciding whether the risks are acceptable or unacceptable.

6.1.2 Meeting the requirements

Where the site is potentially affected by contaminants, a combined geotechnical and geo-environmental investigation should be considered.

6.1.2.1 Hazard identification and assessment

The following hazards shall be considered:

• aggressive substances – including inorganic and organic acids, alkalis, organic solvents and inorganic chemicals such as sulphates and chlorides;	C2.23a
• combustible fill – including domestic waste, colliery spoil, coal, plastics, petrol-soaked ground, etc.	C2.23b
• expansive slags – e.g. blast furnace and steel-making slag;	C2.23c
• floodwater affected by contaminants – substances in the ground, waste matter or sewage.	C2.23d

Although flood resistance is **not** currently a **requirement** of the Building Regulations, as part of their aim to improve the energy-efficiency of buildings and use planning to protect the environment, the policies set out in the revised 2013 National Planning Policy Framework aim to avoid *inappropriate development in areas at risk of flooding, including requiring new development to be flood resilient and resistant, as and where appropriate.* (See https://www.gov.uk/government/uploads/system/uploads/attachment_data/file/6077/2116950.pdf for full details.)

 Further information on flood-resistant and -resilient construction can be found in the Defra 2007 publication *Improving the Flood Performance of New Buildings: Flood Resilient Construction.*

6.1.2.2 Contaminated ground

A preliminary site assessment is required to provide information on the past and present uses of the site and surrounding area that may give rise to contamination (see Table 6.1).

Table 6.1 Examples of possible contaminants

Signs of possible contaminants	Possible contaminant
Vegetation (absence, poor or unnatural growth)	Metals Metal compounds Organic compounds Gases (landfill or natural source)
Surface materials (unusual colours and contours may indicate wastes and residues)	Metals Metal compounds Oily and tarry wastes Asbestos Other mineral fibres Organic compounds, including phenols Combustible material, including coal and coke dust Refuse and waste
Fumes and odours (may indicate organic chemicals)	Volatile organic and/or sulphurous compounds from landfill or petrol/solvent spillage Corrosive liquids Faecal animal and vegetable matter (biologically active)
Damage to exposed foundations of existing buildings	Sulphates
Drums and containers (empty or full)	Various

 The planning authority should be informed prior to any intrusive investigations or if any substance is found which was not identified in a preliminary statement about the nature of the site.

The underlying geology of a potential site has to be considered, as natural contaminants may be present. For example:	C2.3 and 2.4

- naturally occurring heavy metals (e.g. cadmium and arsenic) originating in mining areas;
- gases (e.g. methane and carbon dioxide) originating in coal-mining areas;
- organic-rich soils and sediments such as peat and river silts;
- radioactive radon gas – which can also be a problem in certain parts of the country.

Possible sulphate attack from some strata on concrete floor slabs and oversite concrete needs to be considered.	C2.5

6.1.2.3 Gaseous contaminants

6.1.2.3.1 Radon

All new buildings, extensions and conversions (whether C2.39
residential or non-domestic), which are built in areas where
there may be high radon emissions, may need to incorporate
precautions against radon.

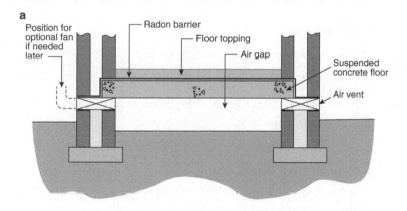

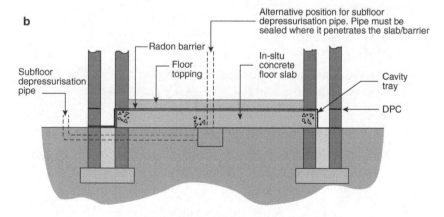

Figure 6.3 Protective measures for concrete floors.

The estimated costs for protecting dwellings against radon are not prohibitive
and should be considered.

Table 6.2 Estimated costs of radon protective measures

Building type	Basic protection	Full protection
Terraced	£318	£418
Semi-detached	£350	£448
Detached	£510	£614
Bungalow	£526	£624
Converted flat	£150	£220
PB low-rise flat	£160	£230
PB high-rise flat	£70	£100

6.1.2.3.2 Landfill gas

Methane is an asphyxiant, will burn, and can explode in air. Carbon dioxide is non-flammable and toxic. Many of the other components of landfill gas are flammable and some are toxic. All will require careful analysis.	C2

6.1.2.3.3 Gaseous risk assessment

A risk assessment should be completed for methane and other gases, particularly:

- on a landfill site or within 250 m of the boundary of a landfill site; C2.28a

- on a site subject to the wide-scale deposition of biodegradable substances (including made ground or fill); C2.28b

- on a site that has been subject to a use that could give rise to petrol, oil or solvent spillages; C2.28c

- in an area subject to naturally occurring methane, carbon dioxide and other hazardous gases (e.g. hydrogen sulphide). C2.28d

During a site investigation for methane and other gases:

- measurements should be taken over a sufficiently long period of time in order to characterize gas emissions fully; C2.30

- measurements should include periods when gas emissions are likely to be higher, e.g. during periods of falling atmospheric pressure. C2.30

Gas risks (i.e. to human receptors) should be considered for:

- gas entering the dwelling through the substructure (and   C2.32
  building up to hazardous levels);
- subsequent householder exposure in garden areas including   C2.32
  outbuildings (e.g. garden sheds and greenhouses),
  extensions and garden features (e.g. ponds).

When land that is affected by contaminants is being developed, 'receptors' (i.e. buildings, building materials and building services, as well as people) are introduced on to the site and it is necessary to break the pollutant linkages. This can be achieved by:

- treating the contaminant (e.g. use of physical, chemical or biological processes to eliminate or reduce the contaminant's toxicity or harmful properties);
- blocking or removing the pathway (e.g. isolating the contaminant beneath protective layers or installing barriers to prevent migration);
- protecting or removing the receptor (e.g. changing the form or layout of the development, using appropriately designed building materials, etc.).

A risk assessment based on the concept of a 'source–pathway–receptor' relationship, or pollutant linkage of a potential site (see Figure 6.4) should be carried out to ensure the safe development of land that is affected by contaminants.

6.1.2.3.4 Ground investigation

The detailed ground investigation:

- must provide sufficient information for the confirmation of a conceptual model for the site, the risk assessment and the design and specification of any remedial works;
- is likely to involve collection and analysis of soil, soil gas, surface and groundwater samples by the use of invasive and/or non-invasive techniques.

During the development of land affected by contaminants the   C2.14
health and safety of both the public and workers should be
considered.

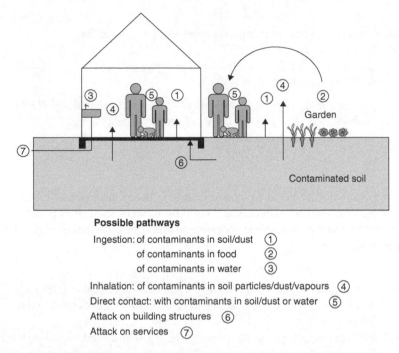

Possible pathways

Ingestion: of contaminants in soil/dust ①
of contaminants in food ②
of contaminants in water ③

Inhalation: of contaminants in soil particles/dust/vapours ④
Direct contact: with contaminants in soil/dust or water ⑤
Attack on building structures ⑥
Attack on services ⑦

Figure 6.4 Conceptual model of a site showing a source–pathway–receptor relationship.

6.1.2.3.5 Remedial measures

If the risks posed by the gas are unacceptable then these need to be managed through appropriate building remedial measures.	C2.36
Site-wide gas control measures may be required if the risks on any land associated with the building are deemed unacceptable.	C2.36
Consideration should be given to the design and layout of buildings to maximize the driving forces of natural ventilation.	C2.37
For non-domestic buildings, expert advice concerning gas control measures should be sought, as the floor area of such buildings can be large and it is important to ensure that gas is adequately dispersed from beneath the floor.	C2.38
There is a need for continued maintenance and calibration of mechanical (as opposed to passive) gas control systems.	C2.38
Subfloor ventilation systems should be carefully designed to ensure adequate performance and should not be modified unless subjected to a specialist review of the design.	C2.38

6.1.2.3.6 Corrective measures

> When building work is undertaken on sites affected by   C2.15
> contaminants where control measures are already in place,
> care must be taken not to compromise these measures.

Depending on the contaminant, three generic types of corrective measure can be considered: treatment, containment and/or removal.

 Note: The containment or treatment of waste may require a waste management licence from the Environment Agency.

6.1.2.3.7 Treatment

> The choice of the most appropriate treatment process for a   C2.16
> particular site is a highly site-specific decision for which
> specialist advice should be sought.

6.1.2.3.8 Containment

> In-ground vertical barriers may also be required to control   C2.17
> lateral migration of contaminants.
>   Cover systems involve the placement of one or more
> layers of materials over the site and may be used to:
>
> • break the pollutant linkage between receptors and   C2.18
>   contaminants;
> • sustain vegetation;
> • improve geotechnical properties;
> • reduce exposure to an acceptable level.
>
> Imported fill and soil for cover systems should be assessed at C2.20
> source to ensure that it is suitable for use.
>
> The size and design of cover systems (particularly soil-based   C2.20
> ones used for gardens) should take account of their long-term
> performance.
>
> Gradual intermixing due to natural effects and activities such C2.20
> as burrowing animals, gardening, etc. needs to be considered.

6.1.2.3.9 Removal

> Imported fill should be assessed at source to ensure that there are no materials that will pose unacceptable risks to potential receptors.   C2.21

6.1.2.3.10 Site preparation

> The effects of roots that are close to the building need to be assessed and any vegetable matter such as turf and roots should be removed from the ground that is going to be covered by the building, sufficient to prevent later growth.   C1.4

 Where mature trees are present (particularly on sites with shrinkable clays; see Table 6.3) the potential damage arising from ground heave to services and floor slabs and oversite concrete should be assessed.

Table 6.3 Volume change potential for some common clays

Clay type	Volume change potential
Glacial till	Low.
London	High to very high.
Oxford and Kimmeridge	High.
Lower lias	Medium.
Gault	High to very high.
Weald	High.
Mercian mudstone	Low to medium.

> Building services such as below-ground drainage should be sufficiently robust or flexible to accommodate the presence of any tree roots.   C1.6
>
> Joints should be made so that roots will not penetrate them.   C1.6
>
> Where roots could pose a hazard to building services, consideration should be given to their removal.   C1.6
>
> On sites previously used for buildings, consideration should be given to the presence of other infrastructure (such as existing foundations, services, buried tanks, etc.) that could endanger   C1.7

persons in and about the building and any land associated with the building.

If the site contains fill or made ground, consideration should be given to its compressibility and its potential to collapse when wet.　　C1.8

6.1.2.4 Foundations and types of soil

Table 6.4 provides guidance on determining the type of soil on which it is intended to lay a foundation.

Table 6.4 Types of subsoil

Type	Applicable field test
Rock (being stronger/ denser than sandstone, limestone or firm chalk)	Requires at least a pneumatic or other mechanically operated pick for excavation.
Compact gravel and/or sand	Requires a pick for excavation. Wooden peg 50 mm square in cross-section hard to drive beyond 150 mm.
Stiff clay or sandy clay	Cannot be moulded with the fingers and requires a pick or pneumatic or other mechanically operated spade for its removal.
Firm clay or sandy clay	Can be moulded by substantial pressure with the fingers and can be excavated with a spade.
Loose sand, silty sand or clayey sand	Can be excavated with a spade. Wooden peg 50 mm square in cross-section can be easily driven.
Soft silt, clay, sandy clay or silty clay	Fairly easily moulded in the fingers and readily excavated.
Very soft silt, clay, sandy clay or silty clay	Natural sample in winter conditions exudes between the fingers when squeezed in fist.

6.1.2.4.1 Subsoil drainage

Where the water table can rise to within 0.25 m of the lowest floor of the building, or where surface water could enter or adversely affect the building, either the ground to be covered by the building should be drained by gravity, or other effective means of safeguarding the building should be taken.　　C3.2

If an active subsoil drain is cut during excavation and if it passes under the building it should be:　　C3.3

• relaid in pipes with sealed joints and have access points outside the building; or
• rerouted around the building; or
• rerun to another outfall (see Figure 6.5).

Where contaminants are present in the ground, consideration   C3.7
should be given to subsoil drainage to prevent the transportation
of water-borne contaminants to the foundations or into the
building or its services.

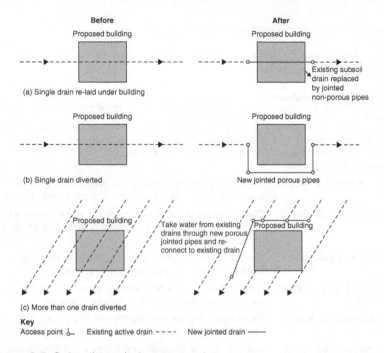

Figure 6.5 Subsoil cut during excavation.

6.1.2.4.2 Ground movement

Known or recorded conditions of ground instability, such as   A1/2 1.9
that arising from landslides, disused mines or unstable strata,
should be taken into account in the design of the building and
its foundations.

6.1.2.4.3 Foundations – plain concrete

There should **not** be:

* non-engineered fill (see BRE Digest 427) or a wide   A1/2 2E1a
  variation in ground conditions within the loaded area;

- weaker or more compressible ground at such a depth   A1/2 2E1b
  below the foundation as could impair the stability of
  the structure.

The foundations should be situated centrally under the wall.   A1/2 2E2a

In non-aggressive soils, concrete should be composed of   A1/2 2E2b
Portland cement to BS EN 197-1 and 197-2 and fine and
coarse aggregate conforming to BS EN 12620, and the mix
should be either:

- 50 kg of Portland cement to not more than 200 kg
  ($0.1 \, m^3$) of fine aggregate and 400 kg ($0.2 \, m^3$) of
  coarse aggregate; or
- Grade ST2 or Grade GEN I concrete to BS 8500-2.

For foundations in chemically aggressive soil conditions,   (8.18)
guidance in BS 8500-1: Part 1 and BRE Special Digest 1
should be followed.

The minimum thickness T of concrete foundation should   A1/2 2E2c
be 150 mm or P, whichever is the greater, where P is   (8.18a)
derived using Table 6.4 and Figure 6.6.

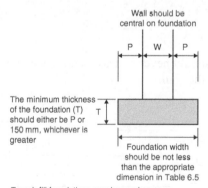

Figure 6.6 Foundation dimensions.

Foundations stepped on elevation should overlap by twice   A1/2 2E2d
the height of the step, by the thickness of the foundation,
or 300 mm, whichever is greater (see Figure 6.7).

Foundations should unite at each
change in level

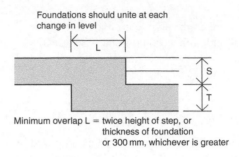

Minimum overlap L = twice height of step, or
thickness of foundation
or 300 mm, whichever is greater

S should not be greater than T

Figure 6.7　Elevation of stepped foundation.

The overlap for trench fill foundations should be twice the height of the step or 1 m, whichever is greater.	A1/2 2E2d
Trench fill foundations may be used as an acceptable alternative to strip foundations.	A1/2 2E2c

Note: Steps in foundations should not be of greater height than the thickness of the foundation (see Figure 6.7).

Foundations for piers, buttresses and chimneys should project as shown in Figure 6.8.	A1/2 2E2f

 The projection X should never be less than the value of P where there is no local thickening of the wall.

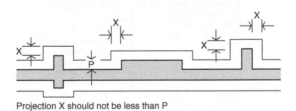

Projection X should not be less than P

Figure 6.8　Piers and chimneys.

6.1.2.4.4　Strip foundations

The recommended minimum widths of strip foundations shall be as indicated in Table 6.5.	A1/2 2E3

Table 6.5 Minimum width of strip footings

Type of ground (including engineered fill)	Condition of ground	Field test applicable	Total load of load-bearing walling not more than (kN/linear metre)					
			Minimum width of strip foundation (mm)					
			20	30	40	50	60	70
I Rock	Not inferior to sandstone, limestone or firm chalk	Requires at least a pneumatic or other mechanically operated pick for excavation.	In each case equal to the width of wall					
II Gravel or sand	Medium dense	Requires pick for excavation. Wooden peg 50mm square in cross-section hard to drive beyond 150mm.	250	300	400	500	600	650
III Clay Sandy clay	Stiff Stiff	Can be indented slightly by thumb.	250	300	400	500	600	650
IV Clay Sandy clay	Firm Firm	Thumb makes impression easily.	300	350	450	600	750	850
V Sand Silty sand Clayey sand	Loose Loose Loose	Can be excavated with a spade. Wooden peg 50mm square in cross-section can be easily driven.	400	600	Note: Foundations on soil types V and VI do not fall within the provisions of this section if the total load exceeds 30kN/m.			
VI Clay Sandy clay Clay or silt	Soft Soft Soft Soft	Finger pushed in up to 10mm.	450	650				
VII Silt Clay Sandy clay Clay or silt	Very soft Very soft Very soft Very soft	Finger easily pushed in up to 25mm.	Refer to specialist advice.					

Note: This depth will commonly need to be increased in areas subject to long periods of frost or in order to transfer the loading on to satisfactory ground.

> In clay soils subject to volume change on drying (i.e.   A1/2 2E4
> 'shrinkable clays' with a plasticity index greater than or
> equal to 10 per cent), strip foundations should be taken to a
> depth where anticipated ground movements (caused by
> vegetation and trees on the ground) will not impair the
> stability of any part of the building.
>
> The depth to the underside of foundations on clay soils   A1/2 2E4
> should not be less than 0.75 m on low-shrinkage clay soils,
> 0.9 m on medium-shrinkage clay soils and 1.0 m on
> high-shrinkage clay soils.

Note: These depths may need to be increased in order to transfer the loading on to satisfactory ground, or where there are trees nearby.

6.1.2.4.5 Underpinning foundations

Underpinning is a construction method to increase the depth of the foundations of a building by excavating the existing foundation and replacing it with new foundation material. The reasons for underpinning are generally because:

- the existing foundations of the building have moved (e.g. caused by poor soil, or changes to the soil conditions through subsidence, etc.); or
- another storey to the building is going to be added and the depth of the existing foundations is inadequate to support the modified building's weight.

Underpinning work will require very careful planning and execution and (if you propose to underpin an existing foundation) Building Regulations approval will normally be required.

In order to avoid excessively undermining the existing foundations, causing further damage to the structure above and running the possibility of the building collapsing (!), the excavations for the underpinning should be carried out to the engineer's instructions and details.

Note: Filling the excavation with concrete will not necessarily guarantee that the underpinning will provide sufficient support to the existing foundation, because of the possibility that cavities will remain between the two. It is, therefore, recommended and usually necessary for a sand and cement packing to be rammed into the void, in stages, to ensure adequate support. The timing of each stage and the specification of the materials to be used will vary on a case-by-case basis and should normally be the subject of a structural engineer's design.

6.1.2.4.6 Maintenance on foundations

Maintenance on foundations generally does not require planning permission. However, if you live in a listed building or designated area (Conservation Area, National Park or Area of Outstanding Natural Beauty) you should check with your local planning authority before carrying out any work.

6.1.2.5 Buildings – size

6.1.2.5.1 Classification of purpose groups

Many of the provisions in the Approved Documents are related to how the building is going to be used, its purpose and who is going to use it. These provisions can apply to a whole building or (where a building is compartmented) to a section of a building. Table 6.6 sets out the purpose group classification.

Table 6.6 Classification of purpose groups

Title	Group	Purpose for which the building or compartment of a building is intended to be used
Residential (dwellings)	1(a)	Flat or maisonette.
	1(b)	House which contains a habitable storey with a floor level which is more than 4.5 m above ground level.
	1(c)	House which does not contain a habitable storey with a floor level which is more than 4.5 m above ground level.
Residential (institutional)	2(a)	Hospital, home, school or other similar establishment.
Other	2(b)	Hotel, boarding house, residential college, hall of residence, hostel, etc.
Office	3	Offices or premises used for the purpose of administration, clerical work, banking and communications.
Shop and business commercial	4	Shops or premises used for a retail trade.
Assembly and recreation	5	Place of assembly, entertainment or recreation.
Industrial	6	Factories and other premises used for generating power or slaughtering livestock.
Storage and other non-industrial	7(a)	Place for the storage or deposit of goods or materials (and any non-residential building not within any of the purpose groups 1 to 6.
	7(b)	Car parks.

6.1.2.6 Maximum floor area

No floor enclosed by structural walls on all sides shall   A1/2 (2C14)
exceed 70 m² (see Figure 6.9).

No floor with a structural wall on one side shall exceed   A1/2 (2C14)
36 m² (see Figure 6.9).

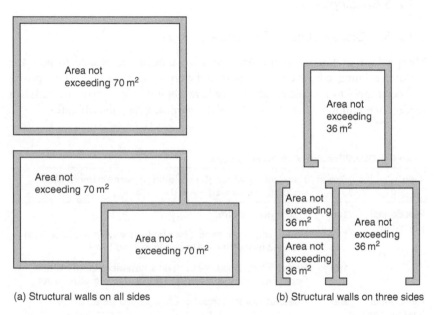

(a) Structural walls on all sides   (b) Structural walls on three sides

Figure 6.9 Maximum floor area that is enclosed by structural walls.

6.1.2.7 Maximum height of buildings

The maximum height of a building shall not exceed the   A1/2 (1C17)
heights given in Figure 6.10 with regard to the relevant
wind speed.

Read map wind speed V from Figure 6.10.1	Find the orographic zone for the site from Figure 6.10.2 and obtain Factor 0 from Table a (or use Figure 6.10.3)	Obtain value of Factor A from **Table b**	Calculate value of Factor S from: S = V×0×A	Obtain maximum allowable building height from **Table c**

Table a Factor 0

Orographic category and average slope of whole hillside, ridge, cliff or escarpment

	Factor 0		
	Zone 1	Zone 2	Zone 3
Category 1: Nominally flat terrain, average slope < 1/20	1.0	1.0	1.0
Category 2: Shallow terrain, average slope < 1/10	1.12	1.07	1.05
Category 3: Moderately steep terrain, average slope < 1/5	1.24	1.13	1.10
Category 4: Steep terrain, average slope > 1/5	1.36	1.20	1.15

Table b Factor A

Site altitude (m)	Factor A
0	1.00
50	1.05
100	1.10
150	1.15
200	1.20
300	1.30
400	1.40
500	1.50

Table c Maximum allowable building height in meters

Factor S	Country sites			Town sites		
	Distance to the coast			Distance to the coast		
	< 2km	2 to 20km	> 50km	< 2km	2 to 20km	> 50km
≤25	15	15	15	15	15	15
26	11.5	13.5	15	15	15	15
27	8	11	14.5	15	15	15
28	5.5	8	11	15	15	15
29	4	6.5	8.5	12.5	15	15
30	3	5	6.5	10	12.5	15
31		4	5.5	8.5	11	13.5
32		3.5	4.5	7	9.5	11.5
33		3	3.5	6	8	10
34			3	5.5	7	8/.5
35				4.5	6.5	7.5
36				4	5.5	6.5
37				3.5	5	6
38				3	4.5	5.5
39					4	5
40					3.5	4.5
41					3	4
42						3.5
43						3.5
44						3

Notes: Table a - Outside of the zones shown in **Table a**, the Factor 0 = 1.0.
Table b - For elevated sites where orography is significant a more accurate assessment of Factor A can be obtained by using the altitude at the base of the topographic feature instead of the attitude at the site. See Figure 6.10.2 or, alternatively, Figure 6.10.3.

Table c - i) Sites in town less than 300mm from the edge of the town should be assumed to be in the country terrain. ii) Where a site is closer than 1km to an inland area of water which extends more than 1km in the wind direction, the distance to the coast should be taken as < 2km.

Interpolation may be used in **Tables b** and **c**.

Figure 6.10 Maximum height of buildings.

Figures 6.10.1, 6.10.2 and 6.10.3 are associated with the requirements in Figure 6.10.

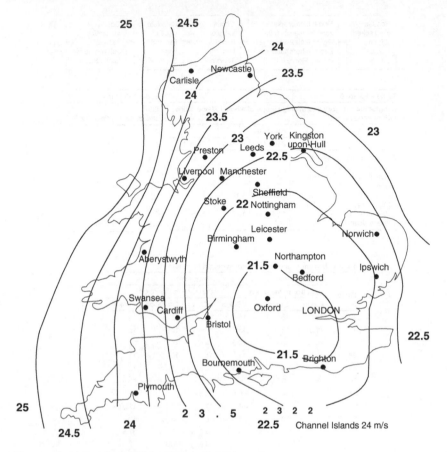

Figure 6.10.1 Map of wind speeds (V) in m/s.

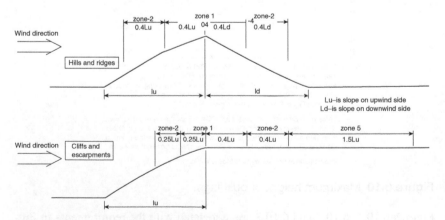

Figure 6.10.2 Orographic zones for Factor 0.

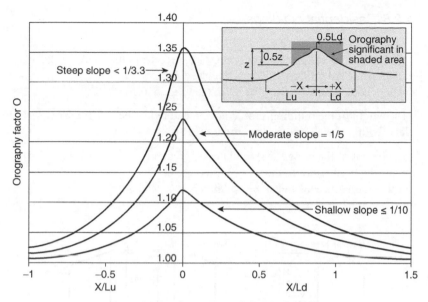

Figure 3A Orography Factor O far hills and ridges

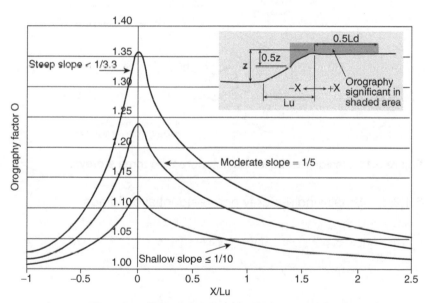

Figure 3B Orography Factor O far cliffs and escarpments
(interpolation between curves may be used)

Figure 6.10.3 Alternative graphical method for determining Orography Factor 0.

6.1.2.8 Residential buildings

> The maximum height of the building measured from the   A1/2 (2C4i)
> lowest finished ground level to the highest point of any
> wall or roof should be less than 15 m (see Figure 6.11).
>
> The height of the building should not exceed twice the   A1/2 (2C4ii)
> least width of the building (see Figure 6.11).
>
> The height of the wing H2 should not be greater than   A1/2 (2C4iii)
> twice the least width of the wing W2 where the
> projection P exceeds twice the width W2.

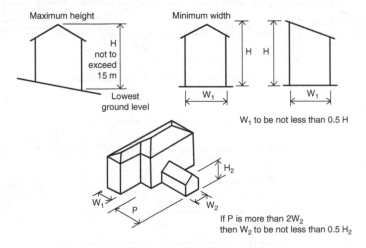

Figure 6.11 Residential buildings not more than three storeys.

6.1.2.8.1 Small single-storey non-residential buildings

> The floor area of the building or annexe shall not   A1/2 2C38(1)a
> exceed 36 m².

 Note: Where the floor area of the building or annexe exceeds 10 m², the walls
shall have a mass of not less than 130 kg/m².

> The height H should not exceed 3 m and the width   A1/2 2C38(1)a
> (or greater length) should not exceed 9 m
> (see Figure 6.12).

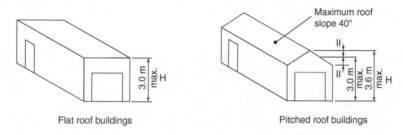

Figure 6.12 Size and proportion of non-residential buildings.

6.1.2.8.2 Annexes

The height H (as variously shown in Figure 6.13) should   A1/2 (2C4b)
not exceed 3 m.

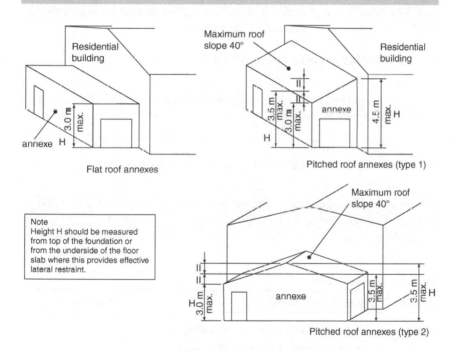

Figure 6.13 Size and proportion of non-residential annexes.

Note: In May 2013 Parliament agreed to increase the size of single-storey rear extensions which can be built under permitted development, and brought into force the 'associated neighbour consultation scheme'. This scheme allows, for a limited period between 30 May 2013 and 30 May 2016, householders to build larger single-storey rear extensions as a 'permitted development'. The size limits have doubled from 4 m^2 to 8 m^2 for detached houses, and from 3 m^2 to 6 m^2 for all other houses.

There is no fee involved but, in order to get permission to build one of these larger extensions, the planning authority still requires a written description of the proposal, a plan of the site, the addresses of any adjoining properties (including those at the rear) and a contact address for the developer.

There is a 42-day consultation period during which time the local authority will serve a notice on adjoining owners or occupiers. If any adjoining neighbour raises an objection during the first 21 days, the local authority will take this into account and make a decision about whether the impact on the amenity of all adjoining properties is acceptable or not. **No** other issues will be considered and the development can go ahead if the local authority gives the developer permission in writing within the 42-day period.

 If the local authority **fails** to notify the developer of its decision within the 42-day determination period then the development may go ahead without restriction. More details are provided in http://www.planningportal.gov.uk/uploads/ neighbour_consultation_scheme_guidance_may13.pdf.

6.1.2.9 Heights of walls and storeys

> The measured height of a wall or a storey should be in accordance with Figure 6.14.   A1/2 2C18

6.1.2.9.1 Imposed loads on roofs, floors and ceilings

> The imposed loads on roofs, floors and ceilings shall not exceed those shown in Table 6.7.   A1/2 (2C15)

Table 6.7 Imposed loads

Element	Distributed loads	Concentrated load
Roofs	1.00 kN/m^2 for spans not exceeding 12 m	
	1.50 kN/m^2 for spans not exceeding 6 m	
Floors	2.00 kN/m^2	
Ceilings	0.25 kN/m^2	0.9 kN/m^2

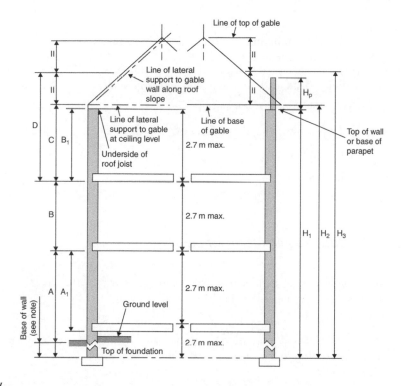

Key

(a) **Measuring Storey Heights**

A_1 is the ground storey height if the ground floor provides effective lateral support to the wall i.e. is adequately tied to the wall or is a suspended floor bearing on the wall.

A is the ground storey height if the ground floor does not provide effective lateral support to the wall.

Note: If the wall is supported adequately and permanently on both sides by suitable compact material, the base of the wall for the purposes of the storey height may be taken as the lower level of this support. (Not greater than 3.7 m ground storey height.)

B is the intermediate storey height.

B_1 is the top storey height for walls which do not include a gable.

C is the top storey height where lateral support is given to the gable at both ceiling level and along the roof slope.

D is the top storey height for the external walls which include a gable where lateral support is given to the gable only along the roof slope.

(b) **Measuring Wall Heights**

H_1 is the height of an external wall that does not include a gable.

H_2 is the height of an internal or separating wall which is built up to the underside of the roof.

H_3 is the height of an external wall which includes a gable.

H_p is the height of a parapet. If Hp is more than 1.2 m add H_p to H_1.

Figure 6.14 Method for measuring the heights of storeys and walls.

6.1.2.10 Structural safety

The safety of a structure depends on the successful combination of design and completed construction, particularly:

- the design – which should also: A1/2 0.2a
  - be based on identification of the hazards (to which the structure is likely to be subjected) and an assessment of the risks;

○	reflect conditions that can reasonably be foreseen during future use;	
•	loading – dead load, imposed load and wind load;	A1/2 0.2b
•	the properties of materials used;	A1/2 0.2c
•	the detailed design and assembly of the structure;	A1/2 0.2d
•	safety factors;	A1/2 0.2e
•	workmanship.	A1/2 0.2f

6.1.2.11 Basic requirements for stability

Adequate provision shall be made to ensure that the building is stable under the likely imposed and wind loading conditions.	A1/2 1A2
The overall size and proportioning of the building shall be limited according to the specific guidance for each form of construction.	A1/2 1A2a
The layout of walls (both internal and external) forming a robust three-dimensional box structure in plan shall be constructed according to the specific guidance for each form of construction.	A1/2 1A2b
The internal and external walls shall be adequately connected either by masonry bonding or by using mechanical connections.	A1/2 1A2c
The intermediate floors and roof shall be constructed so that they: • provide local support to the walls; • act as horizontal diaphragms capable of transferring the wind forces to buttressing elements of the building.	A1/2 1A2d

Note: A traditional cut timber roof (such as one using rafters, purlins and ceiling joists) generally has sufficient built-in resistance to instability and wind forces (e.g. from hipped ends, tiling battens and rigid sarking, etc.). However, the need for diagonal rafter bracing equivalent to that recommended in BS EN 1995-1-1:2004 with its UK National Annex should be considered, **especially** for single-hipped and non-hipped roofs of greater than 40° pitch to detached houses.

6.2 Ventilation

6.2.1 Requirements

There shall be adequate means of ventilation provided for people in the building.

(Approved Document F)

Septic tanks, wastewater treatment systems and cesspools should have adequate ventilation that is kept away from buildings.

(Approved Document H2)

Ventilation is defined in the Building Regulations as *the supply and removal of air (by natural and/or mechanical means) to and from a space or spaces in a building.*

In addition to replacing 'stale' indoor air with 'fresh' outside air, the aim of ventilation is to:

- provide outside air for breathing;
- control thermal comfort;
- limit the accumulation of moisture and pollutants from a building, which could, otherwise, become a health hazard to people living and/or working within that building;
- dilute and remove airborne pollutants (especially odours but also those that are released from materials and products used in the construction, decoration and furnishing of a building, and as a result of the activities of the building's occupants);
- control excess humidity;
- provide air for fuel-burning appliances.

Note: The requirements of Approved Document F have *also* been designed to deal with the products of tobacco smoking.

In general terms, all these aims can be met if the ventilation system:

- disperses residual pollutants and water vapour;
- extracts water vapour from wet areas where it is produced in significant quantities (e.g. kitchens, utility rooms and bathrooms);
- rapidly dilutes pollutants and water vapour produced in habitable rooms, occupiable rooms and sanitary accommodation;
- extracts pollutants from areas where they are produced in significant quantities (e.g. rooms containing processes or activities which generate harmful contaminants);
- is designed, installed and commissioned so that it:

  ○ is not detrimental to the health of the people living and/or working in the building;

- helps maintenance and repair;
- is reasonably secure;

• makes available, over long periods, a minimum supply of outdoor air for the occupants;
• minimizes draughts;
• provides protection against rain penetration.

Ventilation is also a means of controlling thermal comfort.

The aim of Approved Document F is to suggest to the designer the level of ventilation that should be sufficient for a particular situation as opposed to how it should be achieved. The designer is, therefore, free to use whatever ventilation system he or she considers most suitable for a particular building provided that it can be demonstrated that it meets the recommended performance criteria and levels concerning moisture, pollutants and air flow rate standards as shown in Table 6.8.

Table 6.8 Standards for performance-based ventilation

Type	Standard	Part
Intermittent extract fan	BS EN 13141-4	Clause 4
Range hood	BS EN 13141-3	Clause 4
Background ventilator (non-RH controlled)	BS EN 13141-1	Clauses 4.1 and 4.2
Background ventilator (RH controlled)	Pr EN 13141-9	Clauses 4.1 and 4.2
Passive stack ventilator	See Appendix D of Approved Document F	
Continuous mechanical extract ventilation (MEV) system	BS EN 13141-6	Clause 4
Continuous mechanical supply and extract with heat recovery (MVHR)	Pr EN 13141-7	Clauses 6.1, 6.2 and 6.2.2
Single room heat recovery ventilator	Pr EN 13141-8	Clauses 6.1 and 6.2
RH, relative humidity		

 For further details and examples concerning 'performance-based ventilation', see Appendix A to Approved Document F.

 Note: The 2010 edition of Approved Document F, Ventilation included the changes that have been made to the legal requirements of the Building Regulations 2000 and the Building (Approved Inspectors etc.) Regulations 2000. These stipulate that:

• All fixed mechanical ventilation systems, where they can be tested and adjusted, shall be commissioned and a commissioning notice given to the building control body.

- For mechanical ventilation systems installed in new dwellings, air flow rates shall be measured on site and a notice given to the building control body. This shall apply to intermittently used extract fans and cooker hoods, as well as continuously running systems.
- The owner shall be given sufficient information about the ventilation system and its maintenance requirements so that the ventilation system can be operated to provide adequate air flow.

6.2.1.1 Exceptions

Approved Document F provides guidance for those erecting new dwellings or buildings other than dwellings and completing work on existing buildings, with the following exceptions:

- buildings controlled under the Manufacture and Storage of Explosives Regulations 2005;
- buildings controlled under the Nuclear Installations Act 1965;
- detached buildings of less than 30 m² floor area, designed as shelters from nuclear, chemical or conventional weapons;
- buildings included in the schedule of monuments (maintained under section 1 of the Ancient Monuments and Archaeological Areas Act 1979);
- detached buildings into which people do not normally (or only occasionally) go;
- greenhouses (non-retail);
- agricultural buildings – unless part of the building is used as a dwelling;
- temporary buildings not intended to remain in place for more than 28 days;
- ancillary buildings used for the disposal of buildings or building plots on site;
- site construction buildings;
- buildings on the site of mines and quarries which do not contain dwellings or are used as offices or showrooms;
- detached single-storey buildings, with less than 30 m² floor area that do not contain any sleeping accommodation;
- detached buildings of less than 15 m² floor area containing no sleeping accommodation;
- ground-level building extensions (e.g. conservatories, porches, covered yards, covered ways and/or car ports) with a floor area less than 30 m².

6.2.1.2 Notification of work covered by the ventilation requirements

In most cases where it is proposed to carry out notifiable ventilation work on a building, the building control body has to be notified in advance via a full plans application or an initial notice given jointly with the approved inspector.

6.2.1.3 Competent person self-certification schemes

It is not necessary, however, to notify a building control body in advance of work that is going to be carried out by a person registered with a competent person self-certification scheme for that type of work.

There are a number of competent person schemes for the installation of mechanical ventilation and air-conditioning systems in buildings (e.g. BESCA, CORGI, NAPIT and NICEIC); more details can be found at https://www.gov.uk/competent-person-scheme-current-schemes-and-how-schemes-are-authorised#current-schemes.

6.2.1.4 Emergency repairs

Where the work involves an emergency repair, unless the person carrying out the work is registered under an appropriate competent person scheme, a notice has to be provided to the building control body at the earliest opportunity. A completion certificate can then be issued in the normal way.

6.2.1.5 Minor works

Where the work is of a minor nature such as:

*   replacement of parts, or the addition of an output or control device;
*   provision of a self-contained mechanical ventilation or air-conditioning appliance which cannot be adjusted from its factory settings – as in the case of a cooker hood, a bathroom extract fan or a room air-conditioning unit;

the work need **not** be notified to the building control body.

6.2.2 Meeting the requirements

6.2.2.1 External pollution

In urban areas, buildings are exposed to a large number of pollution sources from varying heights and upwind distances (i.e. long, intermediate and short-range). Internal contamination from these pollution sources can have a detrimental effect on the buildings' occupants, and so it is very important to ensure that the ventilation system provided is sufficient and, above all, that the air intake cannot be contaminated.

Typical urban pollutants include:

*   carbon monoxide (CO);
*   nitrogen dioxide, (NO_2);
*   sulphur dioxide (SO_2);
*   ozone (O_3);
*   particulate (PM_{10});
*   benzene;
*   butadiene;
*   polycyclic aromatic hydrocarbons (PAHs);

- ammonia;
- lead.

Typical emission sources include:

- building ventilation system exhaust discharges;
- combustion plants (e.g. heating appliances) running on conventional fuels;
- construction and demolition sites;
- discharges from industrial processes and other sources;
- other combustion-type processes (such as waste incineration, and thermal oxidation abatement schemes);
- road traffic, including traffic junctions and underground car parks;
- uncontrolled discharges from industrial processes and other sources.

6.2.2.2 Indoor air pollutants

The maximum permissible level of indoor air pollutants is listed in Table 6.9.

Table 6.9 Maximum permissible level of indoor air pollutants

Pollutant	Permissible level
Nitrogen dioxide (NO_2)	Not exceeding:
	$288 g/m^3$ (150 ppb) – 1-hour average $40 g/m^3$ (20 ppb) – long-term average
Carbon monoxide (CO)	Not exceeding:
	$100 mg/m^3$ (90 ppm) – 15=minute averaging time $60 mg/m^3$ (50 ppm) – 30-minute averaging time (DOH, 2004) $30 mg/m^3$ (25 ppm) – 1-hour averaging time (DOH, 2004) $10 mg/m^3$ (10 ppm) – 8-hour averaging time (DOH, 2004)
Control of bio-effluents (body odours)	3.5 l/s per person
Total volatile organic compounds	Not exceeding $300 g/m^3$ averaged over 8 hours

Note: Mould growth can occur whether the dwelling is occupied or unoccupied, so the performance criteria for moisture (see Table 6.10) should be met at all times, regardless of occupancy. The other pollutants listed in Table 6.9 are harmful to the occupants only when the dwelling is occupied.

Table 6.10 Mould growth

Moving average speed	Surface water activity
1 month	65%
1 week	75%
1 day	85%

Extract ventilation concerns the removal of air directly from a space or spaces to outside. Extract ventilation may be by natural means such as passive stack ventilation (PSV) or by mechanical means (e.g. by an extract fan or central system).

 Extract fans lower the pressure in a building, which can cause combustion gases from open-flued appliances filling the room instead of going up the flue or chimney.

6.2.2.3 Extract ventilation requirements

All kitchens, utility rooms, bathrooms and sanitary accom-   F5.5
modation shall be provided with extract ventilation to the
outside, which is capable of operating either intermittently or
continuously.

The minimum extract air flow rates should be greater than those shown
in Table 6.11.

Table 6.11 Extract ventilation rates

Room	Minimum intermittent extract rate	Continuous extract	
		Minimum high rate	Minimum low rate
Kitchen	30 l/s (adjacent to hob) or 60 l/s elsewhere	13 l/s	Total extract rate must be at least the whole building ventilation rate shown in Table 6.12
Utility room	30 l/s	8 l/s	
Bathroom	15 l/s	8 l/s	
Sanitary accommodation	6 l/s	6 l/s	

The whole building ventilation rate for habitable rooms   F5.6
in a dwelling should be greater than that shown in
Table 6.12.

The minimum ventilation rate (based on two occupants   F Table 5.1b
in the main bedroom and a single occupant in all other
bedrooms) should be not less than 0.3 l/s per m^2.

Table 6.12 Whole building ventilation rates

	Number of bedrooms				
	1	2	3	4	5
Whole building ventilation rate (l/s)	13	17	21	25	29

Note: For greater occupancy, add 4 l/s per occupant.

Purge ventilation is required in each habitable room and should be capable of extracting a minimum of four air changes per hour (i.e. 4 ach) per room directly to the outside.

> Sufficient information about the ventilation system and its   F4.47
> maintenance requirements must be given to owners so that
> the ventilation system can be operated to provide adequate air
> flow.

6.2.2.4 Ventilation effectiveness

Ventilation effectiveness is (as the term suggests) a measure of how well a ventilation system supplies air to the building's occupants. From an energy-saving perspective, the higher the level of ventilation effectiveness the more efficient the system will be in reducing pollutant levels in the occupant's breathing zone. As this can result in quite significant energy savings, it has to be considered when designing and installing ventilation systems.

As the designer cannot be absolutely certain of the future occupancy and/or use of the building in terms of seating plan, location of computers and printers, etc., a ventilation effectiveness level of 1 (i.e. where the supply air is fully mixed with the room air before it is breathed by the occupants) should be assumed in his or her calculations.

Note: For more details about ventilation effectiveness, see CIBSE Guide A.

6.2.2.4.1 Equivalent ventilator area for dwellings

Designers should use the equivalent ventilator areas shown in Table 6.13 when designing systems using intermittent extract fans and background ventilators for multi-storey dwellings that are more than four storeys above ground level and which have more than one exposed façade.

Note: 'Equivalent area' is defined as *the area of a sharp-edged orifice which air would pass at the same volume flow rate, under an identical applied pressure difference.*

Notes:

(1) For single-storey dwellings up to four storeys above ground level, add 5000 mm^2.
(2) For an occupancy level greater than two persons in the main bedroom and one person in all other bedrooms, assume an extra bedroom for each additional person.
(3) For more than five bedrooms, add an additional 10,000 mm^2 per bedroom.

Table 6.13 Equivalent ventilator area for dwellings

Total floor area (m²)	Equivalent ventilator area (mm²)				
	Number of bedrooms				
	1	2	3	4	5
<50	25,000	35,000	45,000		
51–60	25,000	30,000	40,000		
61–70	30,000	30,000	35,000	45,000	55,000
71–80	35,000	35,000	35,000		
81–90	40,000	40,000	40,000		
91–100	45,000	45,000	45,000		
>100	Add 5000 mm² for every additional 10 m² floor area.				

6.2.2.5 Ventilation air intakes

One method of achieving good indoor air quality is to reduce the amount of water vapour and/or air pollutants that are released into the indoor air, particularly those caused from construction and consumer products.

Air intakes that are located on a less polluted side of the building may be used for fresh air.

Note: Further information about control of emissions from construction products is available in BRE Digest 464.

6.2.2.6 Noise from ventilation systems

As the noise from ventilation systems can disturb the occupants of the building and in doing so affect their work effectiveness, the designer must consider methods of minimizing noise through careful design and use of quieter products such as sound-attenuating ventilation material.

Noise from a ventilation system may also disturb people who are outside the building, and so measures to minimize externally emitted noise should be considered.

The installation and use of ventilation systems in buildings will result in more energy being used (e.g. to heat fresh air taken in from outside or to move air into, out of and/or around the building) and so consideration should always be given to using heat recovery devices, efficient types of fan motor and/or energy-saving control devices in ventilation systems.

If a dwelling is near a busy road or by an airport, etc., where the amount of external noise is likely to be intrusive, a sound-attenuator duct section may be fitted in the roof space just above the ceiling.

6.2.2.7 Control of ventilation

Ventilation should be controllable so that it can maintain reasonable indoor air quality and avoid waste of energy. These controls can be either manual or automatic.

Manually controlled trickle ventilators (the most common type of background ventilators) can be located over the window frames, in window frames, just above the glass or directly through the wall (see Figure 6.23).

Trickle ventilators are intended to be normally left open in occupied rooms in dwellings.

In dwellings, humidity-controlled devices are available to regulate the humidity of the indoor air, and hence minimize the risk of condensation and mould growth. These are best installed as part of an extract ventilator in moisture-generating rooms (e.g. kitchen or bathroom). Humidity control is not appropriate for sanitary accommodation, where the dominant pollutant is normally odour.

Table 6.14 Ventilation systems for basements

Type of basement	Background ventilators and intermittent extract fans	Passive stack ventilation	Continuous mechanical extract	Continuous mechanical supply and extract with heat
Basement connected to the rest of the dwelling by an open stairway	Yes	Yes	Yes	Yes
Basement with a single exposed façade and dwelling above ground with more than one exposed façade		Yes	Yes	
Basements not connected to the rest of the dwelling by an open stairway	Yes	Yes	Yes	Yes
Dwelling above ground has no bedrooms	Yes	Yes	Yes	Yes
Dwelling comprises just a basement	Yes	Yes	Yes	Yes

Other types of *automatic control* (e.g. trickle ventilators, ventilation fans, dampers and air terminal devices) may also be suitable for regulating ventilation devices in dwellings.

Trickle ventilators with automatic controls should also have a manual override, so that the occupant can close the ventilator to avoid draughts. For pressure-controlled trickle ventilators that are fully open at typical conditions (e.g. 1 Pa pressure difference), only a manual close option is recommended.

In buildings other than dwellings, more sophisticated *automatic control* systems such as occupancy sensors (using local passive infrared detectors) or indoor carbon dioxide concentration sensors (using electronic carbon dioxide detectors) can be used as an indicator of occupancy level and, therefore, body odour.

6.2.2.8 Types of ventilation

Buildings are ventilated through a combination of infiltration and purpose-provided ventilation:

- **Infiltration** is the uncontrollable air exchange between the inside and outside of a building through a wide range of air leakage paths in the building structure.
- **Purpose-provided ventilation** is the controllable air exchange between the inside and outside of a building by means of a range of natural and/or mechanical devices.

Through good design and execution, domestic and non-domestic buildings can currently achieve an air permeability down to around 2–4 m^3/h per m^2 of envelope area at 50 pascal (Pa) pressure difference. However, it is envisaged that will improve in the future owing to the UK's commitment to higher energy-efficiency and lower carbon emissions.

The three main controllable ventilation methods are listed in Table 6.15.

Table 6.15 Ventilation methods

Method	Type	Why used	Remarks
Extract ventilation	Intermittent extract fans	In rooms where most water vapour and/or pollutants are released (e.g. cooking, bathing or photocopying).	This extract may be either intermittent or continuous and is aimed at minimizing the spread of vapour and pollutants to the rest of the building.
Whole building ventilation	Trickle ventilators	To provide fresh air to the building, dilute and disperse residual water vapour/pollutants not dealt with by extract ventilation, and remove water vapour and pollutants released by building materials, furnishings, activities and the presence of occupants.	This type of ventilation provides continuous air ventilation exchange, with a ventilation rate that can be reduced or ceased when the building is not occupied. In some cases (e.g. when the building is reoccupied) it may be necessary to purge the air (see below).
Purge ventilation	Windows	To assist in the removal of high concentrations of pollutants and water vapour released from occasional activities (e.g. painting and decorating) or accidental releases (e.g. smoke from burnt food or water spillage).	Purge ventilation is intermittent and may be used to improve thermal comfort and/or overheating in summer (see Approved Documents L1A (New dwellings) and L2A (New buildings other than dwellings)).

6.2.2.8.1 Purge ventilation

Purge ventilation is a manually controlled type of ventilation that is used in rooms and spaces to rapidly dilute pollutants and/or water vapour. It can be achieved by natural means (such as an openable window or an external door) or by mechanical means (e.g. a fan).

 Note: For further guidance on purge ventilations, see BS 5925:1991, *Code of Practice for Ventilation Principles and Designing for Natural Ventilation.*

6.2.2.8.2 Passive stack ventilation

Passive stack ventilation (PSV) is a ventilation device which uses ducts from terminals mounted in the ceiling of rooms to terminals on the roof, to extract air to the outside by a combination of the natural stack effect and the pressure effects of wind passing over the roof of the building.

The so-called 'stack effect' relies on the pressure differential between the inside and the outside of a building caused by differences in the density of the air due to an indoor–outdoor temperature difference.

Table 6.16 Passive stack ventilation

Room	Internal duct diameter (mm)	Internal cross-sectional area (mm)
Kitchen	125	12,000
Utility room	100	8,000
Bathroom	100	8,000
Sanitary accommodation	80	5,000

For sanitary accommodation only, purge ventilation may be used provided that security is not an issue.

6.2.2.8.2.1 PSV DESIGN

In designing PSV systems, the following requirements shall be met.

Dwellings which only have a single exposed façade should be designed so that the habitable rooms are on the exposed façade and are capable of achieving cross-ventilation.	F1 Table 5.2b
PSV extract terminals should be located in the ceiling or, alternatively, on a wall that is less than 400 mm below the ceiling.	F1 Table 5.2b

Figure 6.15 PSV offset requirements.

 Instead of PSV, an open-flued appliance may provide sufficient extract ventilation.

Background ventilators should be located in all rooms   F1 Table 5.2b
with external walls except the rooms where a PSV is
located.

The design and installation of PSV systems are crucial to their operation, and Figure 6.16 shows the preferred option for kitchen and bathroom ducts with ridge terminals.

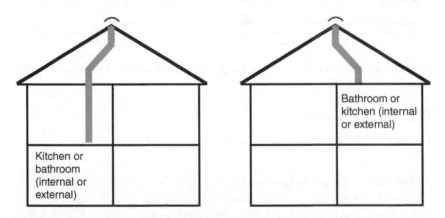

Bathroom or
kitchen (internal
or external)

Kitchen or
bathroom
(internal or
external)

Figure 6.16 Preferred PSV system layouts.

Another option (see Figure 6.17) is to have the kitchen and bathroom ducts penetrating the roof and extend their terminals to ridge height.

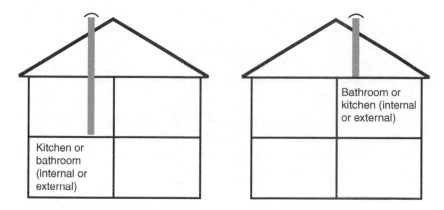

Bathroom or kitchen (internal or external)

Kitchen or bathroom (Internal or external)

Figure 6.17 Alternative PSV system layouts.

6.2.2.8.2.2 LOCATION OF PSVS

PSV extract terminals should be located in the ceiling or, if on a wall, less than 400 mm below the ceiling.	F
Background ventilators should **not** be within the same room as a PSV terminal.	F
If a PSV is located in a protected stairway of a dwelling, it shall not allow smoke or fire to spread into the stairway.	F
The duct length should be just sufficient to fit between the ceiling grille and the outlet terminal.	F
Flexible ducting should be fully extended but not taut.	F

 Note: Allow approximately 300 mm extra to make smooth bends in an offset system.

Ducting should:	F

- be properly supported along its entire length (remembering that flexible ducting generally requires more support than rigid ducting);
- be run straight without any distortion or sagging;
- not have any kinks at bends or connections with ceiling grilles and outlet terminals;
- be securely fixed to the roof outlet terminal so that it cannot sag or become detached.

In roof spaces: F

- ducts should, ideally, be secured to a wooden strut that is securely fixed at both ends;
- flexible ducts should be allowed to curve gently at each end of the strut.

For stability, rigid ducts should be used for any outside part   F
of the PSV system that is above the roof slope. To provide
stability, they should also project down into the roof space far
enough to allow firm support.

6.2.2.8.2.3 PSV controls

All PSV controls: F1 Table 5.2b

- should be set up to operate without occupant intervention;
- should have automatic controls (e.g. sensors for humidity, occupancy/usage, pollutant release);
- should ensure that humidity controls are not used for sanitary accommodation, as odour is the main pollutant.
- in kitchens, should provide sufficient flow during cooking with fossil fuel (e.g. gas) to avoid build-up of combustion products.

Ensure that background ventilators may be either
manually adjustable or automatically controlled.

6.2.2.9 PSV terminals and ducts

Approved Document F includes the following recommendations for terminals
and ducts associated with ventilation:

Common outlet terminals and/or branched ducts shall **not** be used   F
for wet rooms (e.g. the kitchen, bathroom, utility room and/or
WCs).
Ducts should have no more than one offset (i.e. bend) and ideally
these should be 'swept' at an angle of no more than 40° to the
vertical.
If a duct penetrates the roof more than 0.5 m from the roof ridge,
then it must extend above the roof slope to at least the height of the
roof ridge.

If tile ventilators are used on the roof slope they must be positioned no more than 0.5 m from the roof ridge.

Separate ducts shall be taken from the ceilings of wet rooms to separate terminals on the roof.

Ceiling extract grilles should have a free area, not less than the duct cross-sectional area (when in the fully open position if adjustable).

Ducts should be insulated in the roof space and other unheated areas with at least 25 mm of a material having a thermal conductivity of 0.04 W/m K.

If a duct extends above the roof level, then that section of the duct should be insulated or be fitted with a condensation trap just below roof level.

If a conversion fitting is required to connect the duct to the terminal then the duct cross-section area must be maintained (or exceeded) throughout the conversion fitting.

PSVs for dwellings that are situated near a significantly taller building (i.e. more than 50 per cent taller) should be at least five times the difference in height away from the taller building (i.e. if the difference in height is 10 m, then the PSV should not be installed in a dwelling within 50 m of the taller building).

Outlet terminals should have a free area that is not less than the duct cross-section area.

Terminals should not allow ingress of large insects or birds and should be designed so that rain is not likely to enter the duct and run down into the dwelling.

Terminals should be designed so that any condensation forming inside them cannot run down into the dwelling but will run off on to the roof.

 Note: Terminals should have an overall static pressure loss (upstream duct static minus test room static) equivalent to no more than four times the mean duct velocity pressure when measured at a static pressure difference of 10 Pa.

Ventilation ducts supplying or extracting air directly to or from a protected stairway should **not** also serve other areas.	B1 2.17 (V1) B1 2.18 (V2)
Ventilation ducts supplying or extracting air directly to or from a protected stairway or entrance hall should not also serve other areas.	B1 2.18 (V2)
Separate ventilation systems should be provided for each protected stairway.	B1 5.47 (V2)

If a ventilation duct passes through a compartment wall or compartment floor (or is built into a compartment wall) each wall of the duct should have a fire resistance of at least half that of the wall or floor (see Figure 6.18).

B3 7.11 (V1)
B3 8.34 (V2)

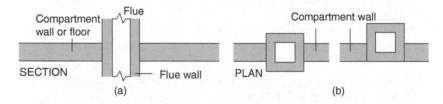

Figure 6.18 Flues penetrating compartment walls or floors.

6.2.2.9.1 General requirements

To ensure good transfer of air throughout the dwelling there shall be an undercut of 7600 mm² (minimum) in all internal doors above the floor finish (equivalent to an undercut of 10 mm for a standard 760 mm width door).

F Tables 5.2a to 5.2d

Adequate replacement air must also be available (e.g. a 10 mm gap under the door or equivalent).

F Tables 5.2a to 5.2d

All ducting that passes through a fire-stopping wall or fire compartment shall meet the requirements of Approved Document B of the Building Regulations.

F Tables 5.2a to 5.2d

Duct runs should be straight, with as few bends and kinks as possible to minimize system resistance.

F Tables 5.2a to 5.2d

Horizontal ducting (including ducting in walls) should be arranged to slope slightly downwards away from the fan to prevent backflow of any moisture.

F Tables 5.2a to 5.2d

Fans and/or ducting placed in, or passing through, an unheated void or loft space should be insulated to reduce the possibility of condensation forming.

F Tables 5.2a to 5.2d

The inner radius of any bend should be greater than or equal to the diameter of the ducting being used (see Figure 6.19).

F Tables 5.2a to 5.2d

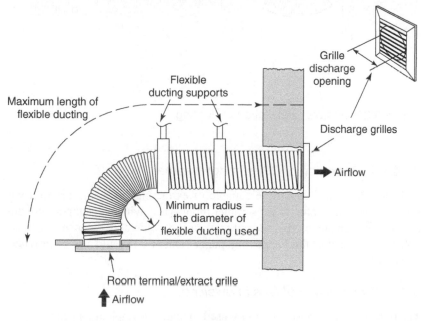

Figure 6.19 Correct installation of ducting.

Vertical duct rises may need to be fitted with a condensation trap in order to prevent the backflow of any moisture.

F Tables 5.2a to 5.2d

The circular profile of a flexible duct should be maintained throughout the full length of the duct run (see Figure 6.19).

F Tables 5.2a to 5.2d

If a back-draught device is used it may be incorporated into the fan itself.

F Tables 5.2a to 5.2d

Flexible ducting should be installed without any peaks or troughs (see Figure 6.20).

F Tables 5.2a to 5.2d

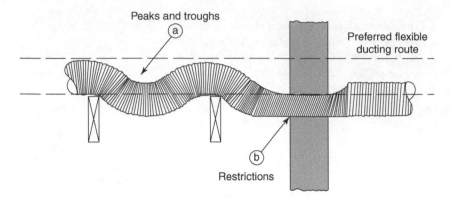

Figure 6.20 Incorrect installation of ducting.

6.2.2.9.2 Operation of PSVs in hot weather

Although PSV units should be capable of extracting sufficient air from wet rooms during the winter, in the summer months (i.e. when the temperature difference between the internal and external air is considerably reduced) they may not. To guard against this happening, purge ventilation should also be provided in wet rooms.

6.2.2.10 Installation of fans in dwellings

The three fan types most commonly used in domestic applications are:

• axial fans;
• centrifugal fans;
• in-line fans.

6.2.2.10.1 Axial fans

The axial fan is the most common form of fan, and can be mounted on the wall, on the window (i.e. through a suitable glazing hole) or in the ceiling (e.g. in a bathroom).

For wall and window mounting applications up to 350 mm thick, use a short length of rigid round duct or a flexible duct pulled taut.

For bathrooms, 100 mm diameter fans can be used as axial fans in the ceiling with a short (1.5 m maximum) length of flexible duct with (a maximum) two 90° bends.

Note: The duct must be pulled taut and the discharge terminal should have at least 85 per cent free area of the duct diameter.

6.2.2.10.2 Centrifugal fans

Centrifugal fans (because they develop greater pressure) permit longer lengths of ducting to be used, and so can be utilized for most wall and/or window applications in high-rise (i.e. above three storeys) buildings or in exposed locations to overcome wind pressure.

Most centrifugal fans are designed with 100 mm diameter outlets, which enables them to be connected to a wide variety of duct types.

Wall- or ceiling-mounted centrifugal fans that are designed to achieve 60 l/s for kitchens should not be ducted more than 3 m and should have no more than one 90° bend. Fans designed to achieve 15 l/s for bathrooms should not be ducted more than 6 m and should have no more than two 90° bends.

6.2.2.10.3 In-line fans

There are two types of in-line fan available:

*   in-line axial fans, which have to be installed with the shortest possible duct length to the discharge terminal;
*   in-line mixed flow fans, which have the characteristics of both axial and centrifugal fans and can, therefore, be used with longer lengths of ducting.

Both types can be used for bathrooms (100 mm diameter), utility rooms (125 mm diameter) and kitchens (150 mm diameter).

6.2.2.10.4 Fan terminals

When installing fans:

> Ensure that the free area of the grill opening of a room terminal   F
> extract grille and/or discharge terminal has a minimum of 85 per
> cent of the free area of the ducting being used.

Note: In these cases (only), the equivalent area may be assumed to be equal to the free area.

6.2.2.11 Background ventilators and intermittent extract fans

The need for background ventilators will depend on the air permeability or air-tightness of a building, where air permeability is defined as *the average volume of air (in cubic metres per hour) that passes through unit area of the building envelope (in square metres) when subject to an internal to external pressure difference of 50 Pa.*

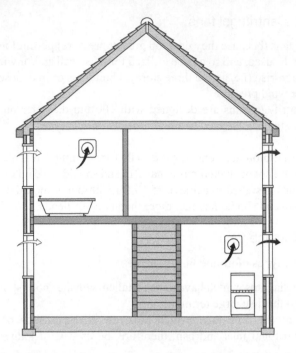

Figure 6.21 Background ventilators and intermittent extract fans.

6.2.2.11.1 Intermittent extract fans

- Intermittent extract fans may be operated manually and/or automatically by a sensor (e.g. humidity, occupancy/usage, pollutant release).   F1 Table 5.2a
- Humidity controls should not be used for *sanitary accommodation*, as odour is the main pollutant.
- In kitchens, any *automatic control* must provide sufficient flow during cooking with fossil fuel (e.g. gas) to avoid build-up of combustion products.
- Any automatic control should have a manual override to allow the occupant to turn the extract on.
- In a room with no openable window (i.e. an internal room) an intermittent extract fan should have a **15-minute overrun**.

In rooms with no natural light, the fans could be controlled by the operation of the main room light switch.

Minimum extract air flow rates for intermittent extract fans should be greater than those shown in Table 6.17.	F Appendix A

Table 6.17 Extract ventilation rates

Room	Minimum intermittent extract rate	Continuous extract	
		Minimum high rate	Minimum low rate
Kitchen	30 l/s (adjacent to hob) or 60 l/s elsewhere	13 l/s	Total extract rate must be at least the whole building ventilation rate shown in Table 6.12
Utility room	30 l/s	8 l/s	
Bathroom	15 l/s	8 l/s	
Sanitary accommodation	6 l/s	6 l/s	

 Note: For sanitary accommodation, a purge ventilation system may be used, and in wet rooms a heat recovery ventilator may be used instead of a conventional fan, provided that it has the same extract rate.

Background ventilators may be either manually adjustable or automatically controlled.	F1 Table 5.2a
Manual controls should be within reasonable reach of the occupants.	F1 Table 5.2a
Fans should be quiet so as not to discourage their use by occupants.	F1 Table 5.2a
Background ventilators for dwellings with a single exposed façade should be located at both high (typically 1.7 m above floor level) and low (i.e. at least 1.0 m below the high ventilators) positions in the façade (see Figure 6.22).	F1 Table 5.2a

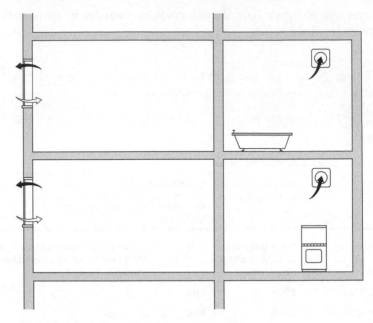

Figure 6.22 Single-sided ventilation.

Dwellings with only a single exposed façade should be designed so that the habitable rooms are on the exposed façade in order to achieve cross-ventilation.	F Table 5.2b
Background ventilators should be at least 0.5 m from an extract fan.	F
Background ventilators should be located so as to avoid draughts (e.g. typically 1.7 m above floor level).	F

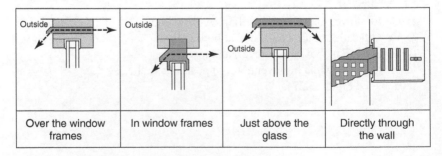

Figure 6.23 Background ventilation systems.

Controllable background ventilators with a minimum equivalent area of 2500 mm^2 shall be fitted in each room (except wet rooms from which air is extracted).	F Table 5.2c
Background ventilators may be manually adjustable or automatically controlled.	F Table 5.2c
Windows with night latches should **not** be used, as they are more liable to draughts as well as being a potential security risk.	F

6.2.2.11.2 Location of background ventilators in rooms

Background ventilators should be located to avoid draughts (e.g. typically 1.7 m above floor level).	F1 Table 5.2a
Background ventilators should be located in all rooms with external walls, with at least 5000 mm^2 equivalent area in each habitable room and 2500 mm^2 equivalent area in each wet room.	F1 Table 5.2a
To ensure good transfer of air throughout the dwelling there should be an undercut of minimum area 7600 mm^2 in all internal doors above the floor finish.	F1 Table 5.2a

6.2.2.12 Trickle ventilators

Manually controlled trickle ventilators are widely used for background ventilation and these can be located as shown in Figure 6.23.

To avoid cold draughts, trickle ventilators are normally positioned 1.7 m above floor level, and usually include a simple control (such as a flap) to allow users to shut off the ventilation according to personal choice or external weather conditions. Nowadays, pressure-controlled trickle ventilators that reduce the air flow according to the pressure difference across the ventilator are available to reduce draught risks during windy weather.

Trickle ventilators are normally left open in occupied rooms in dwellings.

Trickle ventilators that include an automatic control should be capable of being manually overridden so that they can be opened by the occupant when required.	F Table 5.2c
Pressure-controlled trickle ventilators that, under normal conditions, are left open (e.g. 1 Pa pressure difference) should **only** be capable of being manually closed.	F 4.20

| Trickle ventilators etc. should be clearly marked with their equivalent area (measured according to BS EN 13141-1:2004) by means of either a stamp or an indelibly printed self-adhesive label. | F 4.26 |
| All fans should operate quietly at their minimum (i.e. normal) rate so as not to disturb the occupants of the building. | F Table 5.2c |

6.2.2.13 Continuous mechanical extract

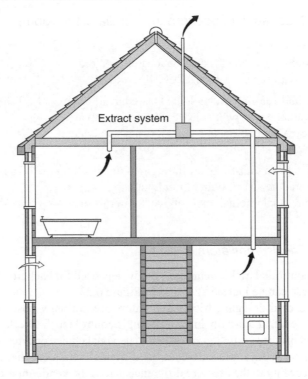

Figure 6.24 Continuous mechanical extract.

This system may consist of either a central extract system or individual room fans, or a combination of both.

To calculate the required extract rate, first determine the **whole building ventilation rate** from Table 6.18.

Table 6.18 Whole building ventilation rates

	Number of bedrooms				
Whole building ventilation rate (l/s)	1	2	3	4	5
	13	17	21	25	29

Then work out the **whole dwelling air extract rate** (at maximum operation) by summing the individual room rates from Table 6.19.

Table 6.19 Whole dwelling air extract rate

Room	Minimum intermittent extract rate (l/s)	Minimum high continuous extract rate (l/s)
Kitchen	30 (adjacent to hob) 60 elsewhere	13
Utility room	30	8
Bathroom	15	8
Sanitary accommodation	6	6

For sanitary accommodation **only**, purge ventilation may be used provided that security is not an issue.

6.2.2.13.1 Continuous mechanical extract – requirements

> Mechanical extract terminals and fans should be installed as high as is practicable and preferably less than 400 mm below the ceiling.   F1 Table 5.2c

 Where ducts, etc. are provided in a dwelling with a protected stairway, precautions may be necessary to avoid the possibility of the system allowing smoke or fire to spread into the stairway.

6.2.2.13.1.1 AIR TRANSFER

> To ensure good transfer of air throughout the dwelling, there should be an undercut of minimum area 7600 mm^2 in all internal doors above the floor finish.   F1 Table 5.2c

6.2.2.13.1.2 CONTROLS

Controls should be set up to operate without occupant intervention, but may have manual or automatic controls to select the boost rate.

> Any manual boost controls should be provided locally to the spaces being served (e.g. bathrooms and kitchen), as provision of a single centrally located switch may result in fans being left in an inappropriate mode of operation.   F1 Table 5.2c

Automatic controls could include sensors for humidity, occupancy/usage and pollutant release.

Humidity controls should not be used for sanitary accommodation, as odour is the main pollutant.

In kitchens, any automatic control must provide sufficient flow during cooking with fossil fuel (e.g. gas) to avoid build-up of combustion products.	F1 Table 5.2c
Where manual controls are provided, they should be within reasonable reach of the occupants.	F1 Table 5.2c

6.2.2.13.1.3 OPERATION

Extract should be from each wet room.	F1 Table 5.2d
Air should normally be supplied to each habitable room.	F1 Table 5.2d
The total supply air flow should usually be distributed in proportion to the habitable room volumes.	F1 Table 5.2d
Recirculation by the system of moist air from the wet rooms to the habitable rooms should be avoided.	F1 Table 5.2d
Cooker hoods should be 650 mm to 750 mm above the hob surface (or follow the manufacturer's instructions).	F1 Table 5.2d
Mechanical extract terminals and fans should be installed as high as is practical and preferably less than 400 mm below the ceiling.	F1 Table 5.2d
Mechanical supply terminals should be located and directed to avoid draughts.	F1 Table 5.2d
Where ducts, etc. are provided in a dwelling with a protected stairway, precautions may be necessary to avoid the possibility of the system allowing smoke or fire to spread into the stairway.	F1 Table 5.2d

The maximum ('boost') rate should be the greater of the whole building ventilation rate and the whole dwelling air extract rate.	F Table 5.2c

The maximum individual room extract rates should be at least those given in Table 6.19.   F Table 5.2c

The minimum air supply rate should be at least the whole building ventilation rate.   F Table 5.2c

Note: Extract terminals located on the prevailing windward façade should be protected against the effects of wind by using ducting to another façade, using a constant-volume flow rate unit or a central extract system.

All fans should operate quietly at their minimum (i.e. normal) rate so as not to disturb the occupants of the building.   F Table 5.2c

Ventilation devices designed to work continuously: F Tables 5.2a to 5.2d

*   shall be set up to operate without occupant intervention;
*   may have a manual control to select maximum 'boost';
*   may have automatic controls such as humidity control (but not if used for sanitary accommodation), occupancy/usage sensor, moisture/pollutant release detector, etc.

Automatic controls for ventilators that are designed to work continuously in kitchens must be capable of providing sufficient flow during cooking with fossil fuels (e.g. gas) so as to avoid the build-up of combustion products.   F Tables 5.2a to 5.2d

6.2.2.14 Continuous mechanical supply and extract with heat recovery

To calculate the air flow rate of a building using continuous mechanical supply and extract with heat recovery (MVHR), first determine the **whole building ventilation rate** from Table 6.18; then, depending on whether it is a multi-storey or single-storey building, subtract the gross internal volume of dwelling heated space (m^2) as follows:

Multi-storey dwelling   Whole building ventilation rate $- 0.04 \times$ gross internal volume

Single-storey dwelling   Whole building ventilation rate $- 0.06 \times$ gross internal volume

Next, work out the whole dwelling air extract rate at maximum operation by summing the individual room rates from Table 6.19.

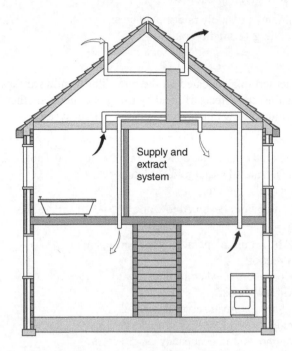

Supply and extract system

Figure 6.25 Continuous mechanical supply – with heat recovery.

6.2.2.15 Single room heat recovery ventilator

If a single room heat recovery ventilator (SRHRV) is used to ventilate a habitable room, to calculate the air flow rate, first determine the **whole building ventilation rate** from Table 6.18, and then work out the room supply rate using the following formula:

$$\frac{\text{Whole building ventilation rate} \times \text{room volume}}{\text{Total volume of all habitable rooms}}$$

When working out the continuous mechanical extract for a whole building, which also includes a room ventilated by an SRHRV, the following formula should be used:

$$\frac{\text{Whole building ventilation rate} \times \text{room volume} - \text{SRHRV rate}}{\text{Total volume of all habitable rooms}}$$

6.2.2.16 Mechanical intermittent extract

As odour is the main pollutant, humidity controls should **not** be used for intermittent extract in sanitary accommodation.	F Tables 5.2a
Ventilators equipped with intermittent extract shall be capable of being operated manually and/or automatically by a sensor (e.g. humidity sensor, occupancy/usage sensor, moisture/pollutant release detector, etc.).	F Tables 5.2a to 5.2d
All ventilator automatic controls must be provided with a manual override to allow the occupant to turn the extract on.	F Tables 5.2a to 5.2d
Automatic controls for ventilators used in kitchens must be capable of providing sufficient flow during cooking with fossil fuels (e.g. gas) so as to avoid the build-up of combustion products.	F Tables 5.2a to 5.2d
If a fan is installed in an internal room without an openable window, then the fan should have a 15-minute overrun.	F Tables 5.2a to 5.2d
In rooms with no natural light, fans could be controlled by the operation of the main room light switch.	F Tables 5.2a to 5.2d

Note: In dwellings, humidistat controls should be available to regulate the humidity of the indoor air and to minimize the risk of condensation and mould growth. Humidistats are normally installed as part of an extract ventilator, especially in moisture-generating rooms such as a kitchen or a bathroom. They should **not** be used for sanitary accommodation, where the dominant pollutant is usually odour.

6.2.2.17 Ventilation systems – dwellings without basements

The following systems may be used in dwellings without basements:

- background ventilators and intermittent extract fans;
- continuous mechanical extract (MEV);
- continuous mechanical supply and extract with heat recovery (MVHR);
- passive stack ventilation (PSV).

6.2.2.18 Ventilation systems – basements

If a basement is connected to the rest of the dwelling by a   F1 5.11
large permanent opening, such as an open stairway, then the
whole dwelling, including the basement, should be treated as
a multi-storey dwelling and ventilated in a similar manner to
dwellings without basements.

If the basement has a single exposed façade, while the rest   F1 5.11
of the dwelling above ground has more than one exposed
façade, then passive stack ventilation or continuous
mechanical extract should be used.

For basements that are **not** connected to the rest of the   F1 5.12
dwelling by a large permanent opening, then:

- the part of the dwelling above ground should be
  considered separately;
- the basement should be treated as if it were a single-
  storey dwelling above ground.

If the part of the dwelling above ground has no bedrooms,   F1 5.12
then for the purpose of ventilation requirements:

- assume that the dwelling has one bedroom; and
- treat the basement as a single-storey dwelling (with one
  bedroom) as if it were above ground.

If a dwelling only compromises a basement, then it should   F1 5.13
be treated as if it were a single-storey dwelling (with one
bedroom) above ground.

6.2.2.19 Ventilation of habitable rooms through another room or a conservatory

Habitable rooms without an openable window shall be either   F1 5.11
ventilated through another habitable room or through a
conservatory.

6.2.2.20 Ventilation through another room

Habitable rooms without an openable window may be   F1 5.15
ventilated through another habitable room (see Figure
6.26) provided that the other room has:

- purge ventilation; and
- an 8000 mm^2 background ventilator; and

there is a permanent opening between the two rooms.

Table 6.20 Ventilation systems for basements

Type of basement	Background ventilators and intermittent extract fans	Passive stack ventilation	Continuous mechanical extract	Continuous mechanical supply and extract with heat
Basement connected to the rest of the dwelling by an open stairway	Yes	Yes	Yes	Yes
Basement with a single exposed façade and dwelling above ground with more than one exposed façade		Yes	Yes	
Basements not connected to the rest of the dwelling by an open stairway	Yes	Yes	Yes	Yes
Dwelling above ground has no bedrooms	Yes	Yes	Yes	Yes
Dwelling comprises just a basement	Yes	Yes	Yes	Yes

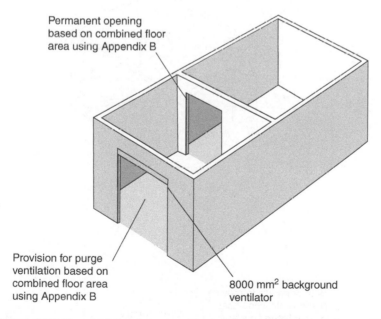

Permanent opening based on combined floor area using Appendix B

Provision for purge ventilation based on combined floor area using Appendix B

8000 mm² background ventilator

Figure 6.26 Two habitable rooms treated as a single room for ventilation purposes.

6.2.2.21 Ventilation through a conservatory

Habitable rooms without an openable window may be   F1 5.16
ventilated through a conservatory (see Figure 6.27) provided
that that conservatory has:

- purge ventilation; and
- an 8000 mm² background ventilator; and

there is a closable opening between the room and the
conservatory that is equipped with:

- purge ventilation; and
- an 8000 mm² background ventilator.

 Note: If a building contains both living accommodation and space to be used
for commercial purposes (e.g. workshop or office), the whole unit should be
treated as a dwelling for the purposes of this Approved Document, as long as
the commercial part could revert to domestic use.

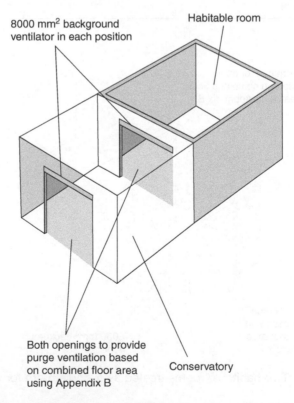

8000 mm² background
ventilator in each position

Habitable room

Both openings to provide
purge ventilation based
on combined floor area
using Appendix B

Conservatory

Figure 6.27 A habitable room ventilated through a conservatory.

6.2.2.22 The effects of solar gains in summer

Solar gains are certainly beneficial in winter, as they reduce the demand for heating. However, they can cause overheating in the summer.

To overcome this problem (and whenever possible) during the summer, the effects of solar gain should be limited by a suitable combination of window size and orientation, shading, solar control measures, ventilation (day and night) and high thermal capacity.

If ventilation is provided using a balanced mechanical system, consideration should be given to providing a summer bypass function for use during warm weather (or allow the dwelling to operate via natural ventilation) so that the ventilation is more effective in reducing overheating.

Although many buildings are equipped with air-conditioning for comfort, they are high energy users and their use should, whenever possible, be limited. This can be achieved by either reducing the need for air-conditioning or reducing the installed capacity of any air-conditioning system that is installed.

6.2.2.23 Ventilation systems – buildings other than dwellings

Ventilation in buildings **other than** dwellings is dependent on occupancy levels, and currently there are some very sophisticated automatic control systems available, such as local passive infrared detectors and electronic carbon dioxide detectors.

> Fresh air supplies should be protected from contaminants that would be injurious to health. F1 6.3

6.2.2.24 Ventilation systems – offices

> All office sanitary accommodation, washrooms and food and beverage preparation areas shall be provided with intermittent air extract ventilation capable of meeting the requirements of Table 6.21. F1 6.10

Table 6.21 Extract ventilation rate

Room	Air extract rate
Rooms containing printers and photocopiers in substantial use (greater than 30 minutes per hour)	20 l/s per machine during use.
Office sanitary accommodation and washrooms	15 l/s per shower/bath 6 l/s per WC/urinal.
Food and beverage preparation areas (not commercial kitchens)	15 l/s with microwave and beverages only; 30 l/s adjacent to the hob with cooker(s); 60 l/s elsewhere with cooker(s).
Specialist buildings and spaces (e.g. commercial kitchens, fitness rooms)	See Table 6.3 of Approved Document F: 2010, pages 36–37.

Extract fans that are located in an internal room which does not have an openable window should have a 15-minute overrun.	F1 Table 6.2c
Extract ventilators should be located as high as practicable and preferably not less than 400 mm below the ceiling level.	F1 Table 6.2b
PSVs should be located in the ceiling of the room.	F1 Table 6.2b
Purge ventilation shall be provided in each office.	F1 6.12
Purged air should be taken directly to outside and should not be recirculated to any other part of the building.	F1 6.12
PSVs can be used as an alternative to a mechanical extract fan for office sanitary, washroom and food preparation areas.	F1 Table 6.2a
PSV controls can be either manual or automatic.	F1 Table 6.2c
The controls for extract fans can be either manual or automatic.	F1 Table 6.2c
Printers and photocopiers that are being used in large numbers and which are in almost constant use (i.e. greater than 30 minutes per hour) shall: • be located in a separate room; • have extract facilities capable of providing an extract rate greater than 20 l/s per machine, during use (see Table 6.21).	F1 6.10
The whole building ventilation rate for the supply of air to the offices should be greater than 10 l/s per person (see Table 6.22).	F1 6.11

Table 6.22 Whole building ventilation rate for air supply to offices

	Air supply rate
Total outdoor air supply rate for offices (no-smoking and no significant pollutant sources)	10 l/s per person

The following air flow rates can mainly be provided by natural ventilation.

6.2.2.25 Ventilation systems – other types of building

The ventilation requirements for other buildings (such as assembly halls, broadcasting studios, computer rooms, factories, hospitals, hotels, museums, schools, sports centres, warehouses, etc.) are listed in Table 6.3 of Approved

Document F, which also provides a link to the relevant controlling Acts of Parliament, statutory instruments, British Standards, and Chartered Institution of Building Services Engineers (CIBSE) and Health and Safety Executive (HSE) standards, practices and recommendations.

6.2.2.26 Ventilation systems – car parks

Ventilation is the important factor and, as heat and smoke cannot be dissipated so readily from a car park that is not open-sided, fewer concessions are made.

Underground car parks, enclosed car parks and multi-storey car parks should be designed to limit the concentration of carbon monoxide to not more than 30 parts per million (ppm) averaged over an 8-hour period and peak concentrations.	F1 6.18a
Ramps and exits shall not go above 90 ppm for periods not exceeding 15 minutes.	F1 6.18b
Naturally ventilated car parks shall have openings at each car parking level: • at least 1/20 of the floor area at that level; • with a minimum of 25 per cent on each of two opposing walls.	F1 6.19a
Mechanically ventilated car parks can have either natural ventilation openings that are not less than 1/40 of the floor area or a mechanical ventilation system capable of at least three air changes per hour (3 ach).	F1 6.2bi
Mechanically ventilated basement car parks shall be capable of at least six air changes per hour (6 ach).	F1 6.2bi
Mechanically ventilated exits and ramps (i.e. where cars queue inside the building with engines running) shall be capable of at least ten air changes per hour (10 ach).	F1 6.2bii

6.2.2.27 Access for maintenance

There should be reasonable access to ventilation systems to enable the changing of filters, the replacing of defective components, the cleaning of duct work and other maintenance activities.	F Tables 5.2a to 5.2d

Where there is a danger of the operator or other person falling through a window above ground-floor level, suitable opening limiters should be fitted or guarding should be provided.

Buildings other than dwellings should include: F 6.6

•   reasonable access for the purpose of replacing filters,
    fans and coils; and
•   availability of access points for cleaning duct work.

Central plant rooms should include adequate space for the   F 6.7
maintenance of the plant (see Figure 6.28).

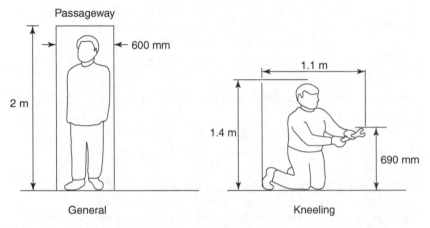

Figure 6.28 Access space in central plant rooms.

6.2.2.27.1 Combustion appliances

A room containing an open-flued appliance may need   J 1.4
permanently open air vents.

If open-flued combustion appliances and extract fans are   F 1.3
going to be installed, then the combustion appliance should
be capable of operating safely – whether or not the fans are
running.

6.2.2.27.2 Exhaust outlets

Exhaust outlets should be located so that re-entry, or ingestion, into the building and/or other nearby buildings is minimized. This can be achieved by ensuring that:

- exhausts are located downstream of air   F Tables 5.2a to 5.2d
  intakes which are located in a prevailing
  wind direction;
- exhausts do not discharge into courtyards,
  enclosures or architectural screens;
- stacks discharge vertically upwards with
  sufficient height to clear surrounding
  buildings and avoid a downwash occurring.

Note: Where possible, pollutants from stacks should be grouped together and discharged vertically upwards.

6.2.2.27.3 Windows

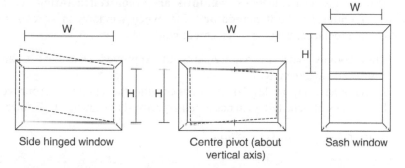

Side hinged window   Centre pivot (about   Sash window
                          vertical axis)

Figure 6.29 Window dimensions.

Note: The window opening area is the dimensions of the open area (i.e. height (H) × width (W)).

The height times the width of the opening part of hinged   F App B
or pivot windows that are designed to open **more** than 30°
and/or sliding sash windows should be at least 1/20 of the
floor area of the room.

The height times the width of the opening part of hinged or   F App B
pivot windows designed to open **less** than 30° should be at
least 1/10 of the floor area of the room.

If a room contains more than one openable window, then the areas of **all** the opening parts may be added together to achieve the required floor area.

If the room contains more than one openable window, the areas of all the opening parts may be added to achieve the required proportion of the floor area.

6.2.2.27.4 Protected shaft

A protected shaft conveying piped flammable gas should have ventilation openings to the outside air at high and low level in the shaft.	B38.41 (V2)

Note: Generally speaking, an external wall of a protected shaft does not need to have fire resistance (but see BS 5588-5:2004 for fire resistance of external walls of firefighting shafts).

6.2.2.28 Work on existing buildings

Under the Building Regulations, **windows are a controlled fitting**. These clauses, therefore, make it **mandatory** that, when windows in an existing building are replaced, then the replacement work:

- **shall** comply with the requirements of Approved Document L and Approved Document N;
- **shall not** have a worse level of compliance with other applicable Approved Documents of Schedule 1 (in particular Approved Documents B, F and J).

6.2.2.29 Replacement windows

All replacement windows should include trickle ventilators or have an equivalent background ventilation opening in the same room.	F 7.4
Ventilation openings should not be smaller than the original opening and they should be controllable.	F 7.6
Where there was no previous ventilation opening, or where the size of the original ventilation opening is not known, the replacement window(s) shall be greater than the minimum requirements shown in Tables 6.23 and 6.24.	F 7.6

Table 6.23 Equivalent areas for replacement windows – dwellings

Type of room	Equivalent area
Habitable rooms	5000 mm²
Kitchen	2500 mm²
Utility room	2500 mm²
Bathroom (without a WC)	2500 mm²

Table 6.24 Equivalent areas for replacement windows – buildings other than dwellings

Type of room	Equivalent area
Occupiable rooms with floor areas <10 m^2	2500 mm^2
Occupiable rooms with floor areas >10 m^2	250 mm^2 per m^2 of floor area
Kitchens (domestic type)	2500 mm^2
Bathrooms and shower rooms	2500 mm^2 per bath or shower
Sanitary accommodation (and/or washing facilities)	2500 mm^2 per WC

6.2.2.30 The addition of a habitable room

The general ventilation rates for an additional habitable room (not including a conservatory) to an existing building may be achieved by using background ventilators, heat recovery ventilators and/or purge ventilation.

A single room heat recovery ventilator may be used to ventilate an additional habitable room (F 7.8b).

6.2.2.30.1 Additional requirements for background ventilators

If the additional room is connected to an existing habitable room which now has no windows opening to outside, then the ventilation opening (or openings) shall be greater than 8000 mm^2 equivalent area.   F 7.8a(i)

If the additional room is connected to an existing habitable room which still has windows opening to outside, but with a total background ventilator equivalent area less than 5000 mm^2 equivalent area, then the ventilation opening (or openings) shall be greater than 8000 mm^2 equivalent area.   F 3.7a(ii)

If the additional room is connected to an existing habitable room which still has windows opening to outside, but with a total background ventilator equivalent area of at least 5000 mm^2 equivalent area, then there should be: F 3.7a(iii)

- background ventilators of at least 8000 mm^2 equivalent area between the two rooms; and
- background ventilators of at least 8000 mm^2 equivalent area between the additional room and outside.

6.2.2.31 The addition of a wet room to an existing building

Internal doors between the wet room and the existing building should have an undercut of at least minimum area 7600 mm^2 (equivalent to an undercut of 10 mm above the floor finish for a standard 760 mm width door).   F 7.13

Whole building and extract ventilation can be provided by: F 7.12

- intermittent extract and a background ventilator of at least 2500 mm^2 equivalent area; or
- single room heat recovery ventilator; or
- passive stack ventilator; or
- continuous extract fan.

6.2.2.32 The addition of a conservatory to an existing building

The general ventilation rate for a conservatory (and   F 7.16–7.18
adjoining rooms) with floor area greater than 30 m^2 can
be achieved by the use of background ventilators.

6.2.2.33 Refurbishing a kitchen or bathroom in an existing dwelling

If any of the work being carried out in the kitchen or bathroom   F 7.23
of an existing building is 'building work' (as defined in
Regulation 3 of the Building Regulations) and there is an
existing extract fan or passive stack ventilator (or cooker hood
extracting to outside in the kitchen) this will suffice (always
assuming that it is in good working order!).

6.2.2.34 Historic buildings

Ventilation systems should **not** introduce new or increased technical risk, or in any other way prejudice the use or character of the building – particularly historic buildings that are:

- listed buildings;
- buildings in Conservation Areas;
- buildings which are of architectural and historical interest;
- buildings which are of architectural and historical interest within National Parks, Areas of Outstanding Natural Beauty, historic parks and gardens, registered battlefields, the curtilages of scheduled ancient monuments, and World Heritage Sites; and
- buildings of traditional construction with permeable fabric that both absorbs and readily allows the evaporation of moisture, as these are exempt from compliance with the requirements of the Building Regulations.

 When undertaking work on, or in connection with, a building that falls within one of the classes listed above, the aim should be to provide adequate ventilation as far as is reasonable and practically possible – without damaging the character of the building or increasing the risk of long-term deterioration of the building fabric or fittings.

Many books have been written about the problems related to restoring historic buildings and, before considering any work of this nature, you would be advised to seek the advice of the local planning authority's conservation officer, particularly if you are contemplating:

- the restoration of a historic building that has been subject to previous inappropriate alteration (e.g. replacement windows, doors and rooflights);
- rebuilding a former historic building following a fire or major demolition;
- making the building's fabric able to 'breathe', in order to control moisture and potential long-term decay.

In **all** cases, the overall aim should be to improve ventilation of a historic building without:

• having a detrimental influence on the character of the building; • increasing the risk of long-term deterioration of the building's fabric or fittings.	F 3.11

6.2.2.35 Fire precautions

In dwellings with three or more storeys and blocks of flats, PSV ducts should not impede fire escape routes.

6.2.2.35.1 Stairs

If a stair serves a place that is a special fire hazard, the lobby or corridor should have not less than $0.4\,m^2$ permanent ventilation or be protected by a mechanical smoke control system.	B1 2.47(V2)

6.2.2.35.2 Ducting

Precautions must be taken where ducting passes through a fire-resisting wall, floor or fire compartment.	FI 4.37

6.2.2.35.3 External escape stairs

Protected lobbies (with less than $0.4\,m^2$ permanent ventilation) should be provided between an escape stairway and a place of special fire hazard.	B1 4.35 (V2)

6.2.2.35.4 Flues

If a flue (or a duct containing a flue and/or ventilation duct) passes through a compartment wall or compartment floor (or is built into a compartment wall) then each wall of the flue or duct should have a fire resistance of at least half that of the wall or floor. (See Figure 6.30.)	B3 10.16 (V2)

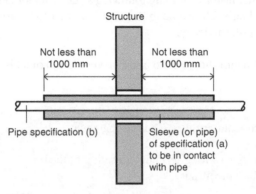

Notes:
1   Make the opening in the structure as small as possible and provide fire-stopping between pipe and structure.

Figure 6.30 Pipes penetrating structure.

6.2.2.35.5 Protected escape routes

Protected lobbies (with less than $0.4\,m^2$ permanent ventilation) should be provided between an escape stairway and a place of special fire hazard.	B1 4.35 (V2)
Ventilation ducts supplying or extracting air directly to or from a protected escape route should **not** also serve other areas.	B1 5.47 (V2)
Separate ventilation systems should be provided for each protected stairway.	B1 5.47 (V2)

6.2.2.35.6 Mechanical ventilation

Smoke control of common escape routes by mechanical ventilation is permitted provided that it meets the requirements of BS EN 12101-6:2005.	B1 2.27 (V2)

Mechanical ventilation systems should be designed to ensure that:	B1 5.46 (V2)
• ductwork does not assist in transferring fire and smoke through the building;	B1 2.18 (V2)
• exhaust points are sited away from final exits, combustible building cladding or roofing materials and openings into the building.	B3 10.2 (V2)

Note: Recirculated air (e.g. from an entrance hall or a stairway) should be designed to shut down on the detection of smoke within the system.

6.3 Drainage

6.3.1 Requirements

Foul water drainage
The foul water drainage system shall:

* *convey the flow-off foul water to a foul water outfall (i.e. sewer, cesspool, septic tank or settlement (i.e. holding) tank),*
* *minimize the risk of blockage or leakage,*
* *prevent foul air from the drainage system from entering the building under working conditions,*
* *be ventilated,*
* *be accessible for clearing blockages,*
* *not increase the vulnerability of the building to flooding.*

(Approved Document H1)

Wastewater treatment systems and cesspools
Wastewater treatment systems shall:

* *have sufficient capacity to enable breakdown and settlement of solid matter in the wastewater from the buildings;*
* *be sited and constructed so as to prevent overloading of the receiving water.*

Cesspools shall have sufficient capacity to store the foul water from the building until they are emptied.

Wastewater treatment systems and cesspools shall be sited and constructed so as not to:

* *be prejudicial to health or a nuisance;*
* *adversely affect water sources or resources;*
* *pollute controlled waters;*
* *be in an area where there is a risk of flooding.*

Septic tanks and wastewater treatment systems and cesspools are constructed and sited so as to:

- *have adequate ventilation;*
- *prevent leakage of the contents and ingress of subsoil water;*
- *have regard to water table levels at any time of the year and rising ground-water levels.*

Drainage fields are sited and constructed so as to:

- *avoid overloading of the soakage capacity, and*
- *provide adequately for the availability of an aerated layer in the soil at all times.*

(Approved Document H2)

Rainwater
Rainwater from roofs shall be carried away from the surface either by a drainage system or by other means.

The rainwater drainage system shall carry the flow of rainwater from the roof to an outfall (e.g. a soakaway, a watercourse, a surface water or combined sewer).

(Approved Document H3)

Rainwater drainage
Rainwater drainage systems shall:

- *minimize the risk of blockage or leakage;*
- *be accessible for clearing blockages;*
- *ensure that rainwater soaking into the ground is distributed sufficiently so that it does not damage foundations of the proposed building or any adjacent structure;*
- *ensure that rainwater from roofs and paved areas is carried away from the surface either by a drainage system or by other means;*
- *ensure that the rainwater drainage system carries the flow of rainwater from the roof to an outfall (e.g. a soakaway, a watercourse, surface water or a combined sewer).*

(Approved Document H3)

Site preparation and resistance to contaminants and water

- *The ground to be covered by the building shall be reasonably free from any material that might damage the building or affect its stability, including vegetable matter, topsoil and pre-existing foundations.*
- *Reasonable precautions shall be taken to avoid danger to health and safety caused by contaminants on or in the ground covered, or to be covered, by the building and any land associated with the building.*
- *Adequate subsoil drainage shall be provided if it is needed to avoid:*

○   *the passage of ground moisture to the interior of the building;*
○   *damage to the building, including damage through the transport of water-borne contaminants to the foundations of the building.*

(Approved Document C1)

 Amendments to the 2010 version of Approved Document C came into force in 2013.

Drains – fire protection
Drains should also provide a degree of fire protection as shown by the following requirement:

- *all openings in fire-separating elements shall be suitably protected in order to maintain the integrity of the continuity of the fire separation,*
- *any hidden voids in the construction shall be sealed and subdivided to inhibit the unseen spread of fire and products of combustion, in order to reduce the risk of structural failure, and the spread of fire.*

(Approved Document B3)

Building over existing sewers
Building or extension or work involving underpinning shall:

- *be constructed or carried out in a manner which will not overload or otherwise cause damage to the drain, sewer or disposal main either during or after the construction;*
- *not obstruct reasonable access to any manhole or inspection chamber on the drain, sewer or disposal main;*
- *in the event of the drain, sewer or disposal main requiring replacement, not unduly obstruct work to replace the drain, sewer or disposal main, on its present alignment;*
- *reduce the risk of damage to the building as a result of failure of the drain, sewer or disposal main.*

(Approved Document H4)

6.3.1.1 Other requirements

You are required by the Building Act 1984 to ensure that all courts, yards and passageways giving access to a house, industrial or commercial building (not maintained at public expense) are capable of allowing satisfactory drainage of its surface or subsoil to a proper outfall.

The local authority can require the owner of any of the buildings to complete such works as may be necessary to remedy the defect.

All plans for building work need to show that drainage of refuse water (e.g. from sinks) and rainwater (from roofs) has been adequately catered for. Failure to do so will mean that these plans will be rejected by the local authority. All plans for buildings must include at least one water or earth closet **unless** the local authority is satisfied that one is not required (e.g. in a large garage separated from the house).

If you propose using an earth closet, the local authority cannot reject the plans unless they consider that there is insufficient water supply to that earth closet.

6.3.1.2 Building Act requirements

The Building Act requires that all drains are either connected with a sewer (unless the sewer is more than 120 ft away or the person carrying out the building work is not entitled to have access to the intervening land) or able to discharge into a cesspool, settlement tank or other tank designed for the reception and/or disposal of foul matter from buildings.

The local authorities view this requirement very seriously and will need to be satisfied that:

*   satisfactory provision has been made for drainage;
*   all cesspools, private sewers, septic tanks, drains, soil pipes, rainwater pipes, spouts, sinks or other appliances are adequate for the building in question;
*   all private sewers that connect directly or indirectly to the public sewer are not capable of admitting subsoil water;
*   the condition of a cesspool is not detrimental to health, or does not present a nuisance;
*   cesspools, private sewers and drains previously used, but no longer in service, do not prejudice health or become a nuisance.

This requirement can become quite a problem if it is not recognized in the early planning stages and so it is always best to seek the advice of the local authority. In certain circumstances, the local authority might even help to pay for the cost of connecting you up to the nearest sewer!

The local authority has the authority to make the owner renew, repair or cleanse existing cesspools, sewers, drains, etc.

6.3.1.3 Can two buildings share the same drainage?

Usually the local authority will require every building to be drained separately into an existing sewer but in some circumstances it may decide that it would be more cost-effective if the buildings were drained in combination. This is common practice on terraced streets, where the houses drain into a common drain which crosses each property (usually at the rear) and then passes down the side of the end of the terrace to join with the sewer. In this case the 'common' pipework is the responsibility of the utilities provider. On occasions, the local authority might even recommend that a private sewer is constructed.

6.3.1.4 What happens if I need to disconnect an existing drain?

If, in the course of your building work, you need to:

*   reconstruct, renew or repair an existing drain that is joined up with a sewer or another drain;

- alter the position of an existing drain that is joined up with a sewer or another drain;
- seal off an existing drain that is joined up with a sewer or another drain,

then, provided that you give 48 hours' notice to the local authority, the person undertaking the reconstruction may break open any street for this purpose.

 You do not need to comply with this requirement if you are demolishing an existing building.

6.3.1.5 Can I repair an existing water closet or drain?

Repairs can be carried out to water closets, drains and soil pipes, but if that repair or construction work is prejudicial to health and/or a public nuisance then the person who completed the installation or repair is liable, on conviction, to a heavy fine.

 Note: In the Greater London area, a 'water closet' can **also** be taken to mean a 'urinal'.

6.3.1.5.1 Can I repair an existing drain?

Only in extreme emergencies are you allowed to repair, reconstruct or alter the course of an underground drain that joins up with a sewer, cesspool or other drainage method (e.g. septic tank).

If you have to carry out repairs, etc. in an emergency, then make sure that you do **not** cover over the drain or sewer without notifying the local authority of your intentions.

6.3.1.6 What about ventilation of soil pipes?

A major requirement of the Building Regulations is that all soil pipes from water closets shall be properly ventilated and that no use shall be made of:

- an existing or proposed pipe designed to carry rainwater from a roof to convey soil and drainage from a sanitary convenience;
- an existing pipe designed to carry surface water from premises to act as a ventilating shaft to a drain or a sewer conveying foul water.

6.3.2 Meeting the requirements

6.3.2.1 General

Drains and sewers should be protected from damage by construction traffic and heavy machinery.	H1 2.57
Heavy materials should not be stored over drains or sewers.	H1 2.57

After laying (including any necessary concrete or other haunching or surrounding and backfiring), gravity drains and private sewers should be tested for watertightness.	H1 2.59
All pipework carrying greywater for reuse should be clearly marked with the word 'GREYWATER'.	H1 (A11)
Material alterations to existing drains and sewers are subject to (and covered by) the Building Regulations.	H1 (B7)
Repairs, reconstruction and alterations to existing drains and sewers should be carried out to the same standards as for new drains and sewers.	H1 (B15)

Note: Separate systems of drains and sewers shall be provided for foul water and rainwater where:

- the rainwater is not contaminated; and
- the drainage is to be connected either directly or indirectly to the public sewer system, which has separate systems for foul water and surface water.

6.3.2.2 Foul water drainage

The capacity of the system should be large enough to carry the expected flow at any point (BS 5572, BS 8301).	H1 0.1
All pipes, fittings and joints should be capable of withstanding an air test of positive pressure of at least 38 mm water gauge for at least 3 minutes.	H1 1.38
Every trap should maintain a water seal of at least 25 mm.	H1 1.38

Some public sewers may carry foul water and rainwater in the same pipe. If the drainage system is also to carry rainwater to such a sewer these combined systems should not be capable of discharging into a cesspool or septic tank.

Foul drainage should be connected to either:	H1 2.1
• a public foul or combined sewer (wherever this is reasonably practicable);	H1 2.3
• an existing private sewer that connects with a public sewer; or	H1 2.6
• a wastewater treatment system or cesspool.	H1 2.7

Combined and rainwater sewers shall be designed to surcharge (i.e. the water level in the manhole rises above the top of the pipe) in heavy rainfall.   H1 2.8

Basements containing sanitary appliances, where the risk of flooding due to surcharge of the sewer is possible, should either use an anti-flooding valve (if the risk is low) or be pumped.   H1 2.9
H1 2.36–2.39
H1 2.10

For other low-lying sites (i.e. not basements) where the risk is considered low, a gully (at least 75 mm below the floor level) can be dug outside the building.

The layout of the drainage system should be kept simple.   H1 2.13

 Pipes should (wherever possible) be laid in straight lines. Changes of direction and gradient should be minimized.

Access points should be provided only if blockages could not be cleared without them.   H1 2.13

Connections should be made using prefabricated components.   H1 2.15

Connection of drains to other drains or private or public sewers and of private sewers to public sewers should be made obliquely, or in the direction of flow.   H1 2.14

The system should be ventilated by a flow of air.   H1 2.18
H1 1.27–1.29

Ventilating pipes should not finish near openings in buildings.   H1 2.18
H1 1.31

Pipes should be laid to even gradients and any change of gradient should be combined with an access point.   H1 2.19
H1 2.49

Pipes should also be laid in straight lines where practicable.   H1 2.20
H1 2.49

6.3.2.2.1 Pumping installations

Where gravity drainage is impracticable, or protection against flooding due to surcharge in downstream sewers is required, a pumping installation will be needed.

Where foul water drainage from a building is to be pumped, the effluent receiving chamber should be sized to contain 24-hour inflow to allow for disruption in service.   H1 2.39

> The minimum daily discharge of foul drainage should be   H1 2.39
> taken as 150 litres per head per day for domestic use.

Although flood resistance is **not** currently a requirement of the Building Regulations, as part of its aim to improve the energy-efficiency of buildings and use planning to protect the environment the revised 2013 National Planning Policy Framework sets out policies that aim to avoid *inappropriate development in areas at risk of flooding, including requiring new development to be flood resilient and resistant, as and where appropriate.* (See https://www.gov.uk/government/uploads/system/uploads/attachment_data/file/6077/2116950.pdf for full details.)

 Further information on flood-resistant and -resilient construction can be found in the Defra 2007 publication *Improving the Flood Performance of New Buildings: Flood Resilient Construction.*

6.3.2.3 Wastewater treatment systems

A notice giving information as to the nature and frequency of maintenance required for the cesspool or wastewater treatment system to continue to function satisfactorily should be displayed within each of the buildings.

The use of non-mains foul drainage, such as wastewater treatment systems, septic tanks or cesspools, should only be considered where connection to mains drainage is **not** practicable.

Any discharge from a wastewater treatment system is likely to require consent from the Environment Agency.

6.3.2.3.1 Septic tanks

Septic tanks with some form of secondary treatment (e.g. from a drainage field/mound, or constructed wetland such as a reed bed) will normally be the most economic means of treating wastewater from small developments (e.g. one to three dwellings). They provide suitable conditions for the settlement, storage and partial decomposition of solids, which need to be removed at regular intervals.

The principle of a septic tank is very simple.

Wastewater flows from the house to the septic tank. The tank is designed to retain wastewater and allow heavy solids to settle to the bottom. These solids are partially decomposed by bacteria to form sludge. Grease and light particles float, forming a layer of scum on top of the wastewater. Baffles installed at the inlet and outlet of the tank help prevent scum and solids from escaping.

A solid pipe leads from the septic tank to a distribution box where the wastewater is channelled into one or more perforated pipes set in trenches of gravel. Here the water slowly seeps into the underlying soil. Dissolved wastes and bacteria in the water are trapped or adsorbed to soil particles or decomposed

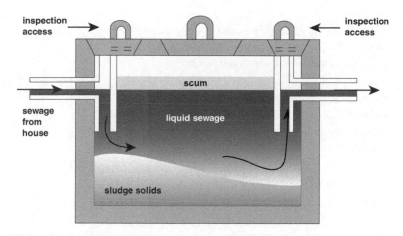

Figure 6.31 Basic septic tank.

by micro-organisms. This process removes disease-causing organisms, organic matter and most nutrients. The purified wastewater then moves to the groundwater or evaporates from the soil. Trench systems are the most common type of system used in new home construction.

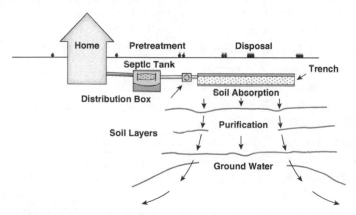

Figure 6.32 Septic tank principles.

Newer septic tanks can have a partial concrete dividing wall in the centre, thus making two compartments. This helps ensure the sludge does not get forced out of the baffle into the drain field. Newer tanks can also have two manhole covers, one above each baffle. Septic tanks should be sited at least 7 m from any habitable parts of buildings, and preferably down a slope.

H2 1.16

Septic tanks should only be used in conjunction with a form of secondary treatment (e.g. a drainage field, drainage mound or constructed wetland).	H2 1.15
Septic tanks should be sited within 30 m of vehicle access to enable the tank to be emptied and cleaned without hazard to the building occupants and without the contents being taken through a dwelling or place of work.	H2 1.17 and 1.64
Septic tanks and settlement tanks should have a capacity below the level of the inlet of at least 2700 litres (2.7 m³) for up to four users. This size should be increased by 180 litres for each additional user.	H2 1.18
Septic tanks may be constructed in brickwork or concrete (roofed with heavy concrete slabs), or factory-manufactured septic tanks (made out of glass-reinforced plastics, polyethylene or steel) can be used.	H2 1.19–20 and 1.65–66
The brickwork should consist of engineering bricks at least 220 mm thick. The mortar should be a mix of 1:3 cement/sand ratio, and in situ concrete should be at least 150 mm thick of C/25/P mix (see BS 5328).	H2 1.20 and 1.66
Septic tanks should be ventilated.	H2 1.21
Septic tanks should incorporate at least two chambers or compartments operating in series.	H2 1.22
Septic tanks should be provided with access for emptying and cleaning.	H2 1.24
A notice should be fixed within the building describing the necessary maintenance.	H2 1.25
Septic tanks should be inspected monthly to check they are working correctly.	H2 (A.11)
Septic tanks should be emptied at least once a year.	H2 (A.13)

6.3.2.4 Cesspools

A cesspool is a watertight tank, installed underground, for the storage of sewage. No treatment is involved.

Cesspools should be sited at least 7 m from any habitable parts of buildings and preferably downslope.	H2 1.58
Cesspools should be provided with access for emptying and cleaning.	H2 1.60

Cesspools should be inspected fortnightly for overflow.	H2 (A.20)
Cesspools should be emptied on a monthly basis by a licensed contractor.	H2 1.60 H2 (A.21)
A filling rate of 150 litres per person per day is assumed, and if the cesspool does not fill within the estimated period the tank should be inspected for leakage.	H2 (A.22)
Cesspools should be ventilated.	H2 1.63
The inlet of a cesspool should be provided with access for inspection.	H2 1.67
Cesspools and settlement tanks (if they are to be desludged using a tanker) should be sited within 30 m of vehicle access.	H2 1.64
Cesspools and settlement tanks should prevent leakage of the contents and ingress of subsoil water.	H2 1.63
Cesspools should have a capacity below the level of the inlet of at least 18,000 litres ($18\,m^3$) for two users, increased by 6800 litres ($6.8\,m^3$) for each additional user.	H2 1.61
Cesspools, septic tanks and settlement tanks may be constructed in brickwork, concrete or glass-reinforced concrete.	H2 1.65–1.66
Factory-made cesspools and septic tanks are available in glass-reinforced plastic, polyethylene or steel.	
The brickwork should consist of engineering bricks at least 220 mm thick. The mortar should be a mix of 1:3 cement/sand ratio, and in situ concrete should be at least 150 mm thick of C/25/P mix (see BS 5328).	H2 1.66
Cesspools should be covered (with heavy concrete slabs) and ventilated.	
Cesspools should have no openings except for the inlet, access for emptying and ventilation.	H2 1.62
Cesspools should be inspected fortnightly for overflow and emptied as required.	H2 (A.20)

6.3.2.4.1 Packaged treatment works

This term is applied to a range of systems designed to treat a given hydraulic and organic load using prefabricated components which can be installed with minimal site work. They are capable of treating effluent more efficiently than septic tank systems and this normally allows the product to be discharged directly to a watercourse.

The discharge from the wastewater treatment plant should be sited at least 10 m away from watercourses and any other buildings.	H2 1.54
Regular maintenance and inspection should be carried out in accordance with the manufacturer's instructions.	H2 (A.17)

6.3.2.4.2 Drainage fields and mounds

Drainage fields (or mounds) serving a wastewater treatment plant or septic tank should be located: • at least 10 m from any watercourse or permeable drain; • at least 50 m from the point of abstraction of any groundwater supply; • at least 15 m from any building; • sufficiently far from any other drainage fields, drainage mounds or soakaways so that the overall soakage capacity of the ground is not exceeded.	H2 1.27
No water supply pipes or underground services other than those required by the disposal system itself should be located within the disposal area.	H2 1.29
No access roads, driveways or paved areas should be located within the disposal area.	H2 1.30
The groundwater table should not rise to within 1 m of the invert level of the proposed effluent distribution pipes.	H2 1.33
An inspection chamber should be installed between the septic tank and the drainage field.	H2 1.43
Constructed wetlands should not be located in the shade of trees or buildings.	H2 1.47
The drainage field/mound should be checked on a monthly basis to ensure that it is not waterlogged and that the effluent is not backing up towards the septic tank.	H2 (A.15)

Under Section 50 (overflowing and leaking cesspools) of the Public Health Act 1936 (as amended in 1961) action could be taken against a builder who had caused the problem, and **not** just against the owner.

Under Section 59 (drainage of building) of the Building Act 1984, local authorities can require either the owner or the occupier to remove (or otherwise make innocuous) any disused cesspool, septic tank or settlement tank.

6.3.2.5 Sewers

The first preference should always be to provide separate foul and surface water sewerage systems. Where 'combined' or 'partially combined' sewerage is unavoidable, then the design and construction of the shared sewer should be in accordance with the *Protocol on Design, Construction and Adoption of Sewers in England and Wales* (see http://archive.defra.gov.uk/environment/quality/water/industry/sewers/documents/sewer-protocol-rev07.pdf for details) so as to prevent the proliferation of private sewers and the problems of ownership and maintenance associated with them.

- Sewers should be laid at an appropriate distance from buildings so as to avoid damage to the foundations. H1 Table C1
- Manholes and chambers should be located so that they are easily accessible either manually or with maintenance equipment such as mini-excavators.
- The last access point on the house drain should be sized to allow man entry and should be located in an accessible position and, as far as practicable, adjacent to the cartilage.
- House 'collector' drains serving each property should normally discharge into the sewer via a single junction or a manhole.
- Sewers should not be laid deeper than necessary.
- Manholes on or near highways or other roads need to be of robust construction.
- Sewers should be laid in straight lines in both vertical and horizontal alignments.

6.3.2.5.1 Building over existing sewers

Where it is proposed to construct a building over or near a drain or sewer shown on any map of sewers, the developer should consult the owner of the drain or sewer. H4 0.3

A building constructed over or within 3 m of any: H4 1.2

- rising main;
- drain or sewer constructed from brick or masonry;
- drain or sewer in poor condition

shall not be constructed in such a position unless special measures are taken.

Buildings or extensions should not be constructed over a manhole or inspection chamber or other access fitting on any sewer (serving more than one property). H4 1.3

A satisfactory diversionary route should be available so that the drain or sewer could be reconstructed without affecting the building.	H4 1.4
The length of drain or sewer under a building should not exceed 6 m except with the permission of the owners of the drain or sewer.	H4 1.5
Buildings or extensions should not be constructed over or within 3 m of any drain or sewer more than 3 m deep, or greater than 225 mm in diameter except with the permission of the owners of the drain or sewer.	H4 1.6
Where a drain or sewer runs under a building at least 100 mm of granular or other suitable flexible filling should be provided round the pipe.	H4 1.9
Where a drain or sewer running below a building is less than 2 m deep, the foundation should be extended locally so that the drain or sewer passes through the wall.	H4 1.10
Where the drain or sewer is more than 2 m deep to invert and passes beneath the foundations, the foundations should be designed with a lintel spanning over the line of the drain or sewer. The span of the lintel should extend at least 1.5 m either side of the pipe and should be designed so that no load is transmitted on to the drain or sewer.	H4 1.12
A drain trench should not be excavated lower than the foundations of any building nearby.	H4 1.13

6.3.2.6 Greywater and rainwater tanks

Greywater and rainwater tanks should:

• prevent leakage of the contents and ingress of subsoil water; • be ventilated; • have an anti-backflow device; • be provided with access for emptying and cleaning.	H2 1.70

6.3.2.7 Rainwater drainage

The capacity of the drainage system should be large enough to carry the expected flow at any point in the system.	H3 0.3
Rainwater or surface water should not be discharged to a cesspool or septic tank.	H3 0.6

6.3.2.7.1 Gutters and rainwater pipes

Although this part of the Building Regulations only actually applies to draining the rainfall from areas of $6\,\mathrm{m}^2$ or more (unless they receive a flow from a rainwater pipe or from paved and/or other hard surfaces), each case should be considered on its own merits separately and a decision made. This particularly applies to small roofs and balconies. Table 6.25 shows the largest effective area that should be drained into the gutter sizes most often used.

For eaves gutters the design rainfall intensity should be $0.021\ \mathrm{l/s/m}^2$. In some cases, eaves drop systems may be used.

Table 6.25 Gutter and outlet sizes

Max effective roof area (m²)	Gutter size (mm diameter)	Outlet size (mm diameter)	Flow capacity (l/s)
6.0	–	–	–
18.07	5	50	0.38
37.0	100	63	0.78
53.0	115	63	1.11
65.0	125	75	1.37
103.0	150	89	2.16

 Gutters should be laid with any fall towards the nearest outlet.

Gutters should be laid so that any overflow in excess of the design capacity (e.g. above-normal rainfall) will be discharged clear of the building.	H3 1.7
Rainwater pipes should discharge into a drain or gully (but may discharge to another gutter or on to another surface if it is drained).	H3 1.8
Any rainwater pipe which discharges into a combined system should do so through a trap.	H3 1.8
The size of a rainwater pipe should be at least the size of the outlet from the gutter.	H3 1.10
A down pipe which serves more than one gutter should have an area at least as large as the combined areas of the outlets.	H3 1.10
On flat roofs, valley gutters and parapet gutters, additional outlets may be necessary.	H3 1.7
Where a rainwater pipe discharges on to a lower roof or paved area, a pipe shoe should be fitted to divert water away from the building.	H3 1.9

Gutters and rainwater pipes should be firmly supported without restricting thermal movement.

The materials used should be of adequate strength and durability, and: H3 1.16

- all gutter joints should remain watertight under working conditions;
- pipework in siphonic roof drainage systems should be able to resist negative pressures in accordance with the design;
- gutters and rainwater pipes should be firmly supported;
- different metals should be separated by non-metallic material to prevent electrolytic corrosion.

6.3.2.8 Surface water drainage

Discharge to a watercourse may require consent from the Environment Agency, which may limit the rate of discharge. Where other forms of outlet are not practicable, discharge should be made to a sewer (H3 3.2–3.3). For design purposes a rainfall interval of $0.014 \, l/s/m^2$ can be assumed as normal.

Some drainage authorities have sewers that carry both foul water and rainwater (i.e. combined systems) in the same pipe. Where they do, they can allow rainwater to discharge into the system if the sewer has enough capacity to take the added flow. Some private sewers (drains serving more than one property) also carry both foul water and rainwater. If a sewer (or private sewer) operated as a combined system does not have enough capacity, the rainwater should be run in a separate system with its own outfall.

Surface water drainage should discharge to a soakaway or other infiltration system where practicable.	H3 3.2
Surface water drainage connected to combined sewers should have traps on all inlets.	H3 3.7
Drains should be at least 75 mm diameter.	H3 3.14
Where any materials that could cause pollution are stored or used, separate drainage systems should be provided.	H3 3.21
On car parks, petrol filling stations or other areas where there is likely to be leakage or spillage of oil, drainage systems should be provided with oil interceptors.	H3 3.22 H3 (A)
Separators should be leak-tight and comply with the requirements of the Environment Agency and BS EN 858:2003.	H3 (A.9–10)

Infiltration devices (including soakaways, swales,   H3 3.23–26
infiltration basins and filter drains) should not be built:

*   within 5 m of a building or road or in areas of
    unstable land;
*   in ground where the water table reaches the bottom
    of the device at any time of the year;
*   sufficiently far from any drainage fields, drainage
    mounds or other soakaways;
*   where the presence of any contamination in the run-
    off could result in pollution of groundwater source or
    resource.

Soakaways should be designed to a return period of once   H3 3.27
in ten years.

Soakaways for areas less than 100 m^2 shall consist of   H3 3.26
square or circular pits, filled with rubble or lined with
dry-jointed masonry or perforated ring units.

Soakaways serving larger areas shall be lined pits or
trench-type soakaways.

The storage volume should be calculated so that, over   H3 3.29
the duration of a storm, it is sufficient to contain the
difference between the inflow volume and the outflow
volume.

Soakaways serving larger areas should be designed in   H3 3.30
accordance with BS EN 752-4.

Under Section 85 (offences concerning the polluting of controlled waters) of
the Water Resources Act 1991 it is an offence to discharge any noxious or polluting material into a watercourse, coastal water or underground water. Most
surface water sewers discharge to watercourses.

Under Section 111 (restrictions on use of public sewers) of the Water Industry
Act 1991 it is an offence to discharge petrol, chemical, waste steam or any matter
likely to damage the drain into any drain or sewer connected to a public sewer.

6.3.2.8.1 Drainage of paved areas

Surface gradients should direct water draining from a   H3 2.2
paved area away from buildings.

Gradients on impervious surfaces should be designed   H3 2.3
to permit the water to drain quickly from the surface.

A gradient of at least 1 in 60 is recommended.

Paths, driveways and other narrow areas of paving should be free-draining to a pervious area such as grassland, provided that:

- the water is not discharged adjacent to buildings where it could damage foundations; and
- the soakage capacity of the ground is not overloaded.

H3 2.6

Where water is to be drained on to the adjacent ground the edge of the paving should be finished above or flush with the surrounding ground to allow the water to run off.

H3 2.7

Where the surrounding ground is not sufficiently permeable to accept the flow, filter drains may be provided.

H3 2.8 and 3.33

Pervious paving should not be used where excessive amounts of sediment are likely to enter the pavement and block the pores.

H3 2.11

Pervious paving should not be used in oil storage areas, or where run-off may be contaminated with pollutants.

H3 2.12

Gullies should be provided at low points where water would otherwise pond.

H3 2.15

Gully gratings should be set approximately 5 mm below the level of the surrounding paved area in order to allow for settlement.

H3 2.16

Provision should be made to prevent silt and grit entering the system, by provision of either gully pots of suitable size or catchpits.

H3 2.17

6.3.2.9 Enclosures for drainage and/or water supply pipes

The enclosure should:

- be bounded by a compartment wall or floor, an outside wall, an intermediate floor or a casing;
- have internal surfaces (except framing members) of Class 0;

B3 7.6–7.9 (V1)

B3 10.7 (V2)

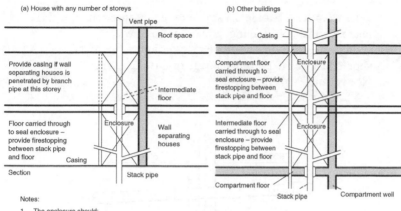

Notes:

1   The enclosure should:
    a   be bounded by a compartment wall or floor, an outside wall, an intermediate floor, or a casing and
    b   have internal surfaces (except framing members) of Class 0, and
    c   not have an access panel which opens into a circulation space or bedroom, and
    d   be used only for drainage, or water supply, or vent pipes for a drainage system
2   The casing should:
    a   be imperforate except for an opening for a pipe or an access panel, and
    b   not be of sheet metal, and
    c   have (including any access pancel) not less than 30 minutes fire resistance
3   The opening for a pipe, either in the structure or the casing, should be as small as possible
    and fire-stopped around the pipe.

Figure 6.33 Enclosure for drainage or water supply pipes.

- not have an access panel which opens into a circulation space or bedroom;
- be used only for drainage, or water supply, or vent pipes for a drainage system.

The casing should: B3 7.6–7.9 (V1)

- be imperforate except for an opening for a pipe   B3 10.7 (V2)
  or an access pane;
- not be of sheet metal;
- have (including any access panel) not less than
  30 minutes' fire resistance.

The opening for a pipe, either in the structure or   B3 7.6–7.9 (V1)
the casing, should be as small as possible and fire-   B3 10.7 (V2)
stopped around the pipe.

6.3.2.10 Protection from settlement

- A drain may run under a building if at least 100 mm of   H1 2.23
  granular or other flexible filling is provided round the
  pipe.

- Where pipes are built into a structure (e.g. inspection chamber, manhole, footing, ground beam or wall) suitable measures (such as using rocker joints or a lintel) should be taken to prevent damage or misalignment (see Figures 6.34 and 6.35).   H1 2.24

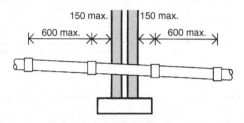

Figure 6.34 Pipe embedded in the wall. Short length of pipe embedded in a wall with joints within 150 mm of either wallface. Additional rocker pipes (max. length 600 mm) with flexible joints are then added.

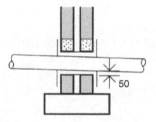

Figure 6.35 Pipe shielded by a lintel. Both sides are masked with rigid sheet material (to prevent entry of fill or vermin) and the void is filled with a compressible sealant to prevent entry of gas.

The depth of cover will usually depend on the levels of the connections to the system, the gradients at which the pipes should be laid and the ground levels.   H1 2.27
H1 2.41–2.45

All drain trenches should not be excavated lower than the foundations of any building nearby (see Figure 6.36).   H1 2.25

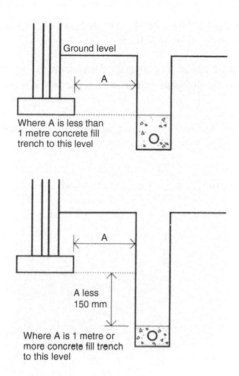

Figure 6.36 Pipe runs near buildings.

6.3.2.11 Bedding and backfill

The choice of bedding and backfill depends on the depth at   H1 2.41
which the pipes are to be laid and the size and strength of
the pipes.

Table 6.26 Materials for below-ground gravity drainage

Material	British Standard
Rigid pipes	BS 7838
Vitrified clay	BS 65, BSEN 295
Concrete	BS 5911
Grey iron	BS 437
Ductile iron	BS EN 598
Flexible pipes	
UPVC (unplasticized polyvinyl chloride)	BS EN 1401
PP (polypropylene)	BS EN 1852
Structured walled plastic pipes	BS EN 13476

Special precautions should be taken to take account of the effects of settlement where pipes run under or near buildings.

The depth of the pipe cover will usually depend on the levels of the connections to the system and the gradients at which the pipes should be laid and the ground levels.

 Pipes need to be protected from damage, particularly pipes which could be damaged by the weight of backfilling.

Rigid pipes should be laid in a trench as shown in Figure 6.37.   H1 2.42

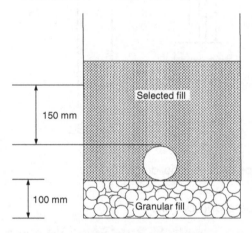

Figure 6.37 Bedding for rigid pipes.

Flexible pipes shall be supported to limit deformation under   H1 2.44
load.

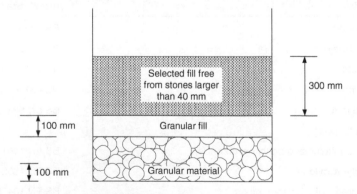

Figure 6.38 Bedding for flexible pipes.

Flexible pipes with very little cover shall be protected   H1 2.42–2.44
from damage by a reinforced cover slab with a flex-
ible filler and at least 75 mm of granular material
between the top of the pipe and the underside of the
flexible filler below the slabs (see Figure 6.39).

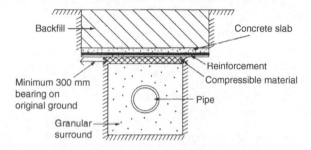

Figure 6.39 Protection of pipes laid in shallow depths.

Trenches may be backfilled with concrete to protect nearby   H1 2.45
foundations. In these cases a movement joint (as shown in
Figure 6.40) formed with a compressible board should be
provided at each socket or sleeve joint.

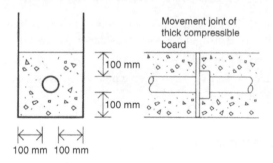

Figure 6.40 Joints for concrete encased pipes.

6.3.2.12 Access points

Access should be provided to long runs.   H1 2.50

Sufficient and suitable access points should be provided for   H1 2.46
clearing blockages from drain runs that cannot be reached
by any other means.

Access points should be provided: H1 2.49

* on or near the head of each drain run;
* at a bend;
* at a change of gradient or pipe size;
* at a junction.

Access points should be either: H1 2.48

* rodding eyes – capped extensions of the pipes;
* access fittings – small chambers on (or an extension of) the pipes but not with an open channel;
* inspection chambers – chambers with working space at ground level;
* manholes – deep chambers with working space at drain level.

Access points should be constructed so as to resist the   H1 2.52
ingress of groundwater or rainwater.

Inspection chambers and manholes should have removable   H1 2.54
non-ventilating covers of durable material (such as cast
iron, cast or pressed steel, precast concrete or UPVC).

Access points to sewers (serving more than one property)   H1 2.51
should be located in places where they are accessible and
apparent for use in an emergency (e.g. highways, public
open space, unfenced front gardens, and shared or unfenced
driveways).

Inspection chambers and manholes in buildings should have   H1 2.54
mechanically fixed airtight covers unless the drain itself has
watertight access covers.

Manholes deeper than 1 m should have metal step irons or   H1 2.54
fixed ladders.

6.3.2.13 Rodent control

If the site has been previously developed, the local authority should be consulted to determine whether any special measures are necessary for control of rodents. Special measures which may be taken include the following:

Sealed drainage – access covers to the pipework should be   H1 2.22a
in the inspection chamber instead of an open channel.

Intercepting traps – should be of the locking type that can   H1 2.22b
be easily removed from the chamber surface and securely
replaced.

Rodent barriers – including enlarged sections on discharge stacks to prevent rats climbing, flexible downward-facing fins in the discharge stack, or one-way valves in underground drainage.	H1 2.22c
Metal cages on ventilator stack terminals – to discourage rats from leaving the drainage system.	H1 2.22d H1 1.31
Covers and gratings to gullies – used to discourage rats from leaving the system.	H1 2.22e
During construction, drains and sewers that are left open should be covered when work is not in progress to prevent entry by rats.	H1 2.56
Disused drains or sewers less than 1.5 m deep that are in open ground should as far as is practicable be removed. Other pipes should be sealed at both ends (and at any point of connection) and grout-filled to ensure that rats cannot gain access.	H1 (B18)

6.3.2.14 Protection of openings for pipes

Pipes which pass through a compartment wall or compartment floor (unless the pipe is in a protected shaft), or through a cavity barrier, should conform to one of the following alternatives:

Proprietary seals (any pipe diameter) that maintain the fire resistance of the wall, floor or cavity barrier.	B3 7.6–7.7 (V1) B3 10.5–10.6 (V1)
Pipes with a restricted diameter should be used where fire-stopping is used around the pipe, keeping the opening as small as possible.	B3 7.6–7.8 (V1) B3 10.5–10.7 (V2)
Sleeving – a pipe of lead, aluminium, aluminium alloy, fibre-cement or UPVC, with a maximum nominal internal diameter of 160 mm, may be used with a sleeving of non-combustible pipe as shown in Figure 6.41.	B3 7.6–7.9 (V1) B3 10.5–10.8 (V2)

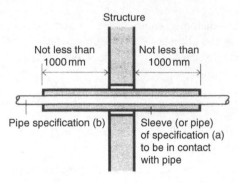

Figure 6.41 Pipes penetrating a structure.

6.3.2.15 Discharge stacks

All stacks should discharge to a drain.	H1 1.26
The bend at the foot of the stack should have as large a radius (i.e. at least 200 mm) as possible.	H1 1.26
Discharge stacks should be ventilated.	H1 1.29
Offsets in the 'wet' portion of a discharge stack should be avoided.	H1 1.27
Stacks serving urinals should be not less than 50 mm.	H1 1.28
Stacks serving closets with outlets less than 80 mm should be not less than 75 mm.	H1 1.28
Stacks serving closets with outlets greater than 80 mm should be not less than 100 mm.	H1 1.28
The internal diameter of the stack should be not less than that of the largest trap or branch discharge pipe.	H1 1.28
Ventilating pipes open to outside air should finish at least 900 mm above any opening into the building within 3 m and should be fitted with a perforated cover or cage (see Figure 6.42) which should be metal if rodent control is a problem.	H1 1.31
Ventilating pipes open to outside air should finish at least 900 mm above any opening into the building within 3 m.	H1 1.31
Air admittance valves complying with BS EN 12380:2002 should be located in areas that have adequate ventilation.	H1 1.33
Air admittance valves should not be used outside buildings or in dust-laden atmospheres.	H1 1.33
Rodding points should be provided in discharge stacks.	H1 1.34

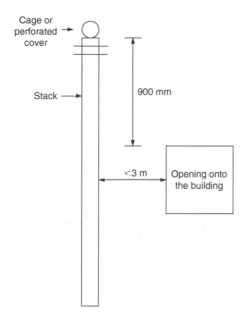

Figure 6.42 Termination of ventilation stacks.

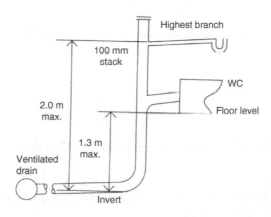

Figure 6.43 Stub stack.

Pipes should be firmly supported without restricting thermal movement.	H1 1.35
Sanitary pipework connected to WCs should not allow light to be visible through the pipe wall, as this is believed to encourage damage by rodents.	H1 1.36

 Drainage serving kitchens in commercial hot food premises should be fitted with a grease separator complying with BS EN 1825-1:2004.

6.3.2.15.1 Branch discharge pipes

> Branch pipes should discharge into either another branch pipe   H1 1.5
> or a discharge stack (unless the appliances discharge into a
> gully on the ground floor or at basement level).

 Note: If the appliances are on the ground floor, the pipe(s) may discharge to a stub stack, discharge stack, direct to a drain or (if the pipe carries only waste-water) to a gully.

> A branch pipe from a ground floor closet should only   H1 1.9
> discharge direct to a drain if the depth from the floor to the
> drain is 1.3 m or less (see Figure 6.44).

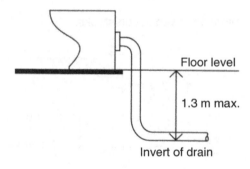

Figure 6.44 Direct connection of ground floor WC to drain.

 Note: A branch pipe serving any ground-floor appliance may discharge direct to a drain or into its own stack.

> A branch pipe should not discharge into a stack in a way   H1 1.10
> which could cause cross-flow into any other branch pipe (see
> Figure 6.45).

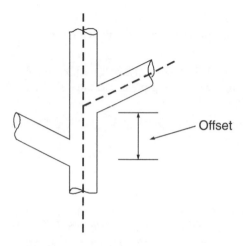

Figure 6.45 Branch connections.

A branch discharge pipe should not discharge into a   H1 1.8
stack lower than 450 mm above the invert of the tail of   H1 (A3, A4)
the bend at the foot of the stack in single dwellings up to H1 1.21
three storeys (see Figure 6.46).

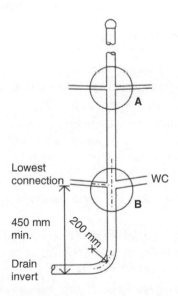

Figure 6.46 Branch discharge stack.

Branch pipes may discharge into a stub stack.	H1 1.12 H1 1.30
A branch pipe discharging to a gully should terminate between the grating or sealing plate and the top of the water seal.	H1 1.13
Bends in branch pipes should be avoided if possible (see Figure 6.47).	H1 1.16

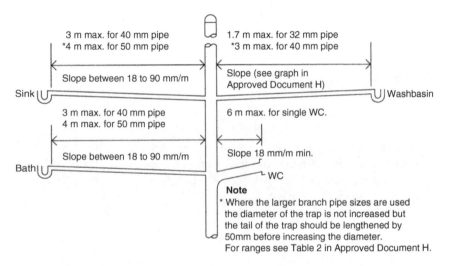

Figure 6.47 Branched connections.

Junctions on branch pipes should be made with a sweep of 25 mm radius or at 45°.	H1 1.17
Rodding points should be provided to give access to any lengths of discharge pipes which cannot be reached by removing traps or appliances with integral traps.	H1 1.25 H1 1.6
A branch pipe discharging to a gully should terminate between the grating or sealing plate and the top of the water seal.	H1 1.13
Condensate drainage from boilers may be connected to sanitary pipework provided the connection is made to an internal stack with a 75 mm condensate trap.	H1 1.14
If the connection is made to a branch pipe, the connection should be made downstream of any sink waste connection.	H1 1.14

All sanitary pipework receiving condensate should be made   H1 1.14
from materials resistant to a pH value of 6.5 or lower and
be installed in accordance with BS 6798.

Pipes serving a single appliance should have at least the same diameter
as the appliance trap (see Table 6.27).

A separate ventilating stack is only likely to be preferred where the numbers of
sanitary appliances and their distance to a discharge stack are large.

Table 6.27 Minimum trap sizes and seal depths

Appliance	Diameter of trap (mm)	Depth of seal (mm of water or equivalent)
Washbasin	32	75
Bidet		
Bath	40	50
Shower		
Food waste disposal unit	40	75
Urinal bowl		
Sink		
Washing machine		
Dishwashing machine		
WC pan (outlet 80 mm)	75	50
WC pan (outlet 80 mm)	100	50

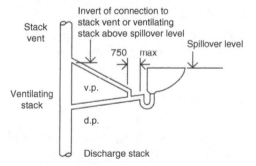

Figure 6.48 Branch ventilation pipes.

Branch ventilation pipes should be connected to the discharge pipe within 750 mm of the trap and should connect to the ventilating stack or the stack vent, above the highest 'spillover' level of the appliances served.	H1 1.22
The ventilating pipe should have a continuous incline from the discharge pipe to the point of connection to the ventilating stack or stack vent.	H1 1.22
Branch ventilating pipes which run direct to outside air should finish at least 900 mm above any opening into the building nearer than 3 m.	H1 1.23
A dry stack may provide ventilation for branch ventilation pipes as an alternative to carrying them to outside air or to a ventilated discharge stack (ventilated system).	H1 (A7 and 1.21)
Ventilation stacks serving buildings with not more than ten storeys and containing only dwellings should be at least 32 mm diameter (for all other buildings see paragraph H1 (1.29)).	H1 (A8) H1 1.21 and 1.29
A separate ventilating stack is only likely to be preferred where the numbers of ventilating pipes and their distance to a discharge stack are large.	H1 1.19 H1 (Table 2)

6.3.2.15.2 Traps

All points of discharge into the system should be fitted with a trap (e.g. a water seal) to prevent foul air from the system entering the building.	H1 1.3–1.4
All traps should be fitted directly over an appliance and should be removable or be fitted with a cleaning eye.	H1 1.6

6.3.2.16 Pipe gradients and sizes

Drains should have enough capacity to carry the anticipated maximum flow (see Table 6.28).	H1 2.29
Sewers (i.e. a drain serving more than one property) should have a minimum diameter of 100 mm when serving ten dwellings or diameter of 150 mm if more than ten.	H1 2.30

Table 6.28 Flow rates from dwellings

Number of dwellings	Flow rate (l/s)
1	2.5
5	3.5
10	4.1
15	4.6
20	5.1
25	5.4
30	5.8

Drains carrying foul water should have an internal diameter of at least 75 mm.	H1 2.33
Drains carrying effluent from a WC or trade effluent should have an internal diameter of at least 100 mm.	H1 2.33

6.3.2.17 Materials for pipes and jointing

To minimize the effects of any differential settlement, pipes should have flexible joints.	H1 2.40
All joints should remain watertight under working and test conditions.	H1 2.40
Nothing in the pipes, joints or fittings should project into the pipe line or cause an obstruction.	H1 2.40
Different metals should be separated by non-metallic materials to prevent electrolytic corrosion.	H1 2.40

6.4 Cellars and basements

6.4.1 Requirements (Building Act 1984 Section 74)

Unless you have the consent of the local authority, you are not allowed to construct a cellar or room *in (or as part of) a house, an existing cellar, a shop, inn, hotel or office if the floor level of the cellar or room is lower than the ordinary level of the subsoil water on, under or adjacent to the site of the house, shop, inn, hotel or office.*

This does not, however apply to:

- the construction of a cellar or room carried out in accordance with plans deposited on an application under the Licensing Act 2003;
- the construction of a cellar or room in connection with a shop, inn, hotel or office that forms part of a railway station.

If the owner of the house, shop, inn, hotel or office allows a cellar (or room forming part of it) to be used in a manner that he or she knows to be in contravention of the Building Regulations, he or she is liable, on summary conviction, to a fine.

Fire precautions
The building shall be provided with:

- *adequate means for venting heat and smoke from a fire in a basement;*
- *sufficient internal fire mains and other facilities to assist firefighters in their tasks.*

(Approved Document B)

Ventilation
There shall be adequate means of ventilation provided for people in the building.

(Approved Document F)

Stairs, ladders and ramps
Any stairs, ramps, floors and balconies and any roof to which people have access, and any light well, basement area or similar sunken area connected to a building, shall be provided with barriers where it is necessary to protect people in or about the building from falling.

(Approved Document K2)

A new issue of Approved Document Part K came into force in 2013.

6.4.2 Meeting the requirements

6.4.2.1 Fire precautions

Owing to the risk that a single stairway may be blocked by smoke from a fire in the basement or ground storey:

• basement storeys that contain a habitable room shall be provided with either:	B1 2.13 (V1) B1 2.6 (V2)
○ an external door or window suitable for egress from the basement; or	
○ a protected stairway leading from the basement to a final exit;	

- final exits shall be sited so that they are clear of any risk from fire or smoke in a basement;

  B1 5.34 (V2)

- in non-residential, purpose group buildings (such as office, shop and commercial, assembly and recreation, industrial, storage, etc.), the following floors shall be constructed as compartment walls and compartment floors:

  B2 8.18c (V2)
  B2 8.18d (V2)

  ○ the floor of every basement storey (except the lowest floor) greater than 10 m below ground level (see Figure 6.49);

  ○ the floor of the ground storey (see Figure 6.50).

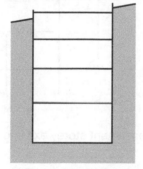

All basement storeys to be separated by compartment floors if any storey is at a depth of more than 10 m

(a) Deep basements

Figure 6.49 Compartment floors – deep basements.

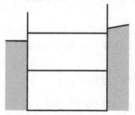

Only the floor of the ground storey need be a compartment floor if the lower basement is at a depth of not more than 10 m

(b) Shallow basement

Figure 6.50 Compartment floors – shallow basements.

6.4.2.2 *Emergency egress windows and external doors*

The window or door should enable the person escaping   B1 2.8 (V1)
to reach a place free from danger of fire (e.g. a courtyard   B1 2.9 (V2)
or back garden which is at least as deep as the dwelling
house is high – see Figure 6.51), and:

• the window should be at least 450 mm high and
  450 mm wide and have an unobstructed openable
  area of at least 0.33 m²; and
• the bottom of the openable area should be not more
  than 1100 mm above the floor.

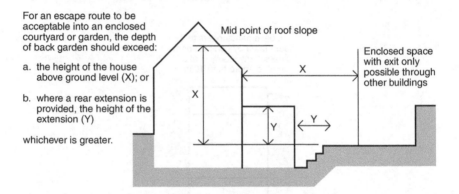

Figure 6.51 Ground or basement storey exit into an enclosed space.

 Notes:

(1) Approved Document K (*Protection from Falling, Collision and Impact*)
    specifies a minimum guarding height of 800 mm, except in the case of a
    window in a roof, where the bottom of the opening may be 600 mm above
    the floor.
(2) Locks (with or without removable keys) and stays may be fitted to egress
    windows, provided that the stay is fitted with a child-resistant release
    catch.
(3) Windows should be designed so that they remain in the open position
    without needing to be held open by the person making an escape.

6.4.2.2.1 Basement stairways

Because basement stairways are more likely to be filled with smoke and heat
than are stairs in ground and upper storeys:

- the flights and landings of an escape stair   B1 5.19 (V2)
  shall be constructed using materials of limited
  combustibility, particularly if it is within a
  basement storey;

- the basement should be served by a separate stair.   B1 2.44 (V2)
                                                        B1 4.42 (V2)

Note: If an escape stair forms part of the **only** escape route from an upper storey of a large building, it should **not** be continued down to serve any basement storey. Other stairs may connect with the basement storey(s) but **only** if there is a protected lobby or a protected corridor between the stair(s) and accommodation at each basement level.

6.4.2.3 Lifts

6.4.2.3.1 Lifts in basements

In basements:

- the lift should be approached only by a protected   B1 5.43 (V2)
  lobby or protected corridor (unless it is within the
  enclosure of a protected stairway);

- lift entrances should be separated from the floor   B1 5.42 (V2)
  area on every storey by a protected lobby;

- lift shafts should not be continued down to serve   B1 5.44 (V2)
  any basement storey if the lift is:

  ○ in a building served by only one escape stair;
  ○ within the enclosure to an escape stair which
    is terminated at ground level.

6.4.2.3.2 Access and facilities for the fire service

Buildings with a basement more than 10 m below the   B5 17.2 (V2)
fire and rescue service vehicle access level should be B5 17.8 (V2)
provided with at least two firefighting shafts containing
firefighting lifts (see Figure 6.52).

Buildings with two or more basement storeys, each   B5 17.4 (V2)
exceeding 900 m² in area, should be provided with
firefighting shaft(s), which need not include firefighting
lifts.

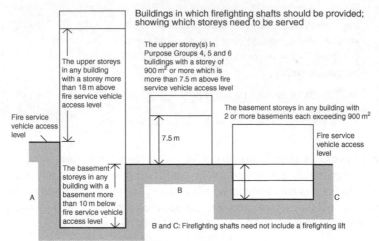

Buildings in which firefighting shafts should be provided;
showing which storeys need to be served

The upper storeys in any building with a storey more than 18 m above fire service vehicle access level

The upper storey(s) in Purpose Groups 4, 5 and 6 bulidings with a storey of 900 m² or more which is more than 7.5 m above fire service vehicle access level

Fire service vehicle access level

The basement storeys in any building with 2 or more basements each exceeding 900 m²

Fire service vehicle access level

7.5 m

The basement storeys in any building with a basement more than 10 m below fire service vehicle access level

A

B

C

B and C: Firefighting shafts need not include a firefighting lift

A. Firefighting shafts should include firefighting lift(s)

Note: Height excludes any top storey(s) consisting exclusively of plant rooms.

Figure 6.52 Provision of firefighting shafts.

6.4.2.4 Venting of heat and smoke from basements

The building should be provided with adequate means for venting heat and smoke from a fire in a basement.	B5 (V2)
Where practicable, each basement space should have one or more smoke outlets (see Figure 6.53).	B5 18.3 (V2)
Outlet ducts or shafts, including any bulkheads over them (see Figure 6.53), should be enclosed in non-combustible construction having a greater fire resistance than the element that they pass through.	B5 18.15 (V2)

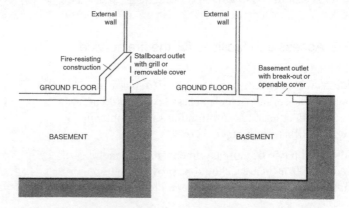

External wall

External wall

Fire-resisting construction

Stallboard outlet with grill or removable cover

Basement outlet with break-out or openable cover

GROUND FLOOR

GROUND FLOOR

BASEMENT

BASEMENT

Figure 6.53 Fire-resisting construction for smoke outlet shafts.

Smoke outlets connected to the open air should be provided from every basement storey, except for a basement in a single-family dwelling.   B5 18.4 (V2)

Smoke outlets (also referred to as smoke vents) should be:

- available so as to provide a route for heat and smoke to escape to the open air from the basement area;   B5 18.2 (V2)

- sited at high level, either in the ceiling or in the wall of the space they serve;   B5 18.7 (V2)
  B5 18.3 (V2)

- evenly distributed around the perimeter to discharge in the open air outside the building.   B5 18.3 (V2)

A system of mechanical extraction may be provided as an alternative to natural venting to remove smoke and heat from basements, **provided** that the basement storey(s) is fitted with a sprinkler system.   B5 18.13 (V2)

Where there are natural smoke outlet shafts from different compartments to the same or different basement storeys, they should be separated from each other by a non-combustible construction.   B5 18.16 (V2)

6.4.2.5 Ventilation

If a basement is connected to the rest of the dwelling by a large permanent opening such as an open stairway, then the whole dwelling including the basement should be treated as a multi-storey dwelling and ventilated in a similar manner to dwellings without basements.   F 5.11

If the basement has a single exposed façade, while the rest of the dwelling above ground has more than one exposed façade, then passive stack ventilation (PSV) or continuous mechanical extract should be used.   F 5.11

If the basement is not connected to the rest of the dwelling by a large permanent opening, then: F 5.12

- the part of the dwelling above ground should be considered separately; and

- the basement should be treated as a single-storey dwelling, as if it were above ground.

If the part of the dwelling above ground has no bedrooms, then for the purpose of ventilation requirements:

> • assume that the dwelling has one bedroom; and   F 5.12
> • treat the basement as a single-storey dwelling (with one
>   bedroom) as if it were above ground.
>
> If a dwelling only comprises a basement, then it should   F 5.13
> be treated as if it were a single-storey dwelling (with one
> bedroom) above ground.
>
> Mechanically ventilated basement car parks shall be capable   F 6.20bi
> of at least six air changes per hour (6 ach).

Table 6.29 Ventilation systems for basements

Type of basement	Background ventilators and intermittent extract fans	Passive stack ventilation	Continuous mechanical extract	Continuous mechanical supply and extract with heat
Basement connected to the rest of the dwelling by an open stairway	Yes	Yes	Yes	Yes
Basement with a single exposed façade and dwelling above ground with more than one exposed façade		Yes	Yes	
Basements not connected to the rest of the dwelling by an open stairway	Yes	Yes	Yes	Yes
Dwelling above ground has no bedrooms	Yes	Yes	Yes	Yes
Dwelling comprises just a basement	Yes	Yes	Yes	Yes

6.4.2.6 Guarding basement stairs

Guarding should be provided whenever it is considered necessary from the point of view of safety to guard any basement stair.

> If a building is likely to be used by children under five   K2 (3.3)
> years old, the guarding should not have horizontal rails,
> the guarding should stop children from easily climbing
> it, and the construction should prevent a 100 mm sphere
> being able to pass through any opening of that guarding.

6.4.2.6.1 Design of guarding

For cellars and basements:

> • guarding must be capable of resisting, as a minimum, the   K2 (3.2)
>   loads given in BS EN 1991-1-1;
> • where glazing is used as part of the guarding, refer also to
>   Approved Document K5.

Figure 6.54 shows typical locations and dimensions for guarding basement or cellar stairs.

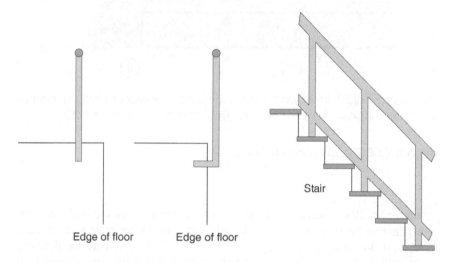

Edge of floor   Edge of floor

Stair

Figure 6.54 Typical locations for guarding

For further guidance on the design of barriers and infill panels, refer to BS 6180.

6.5 Floors

The ground floor of a building is either solid concrete or a suspended timber type. With a concrete floor, a damp-proof membrane is laid between walls.

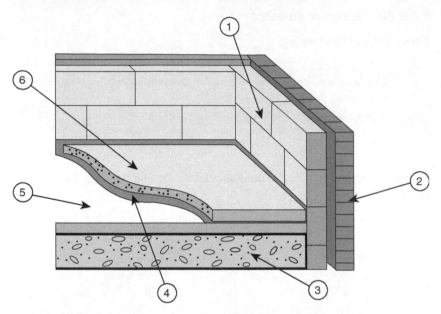

1. Concrete block/inner wall; 2. Outside wall; 3. Hardcore infill; 4. Damp-proof membrane; 5. Sand blinding; 6. Minimum 100 mm concrete.

Figure 6.55 Typical concrete floor.

With timber floors, sleeper walls of honeycomb brickwork are built on over-site concrete between the base brickwork; a timber sleeper plate rests on each wall, and timber joists are supported on them. Their ends may be similarly supported, let into the brickwork or suspended on metal hangers. Floorboards are laid at right angles to joists. First-floor joists are supported by the masonry or hangers.

Similar to the case in a brick-built house, the floors in a timber-framed house are either solid concrete or suspended timber. In some cases, a concrete floor may be screeded or surfaced with timber or chipboard flooring. Suspended timber floor joists are supported on wall plates and surfaced with chipboard.

Guidance on the sizing of certain members in floors and roofs is given in *Span Tables for Solid Timber Members in Floors, Ceilings and Roofs (Excluding Trussed Rafter Roofs) for Dwellings*, published by TRADA, available from Chiltern House, Stocking Lane, Hughenden Valley, High Wycombe, Bucks HP14 4ND.

 Note: Also see BS EN 1995-1-1:2004, *Design of Timber Structures*, and BS 8103-3:2009, *Structural Design of Low-Rise Buildings: Code of Practice for Timber Floors and Roofs for Housing*.

6.5.1 Requirements

Construction
The building shall be constructed so that the combined dead, imposed and wind loads are sustained and transmitted by it to the ground:

- *safely;*
- *without causing such deflection or deformation of any part of the building (or such movement of the ground) as will impair the stability of any part of another building.*

(Approved Document A1)

 Amendments to the 2010 version of Approved Document A came into force in 2013.

The building shall be constructed so that ground movement caused by:

- *swelling, shrinkage or freezing of the subsoil; or*
- *landslip or subsidence (other than subsidence arising from shrinkage)*

will not impair the stability of any part of the building.

(Approved Document A2)

Fire precautions
As a fire precaution, the spread of flame over the internal linings of a building and the amount of heat released from internal linings shall be restricted:

- *all load-bearing elements of structure of the building shall be capable of withstanding the effects of fire for an appropriate period without loss of stability;*
- *ideally the building should be subdivided by elements of fire-resisting construction into compartments;*
- *all openings in fire-separating elements shall be suitably protected in order to maintain the integrity of the continuity of the fire separation;*
- *any hidden voids in the construction shall be sealed and subdivided to inhibit the unseen spread of fire and products of combustion, in order to reduce the risk of structural failure, and the spread of fire.*

(Approved Document B3)

Precautions against moisture
The floors of the building shall adequately protect the building and people who use the building from harmful effects caused by:

- *ground moisture;*
- *precipitation and wind-driven spray;*
- *interstitial and surface condensation; and*
- *spillage of water from or associated with sanitary fittings or fixed appliances.*

All floors next to the ground, walls and roof shall not be damaged by moisture from the ground, rain or snow and shall not carry that moisture to any part of the building that it would damage.

(Approved Document C2)

 Amendments to the 2010 version of Approved Document C came into force in 2013.

Airborne and impact sound
Dwellings shall be designed so that the noise from domestic activity in an adjoining dwelling (or other parts of the building) is kept to a level that:

- *does not affect the health of the occupants of the dwelling;*
- *will allow them to sleep, rest and engage in their normal activities in satisfactory conditions.*

(Approved Document E1)

Dwellings shall be designed so that any domestic noise that is generated internally does not interfere with the occupants' ability to sleep, rest and engage in their normal activities in satisfactory conditions.

(Approved Document E2)

Domestic buildings shall be designed and constructed so as to restrict the transmission of echoes.

(Approved Document E3)

Schools shall be designed and constructed so as to reduce the level of ambient noise (particularly echoing in corridors).

(Approved Document E4)

 Note: The normal way of satisfying Requirement E4 will be to meet the values for sound insulation, reverberation time and internal ambient noise which are given in Section 1 of Building Bulletin 93, *The Acoustic Design of Schools*, produced by Department for Education and Skills (DfES) and published by The Stationery Office (ISBN: 0-11-271105-7).

Ventilation
There shall be adequate means of ventilation provided for people in the building.

(Approved Document F)

Conservation of fuel and power
Reasonable provision shall be made for the conservation of fuel and power in buildings by:

(a)   limiting heat gains and losses

(i)   through thermal elements and other parts of the building fabric; and

(ii) from pipes, ducts and vessels used for space heating, space cooling and hot water services;

(b)   providing fixed building services which

(i)   are energy efficient; and

(ii) have effective controls and are commissioned by testing and adjusting as necessary to ensure they use no more fuel and power than is reasonable in the circumstances.

(Approved Document L)

 A new version of Approved Document L came into force on 6 April 2014.

6.5.1.1 The use of Robust Standards

One of the recommendations of Approved Document E (*Resistance to the Passage of Sound*) involves pre-completion sound testing (PCT) for certain types of home. In an attempt to eliminate the risk of any remedial work being required to completed floor and/or wall constructions (together with the potential for delays in completing the property). In 2003 the Home Builders Federation (HBF) suggested that a series of construction solutions (called Robust Details) should be developed as an alternative to PCT and in 2004 the minister responsible for Building Regulations announced that he would allow Robust Details to be used as an alternative to PCT under a memorandum of understanding (MOU) with a limited company (Robust Details Ltd) which would be responsible for approving, managing and promoting the use of Robust Details as a method of satisfying Building Regulations.

6.5.1.1.1 What is a Robust Detail?

The Robust Details Scheme is an alternative to pre-completion sound testing (PCT) of separating walls and floors in new-build joined houses, bungalows and flats, to demonstrate compliance with the relevant minimum Building Regulation performance standards in England, Wales, Scotland and Northern Ireland.

Robust Details provide builders with a choice of possible construction solutions that have been proven to outperform the standards of Approved Document E, thus eliminating the need for routine PCT!

A Robust Detail is only used in connection with Approved Document E and is defined as *a separating wall or floor (of concrete, masonry, timber, steel or steel-concrete composite) construction, which has been assessed and approved by Robust Details Limited.*

6.5.1.1.2 How are Robust Details approved?

In order to be approved, each Robust Detail must:

- be capable of consistently exceeding the performance standards given in Approved Document E to the Building Regulations;
- be practical to construct on site;
- be reasonably tolerant to workmanship.

6.5.1.1.3 How can Robust Details be used?

Builders are only permitted to use Robust Details instead of PCT **if** the plots concerned have been registered in advance with Robust Details Ltd.

Once a plot has been registered, Robust Details Ltd will provide the relevant registration documentation (which will be accepted by all building control bodies as evidence that the builder is entitled to use Robust Details instead of PCT). The builder will then need to select the Robust Detail specific to the walls and/ or floors he or she wishes to build from the Robust Details handbook (available from Robust Details Ltd) and produce a sitework checklist to show how he or she is going to ensure that building work is carried out exactly in accordance with the Robust Detail specifications.

6.5.1.1.4 Will there be more Robust Details?

Trade associations, manufacturers or other interested parties may submit applications for new Robust Details, and these will be evaluated and, if found acceptable, approved and published.

6.5.1.1.5 Where can I obtain more information?

More information is available from Robust Details Ltd (http://www.robust-details.com) and full contact information is in our 'Useful contact names and addresses' section at the end of this book.

6.5.2 Meeting the requirements

6.5.2.1 Moisture

Floors next to the ground should:	C4.2

- resist the passage of ground moisture to the upper surface of the floor;
- not be damaged by moisture from the ground;
- not be damaged by groundwater;
- resist the passage of ground gases.

Floors next to the ground and floors exposed from below should be designed and constructed so that their structural and thermal performances are not adversely affected by interstitial condensation.	C4.4
All floors should be designed so they do not promote surface condensation or mould growth.	C4.5

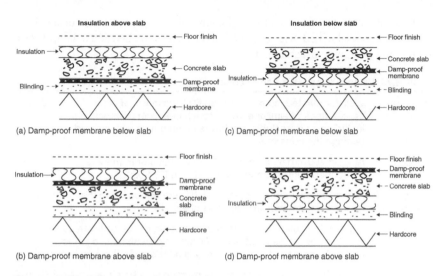

Figure 6.56 Damp-proof courses.

6.5.2.1.1 Ground-supported floors exposed to moisture from the ground

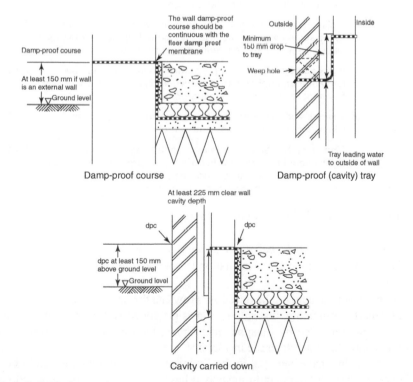

Figure 6.57 Ground-supported floor – construction.

Unless subjected to water pressure, the ground of a ground-supported floor should be covered with dense concrete laid on a hardcore bed and a damp-proof membrane as shown in Table 6.30.

Table 6.30 Ground-supported floor – construction

Hardcore	Well compacted, no greater than 600 mm deep, of clean, broken brick or similar inert material, free from materials including water-soluble sulphates in quantities which could damage the concrete.	C4.7a
Concrete	At least 100 mm thick to mix ST2 in BS 8500 or (if there is embedded reinforcement) to mix ST4 in BS 8500.	C4.7b
Damp-proof membrane	Above or below the concrete which is continuous with the damp-proof courses in walls and piers, etc.	C4.7c
	If below the concrete, the membrane could be formed with a sheet of polyethylene, at least 300 mm thick with sealed joints and laid on a bed of material that will not damage the sheet.	C4.8
	If laid above the concrete, the membrane may be either polyethylene (but without the bedding material) or three coats of cold applied bitumen solution or similar moisture- and water-vapour-resisting material.	C4.9
	In each case it should be protected by either a screed or a floor finish, unless the membrane is pitch mastic or similar material which will also serve as a floor finish.	C4.9
Insulation	If placed beneath floor slabs, it should have sufficient strength to resist the weight of the slab and the anticipated floor loading as well as any possible overloading during construction.	C4.10
	If placed below the damp-proof membrane, it should have low water absorption and (if considered necessary) should be resistant to contaminants in the ground.	C4.10
Timber floor finish	If laid directly on concrete, it may be bedded in a material which can also serve as a damp-proof membrane.	C4.11
	Timber fillets that are laid in the concrete as a fixing for a floor finish should be treated with an effective preservative unless they are above the damp-proof membrane.	C4.11

Note: Suitable insulation may also be incorporated.

Some schools of thought believe that there is a need for an additional damp-proof membrane on top of the insulation to combat interstitial condensation.

This, however, then raises the question 'How can this moisture escape?' Moisture would, presumably, just sit where it is generated and, if interstitial moisture is not controlled by a vapour membrane, then it will surely eventually migrate into the concrete or the insulation!

These points have been put to the Department for Communities and Local Government (DCLG), but unfortunately it has been unable to offer any definite answer – saying that 'the intention of Approved Documents is to provide guidance to the more common building situations and as there may be alternative ways of achieving compliance with the requirements, there is no obligation to adopt any particular solution contained in an Approved Document – if the builder prefers to meet the relevant requirement in some other way'. One of our readers has said that he prefers to employ the insulation below the slab and place a damp-proof membrane between the insulation and the blinding wherever possible (if only for ease of construction), which sounds like a very logical solution.

6.5.2.1.2 Suspended timber ground floors exposed to moisture from the ground

Any suspended timber floor next to the ground should:

- ensure that the ground is covered so as to resist moisture and prevent plant growth;   C4.13a

- have a ventilated air space between the ground covering and the timber;   C4.13b

- have damp-proof courses between the timber and any material which can carry moisture from the ground.   C4.13c

Unless covered with a highly vapour-resistant floor finish, a suspended timber floor next to the ground may be built as shown in Figure 6.58 and as follows:

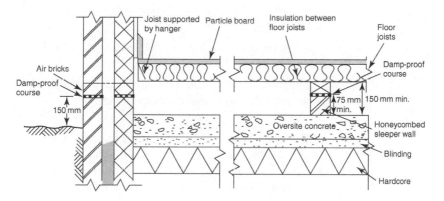

Figure 6.58 Suspended timber floor – construction.

Hardcore   A bed of clean, broken brick or any other inert   C4.14
material free from materials including water-
soluble sulphates in quantities which could
damage the concrete.

Concrete	A ground covering of unreinforced concrete at least 100 mm thick to mix ST 1 in BS 8500 or laid on at least 300 m polyethylene sheet with sealed joints (and itself laid on a bed of material which will not damage the sheet).	C4.14a(i) C4.14a(ii)

Note: To prevent water collecting on the ground covering, either the top should be entirely above the highest level of the adjoining ground or, on sloping sites, consideration should be given to installing drainage on the outside of the upslope side of the building.

Ventilation	There should be a ventilated air space at least 75 mm from the ground (and covering the underside of any wall plates) and at least 150 mm from the underside of the suspended timber floor.	C4.14b
	Two opposing external walls should have ventilation openings placed so that the ventilating air will have a free path between opposite sides and to all parts.	C4.14b
	Ventilating openings should be not less than either 1500 mm²/m run of external wall or 500 mm²/m² of floor area, whichever gives the greater opening area.	C4.14b
	Any pipes needed to carry ventilating air should have a diameter of at least 100 mm.	C4.14b
	Ventilation openings should incorporate suitable grilles to prevent the entry of vermin to the subfloor.	C4.14b
	If floor levels need to be nearer to the ground to provide level access, subfloor ventilation can be provided through offset (periscope) ventilators.	C4.14b
Damp-proof membrane	Damp-proof membranes should be of impervious sheet material, engineering brick or slates in cement mortar or other material which will prevent the passage of moisture.	C4.14c
Timber floor finish	In areas such as kitchens, utility rooms and bathrooms where water may be spilled, any board used as a flooring, irrespective of the storey, should be moisture resistant.	C4.15

In the case of chipboard it should be of one of the grades with improved moisture resistance specified in BS 7331:1990 or BS EN 312 Part 5:1997. — C4.15

Identification marks should be facing upwards. — C4.15

Any softwood boarding should be at least 20 mm thick and from a durable species or treated with a suitable preservative. — C4.15

6.5.2.1.3 Suspended concrete ground floors exposed to moisture from the ground

Concrete suspended floors (including beam and block floors) that are next to the ground should: — C4.17

- adequately prevent the passage of moisture to the upper surface;
- be reinforced to protect against moisture.

There should be a facility for inspecting and clearing out the subfloor voids beneath suspended floors – particularly in localities where flooding is likely.

Hardcore concrete	This is: • in situ concrete at least 100 mm thick containing at least 300 kg of cement for each m³ of concrete; • or precast concrete construction (with or without infilling slabs).
	Reinforcing steel should be protected by a concrete cover of at least 40 mm (if the concrete is in situ) and at least the thickness required for a moderate exposure, if the concrete is precast. — C4.1
Ventilation	There should be a ventilated air space at least 150 mm clear from the ground to the underside of the floor (or insulation if provided). — C4.19b
	Two opposing external walls should have ventilation openings placed so that the ventilating air will have a free path between opposite sides and to all parts. — C4.19b

	Ventilating openings should be not less than either 1500 mm²/m run of external wall or 500 mm²/m² of floor area, whichever gives the greater opening area.	C4.19b
	Any pipes needed to carry ventilating air should have a diameter of at least 100 mm.	C4.19b
	Ventilation openings should incorporate suitable grilles to prevent the entry of vermin to the subfloor.	C4.19b
Damp-proof membrane	A suspended concrete floor should contain a damp-proof membrane (if the ground below the floor has been excavated below the lowest level of the surrounding ground and will not be effectively drained).	C4.19a

6.5.2.1.4 Ground floors and floors exposed from below (resistance to damage from interstitial condensation)

A ground floor (or floor exposed from below such as above an open parking space or passageway (see Figure 6.59) shall be designed in accordance with Clause 8.5 and Appendix D of BS 5250:2002.

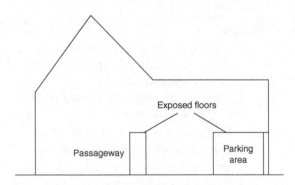

Figure 6.59 Typical floors exposed from below.

6.5.2.1.5 Floors (resistance to surface condensation and mould growth)

Ground floors should be designed and constructed so that the thermal transmittance (U-value) does not exceed 0.7 W/m² K at any point.　　C4.22a

Junctions between elements should be designed in accordance   C4.22b
with robust construction recommendations.

6.5.2.2 Lateral support by floors

Floors should: A1/2 2C33a

- act to transfer lateral forces from walls to buttressing
  walls, piers or chimneys;
- be secured to the supported wall by connections (see
  Figure 6.61).

Intermediate floors and roof shall be constructed so   A1/2 1A2d
that they provide local support to the walls and act as
horizontal diaphragms capable of transferring the wind
forces to buttressing elements of the building.

A wall in each storey of a building should: A1/2 2C32

- extend to the full height of that storey;
- have horizontal lateral supports to restrict movement
  of the wall at right angles to its plane.

Walls should be strapped to floors above ground level, at   A1/2 2C35
intervals not exceeding 2 m and as shown in Figure 6.61
by tension straps conforming to BS EN 845-1.

Where an opening in a floor for a stairway or the like   A1/2 2C37
adjoins a supported wall and interrupts the continuity of
lateral support:

- the maximum permitted length of the opening is to
  be 3 m, measured parallel to the supported wall;
- connections (if provided by means other than by
  anchor) should be throughout the length of each
  portion of the wall situated on each side of the
  opening;
- connections via mild steel anchors should be spaced
  closer than 2 m on each side of the opening to
  provide the same number of anchors as if there were
  no opening;
- there should be no other interruption of lateral
  support.

The maximum span for any floor supported by a wall   A1/2 2C23
is 6 m where the span is measured centre to centre of
bearing (see Figure 6.60).

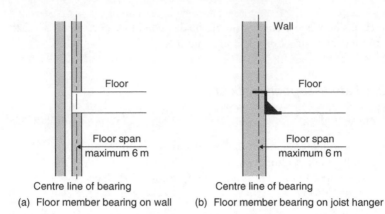

(a) Floor member bearing on wall   (b) Floor member bearing on joist hanger

Figure 6.60 Maximum span of floors.

6.5.2.2.1 Interruption of lateral support

Where an opening in a floor or roof for a stairway or the
like adjoins a supported wall and interrupts the continuity
of lateral support:

- the maximum permitted length of the opening is to be   A1/2 2C37a
  3 m, measured parallel to the supported wall;

- connections (if provided by means other than by   A1/2 2C37b
  anchor) should be throughout the length of each
  portion of the wall situated on each side of the
  opening;

- connections via mild steel anchors should be spaced   A1/2 2C37c
  closer than 2 m on each side of the opening to provide
  the same number of anchors as if there were no
  opening;

- there should be no other interruption of lateral support.   A1/2 2C37d

6.5.2.3 Tension wall straps

Tension straps (conforming to BS EN 845-1) should   A1/2 2C35
be used to strap walls to floors above ground level, at
intervals not exceeding 2 m.

For corrosion resistance purposes, the tension straps   A1/2 2C35
should be material reference 14 or 16.1 or 16.2
(galvanized steel) or other more resistant specifications
including material references 1 or 3 (austenitic stainless
steel).

The declared tensile strength of tension straps should not be less than 8 kN.

Tension straps need **not** be provided:

- in the longitudinal direction of joists in houses of not more than two storeys if the joists: A1/2 2C35a

  ○ are at not more than 1.2 m centres;

  ○ have at least 90 mm bearing on the supported   A1/2 2C35a
     walls or 75 mm bearing on a timber wall plate at each end; or

  ○ are carried on the supported wall by joist   A1/2 2C35b
     hangers (in accordance with BS EN 845-1 and BS 5628 – see Figure 6.61(c)); and

  ○ are incorporated at not more than 2 m centres;   A1/2 2C35b

- when a concrete floor has at least 90 mm bearing on   A1/2 2C35c
  the supported wall (see Figure 6.61(d)); and

- where floors are at or about the same level on each   A1/2 2C35d
  side of a supported wall, and contact between the floors and wall is either continuous or at intervals not exceeding 2 m. Where contact is intermittent, the points of contact should be in line or nearly in line on plan (see Figure 6.61(e)).

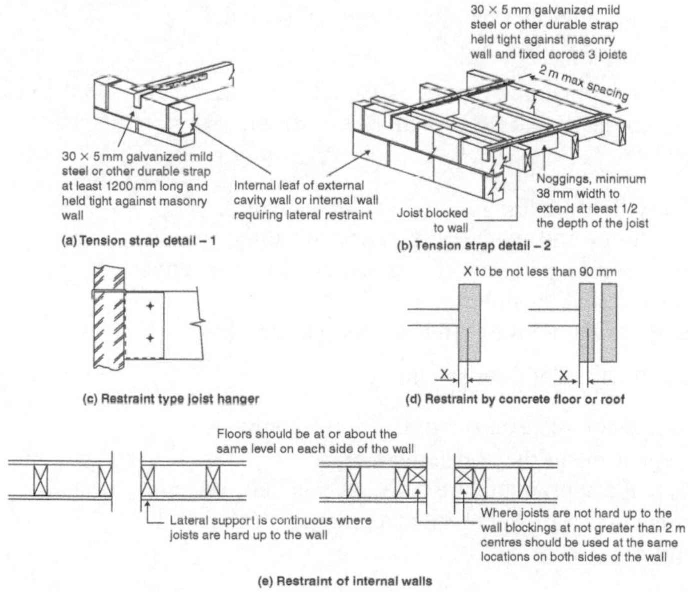

Figure 6.61 Lateral support by floors.

6.5.2.4 *Internal fire spread (structure)*

6.5.2.4.1 Load-bearing elements of structure

All load-bearing elements of a structure shall have a minimum standard of fire resistance.	B3 4.1 (V1) B3 7.1 (V2)
Structural frames, beams, floor structures and gallery structures should have at least the fire resistance given in Appendix A of Approved Document B.	B3 4.2 (V1) B3 7.2 (V2)
When altering an existing two-storey, single-family dwelling house to provide additional storeys, the floor(s), both old and new, shall have the full 30-minute standard of fire resistance.	B3 4.7 (V1)

6.5.2.4.2 Fire resistance – compartmentation

To prevent the spread of fire within a building, whenever possible the building should be subdivided into compartments separated from one another by walls and/or floors of fire-resisting construction.	B3 5.1 (V1) B3 8.1 (V2)
Parts of a building that are occupied mainly for different purposes should be separated from one another by compartment walls and/or compartment floors.	B3 5.3 (V1) B3 8.11 (V2)
The wall and any floor between the garage and the house shall have a 30-minute fire resistance.	B3
In buildings containing flats or maisonettes compartment walls or compartment floors shall be constructed between:	B3 8.13 (V2)

- every floor (unless it is within a maisonette);
- one storey and another within one dwelling;
- every wall separating a flat or maisonette from any other part of the building;
- every wall enclosing a refuse storage chamber.

Every compartment floor should:	B3 5.6 (V1) B3 8.20 (V2)

- form a complete barrier to fire between the compartments they separate; and
- have the appropriate fire resistance as indicated in Appendix A of Approved Document B, Tables A1 and A2.

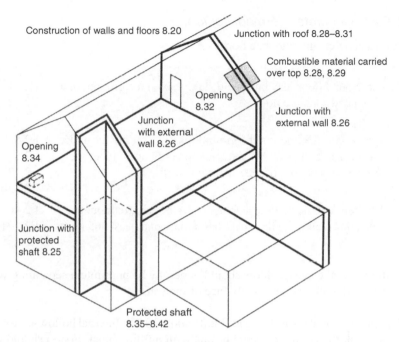

Construction of walls and floors 8.20

Junction with roof 8.28–8.31

Combustible material carried over top 8.28, 8.29

Opening 8.32

Junction with external wall 8.26

Junction with external wall 8.26

Opening 8.34

Junction with protected shaft 8.25

Protected shaft 8.35–8.42

Figure 6.62 Compartment walls and compartment floors with reference to relevant paragraphs in Approved Document B.

Where a compartment wall or compartment floor meets another compartment wall, or an external wall, the junction should maintain the fire resistance of the compartmentation.	B3 5.9 (V1) B3 8.25 (V2)
Junctions between a compartment floor and an external wall that has no fire resistance (e.g. a curtain wall) should be restrained at floor level to reduce the movement of the wall away from the floor when exposed to fire.	B2 8.26 (V2)
Compartment walls should be able to accommodate the predicted deflection of the floor above by either:	B2 8.27 (V2)

- having suitable head detail between the wall and the floor, which can deform but maintain integrity when exposed to a fire; or
- having the wall designed to resist the additional vertical load from the floor above as it sags under fire conditions and thus maintain integrity.

6.5.2.5 Concrete intermediate floor

With a concrete intermediate floor:

The ground floor may be a solid slab, laid on the ground, or a suspended concrete floor.	E2.51 E2.88 E2.126
A concrete slab floor on the ground may be continuous under a solid separating wall but may not be continuous under the cavity masonry core of the separating wall.	E2.51 E2.88 E2.127 E2.130
An internal concrete floor slab may only be carried through a separating wall if the floor base has a mass of at least 365 kg/m².	E2.46 E2.121

 Note: Internal concrete floors should generally be built into a separating wall and carried through to the cavity face of the leaf.

 The cavity should not be bridged (E2.85 and E2.122). Internal hollow-core concrete plank floors (and concrete beams with infilling block floors) should **not** be continuous through or under a separating wall (E2.47, E2.53 and E2.129).

6.5.2.6 Suspended concrete floor

A suspended concrete floor may only pass under a separating wall if the floor has a mass of at least 365 kg/m².	E2.52
A suspended concrete floor should **not** be carried through to the cavity face of the leaf and the cavity should not be bridged	E2.89

6.5.2.7 Separating floors and associated flanking constructions for new buildings

There are three groups of floor, as shown below:

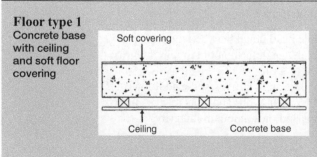

Floor type 1
Concrete base with ceiling and soft floor covering

Soft covering

Ceiling   Concrete base

The resistance to airborne sound depends mainly on the mass per unit area of the concrete base and partly on the mass per unit area of the ceiling. The soft floor covering reduces impact sound at source.

Floor type 2
Concrete base with ceiling and floating floor (three types of floating floor are available, see Table 6.30 on p. 390)

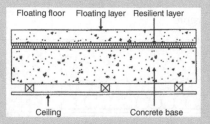

The resistance to airborne and impact sound depends on the mass per unit area of the concrete as well as the mass per unit area and isolation of the floating layer and ceiling. The floating floor reduces impact sound at source.

Floor type 3
Timber frame base with ceiling and platform floor

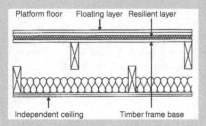

The resistance to airborne and impact sound depends on the structural floor base and the isolation of the platform floor and the ceiling. The platform floor reduces impact sound at source.

6.5.2.7.1 General requirements

Floor types should be capable of achieving the performance standards shown in Table 6.31 on page 405.   E3.1

Care should be taken to correctly detail the junctions between the separating floor and other elements such as external walls, separating walls and floor penetrations.

Note: Where any building element functions as a separating element (e.g. a ground floor that is also a separating floor for a basement flat), then the separating element requirements should take precedence.

6.5.2.7.2 Ceiling treatments

Each floor type should use one of the following three ceiling treatments, which are ranked in order of sound insulation performance from A to C.   E3.17 to E3.18

Note: Use of a better-performing ceiling than that described in the guidance should improve the sound insulation of the floor, provided there is no significant flanking transmission.

Ceiling treatment A
Independent ceiling with absorbent material

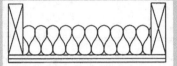

- At least two layers of plasterboard with staggered joints;
- minimum total mass per unit area of plasterboard 20 kg/m²;
- an absorbent layer of mineral wool (minimum thickness 100 mm, minimum density 10 kg/m³) laid in the cavity formed above the ceiling.

The type of ceiling support depends on the floor type

For floor types 1, 2 and 3
Use independent joists fixed only to the surrounding walls.

For floor type 3
Use independent joists fixed to the surrounding walls with additional support provided by resilient hangers attached directly to the floor.

Always ensure:

- that you seal the perimeter of the independent ceiling with tape or sealant;
- you do not create a rigid or direct connection between the independent ceiling and the floor base.

Ceiling treatment B
Plasterboard on proprietary resilient bars with absorbent material

- Single layer of plasterboard, minimum mass per unit area of plasterboard 10 kg/m²;
- fixed using proprietary resilient metal bars;
- an absorbent layer of mineral wool (minimum density 10 kg/m³) should fill the ceiling void.

Ceiling treatment C
Plasterboard on timber battens or proprietary resilient channels with absorbent material

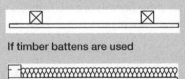

If timber battens are used

If resilient channels are used

- Single layer of plasterboard, minimum mass per unit area 10 kg/m²;
- fixed using timber battens or proprietary resilient channels;
- if resilient channels are used, incorporate an absorbent layer of mineral wool (minimum density 10 kg/m³) that fills the ceiling void.

Notes:

(1) Electrical cables give off heat when in use, and special precautions may be required when they are covered by thermal insulating materials. See BRE BR 262, *Thermal Insulation: Avoiding Risks*, Section 2.3.
(2) Installing recessed light fittings in ceiling treatments A to C can reduce their resistance to the passage of airborne and impact sound.

Extensive cavities in floor voids should be subdivided with cavity barriers.	B3

6.5.2.8 Floors – general

Floors that separate a dwelling from another dwelling (or part of the same building) shall resist the transmission of airborne sounds.	E
Floors above a dwelling that separate it from another dwelling (or another part of the same building) shall resist: • the transmission of impact sound (such as speech, musical instruments and loudspeakers and impact sources such as footsteps and furniture moving); • the flow of sound energy through walls and floors; • the level of airborne sound.	E
Air paths, including those due to shrinkage, must be avoided – porous materials and gaps at joints in the structure must be sealed.	E
The possibility of resonance in parts of the structure (e.g. a dry lining) should be avoided.	E
Flanking transmission (i.e. the indirect transmission of sound from one side of a wall or floor to the other side) should be minimized.	E

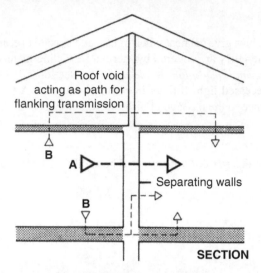

Roof void acting as path for flanking transmission

Separating walls

SECTION

Openings within 700 mm of junctions reduce dimensions of flanking elements and reduce flanking transmission

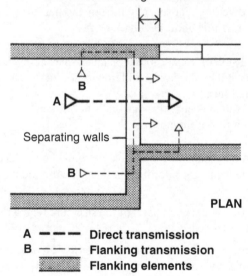

Separating walls

PLAN

A — — — **Direct transmission**
B — — — **Flanking transmission**
        Flanking elements

Figure 6.63 Direct and flanking transmission.

Note: For clarity, not all flanking paths have been shown.

6.5.2.8.1 Requirement E1

Figure 6.64 illustrates the relevant parts of the building that should be protected from airborne and impact sound in order to satisfy Requirement E1.

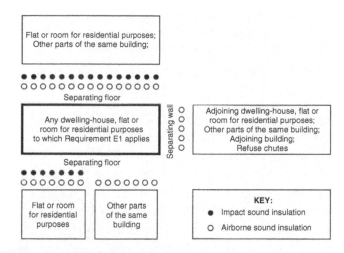

Figure 6.64 Requirement E1 – resistance to sound.

> Floors for rooms for residential purposes and for dwelling   E0.1
> houses and flats that have a separating function should achieve
> the sound insulation values as set out in Table 6.31.

All new floors constructed within a dwelling house (flat or room used for residential purposes) – whether purpose-built or formed by a material change of use – shall meet the laboratory sound insulation values set out in Table 6.31.

Table 6.31 Dwelling houses and flats – performance standards for separating floors and stairs that have a separating function

Type	Airborne sound insulation, DnT,w+Ctr (dB) (minimum values)	Impact sound insulation, LnT,w (dB) (maximum values)
Purpose-built rooms for residential purposes	45	62
Purpose-built dwelling houses and flats	45	62
Rooms for residential purposes formed by material change of use	43	64
Dwelling houses and flats formed by material change of use	43	64

Notes:

(1) The sound insulation values in Table 6.31 include a built-in allowance for 'measurement uncertainty', and so, if any of these test values are not met, that particular test will be considered as failed.

(2) Occasionally, a higher standard of sound insulation may be required between spaces used for normal domestic purposes and noise generated in and to an adjoining communal or non-domestic space.

In these cases it would be best to seek specialist advice before committing yourself.

Flanking transmission from walls and floors connected to the separating wall shall be controlled.	E2
Tests should be carried out between rooms or spaces that share a common area formed by a separating wall or separating floor.	E1
Impact sound insulation tests should be carried out without a soft covering (e.g. carpet, foam-backed vinyl, etc.) on the floor.	E1
When a separating floor is used, a minimum mass per unit area of 120 kg/m^2 (excluding finish) shall always apply, irrespective of the presence or absence of openings.	E2
Care should be taken to correctly detail the junctions between the separating floor and other elements such as external walls, separating walls and floor penetrations.	E3
Spaces between floor joists should be sealed with full depth timber blocking.	E2
If the floor joists are to be supported on the separating wall then they should be supported on hangers and should not be built in.	E2
If the joists are at right angles to the wall, spaces between the floor joists should be sealed with full depth timber blocking.	E3
The floor base (excluding any screed) should be built into a cavity masonry external wall and carried through to the cavity face of the inner leaf.	E
The floor base should be continuous or above an internal masonry wall.	E3
All pipes and ducts that penetrate a floor separating habitable rooms in different flats should: • be enclosed for their full height in each flat; • have fire protection to satisfy Approved Document B (Fire safety).	E3

Notes:

(1) Where any building element functions as a separating element (e.g. a ground floor that is also a separating floor for a basement flat) then the separating element requirements should take precedence.

(2) In some circumstances (e.g. when a historic building is undergoing a material change of use) it may not be practical to improve the sound insulation to the standards set out in Approved Document E1, particularly if the

special characteristics of such a building need to be recognized. In these circumstances the aim should be to improve sound insulation to the 'extent that it is practically possible'.

(3) BS 7913:1998, *The Principles of the Conservation of Historic Buildings*, provides guidance on the principles that should be applied when proposing work on historic buildings.

6.5.2.8.2 Requirement E2

Constructions for new floors within a dwelling-house (flat or   E0.9
room for residential purposes) – whether purpose-built or formed
by a material change of use – shall meet the laboratory sound
insulation values set out in Table 6.32.

Table 6.32 Laboratory values for new internal walls within dwelling houses, flats and rooms for residential purposes – whether purpose-built or formed by a material change of use

	Airborne sound insulation, RW (dB) (minimum values)
Floors in purpose-built dwelling houses and flats	40

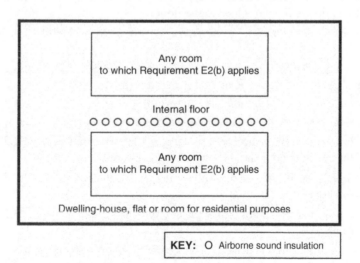

Figure 6.65 Requirement E2(b) – internal floors.

6.5.2.8.3 Requirement E3

Sound absorption measures described in Section 7 of Approved   E0.11
Document N shall be applied.

6.5.2.8.4 Requirement E4

> The values for sound insulation, reverberation time and indoor   E0.12
> ambient noise as described in Section 1 of Building Bulletin
> 93, *The Acoustic Design of Schools* (produced by Department
> for Education and Skills (DFES) and published by The
> Stationery Office (ISBN 0-11-271105-7)) shall be satisfied.

6.5.2.9 Floor type 1: concrete base with ceiling and soft floor covering

A floor of type 1 consists of a concrete floor base with a soft floor covering and a ceiling. Its resistance to airborne sound mainly depends on:

- the mass per unit area of the concrete base;
- the mass per unit area of the ceiling;
- the soft floor covering (which helps to reduce the source of the impact sound).

6.5.2.9.1 General requirements

> To allow for future replacements, the soft floor covering   E3.27a
> should not be fixed or glued to the floor.
>
> To avoid air paths all joints between parts of the floor should   E3.27b
> be filled.
>
> To reduce flanking transmission, air paths should be avoided   E3.27c
> at all points where a pipe or duct penetrates the floor.
>
> A separating concrete floor should be built into the walls   E3.27d
> (around its entire perimeter) if the walls are masonry.
>
> All gaps between the head of a masonry wall and the underside   E3.27e
> of the concrete floor should be filled with masonry.
>
> Flanking transmission from walls connected to the separating   E3.27f
> floor should be controlled.
>
> The floor base shall not bridge the cavity in a two-cavity   E3.27a
> masonry wall.
>
> Non-resilient floor finishes (such as ceramic floor tiles and   E3.27b2
> wood block floors that are rigidly connected to the floor
> base) shall not be used.
>
> Any soft floor covering that is used should be of resilient   E3.28a
> material with an overall uncompressed thickness of at least
> 4.5 mm (also see BS EN ISO 140-8:1998).

Two floor types will meet these requirements:

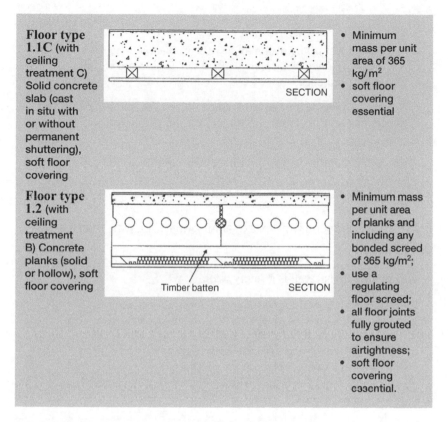

| Floor type 1.1C (with ceiling treatment C) Solid concrete slab (cast in situ with or without permanent shuttering), soft floor covering | SECTION | • Minimum mass per unit area of 365 kg/m² • soft floor covering essential |

| Floor type 1.2 (with ceiling treatment B) Concrete planks (solid or hollow), soft floor covering | Timber batten   SECTION | • Minimum mass per unit area of planks and including any bonded screed of 365 kg/m²; • use a regulating floor screed; • all floor joints fully grouted to ensure airtightness; • soft floor covering essential. |

6.5.2.9.2 Junction requirements for floor type 1

6.5.2.9.2.1 JUNCTIONS WITH AN EXTERNAL CAVITY WALL WITH MASONRY INNER LEAF

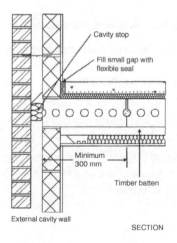

Figure 6.66 Junctions with an external cavity wall with masonry inner leaf.

If the external wall is a cavity wall: E3.31

- the outer leaf of the wall may be of any construction;
- the cavity should be stopped with a flexible closer to ensure adequate drainage.

The masonry inner leaf of an external cavity wall should have a   E3.32
mass per unit area of at least 120 kg/m^2 excluding finish.

The floor base (excluding any screed) should be built into a cav-   E3.33
ity masonry external and carried through to the cavity face of
the inner leaf.

6.5.2.9.2.2 JUNCTIONS WITH AN EXTERNAL CAVITY WALL WITH TIMBER FRAME INNER LEAF

Where the external wall is a cavity wall: E3.36

- the outer leaf of the wall may be of any construction;
- the cavity should be stopped with a flexible closer;
- the wall finish of the inner leaf of the external wall should be two layers of plasterboard, each sheet of plasterboard to be a minimum mass per unit area 10 kg/m^2;
- all joints should be sealed with tape or caulked unenclosed.

6.5.2.9.2.3 JUNCTIONS WITH AN EXTERNAL SOLID MASONRY WALL

No official guidance is currently available about junctions with a solid masonry
wall and so it would always be best to seek specialist advice.

6.5.2.9.2.4 JUNCTIONS WITH INTERNAL FRAMED WALLS

There are no restrictions on internal walls meeting a type 1 separating floor.

6.5.2.9.2.5 JUNCTIONS WITH INTERNAL MASONRY WALLS

The floor base should be continuous through or above an   E3.39
internal masonry wall.

The mass per unit area of any load-bearing internal wall (or   E3.40
any internal wall rigidly connected to a separating floor) should
be at least 120 kg/m^2 excluding finish.

6.5.2.9.2.6 JUNCTIONS WITH FLOOR PENETRATIONS (EXCLUDING GAS PIPES)

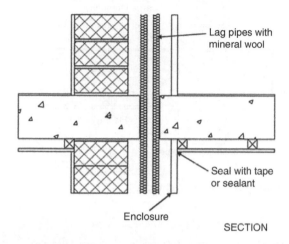

Figure 6.67 Floor type 1 – floor penetrations.

Pipes and ducts should be in an enclosure (both above and below the floor). In all cases:

The enclosure should be constructed of material having a mass per unit area of at least 15 kg/m².	E3.32
Either the enclosure should be lined or the duct (or pipe) within the enclosure wrapped with 25 mm unfaced mineral fibre.	E3.42
Penetrations through a separating floor by ducts and pipes should have fire protection to satisfy Approved Document B (Fire safety).	E3.43
Fire-stopping should be flexible to prevent a rigid contact between the pipe and the floor.	E3.43
Gas pipes may be contained in a separate (ventilated) duct or can remain unenclosed.	E3.43
If a gas service is installed it shall comply with the Gas Safety (Installation and Use) Regulations 1998, SI 1998 No. 2451.	E3.43
If the pipes and ducts penetrate a floor separating habitable rooms in different flats, they should be enclosed for their full height in each flat.	E3.41

6.5.2.9.2.7 JUNCTIONS WITH A SEPARATING WALL TYPE 1 – SOLID MASONRY

For floor types 1.1C and 1.2C, two possibilities exist:

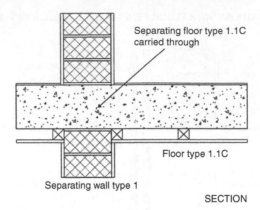

Figure 6.68 Floor type 1.1C – wall type 1.

For a separating floor type 1.2B, the base (excluding any   E3.44
screed) should **not** pass through a separating wall type 1.

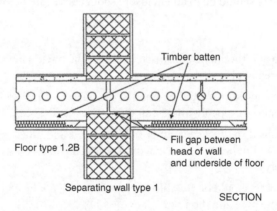

Figure 6.69 Floor type 1.2B – wall type 1.

 It should be noted that for both types of floor, the base (excluding any screed) should **not** pass through a separating wall type 1, but there are no restrictions on internal walls meeting a type 1 separating floor.

6.5.2.9.2.8 JUNCTIONS WITH A SEPARATING WALL TYPE 2 – CAVITY MASONRY

The mass per unit area of any leaf that is supporting or   E3.46
adjoining the floor should be at least 15 kg/m^2 excluding finish.

The floor base (excluding any screed) should be carried through to the cavity face of the leaf.   E3.47

The wall cavity should not be bridged.   E3.47

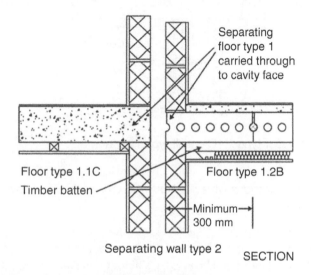

Figure 6.70 Floor types 1.1C and 1.2B – wall type 2.

Where floor type 1.2B is used (and the planks are parallel to the separating wall) the first joint should be a minimum of 300 mm from the inner face of the adjacent cavity leaf.   E3.48

6.5.2.9.2.9 JUNCTIONS WITH SEPARATING WALL TYPES 3.1 AND 3.2 – SOLID MASONRY CORE

A separating floor type 1.1C base (excluding any screed) should pass through separating wall types 3.1 and 3.2.   E3.49

A separating floor type 1.2B base (excluding any screed) should not be continuous through a separating wall type 3.   E3.50

Where a separating wall type 3.2 is used with floor type 1.2B (and the planks are parallel to the separating wall) the first joint should be a minimum of 300 mm from the centreline of the masonry core.   E3.51

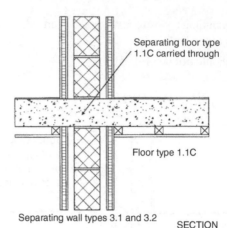

Separating floor type
1.1C carried through

Floor type 1.1C

Separating wall types 3.1 and 3.2   SECTION

Figure 6.71 Floor type 1.1C – wall types 3.1 and 3.2.

6.5.2.9.2.10 JUNCTIONS WITH A SEPARATING WALL TYPE 3.3 – CAVITY
MASONRY CORE

The mass per unit area of any leaf that is supporting or adjoining the floor should be at least 120 kg/m^2 excluding finish.	E3.52
The floor base (excluding any screed) should be carried through to the cavity face of the leaf of the core.	E3.53
The cavity should not be bridged.	E3.53
Where floor type 1.2B is used (and the planks are parallel to the separating wall) the first joint should be a minimum of 300 mm from the inner face of the adjacent cavity leaf of the masonry core.	E3.54

6.5.2.9.2.11 JUNCTIONS WITH A SEPARATING WALL TYPE 4 – TIMBER FRAMES
WITH ABSORBENT MATERIAL

At the time of publication, there is no official guidance available and so it would be advisable to seek specialist advice for this type of situation

6.5.2.10 Floor type 2: concrete base with ceiling and floating floor

A floor of type 2 consists of a concrete floor base with a floating floor (which in turn consists of a floating layer and a resilient layer) and a ceiling. Its resistance to airborne and impact sound depends on:

- the mass per unit area of the concrete base;
- the mass per unit area and isolation of the floating layer and the ceiling;
- the floating floor (which reduces impact sound at source).

6.5.2.10.1 General requirements

All joints between parts of the floor should be filled to avoid air paths.	E3.61a
To reduce flanking transmission, air paths should be avoided at all points where a pipe or duct penetrates the floor.	E3.61b
A separating concrete floor should be built into the walls (around its entire perimeter) if the walls are masonry.	E3.61c
All gaps between the head of a masonry wall and the underside of the concrete floor should be filled with masonry.	E3.61d
Flanking transmission from walls connected to the separating floor should be controlled.	E3.61e
The floor base shall not bridge a cavity in a cavity masonry wall.	E3.61

Two floor types (consisting of a floating layer and resilient layer – see below) will meet these requirements. A performance-based approach (type C) is also available.

Floating floor (a) Timber raft floating layer with resilient layer	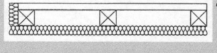	• Timber raft of board material (with bonded edges, e.g. tongued and grooved); • minimum mass per unit area 12 kg/m²; • fixed to 45 mm × 45 mm battens laid loose on the resilient layer (but not along any joints in the resilient layer); • resilient layer of mineral wool (which may be paper faced on the underside) with density 36 kg/m³ and minimum thickness 25 mm.

Floating floor (b) Sand cement screed floating layer with resilient layer

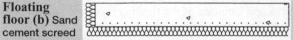

Floating layer: of 65 mm sand–cement screed with a mass per unit area of at least 80 kg/m². Resilient layer: protected while the screed is being laid (e.g. by a 20–50 mm wire mesh) and consisting of either:

- a layer of mineral wool of minimum thickness 25 mm with density 36 kg/m³ (paper faced on the upper side);
- an alternative type of resilient layer with maximum dynamic stiffness of 15 kg/m³;
- an alternative type of resilient layer with minimum thickness of 5 mm (see BS EN ISO 29052-1:1992).

Floating floor (c) Performance-based approach

Floating floors should meet the following specification:

- Rigid boarding above a resilient and/or damping layer;
- weighted reduction in impact sound pressure level of not less than 29 dB (see BS EN ISO 717-2:1997 and BS EN ISO 140-8:1998).

A small gap filled with a flexible sealant should be left between the floating layer and wall at all room edges.   E3.63a

A small gap (approx. 5 mm and filled with a flexible sealant) should be left between skirting and floating layer.   E3.63b

Resilient materials should be laid in rolls or sheets either with lapped joints or with joints tightly butted and taped. — E3.63c

Paper facing should be used on the upper side of fibrous materials to prevent screed entering the resilient layer. — E3.63d

The floating layer and the base or surrounding walls shall not be bridged (e.g. with services or fixings that penetrate the resilient layer). — E3.63a2

The floating screed shall create a bridge (e.g. through a gap in the resilient layer) to the concrete floor base or surrounding walls. — E3.63b2

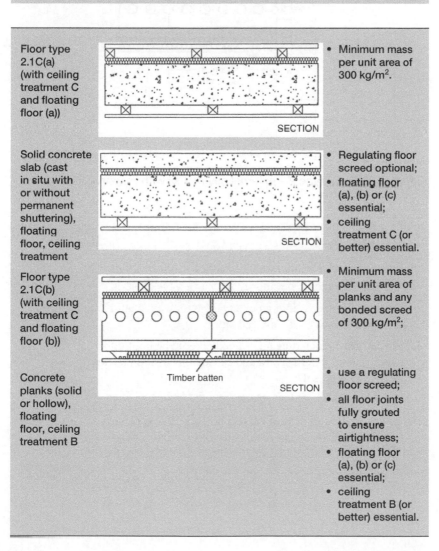

Floor type 2.1C(a) (with ceiling treatment C and floating floor (a))

Solid concrete slab (cast in situ with or without permanent shuttering), floating floor, ceiling treatment

SECTION

- Minimum mass per unit area of 300 kg/m^2.

- Regulating floor screed optional;
- floating floor (a), (b) or (c) essential;
- ceiling treatment C (or better) essential.

Floor type 2.1C(b) (with ceiling treatment C and floating floor (b))

Concrete planks (solid or hollow), floating floor, ceiling treatment B

Timber batten

SECTION

- Minimum mass per unit area of planks and any bonded screed of 300 kg/m^2;

- use a regulating floor screed;
- all floor joints fully grouted to ensure airtightness;
- floating floor (a), (b) or (c) essential;
- ceiling treatment B (or better) essential.

6.5.2.10.2 Junction requirements for floor type 2

6.5.2.10.2.1 JUNCTIONS WITH AN EXTERNAL CAVITY WALL WITH TYPE 2 TIMBER FRAME INNER LEAF

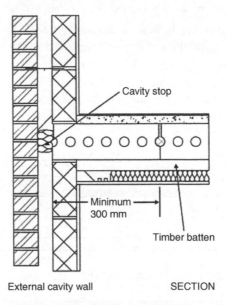

External cavity wall   SECTION

Figure 6.72 Floor type 2 – external cavity wall with masonry internal leaf.

Where the external wall is a cavity wall:	E3.69
• the outer leaf of the wall may be of any construction; • the cavity should be stopped with a flexible closer.	
The masonry inner leaf of an external cavity wall should have a mass per unit area of at least 120 kg/m².	E3.70
The floor base (excluding any screed) should be built into a cavity masonry external wall and carried through to the cavity face of the inner leaf.	E3.71
The cavity should not be bridged.	E3.71
If a floor 2.2B is used (and the planks through, or above, an internal masonry wall are parallel to the external wall) the first joint should be a minimum of 300 mm from the cavity face of the inner leaf.	E3.72

6.5.2.10.2.2 JUNCTIONS WITH AN EXTERNAL CAVITY WALL WITH TIMBER FRAME INNER LEAF

Where the external wall is a cavity wall: E3.74

- the outer leaf of the wall may be of any construction;
- the cavity should be stopped with a flexible closer;
- the wall face of the inner leaf of the external wall should be two layers of plasterboard;
- each sheet of plasterboard should be of minimum mass per unit area 10 kg/m^2;
- all joints should be sealed or caulked with sealant.

6.5.2.10.2.3 JUNCTIONS WITH AN EXTERNAL SOLID MASONRY WALL

No official guidance is currently available and so it is best to seek specialist advice.

6.5.2.10.2.4 JUNCTIONS WITH INTERNAL FRAMED WALLS

There are no restrictions on internal walls meeting a type 4 separating wall.

6.5.2.10.2.5 JUNCTIONS WITH INTERNAL MASONRY WALLS

The floor base should be continuous or above an internal   E3.77
masonry wall.

The mass per unit area of any load-bearing internal wall   E3.78
or any internal wall rigidly connected to a separating floor
should be at least 120 kg/m^2 excluding finish.

6.5.2.10.2.6 JUNCTIONS WITH FLOOR PENETRATIONS (EXCLUDING GAS PIPES)

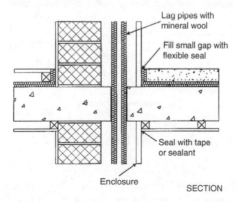

Lag pipes with mineral wool

Fill small gap with flexible seal

Seal with tape or sealant

Enclosure

SECTION

Figure 6.73 Floor type 2 – floor penetrations.

Pipes and ducts that penetrate a floor separating habitable rooms in different flats should be enclosed for their full height in each flat.	E3.79
The enclosure should be constructed of material having a mass per unit area of at least 15 kg/m².	E3.80
Either line the enclosure or wrap the duct or pipe within the enclosure with 25 mm unfaced mineral wool.	E3.80
A small gap (sealed with sealant or neoprene) of about 5 mm should be left between the enclosure and the floating layer.	E3.81
Where floating floor (a) or (b) is used the enclosure may go down to the floor base (provided that the enclosure is isolated from the floating layer).	E3.81
Penetrations through a separating floor by ducts and pipes should have fire protection to satisfy Approved Document B (Fire safety).	E3.82
Gas pipes may be contained in a separate (ventilated) duct or can remain unenclosed.	E3.82
If a gas service is installed it shall comply with the Gas Safety (Installation and Use) Regulations 1998, S1 1998 No. 2451.	E3.82

6.5.2.10.2.7 JUNCTIONS WITH A SEPARATING WALL
TYPE 1 – SOLID MASONRY

A separating floor type 2.1C base (excluding any screed) should pass through a separating wall type 1.	E3.84

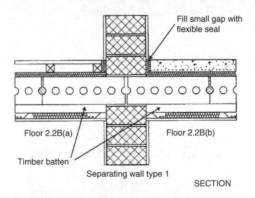

Figure 6.74 Floor type 2.1C – wall types 3.1 and 3.2.

A separating floor type 2.2B base (excluding any screed) should **not** be continuous through a separating wall type 1.	E3.84

6.5.2.10.2.8 JUNCTIONS WITH A SEPARATING WALL TYPE 2 – CAVITY MASONRY

The floor base (excluding any screed) should be carried through to the cavity face of the leaf.   E3.85

The cavity should not be bridged.   E3.85

If a floor type 2.2B is used (and the planks are parallel to the separating wall) the first joint should be a minimum of 300 mm from the cavity face of the leaf.   E3.86

6.5.2.10.2.9 JUNCTIONS WITH SEPARATING WALL TYPES 3.1 AND 3.2 – SOLID MASONRY CORE

A separating floor type 2.1C base (excluding any screed) should pass through separating wall types 3.1 and 3.2.   E3.87

A separating floor type 2.2B base (excluding any screed) should not be continuous through a separating wall type 3.   E3.88

If a separating wall type 3.2 is used with floor type 2.2B (and the planks are parallel to the separating wall) the first joint should be a minimum of 300 mm from the centreline of the masonry core   E3.89

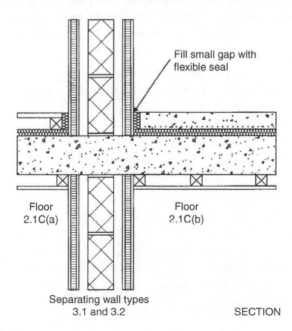

Figure 6.75 Floor type 2.1C – wall types 3.1 and 3.2.

6.5.2.10.2.10 JUNCTIONS WITH A SEPARATING WALL TYPE 3.3 – CAVITY
MASONRY CORE

The mass per unit area of any leaf that is supporting or adjoining the floor should be at least 120 kg/m² excluding finish.	E3.90
The floor base (excluding any screed) should be carried through to the cavity face of the leaf of the core.	E3.91
The cavity should not be bridged.	E3.91
If a floor type 2.2B is used (and the planks are parallel to the separating wall) the first joint should be a minimum of 300 mm from the inner face of the adjacent cavity leaf of the masonry core.	E3.92

6.5.2.10.2.11 JUNCTIONS WITH A SEPARATING WALL TYPE 4 – TIMBER
FRAMES WITH ABSORBENT MATERIAL

Currently there is no official guidance available and so it would be best to seek specialist advice.

6.5.2.11 Floor type 3: timber frame base with ceiling and platform floor

A floor of type 3 consists of a timber frame structural floor base with a deck, platform floor (consisting of a floating layer and a resilient layer) and ceiling treatment. Its resistance to airborne and impact sound depends on:

- the structural floor base;
- the isolation of the platform floor and the ceiling;
- the platform floor (which reduces impact sound at source).

6.5.2.11.1 General requirements

To reduce flanking transmission, air paths should be avoided at all points where the floor is penetrated.	E3.99a
Flanking transmission from walls connected to the separating floor should be as described in the following junction requirements for floor type 3.	E3.99b
There should be no bridge (e.g. formed by services or fixings that penetrate the resilient layer) between the floating layer and the base or surrounding walls.	E3.99
For the platform floor, ensure that:	E3.99

- the correct density of resilient layer is used;
- the layer can carry the anticipated load;
- during construction a gap is maintained between the wall and the floating layer (filled with a flexible sealant, expanded or extruded polystyrene strip);
- resilient materials are laid in sheets with all joints tightly butted and taped.

Floor type 3.1A will meet these requirements.

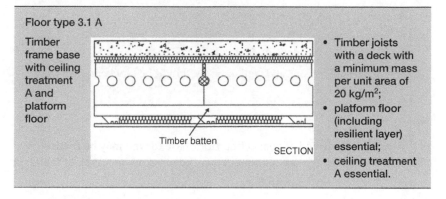

Floor type 3.1 A

Timber frame base with ceiling treatment A and platform floor

Timber batten

SECTION

- Timber joists with a deck with a minimum mass per unit area of 20 kg/m^2;
- platform floor (including resilient layer) essential;
- ceiling treatment A essential.

6.5.2.11.2 Platform floor

The floating layer should: E3.101

- be a minimum of two layers of board material;
- be minimum total mass per unit area 25 kg/m^2;
- have layers of minimum thickness 8 mm;
- be fixed together with joints staggered;
- be laid loose on a resilient layer.

6.5.2.11.3 Resilient layer

The resilient layer should be of mineral wool: E3.102

- minimum thickness 25 mm;
- density 60 kg/m^3 to 100 kg/m^3;
- paper faced on the underside.

6.5.2.11.4 Junction requirements for floor type 3

6.5.2.11.4.1 JUNCTIONS WITH AN EXTERNAL CAVITY WALL WITH MASONRY INNER LEAF

Where the external wall is a cavity wall: • the outer leaf of the wall may be of any construction; • the cavity should be stopped with a flexible closer.	E3.103
The masonry inner leaf of a cavity wall should be lined with an independent panel.	E3.104
The ceiling should be taken through to the masonry.	E3.105
The junction between the ceiling and the independent panel should be sealed with tape or caulked with sealant.	E3.105
Air paths between floor and wall cavities should be blocked.	E3.106

Notes:

(1) Any normal method of connecting floor base to wall may be used.
(2) Independent panels are not required if the mass per unit area of the inner leaf is greater than 375 kg/m^2.

6.5.2.11.4.2 JUNCTIONS WITH AN EXTERNAL CAVITY WALL WITH TIMBER FRAME INNER LEAF

Where the external wall is a cavity wall: • the outer leaf of the wall may be of any construction; • the cavity should be stopped with a flexible closer.	E3.109
The wall finish of the inner leaf of the external wall should: • be two layers of plasterboard; • be each sheet of plasterboard of minimum mass per unit area 10 kg/m^2; • have all joints sealed with tape or caulked with sealant.	E3.110
Any normal method of connecting floor base to wall may be used.	E3.111
If the joists are at right angles to the wall, spaces between the floor joists should be sealed with full depth timber blocking.	E3.112
The junction between the ceiling and wall lining should be sealed with tape or caulked with sealant.	E3.113

6.5.2.11.4.3 JUNCTIONS WITH AN EXTERNAL SOLID MASONRY WALL

Currently there is no official guidance available and so it would be best to seek specialist advice.

6.5.2.11.4.4 JUNCTIONS WITH INTERNAL FRAMED WALLS

The spaces between joists are at right angles and should be sealed with full depth timber blocking.	E3.114
The junction between the ceiling and the internal framed wall should be sealed with tape or caulked with sealant.	E3.115

6.5.2.11.4.5 JUNCTIONS WITH INTERNAL MASONRY WALLS

Currently there is no official guidance available and so it would be best to seek specialist advice.

6.5.2.11.4.6 JUNCTIONS WITH FLOOR PENETRATIONS (EXCLUDING GAS PIPES)

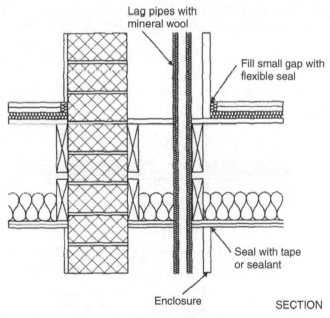

Figure 6.76 Floor type 3 – floor penetrations.

Pipes and ducts that penetrate a floor separating habitable rooms in different flats should be enclosed for their full height in each flat.	E3.117

The enclosure should:

- be constructed of material having a mass per unit area of   E3.118
  at least 15 kg/m²;

- have a small, sealed (with sealant or neoprene) 5 mm gap   E3.119
  between the enclosure and floating layer;

- go down to the floor base;   E3.119

- be isolated from the floating layer.   E3.119

The duct or pipe within the enclosure should be lined or   E3.118
wrapped with 25 mm unfaced mineral wool.

Penetrations through a separating floor by ducts and pipes   E3.120
should have fire protection to satisfy Approved Document B
(Fire safety).

Fire-stopping should be flexible and also prevent rigid contact   E3.121
between the pipe and floor.

Gas pipes may be contained in a separate (ventilated) duct or   E3.120
can remain unenclosed.

If a gas service is installed it shall comply with the Gas   E3.120
Safety (Installation and Use) Regulations 1998, S1 1998 No.
2451.

6.5.2.11.4.7 JUNCTIONS WITH A SEPARATING WALL TYPE 1 – SOLID MASONRY

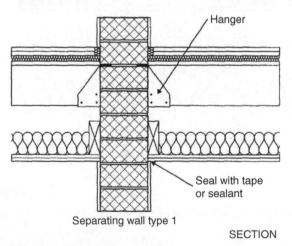

Figure 6.77 Floor type 3 – wall type 1.

| Floor joists supported on a separating wall should be supported on hangers as opposed to being built in. | E3.121 |
| The junction between the ceiling and wall should be sealed with tape or caulked with sealant. | E3.122 |

Note: The above is particularly relevant for flats where there are separating walls.

6.5.2.11.4.8 JUNCTIONS WITH A SEPARATING WALL TYPE 2 – CAVITY MASONRY

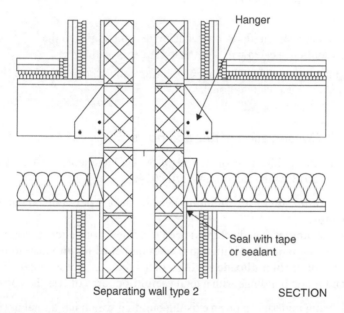

Separating wall type 2   SECTION

Figure 6.78 Floor type 3 – wall type 2.

Floor joists that are supported on a separating wall should be supported on hangers and not built in.	E3.123
The adjacent leaf of a cavity separating wall should be lined with an independent panel.	E3.124
The ceiling should be taken through to the masonry.	E3.125
The junction between the ceiling and the independent panel should be sealed with tape or caulked with sealant.	E3.125

Note: Independent panels are not required if the mass per unit area of the inner leaf is greater than 375 kg/m^2.

6.5.2.11.4.9 JUNCTIONS WITH A SEPARATING WALL TYPE 3 – MASONRY BETWEEN INDEPENDENT PANELS

Floor joists that are supported on a separating wall should be supported on hangers and not built in.	E3.127
The ceiling should be taken through to the masonry.	E3.128
The junction between the ceiling and the independent panel should be sealed with tape or caulked with sealant.	E3.128

6.5.2.11.4.10 JUNCTIONS WITH A SEPARATING WALL TYPE 4 – TIMBER FRAMES WITH ABSORBENT MATERIAL

Spaces between the floor joists that are at right angles to the wall should be sealed with full depth timber blocking.	E3.129
The junction of the ceiling and wall lining should be sealed with tape or caulked with sealant.	E3.130

6.5.2.12 Conservation of fuel and power

Following the publication of new Approved Documents L1A and L1B in November 2013, with effect from April 2014:

- all **new non-domestic buildings** need to make **9 per cent carbon dioxide savings**;
- all **new homes** (i.e. dwellings) need to achieve, or better, a fabric energy-efficiency target in addition to the strengthened requirement to deliver **6 per cent carbon dioxide savings**. (This requirement is not applicable to stand-alone buildings with a total useful floor area of less than $50\,m^2$.

If work being undertaken on an existing building which has a total useful floor area of over $1000\,m^2$, then, in addition to the principal works (which must still comply with the energy-efficiency requirements detailed in Approved Document L in the normal way), consequential improvements (where technically, functionally and economically feasible) will have to be completed, and

If upgrades are proposed to the existing dwelling, then such upgrades should be implemented to a standard that is no worse than that shown in column (b) of Table 6.33.	L1B 4.7

Table 6.33 Upgrading retained thermal elements

Element	(a) Threshold U-value (W/m² K)	(b) Improved U-value (W/m² K)
Floor	0.70	0.25

In addition, the standards for new thermal elements should be no worse than those shown in Table 6.34.

Table 6.34 Upgrading required for new thermal elements

Location	Element	Standard (W/m² K)
Conservatories and porches	Floor	0.26
Fabric standards	Floor	0.20
Buildings (other than dwellings)	Floor	0.25

6.5.2.12.1 New domestic buildings

If work is being undertaken in a **new** domestic building, then the requirements of LIA prevail:

Element or system	Values
Opening areas (windows and doors)	Same as actual dwelling up to a maximum proportion of 25% of total floor area
Floor	0.13 W/m² K

and the limiting standards for a domestic building's fabric shall be:

Floor	0.25 W/m² K

6.5.2.12.2 New homes

- If work is being undertaken in a new home, with a total useful floor area of less than 50 m², then the requirements of L2A must be met:

Element or system	Values
Floor	0.22 W/m² K

6.5.2.12.3 Energy meters

Energy metering systems should be capable of enabling and ensuring that:

• in buildings with a total useful floor area greater than 1000m², automatic meter reading and data collection facilities are available.	L2A 2.47

and the limiting standards for a domestic building's fabric shall be:

Floor	0.25 W/m² K

6.6 Walls

In a brick-built house, the external walls are load-bearing elements that support the roof, floors and internal walls. These walls are normally cavity walls comprising two leaves braced with metal ties, but older houses will have solid walls, at least 225 mm (9 in.) thick. Bricks are laid with mortar in overlapping bonding patterns to give the wall rigidity and a damp-proof course is laid just above ground level to prevent the moisture rising. Window and door openings are spanned above with rigid supporting beams called 'lintels'. The internal walls of a brick-built house are either non-load-bearing divisions (made from lightweight blocks, manufactured boards or timber studding) or load-bearing structures made of brick or block.

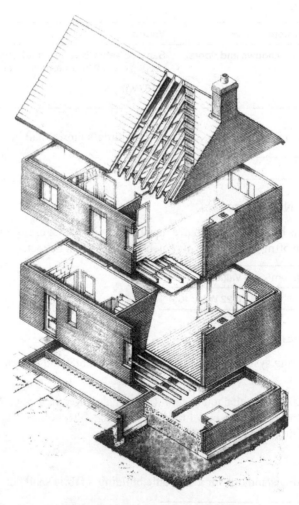

Figure 6.79 Brick-built house – typical components.

Modern timber-framed house walls are constructed of vertical timber studs with horizontal top and bottom plates nailed to them. The frames, which are erected on a concrete slab or a suspended timber platform supported by cavity brick walls, are faced on the outside with plywood sheathing to stiffen the structure. Breather paper is fixed over the top to act as a moisture barrier. Insulation quilt is used between studs. Rigid timber lintels at openings carry the weight of the upper floor and roof. Brick cladding is typically used to cover the exterior of the frame. It is attached to the frame with metal ties. Weatherboarding often replaces the brick cladding on upper floors.

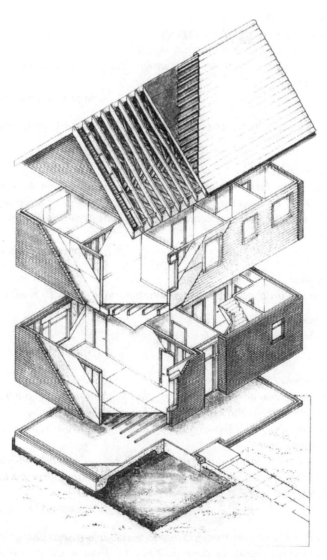

Figure 6.80 Timber-framed house – typical components.

Note: When reading this section, you will probably notice that a few of the requirements are also covered in Sections 6.5 (Floors) and 6.7 (Ceilings). This has been done in order to save the reader having to constantly turn back and reread a previous page.

6.6.1 Requirements

Moisture
The walls of the building shall adequately protect the building and people who use the building from harmful effects caused by:

- *ground moisture;*
- *precipitation and wind-driven spray;*
- *interstitial and surface condensation; and*
- *spillage of water from or associated with sanitary fittings or fixed appliances.*

(Approved Document C2)

Amendments to the 2010 version of Approved Document C came into force in 2013.

Cavity insulation
Fumes given off by insulating materials such as by urea formaldehyde (UF) foams should not be allowed to penetrate occupied parts of buildings to an extent where it could become a health risk to persons in the building by becoming an irritant concentration.

(Approved Document D)

Fire precautions
Materials and/or products used for the internal linings of walls shall restrict:

- *the spread of flame;*
- *the amount of heat released.*

(Approved Document B2)

Internal fire spread (structure)
A wall common to two or more buildings shall be designed and constructed so that it adequately resists the spread of fire between those buildings.

The building should be sub-divided by elements of fire-resisting construction into compartments.

Any hidden voids in the construction shall be sealed and sub-divided to inhibit the unseen spread of fire and products of combustion, in order to reduce the risk of structural failure, and the spread of fire.

(Approved Document B3)

External walls shall be constructed so as to have a low rate of heat release and thereby be capable of reducing the risk of ignition from an external source and the spread of fire over their surfaces.

The amount of unprotected area in the sides of the building shall be restricted so as to limit the amount of thermal radiation that can pass through the wall.

(Approved Document B4)

Airborne and impact sound
Dwellings shall be designed so that the noise from domestic activity in an adjoining dwelling (or other parts of the building) is kept to a level that:

- *does not affect the health of the occupants of the dwelling;*
- *will allow them to sleep, rest and engage in their normal activities in satisfactory conditions.*

(Approved Document E1)

Dwellings shall be designed so that any domestic noise that is generated internally does not interfere with the occupants' ability to sleep, rest and engage in their normal activities in satisfactory conditions.

(Approved Document E2)

Domestic buildings shall be designed and constructed so as to restrict the transmission of echoes.

(Approved Document E3)

Schools shall be designed and constructed so as to reduce the level of ambient noise (particularly echoing in corridors).

(Approved Document E4)

Conservation of fuel and power
Reasonable provision shall be made for the conservation of fuel and power in buildings by:

(a) limiting heat gains and losses

  (i) through thermal elements and other parts of the building fabric; and
  (ii) from pipes, ducts and vessels used for space heating, space cooling and hot water services;

(b) providing fixed building services which

  (i) are energy efficient; and
  (ii) have effective controls and are commissioned by testing and adjusting as necessary to ensure they use no more fuel and power than is reasonable in the circumstances.

(Approved Document L)

 A new version of Approved Document L came into force on 6 April 2014.

6.6.2 Meeting the requirements

6.6.2.1 General

Walls should comply with the relevant requirements of BS EN 1996-2:2006 (except those walls that meet the conditions given in paragraphs 2C4 and 2C14 to 2C38 of Approved Document A).

 Note: Also see BS EN 1996-1-1:2005 if design strengths (and suitability) of walls using masonry units with different compressive strengths are being considered.

6.6.2.1.1 Basic requirements for stability

The layout of walls (both internal and external) shall:	A1/2 1A2b
• form a robust three-dimensional box structure in plan; • be constructed according to the specific guidance for each form of construction.	
Internal and external walls shall be adequately connected either by masonry bonding or by using mechanical connections.	A1/2 1A2c

6.6.2.2 Building height

For residential buildings, the maximum height of the building measured from the lowest finished ground level adjoining the building to the highest point of any wall or roof should not be greater than 15 m.	A1/2 2C4i
Types of wall shown in Table 6.35 must extend to the full storey height.	A1/2 2C2

Table 6.35 Wall types

Residential buildings of up to three storeys	Small, single-storey, non-residential buildings and annexes
External walls	External walls
Internal load-bearing walls	Internal load-bearing walls
Compartment walls	
Separating walls	

6.6.2.3 Small, single-storey, non-residential buildings and annexes

6.6.2.3.1 General construction

The walls shall be solidly constructed in brickwork or blockwork.	A1/2 2C38(i)b
Where the floor area of the building or annexe exceeds $10\,m^2$, the walls shall have a mass of not less than 130 kg/m^2.	A1/2 2C38(i)c
The only lateral loads are wind loads.	A1/2 2C38(i)e
The maximum length or width of the building or annexe shall not exceed 9 m.	A1/2 2C38(i)
The height of the building or annexe shall not exceed the lower value derived from Figure 6.81.	A1/2 2C38(i)

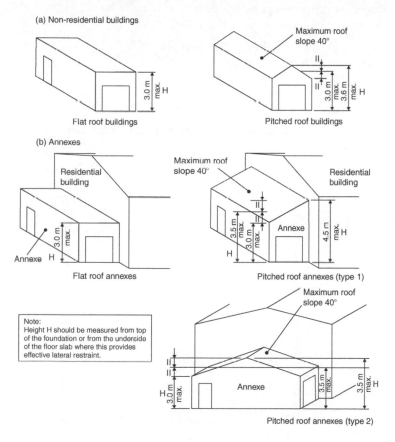

Figure 6.81 Size and proportions of non-residential buildings and annexes.

> Walls shall be tied to the roof structure vertically and   A1/2 2C38(i)i
> horizontally and have a horizontal lateral restraint at
> roof level.

 Note: The roof should be braced at rafter level, horizontally at eaves level and at the base of any gable by roof decking, rigid sarking or diagonal timber bracing, as appropriate, in accordance with BS EN 1995-1-1:2004.

6.6.2.3.2 Size and location of openings

> One or two major openings not more than 2.1 m in   A1/2 2C38(ii)
> height are permitted in one wall of the building or
> annexe only.
>
> The width of a single opening or the combined width of   A1/2 2C38(ii)
> two openings should not exceed 5 m.
>
> The only other openings permitted in a building or   A1/2 2C38(ii)
> annexe are for windows and a single-leaf door.
>
> The size and location of these openings should be in   A1/2 2C38(ii)
> accordance with Figure 6.82.

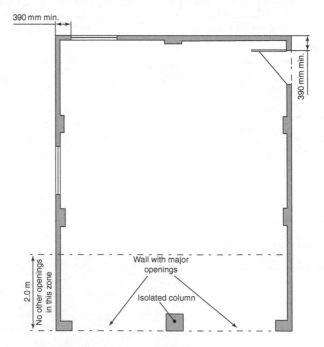

Figure 6.82 Size and location of openings.

Major openings should be restricted to one wall only. Their aggregate width should not exceed 0.5 m and their height should not be greater than 2.1 m.	A1/2 2C38(ii)
There should be no openings within 2.0 m of a wall containing a major opening.	A1/2 2C38(ii)
The aggregate size of the openings in a wall not containing a major opening should not exceed 2.4 m.	A1/2 2C38(ii)
There should not be more than one opening between piers.	A1/2 2C38(ii)
Unless there is a corner pier, the distance from a window or a door to a corner should not be less than 390 mm.	A1/2 2C38(ii)

6.6.2.4 Thickness of walls

The thickness of the wall depends on the general conditions relating to the building of which the wall forms a part (e.g. floor area, roof loading, wind speed, etc.) and the design conditions relating to the wall (e.g. type of materials, loading, end restraints, openings, recesses, overhangs, lateral floor support requirements, etc.).

 Note: Where walls are constructed of bricks or blocks, they shall be in accordance with BS EN 1991-1-4:2005.

6.6.2.5 Chases

Vertical chases should not be deeper than 1/3 of the wall thickness.	A1/2 2C30a

 Note: Or, in cavity walls, 1/3 of the thickness of the leaf.

Horizontal chases should not be deeper than 1/6 of the thickness of the leaf of the wall.	A1/2 2C30b
Chases should not be so positioned as to impair the stability of the wall (particularly where hollow blocks are used).	A1/2 2C30c

6.6.2.6 Overhangs

The amount of any projection should not impair the stability of the wall.	A1/2 2C31

6.6.2.7 Wall cladding

Wall cladding presents a hazard if it becomes detached from the building. An acceptable level of safety can be achieved depending on the type and location of the cladding.

 The guidance given below relates to **all** forms of cladding, including curtain walling and glass façades.

Cladding shall be capable of safely sustaining and transmitting (to the supporting structure of the building) all dead, imposed and wind loads.	A1/2 3.2a
Provision shall be made, where necessary, to accommodate differential movement of the cladding and the supporting structure of the building.	A1/2 3.2
Wind loading on the cladding should be derived from BS EN 1991-1-4:2005.	A1/2 3.3
Due consideration shall be given to local increases in wind suction arising from funnelling of the wind through gaps between buildings.	A1/2 3.3

 Note: Guidance on funnelling effects is given in BRE Digest 436, *Wind Loading on Buildings: Brief Guidance for Using BS 6399-2:1997*, available from BRE, Bucknalls Lane, Garston, Watford, Herts WD2 7JR.

The cladding shall be securely fixed to, and supported by, the structure of the building using both vertical support and horizontal restraint.	A1/2 3.2b
The cladding and its fixings (including any support components) shall be of durable materials.	A1/2 3.2d
The design life of the fixings shall not be less than that of the cladding.	A1/2 3.2d
Fixings shall be corrosion resistant and of a material type appropriate for the local environment.	A1/2 3.2d
Fixings for supporting cladding should be determined from a consideration of the proven performance of the fixing and the risks associated with the particular application.	A1/2 3.7

The strength of fixings should be derived from tests using materials representative of the material into which the fixing is to be anchored, taking account of any inherent weaknesses that may affect the strength of the fixing (e.g. cracks in concrete due to shrinkage and flexure, or voids in masonry construction).	A1/2 3.8
Where the cladding is required to support other fixtures (e.g. handrails or fittings such as antennas and signboards) account should be taken of the loads and forces arising from such fixtures and fittings.	A1/2 3.4
Where the wall cladding is required to function as pedestrian guarding to stairs, ramps, vertical drops of more than 600 mm in dwellings or more than the height of two risers (or 380 mm if not part of a stair) in other buildings or as a vehicle barrier, account should be taken of the additional imposed loading as stipulated in Approved Document K.	A1/2 3.5
Where the wall cladding is required to safely withstand lateral pressures from crowds, an appropriate design loading is given in BS EN 1991-1-1:2002 with its UK National Annex and the *Guide to Safety at Sports Grounds* (4th edition, 1997).	A1/2 3.6
Applications should be designated as being either non-redundant (where the failure of a single fixing could lead to the detachment of the cladding) or redundant (where failure or excessive movement of one fixing results in load sharing by adjacent fixings) and the required reliability of the fixing determined accordingly.	A1/2 3.7
All cladding (used to protect the building from rain or snow) shall be jointless or have sealed joints.	C4 5.1–5.6

 Note: Large glass panels in cladding of walls and roofs (where the cladding is not divided into small areas by load-bearing framing) need special consideration. Guidance is given in the Institution of Structural Engineers' report *Structural Use of Glass in Buildings*, 1999, available from 11 Upper Belgrave Street, London SW1X 8BH, and *Nickel Sulfide in Toughened Glass*, published by the Centre for Window Cladding and Technology, 2000.

Further guidance on cladding is also provided in the following documents:

*   The Institution of Structural Engineers' report *Aspects of Cladding*, 1995;
*   The Institution of Structural Engineers' report *Guide to the Structural Use of Adhesives*, 1999;

- BS 8297 *Code of Practice for the Design and Installation of Non-Load-bearing Precast Concrete Cladding*;
- BS 8298 *Code of Practice for the Design and Installation of Natural Stone Cladding and Lining.*

6.6.2.8 Masonry units

Walls should be properly bonded and solidly put together with mortar and constructed of masonry units conforming to the following standards:

> - clay bricks or blocks to BS EN 771-1;   A1/2 2C20
> - calcium silicate bricks or blocks to BS EN 771-2;
> - concrete bricks or blocks to BS EN 771-3 or BS EN 771-4;
> - manufactured stone to BS EN 771-5;
> - square dressed natural stone to the appropriate requirements described in BS EN 771-6.

 Note: See BS 3921, BS 6073-1, BS 187, BS 5390 and BS 6649 for further details about the minimum compressive strength requirements for masonry units.

6.6.2.9 Mortar

> Mortar should be one of the following: A1/2.2C22
>
> - mortar designation (iii) according to BS EN 1996-1-1:2005;
> - strength class M4 according to BS EN 998-2:2010;
> - 1:1:5 to 6 CEM I, lime, and fine aggregate measured by volume of dry materials; or
> - of equivalent or greater strength and durability to the specifications in a. above.

6.6.2.10 Tension straps

> Tension straps (conforming to BS EN 845-1) should be   A1/2 2C35
> used to strap walls to floors above ground level, at intervals
> not exceeding 2 m and as shown in Figure 6.83.

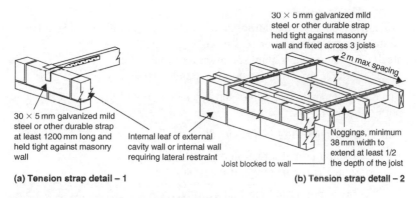

(a) Tension strap detail – 1

(b) Tension strap detail – 2

Figure 6.83 Lateral support by floors.

Gable walls should be strapped to roofs as shown in Figure   A1/2 2C36
6.84(a) and (b) by tension straps.

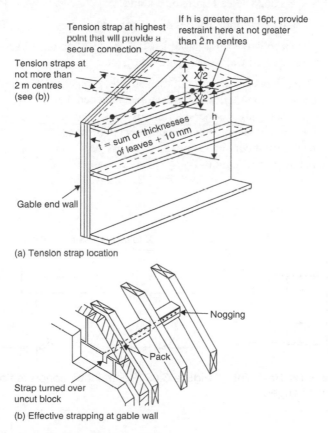

(a) Tension strap location

(b) Effective strapping at gable wall

Figure 6.84 Lateral support at roof level.

For corrosion resistance purposes, tension straps should be material reference 14 or 16.1 or 16.2 (galvanized steel) or other more resistant specifications including material references 1 or 3 (austenitic stainless steel).   A1/2 2C35

The declared tensile strength of tension straps should not be less than 8 kN.

Tension straps need **not** be provided: A1/2 2C35

- in the longitudinal direction of joists in houses of not more than two storeys if the joists:
  - are at not more than 1.2 m centres;   A1/2 2C35a
  - have at least 90 mm bearing on the supported walls or 75 mm bearing on a timber wall plate at each end; or   A1/2 2C35a
  - are carried on the supported wall by joist hangers (in accordance with BS EN 845-1 and BS 5628 – see Figure 6.85); and   A1/2 2C35b
  - are incorporated at not more than 2 m centres;   A1/2 2C35b
- when a concrete floor has at least 90 mm bearing on the supported wall (see Figure 6.86); and   A1/2 2C35c
- where floors are at or about the same level on each side of a supported wall, and contact between the floors and wall is either continuous or at intervals not exceeding 2 m. Where contact is intermittent, the points of contact should be in line or nearly in line on plan (Figure 6.87).   A1/2 2C35d

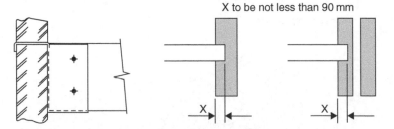

X to be not less than 90 mm

Figure 6.84 Restraint type joist hanger.

Figure 6.85 Restraint by concrete floor or roof.

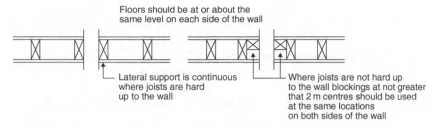

Floors should be at or about the
same level on each side of the wall

— Lateral support is continuous
where joists are hard
up to the wall

— Where joists are not hard up
to the wall blockings at not greater
that 2 m centres should be used
at the same locations
on both sides of the wall

Figure 6.87 Restraint of internal walls.

6.6.2.11 Vertical lateral restraint to walls

The ends of every wall should be bonded or otherwise   A1/2 2C25
securely tied throughout their full height to a buttressing
wall, pier or chimney.

Long walls may be provided with intermediate buttressing A1/2 2C25
walls, piers or chimneys dividing the wall into distinct
lengths within each storey.

Note: Each distinct length is considered to be a
supported wall for the purposes of the Building
Regulations.

Intermediate buttressing walls, piers or chimneys should A1/2 2C25
provide lateral restraint to the full height of the supported
wall.

They may be staggered at each storey.

A wall in each storey of a building should: A1/2 2C32

*   extend to the full height of that storey;

*   have horizontal lateral supports to restrict movement
    of the wall at right angles to its plane.

The requirements for lateral restraint are shown in Table A1/2 2C34
6.36.

Table 6.36 Lateral support for walls

Wall type	Wall length	Lateral support required
Solid or cavity: external compartment separating	Any length	Roof lateral support by every roof forming a junction with the supported wall.
	Greater than 3 m	Floor lateral support by every floor forming a junction with the supported wall.
Internal load-bearing wall (not being a compartment or separating wall)	Any length	Roof or floor lateral support at the top of each storey.

Walls should be strapped to floors above ground level, at   A1/2 2C35
intervals not exceeding 2 m and, as shown in Figure 6.92,
by tension straps conforming to BS EN 845-1.

6.6.2.12 Internal load-bearing walls in brickwork or blockwork

The maximum span for any floor supported by a wall   A1/2 2C23
is 6 m, where the span is measured centre to centre of
bearing (Figure 6.88).

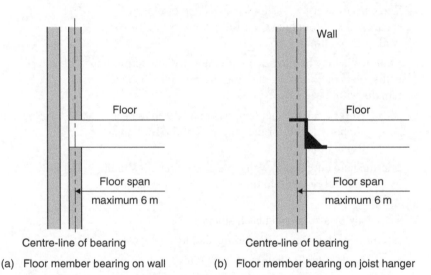

(a)   Floor member bearing on wall   (b)   Floor member bearing on joist hanger

Figure 6.88 Maximum span of floors.

Vertical loading on walls should be distributed.   A1/2 2C23a

Differences in level of ground or other solid   A1/2 2C23b
construction between one side of the wall and the
other should be less than four times the thickness of
the wall, as shown in Figure 6.89.

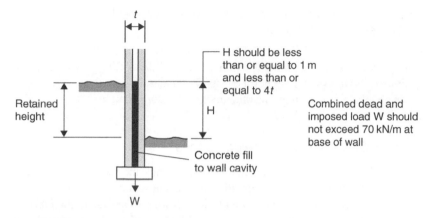

Figure 6.89 Combined and imposed dead load.

Dead load, imposed load and wind load should be in accordance with current codes of practice.	A1/2 0.2b
Loads used in calculations should allow for possible dynamic, concentrated and peak load effects that may occur.	A1/2 0.2
All walls (except compartment and/or separating walls) should have a thickness not less than:	A1/2 2C10

$$\frac{\text{Specified thickness from Table 6.32} - 5\text{mm}}{2}$$

Note: Except for a wall in the lowest storey of a three-storey building, carrying load from both upper storeys, walls should have a thickness as determined by the equation or 140 mm, whichever is the greater.

6.6.2.13 Solid external walls

6.6.2.13.1 Compartment walls and separating walls in coursed brickwork or blockwork

Solid walls constructed of coursed brickwork or blockwork should be at least as thick as 1/16 of the storey height.	A1/2 2C6

6.6.2.13.2 Compartment walls and separating walls in uncoursed stone, flints, etc.

> The thickness of walls constructed in uncoursed stone, flints,   A1/2 2C7
> clunches or bricks or other burnt or vitrified material should
> not be less than 1.33 times the thickness of the storey height.

6.6.2.14 Cavities and concealed spaces

Concealed spaces or cavities in walls, floors, ceilings and roofs will provide an easy route for smoke and flame spread which, because it is hidden, will present a greater danger than would be more obvious from a weakness in the fabric of the building. To overcome this danger, buildings shall be designed and constructed so that the unseen spread of fire and smoke within concealed spaces in their structure and fabric is prevented.

Note: With the introduction of the 2013 (amended) Approved Document B, window and door frames are now **only** suitable for use as cavity barriers if they are constructed of steel or timber of an appropriate thickness.

6.6.2.14.1 Provision of cavity barriers

> Cavity barriers should be provided:
> *   at the edges of cavities;   B3 6.3 (V1)
> *   around window and door openings;   B3 9.3 (V2)
> *   at the junction between an external cavity wall and a
>   compartment wall;
> *   at the junction between an external cavity wall and a
>   compartment floor;
> *   at the top of such an external cavity wall;
> *   at the junction between an internal cavity wall and
>   any assembly which forms a fire-resisting barrier;
>
> *   above the enclosures to a protected stairway in a   B1 2.14
>   dwelling house with a floor more than 4.5 m above
>   ground level.
>
> Cavity barriers need not be provided between
>   double-skinned corrugated or profiled insulated   B3 6.4 (V1)
>   roof sheeting, if the sheeting is a material of limited   B3 9.5 (V2)
>   combustibility.

Note: Separate rules exist for bedrooms in institutional and other residential buildings (see B3 9.7 (V2)).

6.6.2.15 Construction and fixings for cavity barriers

Every cavity barrier should be constructed to provide at least 30 minutes' fire resistance. — B3 6.5 (V1) / B3 9.13 (V2)

A cavity barrier should, wherever possible, be tightly fitted to a rigid construction and mechanically fixed in position. — B3 6.6 (V1) / B3 9.14 (V2)

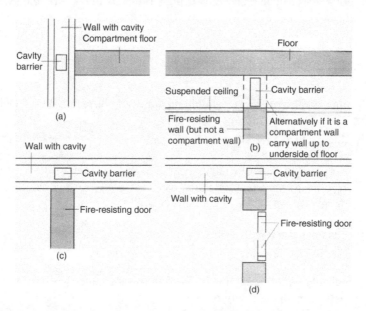

Figure 6.90 Interrupting concealed spaces and cavities. (a, b) Sections. (c, d) Plans.

Cavity barriers should be fixed so that their performance is unlikely to be made ineffective by: — B3 6.7 (V1)

- movement of the building due to subsidence, shrinkage or temperature change; — B3 9.15 (V2)
- movement of the external envelope due to wind;
- collapse in a fire of any services penetrating them;
- failure in a fire of their fixings;
- failure in a fire of any material or construction which they abut.

Cavity barriers in a stud wall or partition (or provided around openings) may be made of:

- steel at least 0.5 mm thick; or
- timber at least 38 mm thick; or

- polythene-sleeved mineral wool, or mineral-wool slab (in either case under compression);
- calcium silicate, cement-based or gypsum-based boards at least 12 mm thick.

Extensive cavities in floor voids should be subdivided with cavity barriers.   B3 (V2)

The need for cavity barriers in some concealed floor or roof spaces can be reduced by using a fire-resisting ceiling below the cavity.   B2 3.6 (V1)
B2 6.6 (V2)

6.6.2.15.1 Openings in cavity barriers

Openings in a cavity barrier should be limited to those for: B3 6.8 (V1)

- doors which have at least 30 minutes' fire resistance;   B3 9.13 (V2)
- the passage of pipes which meet the provisions in Approved Document P Section 7;
- the passage of cables or conduits containing one or more cables;
- openings fitted with a suitably mounted automatic fire damper;
- ducts which are fire-resisting or are fitted with a suitably mounted automatic fire damper where they pass through the cavity barrier.

6.6.2.16 Cavity walls in coursed brickwork or blockwork

All cavity walls should have: A1/2 2C8

- leaves at least 90 mm thick and cavities at least 50 mm wide;
- wall ties with a horizontal spacing of 900 mm and a vertical spacing of 450 mm (or alternatively spaced so that the number of wall ties per square metre is not less than 2.5 ties/m^2);
- wall ties spaced not more than 300 mm apart vertically, within a distance of 225 mm from the vertical edges of all openings, movement joints and roof verges.

 Note: A selection of wall ties for use in a range of cavity widths is shown in Table 6.37.

Table 6.37 Minimum thickness of certain external walls, compartment walls and separating walls

Height of wall	Length of wall	Minimum thickness of wall
Not exceeding 3.5 m	Not exceeding 12 m	190 mm for whole of its height
Exceeding 3.5 m but not exceeding 9 m	Not exceeding 9 m	190 mm for whole of its height
	Exceeding 9 m	290 mm from the base for the height of one storey and 190 mm for the rest of its height
Exceeding 9 m but not exceeding 12 m	Not exceeding 9 m	290 mm from the base for the height of one storey and 190 mm for the rest of its height
	Exceeding 9 m but not exceeding 12 m	290 mm from the base for the height of two storeys and 190 mm for the rest of its height

Wall ties should comply with BS EN 845-1.	A1/2 2C19
Wall ties should have a horizontal spacing of 900 mm and a vertical spacing of 450 mm, equivalent to 2.5 ties per square metre.	A1/2 2C8
Wall ties should also be provided, spaced not more than 300 mm apart vertically, within a distance of 225 mm from the vertical edges of all openings, movement joints and roof verges.	A1/2 2C8
For external walls, compartment walls and separating walls in cavity construction, the combined thickness of the two leaves plus 10 mm should be at least as thick as 1/16 of the storey height.	A1/2 2C8

6.6.2.17 Ventilation of rooms containing openable windows

Table 6.38 Ventilation of rooms containing openable windows (i.e. located on an external wall)

Room	Rapid ventilation (e.g. opening windows)	Background ventilation (mm^2)	Extract ventilation fan rates or passive stack ventilation (PSV)
Habitable room	1/20 of floor area	8000	
Kitchen	Opening window (no minimum size)	4000	30 l/s adjacent to a hob or 60 l/s elsewhere or PSV
Utility room	Opening window (no minimum size)	4000	30 l/s or PSV
Bathroom (with or without WC)	Opening window (no minimum size)	4000	15 l/s or PSV
Sanitary accommodation (separate from bathroom)	1/20 of floor area or mechanical extract at 6 l/s	4000	

6.6.2.18 Walls providing vertical support to other walls

Irrespective of the material used in the construction, a wall   A1/2 2C9
should not be less than the thickness of any part of the wall
to which it gives vertical support.

6.6.2.19 Parapet walls

The minimum thickness and maximum height of parapet   A1/2 2C11
walls should be as shown in Figure 6.91.

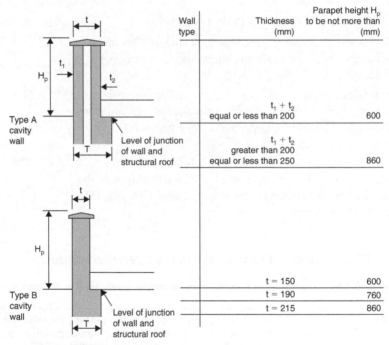

Wall type	Thickness (mm)	Parapet height H_p to be not more than (mm)
Type A cavity wall	$t_1 + t_2$ equal or less than 200	600
	$t_1 + t_2$ greater than 200 equal or less than 250	860
Type B cavity wall	$t = 150$	600
	$t = 190$	760
	$t = 215$	860

Note: t should be less than or equal to T

Figure 6.91 Height of parapet walls.

6.6.2.20 Single leaves of certain external walls

The single leaf of external walls of small, single-storey,   A1/2 2C12
non-residential buildings and of annexes need be only
90 mm thick.

6.6.2.21 Buttressing walls

If the buttressing wall is not itself a supported wall, its
thickness T2 should not be less than:

- half the thickness required for an external or   A1/2 2C26a
  separating wall of similar height and length less
  5 mm; or

- 75 mm if the wall forms part of a dwelling house   A1/2 2C26b
  and does not exceed 6 m in total height and 10 m in
  length; and

- 90 mm in other cases.   A1/2 2C26c

The length of the buttressing wall should be:

- at least 1/6 of the overall height of the supported
  wall;
- bonded or securely tied to the supporting wall and at
  the other end to a buttressing wall, pier or chimney.

The size of any opening in the buttressing wall should be   A1/2 2C26c
restricted as shown in Figure 6.92.

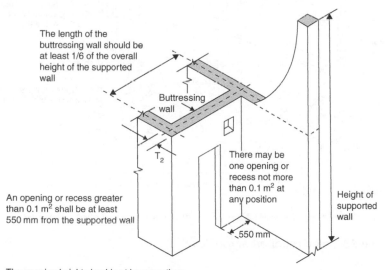

The length of the
buttressing wall should be
at least 1/6 of the overall
height of the supported
wall

Buttressing
wall

T_2

There may be
one opening or
recess not more
than 0.1 m² at
any position

Height of
supported
wall

An opening or recess greater
than 0.1 m² shall be at least
550 mm from the supported wall

550 mm

The opening height should not be more than
0.9 times the floor to ceiling height and the
depth of the lintel including any masonry
over the opening should be not less than 150 mm

Figure 6.92 Openings in a buttressing wall.

6.6.2.22 Gable walls

Gable walls should be strapped to roofs as shown in Figure 6.93(a) and (b) by tension straps.	A1/2 2C36
Vertical strapping at least 1 m in length should be provided at eaves level at intervals not exceeding 2 m as shown in Figure 6.93(c) and (d).	A1/2 2C36
Vertical strapping may be omitted if the roof:	A1/2 2C36a–d

- has a pitch of 15° or more; and
- is tiled or slated; and
- is of a type known by local experience to be resistant to wind gusts; and
- has main timber members spanning on to the supported wall at not more than 1.2 m centres.

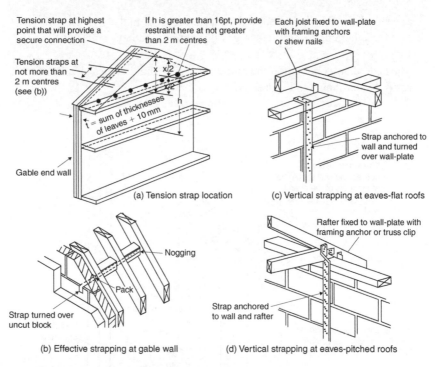

Figure 6.93 Lateral support at roof level.

6.6.2.23 Piers

Piers should have a minimum width of 190 mm (Figure 6.96).	A1/2 2C27a

Piers should measure at least three times the thickness of the supported wall.   A1/2 2C27a

6.6.2.23.1 Wall thicknesses and recommendations for piers

The walls should have a minimum thickness of 90 mm.   A1/2 2C38(iii)

Walls that do not contain a major opening but exceed 2.5 m in length or height should be bonded or tied to piers for their full height at not more than 3 m centres as shown in Figure 6.94.   A1/2 2C38(iii)

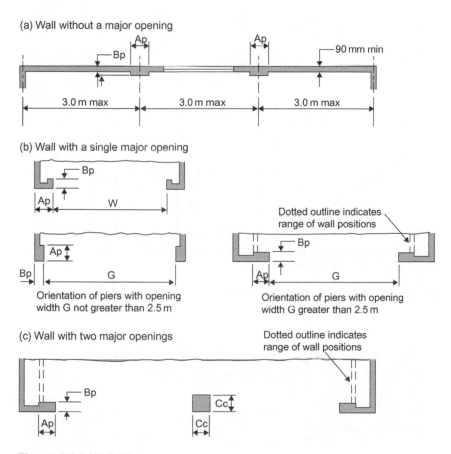

Figure 6.94 Wall thicknesses.

Walls that contain one or two major openings should in   A1/2 2C38(iii)
addition have piers as shown in Figure 6.81(b) and (c).

Where ties are used to connect piers to walls they   A1/2 2C38(iii)
should be:

- flat;
- 20 mm × 3 mm in cross-section;
- stainless steel;
- placed in pairs;
- spaced at not more than 300 mm centres vertically.

Walls should be tied horizontally at no more than 2 m   A1/2 2C38(iv)
centres to the roof structure at eaves level, base of
gables and along roof slopes (as shown in Figure 6.95)
with straps.

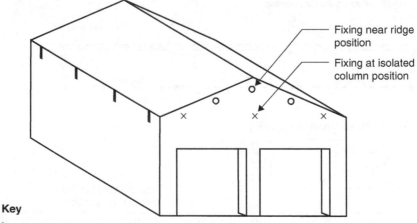

Key

❙ denotes fixings at eaves level.   × denotes fixings at base of gable.
○ denotes fixings along roof slope.

Figure 6.95 Lateral restraint at roof level.

Where straps cannot pass through a wall, they should   A1/2 2C38(iv)
be adequately secured to the masonry using suitable
fixings.

Isolated columns should also be tied to the roof   A1/2 2C38(iv)
structure (Figure 6.95).

6.6.2.24 Chimneys

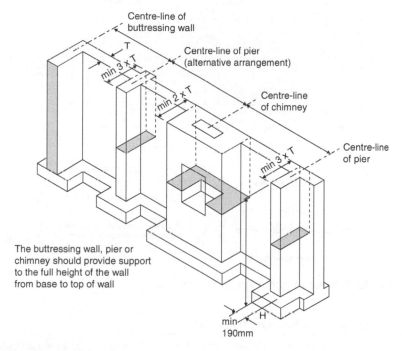

Figure 6.96 Buttressing.

Chimneys should measure at least twice the thickness,   A1/2 2C27a
measured at right angles to the wall (see Figure 6.96).

- The sectional area on plan of chimneys (excluding openings for fireplaces and flues) should be not less than the area required for a pier in the same wall.
- The overall thickness should not be less than twice the required thickness of the supported wall (Figure 6.96).

The number, size and position of openings and recesses should not impair the stability of a wall or the lateral restraint afforded by a buttressing wall to a supported wall.	A1/2 2C28
Construction over openings and recesses should be adequately supported.	A1/2 2C28
No openings should be provided in walls below ground floor except for small holes for services, ventilation, etc., which should be limited to a maximum area of 0.1 m² at not less than 2 m centres (Figure 6.97 and Table 6.39).	A1/2 2C29

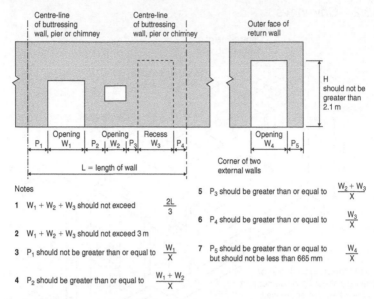

Figure 6.97 Sizes of openings and recesses.

Table 6.39 Value of X factor for Figure 6.97

Nature of roof span	Maximum roof span (m)	Minimum thickness of wall inner (mm)	Span of floor is parallel to wall	Span of timber floor into wall		Span of concrete floor into wall	
				Max. 4.5 m	Max. 6.0 m	Max. 4.5 m	Max. 6.0 m
				Value of factor X			
Roof spans parallel to wall	Non-applicable	100	6	6	6	6	6
		90	6	6	6	6	5
Timber roof spans into wall	9	100	6	6	5	4	3
		90	6	4	4	3	3

6.6.2.25 Moisture

6.6.2.25.1 Resistance to the passage of moisture

Walls should:

- resist the passage of moisture from the ground to the inside of the building;   C5.2

- not be damaged by moisture from the ground;   C5.2b

- not carry moisture from the ground to any part that would be damaged by moisture.   C5.2b

External walls should:

- resist rain penetrating components of the structure that could be damaged by moisture; C5.2c
- resist rain penetrating to the inside of the building; C5.2d
- be designed and constructed so that their structural and thermal performance is not adversely affected by interstitial condensation; C5.2e
- not promote surface condensation or mould growth. C5.2f

 For buildings that are used wholly for storing goods and/or provisions, this requirement may not apply.

6.6.2.25.2 Foundations

A wall shall be erected to prevent undue moisture from the ground reaching the inside of the building and (if it is an outside wall) adequately resisting the penetration of rain and snow to the inside of the building (Figure 6.98). C3

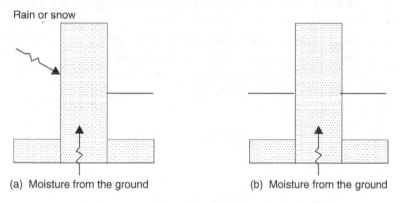

(a) Moisture from the ground   (b) Moisture from the ground

Figure 6.98 Resistance to moisture. (a) External wall. (b) Internal wall.

6.6.2.25.3 Internal and external walls exposed to moisture from the ground

Internal and external walls (that are subject to moisture from the ground) shall have a damp-proof course of bituminous material, polyethylene, engineering bricks or slates in cement mortar, or any other material that will prevent the passage of moisture. C5.5a

The damp-proof course should be continuous with any damp-proof membrane in the floors.   C5.5a

If the wall is an external wall, the damp-proof course should be at least 150 mm above the level of the adjoining ground (Figure 6.99) unless the design is such that a part of the building will protect the wall.   C5.5b

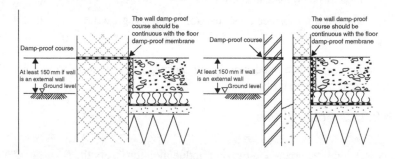

Figure 6.99 Damp-proof courses.

If the wall is an external cavity wall (Figure 6.100), either:

- the cavity should be taken down at least 225 mm below the level of the lowest damp-proof course; or   C5.5c

- a damp-proof tray should be provided so as to prevent precipitation passing into the inner leaf (Figure 6.101), with weep holes every 900 mm to assist in the transfer of moisture through the external leaf.   C5.5c

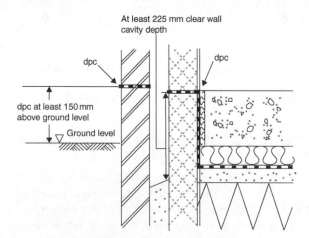

Figure 6.100 Cavity carried down.

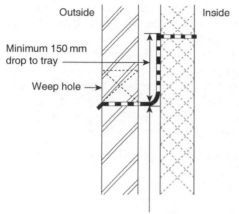

Outside   Inside

Minimum 150 mm
drop to tray ──────

Weep hole →

Tray leading water to outside of wall

Figure 6.101 Damp-proof (cavity) tray.

Where the damp-proof tray does not extend the full length of   C5.5c
the exposed wall (i.e. above an opening), stop ends and at least
two weep holes should be provided.

As well as giving protection against moisture from the ground,   C5.7
an external wall should give protection against precipitation.

6.6.2.25.4 Solid external walls

Solid walls should be capable of holding moisture arising from rain and snow
until it can be released in a dry period without penetrating to the inside of the
building or causing damage to the building.

Solid external walls exposed to **very severe** conditions   C5.9
should be protected by external impervious cladding.

Solid external walls exposed to **severe** conditions may be   C5.9a
built with:

*   brickwork (or stonework) at least 328 mm thick;
*   dense aggregate concrete blockwork at least 250 mm
    thick; or
*   lightweight aggregate (aerated autoclaved concrete
    blockwork) at least 215 mm thick.

Solid external walls exposed to **severe** conditions may be   C5.9b
built, providing:

*   the rendering is in two coats with a total thickness of at
    least 20 mm and has a scraped or textured finish;

- the strength of the mortar is compatible with the strength of the bricks or blocks;
- the joints (if the wall is to be rendered) are raked out to a depth of at least 10 mm;
- the rendering mix is 1 part of cement, 1 part of lime and 6 parts of well-graded sharp sand (nominal mix 1:1:6) unless the blocks are of dense concrete aggregate, in which case the mix may be 1:½.

Adequate protection should be provided at the top of walls, etc. (Figure 6.102).   C5.9c

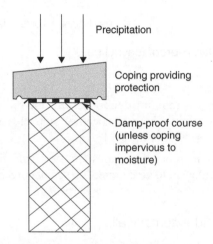

Figure 6.102 Projection of wall head from precipitation.

Unless the protection and joints are a complete barrier to   C5.9c
moisture, a damp-proof course should also be provided.

Damp-proof courses, cavity trays and closers should be
provided and designed to ensure that water drains outwards:

- where the downward flow will be interrupted by an   C5.9d(i)
  obstruction (e.g. from some types of lintel);

- under openings – unless there is a sill and the sill and   C5.9d(ii)
  its joints will form a complete barrier;

- at abutments between walls and roofs.   C5.9d(iii)

A solid external wall may be insulated on the inside or on   C5.10
the outside.

Where the insulation is on the inside, a cavity should be   C5.10
provided to give a break in the path for moisture.

Where the insulation is on the outside, it should provide   C5.10
some resistance to the ingress of moisture to ensure the wall
remains relatively dry (see Figure 6.103).

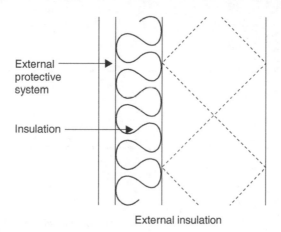

External insulation

Figure 6.103 Insulated (solid) external wall.

6.6.2.25.5 Cavity insulation

 The outer leaf of cavity external walls shall be separated from the inner leaf by
a drained air space (or in any other way which will prevent precipitation from
being carried to the inner leaf).

The suitability of the wall for installing insulation   C5.15a and d
material(s) is to be assessed before the work is carried
out.

When the cavity of an existing house is being filled,   C5.15e
attention should be given to the condition of the external
leaf of the wall, e.g. its state of repair and type of
pointing.

A full or partial fill insulating material may be placed   C5.15a
in the cavity between the outer leaf and an inner leaf of
masonry, subject to the suitability of a wall for installing
insulation into the cavity (Table 6.40).

Table 6.40 Maximum recommended exposure zones for insulated masonry walls

Insulation method	Wall construction								
	Minimum width of filled or clear cavity (mm)	Maximum recommended exposure zone for each construction							
		Impervious cladding		Rendered finish		Facing masonry			Flush sills and copings
		Full height of wall	Above facing masonry	Full height of wall	Above facing masonry	Tooled flush joints	Recessed mortar joints		
Built-in full fill	50	4	3	3	3	2	1		1
	75	4	3	4	3	3	1		1
	100	4	4	4	3	3	1		2
	125	4	4	4	3	3	1		2
	150	4	4	4	4	4	1		2
Injected fill, not UF foam	50	4	2	3	2	2	1		1
	75	4	3	4	3	3	1		1
	100	4	3	4	3	3	1		1
	125	4	4	4	3	3	1		2
	150	4	4	4	4	4	1		2
Injected fill, UF foam	50	4	2	3	2	1	1		1
	75	4	2	3	2	2	1		1
	100	4	2	3	2	2	1		1

Table 6.40 Maximum recommended exposure zones for insulated masonry walls

Insulation method	Minimum width of filled or clear cavity (mm)	Wall construction						
		Maximum recommended exposure zone for each construction						
		Impervious cladding		Rendered finish		Facing masonry		Flush sills and copings
		Full height of wall	Above facing masonry	Full height of wall	Above facing masonry	Tooled flush joints	Recessed mortar joints	
Partial fill:								
Residual 50 mm cavity	50	4	4	4	4	3	1	1
Residual 75 mm cavity	75	4	4	4	4	4	1	1
Residual 100 mm cavity	100	4	4	4	4	4	2	1
Internal insulation:								
Clear cavity 50 mm	50	4	3	4	3	3	1	1
Clear cavity 100 mm	100	4	4	4	4	4	2	2
Fully filled:								
Cavity 50 mm	50	4	3	3	3	2	1	1
Cavity 100 mm	100	4	4	4	3	3	1	2

The insulating material should be the subject of current certification from an appropriate body or a European technical approval.	C5.15c
When partial fill materials are used, the residual cavity should not be less than 50 mm nominal.	C5.15b
Rigid (board or batt) thermal insulating material built into the wall must be certified as being in conformance by an approved installer.	C5.15b
Urea formaldehyde foam inserted into the cavity should be: • in accordance with BS 5617:1985; • installed in accordance with BS 5618:1985.	C5.15d
The person undertaking installation work should operate under an approved installer scheme.	C5.15c

6.6.2.25.6 Framed external walls

The cladding shall be separated from the insulation or sheathing by a vented and drained cavity with a membrane that is vapour open, but resists the passage of liquid water, on the inside of the cavity (see Figure 6.104).	C5.17

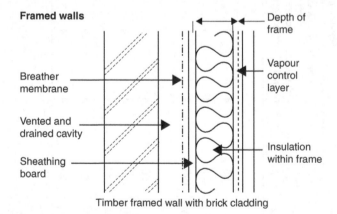

Figure 6.104 Insulated framed wall.

6.6.2.25.7 Cracking of external walls

 The possibility of severe rain penetrating through cracks in masonry external walls should be taken into account when designing a building.

6.6.2.25.8 Impervious cladding systems for walls

Cladding systems for walls should:

- resist the penetration of precipitation to the inside of the building;   C5.19a

- not be damaged by precipitation;   C5.19b

- not carry precipitation to any part of the building that would be damaged by it.   C5.19b

Cladding that is designed to protect a building from precipitation shall be:

- jointless or have sealed joints;   C5.21a

- impervious to moisture.   C5.21a

If the cladding has overlapping dry joints it shall be:

- impervious or weather resisting;   C5.21b

- backed by a material which will direct precipitation that enters the cladding towards the outer face.   C5.21b

Materials that can deteriorate rapidly without special care should only be used as the weather-resisting part of a cladding system.   C5.22

Cladding may be:

- impervious (e.g. metal, plastic, glass and bituminous products);   C5.23a

- weather-resisting (e.g. natural stone or slate, cement-based products, fired clay and wood);   C5.23b

- moisture-resisting (e.g. bituminous and plastic products lapped at the joints);   C5.23c

- jointless materials and sealed joints (i.e. to allow for structural and thermal movement).   C5.23d

Dry joints between cladding units should be designed so that:

- precipitation will not pass through them;   C5.24

- precipitation which enters the joints will be directed towards the exposed face without it penetrating beyond the back of the cladding.   C5.24

 Note: Whether dry joints are suitable will depend on the design of the joint or the design of the cladding and the severity of the exposure to wind and rain.

Each sheet, tile and section of cladding should be securely   C5.25
fixed (according to guidance contained in BS 8000-6:1990).

Particular care should be taken with designing and installation   C5.25
at the junctions between cladding and window and door
openings, as they are vulnerable to moisture ingress.

Insulation may be incorporated into the construction provided   C5.26
that it is either protected from moisture or unaffected by it.

Where cladding is supported by timber components (or is on   C5.27
the façade of a timber-framed building) the space between the
cladding and the building should be ventilated to ensure rapid
drying of any water that penetrates the cladding.

6.6.2.25.9 Joints between walls and doors/window frames

Joints between walls and doors and window frames should:

- resist the penetration of precipitation to the inside of the   C5.29a
  building;
- not be damaged by precipitation;   C5.29b
- not permit precipitation to reach any part of the building   C5.29
  that would be damaged by it.

Damp-proof courses should be provided to direct moisture
towards the outside, particularly:

- where the downward flow of moisture would be   C5.30a
  interrupted at an obstruction, e.g. at a lintel;
- where sill elements (including joints) do not form a   C5.30b
  complete barrier to the transfer of precipitation (e.g.
  under openings, windows and doors);
- where reveal elements, including joints, do not form a   C5.30c
  complete barrier to the transfer of rain and snow (e.g. at
  openings, windows and doors).

Direct plastering of the internal reveal of any window frame   C5.31
should only be used with a backing of expanded metal lathing
or similar.

In areas of the country that are exposed to very severe driving
rain:

- checked rebates should be used in all window and door   C5.32
  reveals;
- the frame should be set back behind the outer leaf of   C5.32
  masonry as shown in Figure 6.105;
- alternatively an insulated finned cavity closer may be used.   C5.32

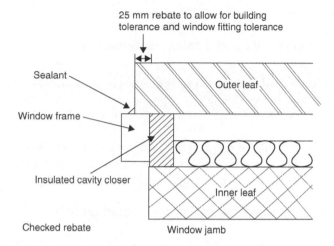

Figure 6.105 Window reveals for use in areas subject to very severe driving rain.

6.6.2.25.10 Door thresholds

Where an accessible threshold is provided to allow unimpeded access (as specified in Approved Document M):

- the external landing (Figure 6.106) should be laid to a fall   C5.33a
  between 1 in 40 and 1 in 60 in a single direction away
  from the doorway;

- the sill leading up to the door threshold should have a   C5.33b
  maximum slope of 15°.

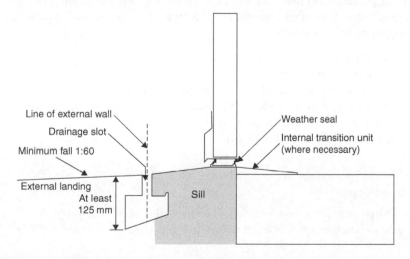

Figure 6.106 Accessible threshold for use in exposed areas.

6.6.2.26 Damp

6.6.2.26.1 Interstitial condensation (external walls)

> External walls shall be designed and constructed in accordance   C5.34
> with Clause 8.3 of BS 5250:2002.

 Specialist advice should be sought when designing swimming pools and other buildings where interstitial condensation in the walls (caused by high internal temperatures and humidities) can cause high levels of moisture being generated.

6.6.2.26.2 Surface condensation and mould growth (external walls)

External walls shall be designed and constructed so that the:

> - thermal transmittance (U-value) does not exceed 0.7 W/   C5.36a
>   m² K at any point;
> - junctions between elements and details of openings (such   C5.36b
>   as doors and windows) meet with the recommendations
>   in the report on robust construction details.

6.6.2.27 Fire

6.6.2.27.1 Fire resistance

Proprietary fire-stopping and sealing systems (including those designed for service penetrations) that have been shown by test to maintain the fire resistance of the wall or other element are available and may be used. Other fire-stopping materials include:

- cement mortar;
- gypsum-based plaster;
- cement or gypsum-based vermiculite/perlite mixes;
- glass fibre, crushed rock, blast furnace slag or ceramic-based products (with or without resin binders); and
- intumescent mastics (B3 11.14).

> Joints between fire-separating elements should be   B3 7.12a (V1)
> fire-stopped.   B3 10.17a (V2)
>
> All openings for pipes, ducts, conduits or cables to   B3 7.12b (V1)
> pass through any part of a fire-separating element   B3 10.17b (V2)
> should be:

- kept as few in number as possible;
- kept as small as practicable;
- fire-stopped (which in the case of a pipe or duct, should allow thermal movement).

To prevent displacement, materials used for fire-stopping should be reinforced with (or supported by) materials of limited combustibility.   B3 7.13 (V1) / B3 10.18 (V2)

Where a compartment wall meets another compartment wall, the junction should maintain the fire resistance of the compartmentation.   B3 5.9 (V1) / B3 8.25 (V2)

At the junction of a compartment floor with an external wall that has no fire resistance (such as a curtain wall) the external wall should be restrained at floor level to reduce the movement of the wall away from the floor when exposed to fire.   B3 5.10 (V1)

6.6.2.27.2 Internal fire spread (linings)

The choice of materials for walls and ceilings can significantly affect the spread of a fire and its rate of growth, even though they are not likely to be the materials first ignited. Although furniture and fittings can have a major effect on fire spread it is not possible to control them through Building Regulations.

The surface linings of walls should meet the classifications shown in Table 6.41.

Table 6.41 Classification of linings

Location	Class*
Small rooms with an area of not more than 4 m² (in residential accommodation) or 30 m² (in non-residential accommodation)	3
Other rooms (including garages)	1
Circulation spaces within buildings	1
Other circulation spaces (including the common area of flats and maisonettes)	0

* **Classifications are based on tests as per BS 476 and as described in Appendix A of Approved Document B.**

Any flexible membrane covering a structure (other than an air-supported structure) should comply with the recommendations given in Appendix A of BS 7157.	B2 6.8 (V2)
The wall and any floor between the garage and the house shall have a 30-minute standard of fire resistance.	B2

6.6.2.27.3 Walls adjacent to hearths

Walls that are not part of a fireplace recess or a prefabricated appliance chamber but are adjacent to hearths or appliances also need to protect the building from catching fire. A way of achieving the requirement is shown in Figure 6.107.	J 2.31

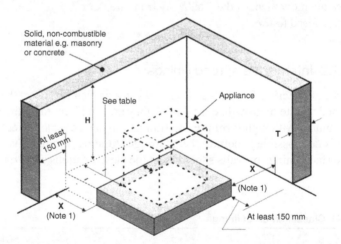

Location of hearth or appliance	Solid, non-combustible material	
	Thickness (T)	Height (H)
Where the hearth abuts a wall and the appliance is not more than 50 mm from the wall	200 mm	At least 300 mm above the appliance and 1.2 m above the hearth
Where the hearth abuts a wall and the appliance is more than 50 mm but not more than 300 mm from the wall	75 mm	At least 300 mm above the appliance and 1.2 m above the hearth
Where the hearth does not abut a wall and is no more than 150 mm from the wall (see Note 1)	75 mm	At least 1.2 m above the hearth
Note 1: There is no requirement for protection of the wall where X is more than 150 mm		

Figure 6.107 Walls adjacent to hearths.

6.6.2.27.4 Load-bearing elements of structure

All load-bearing elements of a structure shall have a minimum standard of fire resistance.	B3 4.1 (V1) B3 7.1 (V2)
Structural frames, beams, columns, load-bearing walls (internal and external), floor structures and gallery structures should have at least the fire resistance given in Appendix A of Approved Document B.	B3 4.2 (V1) B3 7.2 (V2)

6.6.2.27.5 Compartmentation

To prevent rapid fire spread and to reduce the chance of fires becoming out of control, the spread of fire within a building can be restricted by subdividing that building into compartments that are separated from one another by walls and/or floors of fire-resisting construction.

The appropriate degree of subdivision depends on:

* the use of and fire load in the building;
* the height to the floor of the top storey in the building; and
* the availability of a sprinkler system.

To prevent the spread of fire within a building, whenever possible:

• the building should be subdivided into compartments separated from one another by walls and/or floors of fire-resisting construction;	B3 5.1 (V1) B3 8.1 (V2)
• walls separating semi-detached houses, or houses in terraces, should be constructed as a compartment wall and the houses should be considered as separate buildings.	B3 5.3 (V1)
Compartment walls that are common to two or more buildings should:	B3 8.10 (V2)
• be constructed as a compartment wall;	B3 5.7 (V1)
• run the full height of the building in a continuous vertical plane.	

Note: The lowest floor in a building does not need to be constructed as a compartment floor.

Compartment walls should:	B3 5.6 (V1)
• form a complete barrier to fire between the compartments they separate; and	B2 8.20a (V2)
• have the appropriate fire resistance as indicated in Approved Document B, Appendix A, Tables A1 and A2.	B3 5.6 (V1) B2 8.20b (V2)

Note: Adjoining buildings should only be separated by walls, not floors.

Junction of compartment wall with other walls

6.6.2.27.6 Junction of compartment wall with roof

If a fire penetrates a roof near a compartment wall there is a risk that it will spread over the roof to the adjoining compartment. To reduce this risk the wall should be:

- taken up to meet the underside of the roof;
- covered with fire-stopping material (where   B3 8.28 (V2)
  necessary) at the wall/roof junction, to maintain the
  continuity of fire resistance;
- continued across any eaves;

- extended up through the roof for a height of at least   B3 5.12 (V1)
  375 mm above the top surface of the adjoining roof   B3 8.29 (V2)
  covering, or a 1500 mm wide zone on either side of   B3 8.30 (V2)
  the wall should have a suitable covering (see Figure
  6.108).

Compartment walls in a top storey beneath a roof   B3 5.8 (V1)
should be continued through the roof space.   B3 8.24 (V2)

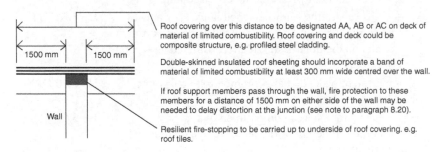

Roof covering over this distance to be designated AA, AB or AC on deck of material of limited combustibility. Roof covering and deck could be composite structure, e.g. profiled steel cladding.

Double-skinned insulated roof sheeting should incorporate a band of material of limited combustibility at least 300 mm wide centred over the wall.

If roof support members pass through the wall, fire protection to these members for a distance of 1500 mm on either side of the wall may be needed to delay distortion at the junction (see note to paragraph 8.20).

Resilient fire-stopping to be carried up to underside of roof covering. e.g. roof tiles.

Figure 6.108 Junction of compartment wall with roof.

6.6.2.27.7 Openings in compartment walls separating buildings or occupancies

Any openings in a compartment wall that is common to two or more buildings should be limited to those for a door which is providing 'means of access' in case of fire (and which has the same fire resistance as that required for the wall). All other openings in compartment walls or compartment floors should be limited to those for:

- doors that have the appropriate fire resistance; B3 8.34 (V2)
- the passage of pipes, ventilation ducts, service cables, chimneys, appliance ventilation ducts or ducts encasing one or more flue pipes;
- refuse chutes of non-combustible construction;
- atria designed in accordance with BS 5588-7:1997; and
- protected shafts (see B3 8.35 V2 for details of the relevant requirements).

6.6.2.27.8 Separation between garage and dwelling house

Compartment walls and compartment floors should be B3 5.4 (V1) provided if a domestic garage is attached to (or forms an integral part of) a dwelling house and the garage should be separated from the rest of the dwelling house, as shown in Figure 6.109.

Note: The wall and any floor between the garage and the house shall have a 30-minute standard of fire resistance. Any opening in the wall should be at least 100 mm above the garage floor level with an FD30 door.

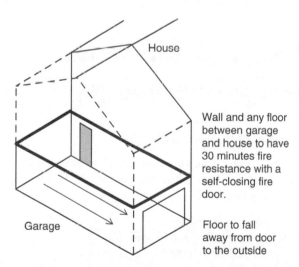

Figure 6.109 Separation between garage and dwelling-house.

If a door is provided between a dwelling house and the garage, the floor of the garage should: B3 5.5 (V1)

- be laid so as to allow fuel spills to flow away from the door to the outside; or
- the door opening should be positioned at least 100 mm above garage floor level.

Under new Building Regulations, fire doors **only** need to be provided with self-closing devices if they are between a dwelling house and an integral garage.

6.6.2.27.9 All-purpose groups

Parts of a building that are used and/or occupied for different purposes should be separated from one another by compartment walls and/or compartment floors (also see Appendix D to Approved Document B (V2)).   B2 8.11 (V2)

Walls that are common to two or more buildings should be constructed as compartment walls.   B2 8.10 (V2)

6.6.2.27.10 Flats

In buildings containing flats, the following should be constructed as compartment walls:

- every wall separating a flat from any other part of the building; and
- every wall enclosing a refuse storage chamber.

6.6.2.27.11 Non-residential buildings

In non-residential, purpose group buildings (such as office, shop, commercial, assembly, recreational facility, industrial, storage, etc. buildings), the following walls should be constructed as compartment walls:

- walls that are required to subdivide buildings in order to meet the size limits on compartments given in Table 12 of Volume 2 of Approved Document B;
- walls of a building that form part of a shopping complex;   B2 8.18e (V2)

- walls that divide a building into separate occupancies (i.e. spaces used by different organizations whether they fall within the same purpose group or not).   B2 8.18f (V2)

6.6.2.27.12 Construction of compartment walls

Note: Adjoining buildings should only be separated by walls, not floors.

Compartment walls should be able to accommodate the predicted deflection of the floor above by either: • having a suitable head detail between the wall and the floor that can deform (but still maintain its integrity) when exposed to a fire; or • the wall being designed to resist the additional vertical load from the floor above as it sags under fire conditions and thus maintain integrity.	B2 8.27 (V2)
Compartment walls that are common to two or more buildings should run the full height of the building in a continuous vertical plane.	B2 8.21 (V2)
Compartment walls used to form a separated part of a building should run the full height of the building in a continuous vertical plane.	B2 8.22 (V2)
If trussed rafters bridge the wall, they should be designed so that failure of any part of the truss due to a fire in one compartment will not cause failure of any part of the truss in another compartment.	B3 5.6 (V1) B3 8.20 (V2)
Junctions between a compartment floor and an external wall that has no fire resistance (such as a curtain wall) should be restrained at floor level to reduce the movement of the wall away from the floor when exposed to fire.	B2 8.26 (V2)
Load-bearing walls (internal and external) should have at least the fire resistance given in Appendix A, Table A1 of Approved Document B.	B3 7.2 (V2)
Timber beams, joists, purlins and rafters may be built into or carried through a masonry or concrete compartment wall if the openings for them are kept as small as practicable and then fire-stopped.	B3 5.6 (V1) B3 8.20 (V2)
There should be continuity at the junctions of the fire-resisting elements enclosing a compartment.	B3 8.6 (V2)

Note: Generally speaking, an external wall of a protected shaft does not need to have fire resistance (but see BS 5588-5:2004 for fire resistance of external walls of firefighting shafts).

6.6.2.27.13 Garages

If a domestic garage is attached to (or forms an integral part of) a house:	B3 5.4 (V1)
• the wall between the garage and the house shall have a 30-minute standard of fire resistance;	
• any opening in the wall should be at least 100 mm above the garage floor level with an FD30 door.	

6.6.2.27.14 Ventilation ducts and flues, etc.

Air-circulation-system transfer grilles should not be fitted in any wall enclosing a protected stairway.	B3 7.10 (V1) B1 2.17 (V1)
If a flue (or a duct containing flues and/or ventilation duct(s)), passes through a compartment wall, or is built into a compartment wall, each wall of the flue or duct should have a fire resistance of at least half that of the wall or floor (Figure 6.110).	B1 7.11 (V1)

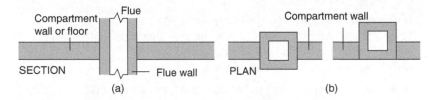

Figure 6.110 Flues penetrating compartment walls or floors.

6.6.2.27.15 Construction of an external wall

Where a portal framed building is near a relevant boundary, the external wall near the boundary may need fire resistance to restrict the spread of fire between buildings.	B4 12.4 (V2)
In cases where the external wall of the building cannot be wholly unprotected, the rafter members of the frame, as well as the column members, may need to be fire-protected.	B4 12.4 (V2)
The external surfaces of walls should meet the provisions shown in Figure 6.111.	B4 8.4 (V1) B4 12.4 (V2)

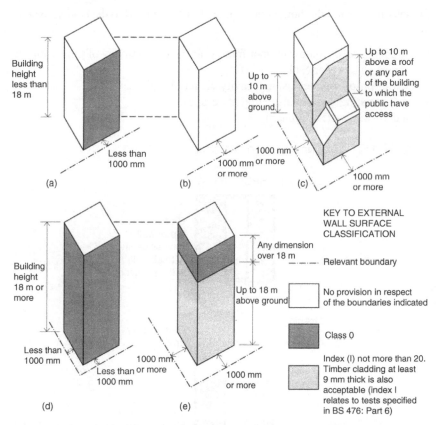

Figure 6.111 Provisions for external surfaces of walls. (a), (d), (e) Any building. (b) Any building other than (c). (c) Assembly or recreation building of more than one storey.

It should be noted that the use of combustible materials for cladding framework, or the use of combustible thermal insulation as an overcladding may be risky in tall buildings, even though the provisions for external surfaces in Figure 6.111 may have been satisfied.

The external envelope of a building should **not** provide a medium for fire spread if it is likely to be a risk to health or safety.	B4 12.5 (V2)
In a building with a storey 18 m or more above ground level, insulation material used in ventilated cavities in the external wall construction should be of limited combustibility (this restriction does not apply to masonry cavity wall construction).	B4 12.5 (V2)

Combustible materials should not be placed in or exposed to the cavity, except for:

• timber lintels, window or door frames, or the end stairway of timber joists;
• pipes, conduits or cables;
• damp-proof course, flashing, cavity closer or wall ties;
• fire-resisting thermal insulating material;
• a domestic meter cupboard.

6.6.2.27.15.1 MASONRY WALL CONSTRUCTION

SECTION THROUGH CAVITY WALL

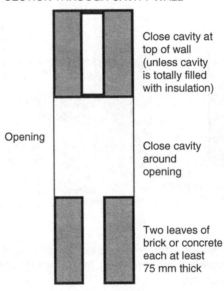

Close cavity at top of wall (unless cavity is totally filled with insulation)

Opening

Close cavity around opening

Two leaves of brick or concrete each at least 75 mm thick

Figure 6.112 Masonry cavity walls excluded from the previous for cavity barriers.

6.6.2.28 Airborne sound

• The flow of sound energy through walls should be restricted.
• Walls should reduce the level of airborne sound.
• Walls that separate a dwelling from another building (or another dwelling) shall resist the transmission of airborne sound.
• Habitable rooms (or kitchens) within a dwelling shall resist the transmission of airborne sound.
• Air paths, including those due to shrinkage, must be avoided.
• Porous materials and gaps at joints in the structure must be sealed.
• Flanking transmission (i.e. the indirect transmission of sound from one side of a wall to the other side) should be minimized.
• The possibility of resonance in parts of the structure (such as a dry lining) should be avoided.

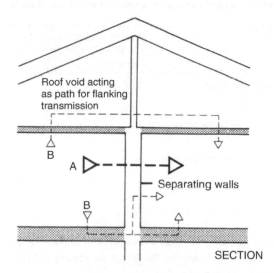

Roof void acting as path for flanking transmission

B

A

Separating walls

B

B

SECTION

Openings within 700 mm of junctions reduce dimensions
of flanking elements and reduce flanking transmission

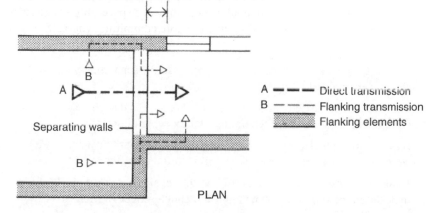

B

A

Separating walls

B

A ━ ━ ━ ━ Direct transmission
B ─ ─ ─ ─ Flanking transmission
░░░ Flanking elements

PLAN

Figure 6.113 Direct and flanking transmission.

6.6.2.28.1 Separating walls (new buildings)

All new walls constructed within a dwelling house (flat or room used for residential purposes) – whether purpose-built or formed by a material change of use – shall meet the laboratory sound insulation values set out in Table 6.42.	E0.9
Walls for rooms for residential purposes, dwelling houses and flats that have a separating function should achieve the sound insulation values, as set out in Table 6.42.	E0.1

Table 6.42 Dwelling houses and flats – performance standards for walls that have a separating function

Type	Airborne sound insulation $D_{nT,W} + C_{tr}$ (dB) (minimum values)
Purpose-built rooms for residential purposes	43
Purpose-built dwelling houses and flats	45
Rooms for residential purposes formed by material change of use	43
Dwelling houses and flats formed by material change of use	43

Notes:

(1) The sound insulation values in Table 6.42 include a built-in allowance for 'measurement uncertainty', and so, if any of these test values are not met, that particular test will be considered as failed.

(2) Occasionally, a higher standard of sound insulation may be required between spaces used for normal domestic purposes and noise generated in and to an adjoining communal or non-domestic space. In these cases it would be best to seek specialist advice before committing yourself.

Flanking transmission from walls connected to the separating wall shall be controlled.	E2
Tests should be carried out between rooms or spaces that share a common area formed by a separating wall or separating floor.	E1
Impact sound insulation tests should be carried out without a soft covering (e.g. carpet, foam-backed vinyl, etc.) on the floor.	E1
If the floor joists are to be supported on the separating wall then they should be supported on hangers and should not be built in.	E2
If the joists are at right angles to the wall, spaces between the floor joists should be sealed with full-depth timber blocking.	E3
The floor base (excluding any screed) should be built into a cavity masonry external wall and carried through to the cavity face of the inner leaf.	E
Walls that separate a dwelling from another dwelling (or part of the same building) shall resist: • level (and transmission) of airborne sounds; • the transmission of impact sound (such as speech, musical instruments and loudspeakers, and impact sources such as footsteps and furniture moving); • flow of sound energy through walls and floors.	E

6.6.2.28.2 Requirement E1

Figure 6.114 illustrates the relevant parts of the building that should be protected from airborne and impact sound in order to satisfy Requirement E1.

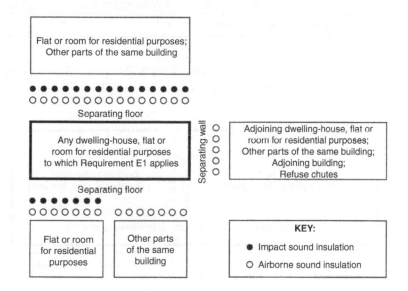

Figure 6.114 Requirement E1 – resistance to sound.

In some circumstances (for example, when a historic building is undergoing a material change of use) it may not be practical to improve the sound insulation to the standards set out in Approved Document E1, particularly if the special characteristics of such a building need to be recognized. In these circumstances the aim should be to improve sound insulation to the *extent that it is practically possible.*

 Note: BS 7913:1998, *The Principles of the Conservation of Historic Buildings*, provides guidance on the principles that should be applied when proposing work on historic buildings.

 Any building work involving a historic building is a virtual minefield and you would be well advised to seek professional advice well before embarking on any construction or reconstruction work.

6.6.2.28.3 Requirement E2

Constructions for new walls within a dwelling house (flat   E0.9
or room for residential purposes) – whether purpose-built or
formed by a material change of use – shall meet the laboratory
sound insulation values set out in Table 6.43.

Table 6.43 Laboratory values for new internal walls within dwelling houses, flats and rooms for residential purposes – whether purpose-built or formed by a material change of use

	Airborne sound insulation RW (dB) (minimum values)
Purpose-built dwelling houses and flats	40

Figure 6.115 illustrates the relevant parts of the building that should be protected from airborne and impact sound in order to satisfy Requirement E2a.

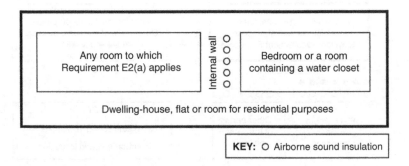

Figure 6.115 Requirement E2a – internal walls.

6.6.2.28.4 Requirement E3

Sound absorption measures described in Section 7 of Approved Document N shall be applied.

6.6.2.28.5 Requirement E4

The values for sound insulation, reverberation time and indoor ambient noise as described in Section 1 of Building Bulletin 93, *The Acoustic Design of Schools* (produced by DfES and published by The Stationery Office (ISBN 0-11-271105-7)), shall be satisfied.

6.6.2.28.6 Types of wall

As shown in Figure 6.116 there are four main types of separating walls that can be used in order to achieve the required performance standards shown in Table 6.44.

 Other designs, materials and/or products may also be available and so it is always worth talking to the manufacturers and/or suppliers first.

The resistance to airborne sound depends mainly on the mass of the wall.

Solid masonry
(Wall type 1)

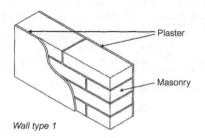

Plaster

Masonry

Wall type 1

The resistance to
airborne sound
depends mainly
on the mass per unit
area of the wall.

Cavity masonry
(Wall type 2)

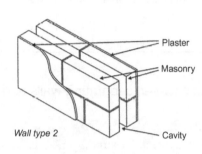

Plaster

Masonry

Wall type 2

Cavity

The resistance to
airborne sound
depends on the
mass per unit
area of the leaves
and on the degree
of isolation
achieved. The
isolation is
affected by
connections
(such as wall ties
and foundations)
between the wall
leaves and by the
cavity width.

**Masonry
between
independent
panels**
(Wall type 3)

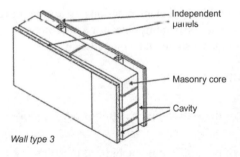

Independent
panels

Masonry core

Cavity

Wall type 3

The resistance
to airborne
sound depends
partly on the
type and mass
per unit area
of the core,
and partly
on the
isolation and
mass per unit
area of the
independent
panels.

**Framed wall
absorbent
with
material**
(Wall type 4)

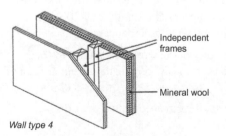

Independent
frames

Mineral wool

Wall type 4

The resistance
to airborne
sound depends
on the mass per
unit area of the
leaves, the
isolation of
the frames,
and the
absorption in
the cavity
between
the frames.

Figure 6.116 Types of separating walls.

Table 6.44 Dwelling houses and flats – performance standards for separating walls, separating floors, and stairs that have a separating function

	Airborne sound insulation $D_{nT,w}$ C_{tr} dB (minimum values)	Impact sound insulation $L^9{}_{nT,w}$ dB (maximum values)
Purpose-built dwelling houses and flats:		
Walls	45	–
Floors and stairs	45	62
Dwelling houses and flats formed by material change of use:		
Walls	43	–
Floors and stairs	43	64

6.6.2.28.7 Junctions between separating walls and other building elements

Care should be taken to correctly detail the junctions between the separating wall and other elements, such as floors, roofs, external walls and internal walls.	E2.9

 Note: Where any building element functions as a separating element (e.g. a ground floor that is also a separating floor for a basement flat) then the separating-element requirements should take precedence.

6.6.2.28.8 Cavity insulation

The outer leaf of the wall should be built of masonry or concrete.	D1 1.1–1.2
The inner leaf of the wall should be built of masonry (bricks or blocks).	D1 1.1–1.2
The wall being insulated with urea formaldehyde (UF) shall be assessed (in accordance with BS 8208) for suitability before any work commences.	D1 1.1–1.2
The person carrying out the work needs to hold (or operate under) a current BSI certificate of registration of assessed capability for the work he or she is doing.	D1 1.1–1.2
The installation shall be in accordance with BS 5618:1985.	D1 1.1–1.2
The material shall be in accordance with the relevant recommendations of BS 5617:1985.	D1 1.1–1.2

6.6.2.28.9 Mass per unit area of walls

The mass per unit area of a wall is expressed in kilograms per square metre (kg/m^2) and is equivalent to:

$$\text{mass per unit area of a wall} = \frac{\text{mass of coordinating area}}{\text{coordinating area}} \qquad (6.1)$$

Mass per unit area of a wall can be calculated as follows:

$$\text{mass per unit area of a wall} = \frac{M_B + \rho_m\,[Td(1+h-d)+V]}{LH}\ kg/m^2 \qquad (6.2)$$

Where:
M_B = brick/block mass (kg) at appropriate moisture content
ρ_m = density of mortar (kg/m^3) at appropriate mortar content
T = the brick/block finish without surface finish (m)
d = mortar thickness (m)
L = coordinating length (m)
H = coordinating height (m)
V = volume of any frog/void filled with mortar (m^3).

Note: The method for calculating mass per unit area is provided in Annex A to Approved Document E of the regulations, together with some examples.

6.6.2.28.10 Density of the materials

The density of the materials used (and on which the mass per unit area of the wall depends) is expressed in kilograms per cubic metre (kg/m^3).

6.6.2.28.10.1 PLASTERBOARD LININGS ON SEPARATING AND EXTERNAL MASONRY WALLS

Wherever plasterboard is recommended (or the finish is not specified) a dry lining laminate of plasterboard with mineral wool may be used.	E2.15
Plasterboard linings should be fixed according to the manufacturer's instructions.	E2.16

Note: Recommended cavity widths in separating cavity masonry walls are minimum values.

6.6.2.28.11 Wall ties in separating and external cavity masonry walls

There are two types of wall tie that can be used in masonry cavity walls: type A (butterfly ties), which are normal; and type B (double-triangle ties), which are used only in external masonry cavity walls where tie type A does not satisfy the requirements of Approved Document A (Structure).

Notes:

(1) Recommended cavity widths in separating cavity masonry walls are minimum values.
(2) In external cavity masonry walls, tie type B may decrease the airborne sound insulation due to flanking transmission via the external wall leaf compared to tie type A.

Stainless steel cavity wall ties are specified for all houses regardless of their location.	A1/2
Wall ties should have a horizontal spacing of 900 mm and a vertical spacing of 450 mm.	A1/2 2C8

This is equivalent to 2.5 ties per square metre.

Wall ties should be spaced not more than 300 mm apart vertically, within a distance of 225 mm from the vertical edges of all openings, movement joints and roof verges.	A1/2 2C8
Wall ties should comply with BS 1243, DD 140 or BS EN 845-1.	A1/2 2C19
Wall ties should be selected in accordance with Table 6.45.	A1/2 2C19

Table 6.45 Cavity wall ties

Nominal cavity width mm (Note 1)	Tie length mm (Note 2)	BS EN 845-1 tie
50 to 75	200	Type 1, 2, 3 or 4 to BSI PD 6697:2010 and selected on the basis of the design loading and design cavity width.
76 to 100	225	
101 to 125	250	
126 to 150	275	
151 to 175	300	
176 to 300	(See Note 3)	

Notes:
1. Where face insulated blocks are used the cavity width should be measured from the face of the masonry unit.
2. The embedment depth of the tie should not be less than 50 mm in both leaves.
3. For cavities wider than 175 mm calculate the length as the nominal cavity width plus 125 mm and select the nearest stock length. For wall ties requiring embedment depths in excess of 50 mm, increase the calculated tie length accordingly.

The leaves of a cavity masonry wall construction should be connected by either butterfly ties or double-triangle ties spaced as per BS 5628-3:2001, which limits this tie type and spacing to cavity widths of 50 mm to 75 mm with a minimum masonry leaf thickness of 90 mm.   E2.19

Note: Wall ties may be used provided that they have the measured dynamic stiffness for the cavity width (see E2.20 and E2.21 for details of the relevant formula for measuring the dynamic stiffness).

In conditions of severe exposure, austenitic stainless steel or suitable non-ferrous ties should be used.   A1/2 (1C20)

The number of ties per square metre, n, shall be calculated from the horizontal (S_x) and vertical (S_y) tie spacing distances (in metres) using the formula $n = 1/(S_x \cdot S_y)$.   E2.22

All wall ties and spacings specified using the dynamic stiffness parameter should also satisfy the requirements of Approved Document A (Structure).   E2.24

6.6.2.28.12 Corridor walls and doors

Separating walls should be used between corridors and rooms in flats, in order to control flanking transmission and to provide the required sound insulation.   E2.25

Note: It is highly likely that the amount of sound insulation gained by using a separating wall will be reduced by the presence of a door.

Noisy parts of the building should preferably have a lobby, double door or high-performance doorset to contain the noise.   E2.27

All corridor doors shall have a good perimeter sealing (including the threshold where practical).   E2.26

All corridor doors shall have a minimum mass per unit area of 25 kg/m².   E2.26

All corridor doors shall have a minimum sound reduction index of 29 dB Rw (measured according to BS EN ISO 140-3:1995 and rated according to BS EN ISO 717-1:1997).   E2.26

All corridor doors shall meet the requirements for fire safety (see Approved Document B (Fire safety)).   E2.26

6.6.2.28.13 Refuse chutes

A wall separating a habitable room (or kitchen) from a refuse chute should have a mass per unit area (including any finishes) of at least 1320 kg/m².	E2.28
A wall separating a non-habitable room from a refuse chute should have a mass per unit area (including any finishes) of at least 220 kg/m².	E2.28

6.6.2.29 Wall type 1 (solid masonry)

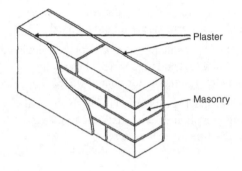

Figure 6.117 Wall type 1 (solid masonry).

When using a solid masonry wall, the resistance to airborne sound depends mainly on the mass per unit area of the wall. As shown in Table 6.46, there are three different categories of solid masonry walls.

Table 6.46 Wall type 1 – categories

Wall type 1		Minimum mass per unit area (including plaster) 415 kg/m²
Category 1.1 Solid masonry		Plaster on both room faces
Dense aggregate concrete block, plaster on both room faces		Blocks laid flat to the full thickness of the wall
		For example:
		Size · · · · · · · · · 215 mm laid flat Density · · · · · · · 1840 kg/m³ Coursing · · · · · · 110 mm Plaster · · · · · · · 13 mm lightweight

Table 6.46 Wall type 1 – categories (Continued)

Wall type 1

Category 1.2 Dense aggregate concrete

Dense aggregate concrete, cast in situ, plaster on both room faces

Minimum mass per unit area (including plaster) 415 kg/m²

Plaster on both room faces

For example:

Concrete	190 mm
Density	2200 kg/m³
Plaster lightweight	13 mm

Wall type 1

Category 1.3 Brick

Brick, plaster on both room faces

Minimum mass per unit area (including plaster) 375 kg/m²

Bricks to be laid frog up, coursed with headers

For example:

Size	215 mm laid flat
Density	1610 kg/m³
Coursing	75 mm
Plaster lightweight	13 mm

6.6.2.29.1 Wall type 1 – general requirements

Fill and seal all masonry joints with mortar.	E2.32a
Lay bricks frog up to achieve the required mass per unit area and avoid air paths.	E2.32b
Use bricks/blocks that extend to the full thickness of the wall.	E2.32c
Ensure that an external cavity wall is stopped with a flexible closer at the junction with a separating wall.	E2.32d

 It is not necessary to stop the cavity wall with a flexible closer if the cavity is fully filled with mineral wool or expanded polystyrene beads.

Control flanking transmission from walls and floors connected
to the separating wall (see guidance on junctions).

Deep sockets and chases should **not** be used in separating   E2.32
walls.

Stagger the position of sockets on opposite sides of the
separating wall.

Ensure flue blocks:

*   will not adversely affect the sound insulation;
*   use a suitable finish.

A cavity separating wall may **not** be changed into a solid   E2.32
masonry (i.e. type 1) wall by filling in the cavity with mortar
and/or concrete.

When the cavity wall is bridged by the solid wall, ensure that   E2.32
there is no junction between the solid masonry wall and a
cavity wall.

6.6.2.29.2 Wall type 1 – junction requirements

6.6.2.29.2.1 JUNCTIONS WITH AN EXTERNAL CAVITY WALL WITH MASONRY INNER LEAF

Where the external wall is a cavity wall:

*   the outer leaf of the wall may be of any construction;   E2.36a
*   the cavity should be stopped with a flexible closer.   E2.36b

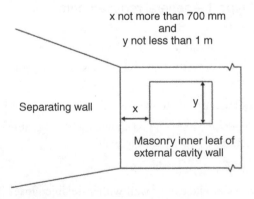

Figure 6.118 Wall type 1 – position of openings in a masonry inner leaf of
an external cavity wall.

The masonry inner leaf should have a mass per unit area of at least 120 kg/m^2 excluding finish unless there are openings in the external wall (Figure 6.118) that are: E2.38a

- not less than 1 m high;   E2.38b

- on both sides of the separating wall at every storey;   E2.38c

- not more than 700 mm from the face of the separating wall on both sides.   E2.39

Note: If there is also a separating floor, then the minimum mass per unit area of 120 kg/m^2 (excluding finish) will always apply, irrespective of the presence or absence of openings.

The separating wall should be joined to the inner leaf of the external cavity wall by one of the methods shown in Figure 6.120.

6.6.2.29.2.2 JUNCTIONS WITH AN EXTERNAL CAVITY WALL WITH TIMBER FRAME INNER LEAF

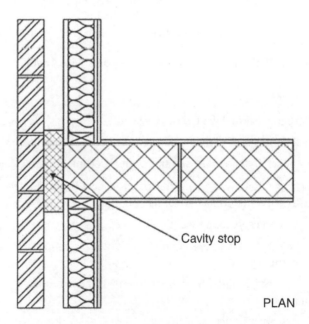

Cavity stop

PLAN

Figure 6.119 Junctions with an external cavity wall with timber frame inner leaf.

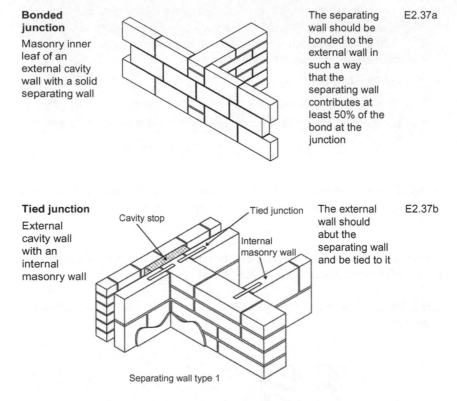

Bonded junction

Masonry inner leaf of an external cavity wall with a solid separating wall

The separating wall should be bonded to the external wall in such a way that the separating wall contributes at least 50% of the bond at the junction

E2.37a

Tied junction

External cavity wall with an internal masonry wall

Cavity stop

Tied junction

Internal masonry wall

Separating wall type 1

The external wall should abut the separating wall and be tied to it

E2.37b

Figure 6.120 Separating wall junctions for a type 1 wall.

Where the external wall is a cavity wall:

- the outer leaf of the wall may be of any construction;   E2.40a
- the cavity should be stopped with a flexible closer.   E2.40b

Where the inner leaf of an external cavity wall is of framed construction, the framed inner leaf should:

- abut the separating wall;   E2.41a1
- be tied to it with ties at no more than 300 mm centres vertically.   E2.41b1

The wall finish of the framed inner leaf of the external wall should be:

- one layer of plasterboard; or   E2.41a2
- two layers of plasterboard where there is a separating floor;   E2.41b2

and

- each sheet of plasterboard should be of minimum mass   E2.41c
  per unit area 10 kg/m²;

- all joints should be sealed with tape or caulked with   E2.41d
  sealant.

6.6.2.29.2.3 JUNCTIONS WITH INTERNAL TIMBER FLOORS

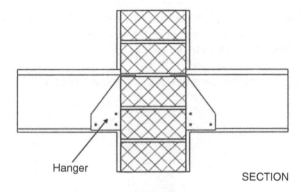

Hanger

SECTION

Figure 6.121 Junctions with internal timber floors.

If the floor joists are to be supported on a type 1 separating   E2.45
wall then they should be supported on hangers as opposed to
being built in.

6.6.2.29.2.4 JUNCTIONS WITH INTERNAL CONCRETE FLOORS

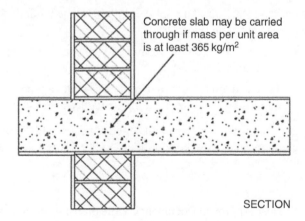

Concrete slab may be carried
through if mass per unit area
is at least 365 kg/m²

SECTION

Figure 6.122 Junctions with internal concrete floors.

An internal concrete floor slab may only be carried through a type 1 separating wall if the floor base has a mass per unit area of at least 365 kg/m².	E2.46
Internal hollow-core concrete plank floors and concrete beams with infilling block floors should not be continuous through a type 1 separating wall.	E2.47

Note: For internal floors of concrete beams with infilling blocks, avoid beams built into the separating wall unless the blocks in the floor fill the space between the beams where they penetrate the wall.

6.6.2.29.2.5 JUNCTIONS WITH CONCRETE GROUND FLOORS

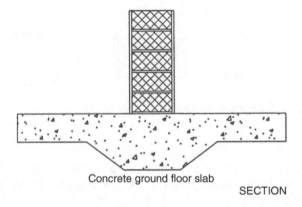

Concrete ground floor slab

SECTION

Figure 6.123 Junctions with concrete ground floors.

The ground floor may be a solid slab, laid on the ground, or a suspended concrete floor.	E2.51
A concrete slab floor on the ground may be continuous under a type 1 separating wall.	E2.51
A suspended concrete floor may only pass under a type 1 separating wall if the floor has a mass of at least 365 kg/m².	E2.52
Hollow-core concrete plank and concrete beams within filling block floors should not be continuous under a type 1 separating wall.	E2.53

Note: See also Approved Document C (Site preparation and resistance to moisture) and Approved Document L (Conservation of fuel and power).

6.6.2.29.2.6 JUNCTIONS WITH CEILING AND ROOF

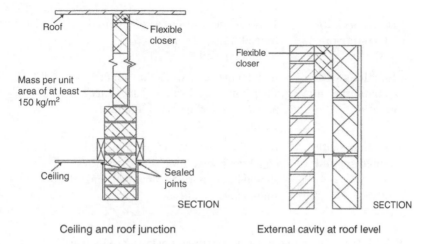

Ceiling and roof junction   External cavity at roof level

Figure 6.124 Junctions with ceiling and roof.

Where a type 1 separating wall is used it should be continuous to the underside of the roof.	E2.55
The junction between the separating wall and the roof should be filled with a flexible closer which is also suitable as a fire stop.	E2.56
Where the roof or loft space is not a habitable room (and there is a ceiling with a minimum mass per unit area of 10 kg/m^2 with sealed joints), the mass per unit area of the separating wall above the ceiling may be reduced to 150 kg/m^2.	E2.57
If lightweight aggregate blocks of density less than 1200 kg/m^3 are used above ceiling level, one side should be sealed with cement paint or plaster skim.	E2.58
Where there is an external cavity wall, the cavity should be closed at eaves level with a suitable flexible material (e.g. mineral wool).	E2.59
A rigid connection between the inner and external wall leaves should be avoided.	Ep23
If a rigid material is used, then it should only be rigidly bonded to one leaf.	Ep23

6.6.2.29.2.7 GUIDANCE FOR OTHER TYPES OF WALL TYPE 1 JUNCTIONS

Junctions with internal framed walls	There are no restrictions on internal framed walls meeting a type 1 separating wall.	E2.43
Junctions with internal masonry walls	Internal masonry walls that abut a type 1 separating wall should have a mass per unit area of at least 120 kg/m^2 excluding finish.	E2.44
Junctions with timber ground floors	If the floor joists are to be supported on a type 1 separating wall then they should be supported on hangers and should not be built in.	E2.49

6.6.2.29.2.8 JUNCTIONS WITH AN EXTERNAL SOLID MASONRY WALL

Currently there is no guidance available in the Approved Documents and it would be better to seek specialist advice if this is part of your building work.

6.6.2.30 Wall type 2 (cavity masonry)

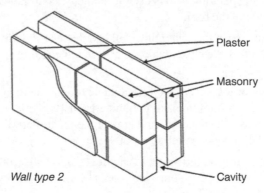

Figure 6.125 Wall type 2 (cavity masonry).

When using a cavity masonry wall, the resistance to airborne sound depends on the mass per unit area of the leaves and on the degree of isolation achieved. The isolation is affected by connections (e.g. wall ties and foundations) between the wall leaves and by the cavity width.

As shown in Table 6.47, there are four different categories of cavity masonry wall.

Table 6.47 Wall type 2 – categories

Wall type 2

Category 2.1

Two leaves of dense aggregate concrete block with 50 mm cavity incorporated at the separating wall

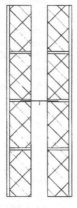

SECTION

Minimum mass per unit area (including plaster) 415 kg/m^2

Plaster on both room faces

Minimum cavity width 50 mm

For example:

Block leaves	100 mm
Density	1990 kg/m^3
Coursing	225 mm
Plaster	13 mm
lightweight	

Wall type 2

Category 2.2

Two leaves of lightweight aggregate block with 75 mm cavity, plaster on both room faces

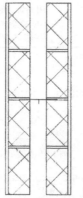

SECTION

Minimum mass per unit area (including plaster) 300 kg/m^2

Plaster on both room faces

Minimum cavity width of 75 mm

For example:

Block leaves	100 mm
Density	1375 kg/m^3
Coursing	225 mm
Plaster	13 mm
lightweight	

Wall type 2

Category 2.3

Two leaves of lightweight aggregate block with 75 mm cavity and step/ stagger plasterboard on both room faces

Wall type 2.3 should only be used where there is a step and/or stagger of at least 300 mm

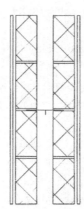

SECTION

Minimum mass per unit area (including plaster) 290 kg/m^2

Lightweight aggregate blocks should have a density in the range 1350 to 1600 m^3

Minimum cavity width of 75 mm
Plasterboard (lightweight) each sheet of minimum mass per unit area 10 kg/m^2 on both room faces

For example:

Block leaves	100 mm
Density	1375 kg/m^3
Coursing	225 mm

Increasing the size of the step or stagger in the separating wall tends to increase the airborne sound insulation

Lightweight plasterboard (minimum mass per unit area 10 kg/m^2) on both room faces

Table 6.47 Wall type 2 – categories (Continued)

Wall type 2		
Category 2.4 Two leaves of Aircrete block with 75 mm cavity and step/stagger plasterboard or plaster on both room faces Wall type 2.4 should only be used where there is a step and/or stagger of at least 300 mm Increasing the size of the step or stagger in the separating wall tends to increase the airborne sound insulation	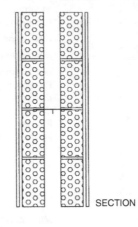 SECTION	Minimum mass per unit area (including plaster) 150 kg/m^2 Lightweight aggregate blocks should have a density in the range 1350 to 1600 kg/m^3 Minimum cavity width of 75 mm Plasterboard (lightweight) minimum mass per unit area 10 kg/m^2 on both room faces or 13 mm plasterboard on both faces For example: Aircrete block leaves 100 mm Density 650 kg/m^3 Coursing 225 mm Plaster (lightweight) minimum mass per unit area 10 kg/m^2 on both room faces

6.6.2.30.1 Wall type 2 – general requirements

Fill and seal all masonry joints with mortar.	E2.65a
Keep the cavity leaves separate below ground-floor level.	E2.65b
Ensure that any external cavity wall is stopped with a flexible closer at the junction with the separating wall.	E2.65c
Control flanking transmission from walls and floors connected to the separating wall.	E2.65d
Stagger the position of sockets on opposite sides of the separating wall.	E2.65e
Ensure that flue blocks will not adversely affect the sound insulation and that a suitable finish is used over the flue blocks.	E2.65f
The cavity separating wall should **not** be converted to a type 1 (solid masonry) separating wall by inserting mortar or concrete into the cavity between the two leaves.	E2.65a2
A solid wall construction in the roof space should **not** be changed.	E2.65b2
Cavity walls should **not** be built off a continuous solid concrete slab floor.	E2.65c2

Deep sockets and chases should **not** be used in a separating wall.	E2.65d2
Deep sockets and chases in a separating wall should **not** be placed back to back.	E2.65d2
Wall ties used to connect the leaves of a cavity masonry wall should be tie type A.	E2.66

6.6.2.30.2 Wall type 2 – junction requirements

6.6.2.30.2.1 JUNCTIONS WITH AN EXTERNAL CAVITY WALL WITH MASONRY INNER LEAF

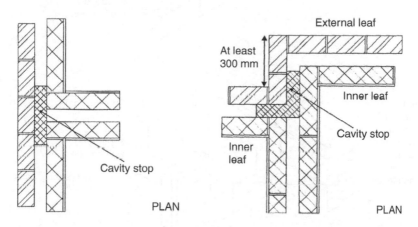

Wall types 2.1 and 2.2 – external cavity wall with masonry inner leaf

Wall types 2.3 and 2.4 – external cavity wall with masonry inner leaf – stagger

Figure 6.126 Junctions with an external cavity wall with masonry inner leaf.

Where the external wall is a cavity wall:	
• the outer leaf of the wall may be of any construction;	E2.73a
• the cavity should be stopped with a flexible closer.	E2.73b
The separating wall should be joined to the inner leaf of the external cavity wall.	E2.74
The masonry inner leaf should have a mass per unit area of at least 120 kg/m² excluding finish.	E2.75

There is no minimum mass requirement where separating wall   E2.76
type 2.1, 2.3 or 2.4 is used **unless** there is also a separating
floor.

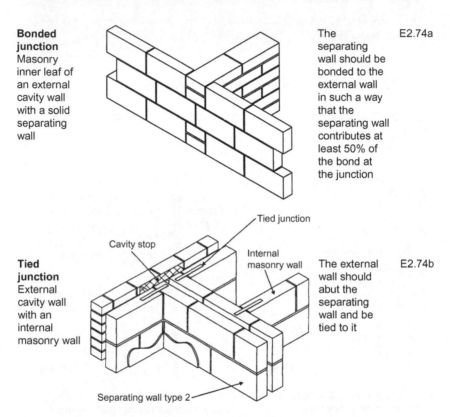

Bonded junction
Masonry inner leaf of an external cavity wall with a solid separating wall

The separating wall should be bonded to the external wall in such a way that the separating wall contributes at least 50% of the bond at the junction   E2.74a

Tied junction
External cavity wall with an internal masonry wall

Cavity stop

Tied junction

Internal masonry wall

Separating wall type 2

The external wall should abut the separating wall and be tied to it   E2.74b

Figure 6.127 Separating wall junctions for a type 2 wall.

6.6.2.30.2.2 JUNCTIONS WITH AN EXTERNAL CAVITY WALL WITH TIMBER
FRAME INNER LEAF

Where the external wall is a cavity wall:

- the outer leaf of the wall may be of any construction;   E2.77a
- the cavity should be stopped with a flexible closer.   E2.77b

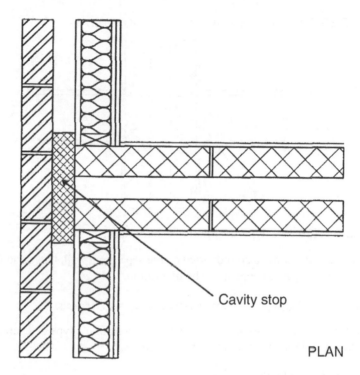

Figure 6.128 Wall type 2 – external cavity wall with timber frame inner leaf.

Where the inner leaf of an external cavity wall is of framed construction, the framed inner leaf should abut the separating wall and be tied to it with ties at no more than 300 mm centres vertically and the wall finish of the inner leaf of the external wall should be:

• one layer of plasterboard;		E2.98a2
• two layers of plasterboard where there is a separating floor;		E2.78b2
Each sheet of plasterboard to be of minimum mass per unit area 10 kg/m^2;		E2.78c2
All joints should be sealed with tape or caulked with sealant.		E2.78d2

6.6.2.30.2.3 JUNCTIONS WITH INTERNAL TIMBER FLOORS

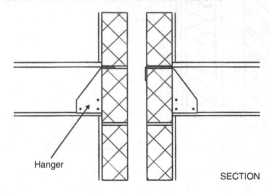

Hanger

SECTION

Figure 6.129 Wall type 2 – internal timber floor.

If the floor joists are to be supported on the separating wall, they should be supported on hangers as opposed to being built in.

6.6.2.30.2.4 JUNCTIONS WITH INTERNAL CONCRETE FLOORS

Internal concrete floors should generally be built into a type 2 separating wall and carried through to the cavity face of the leaf.

 The cavity should not be bridged.

6.6.2.30.2.5 JUNCTIONS WITH CONCRETE GROUND FLOORS

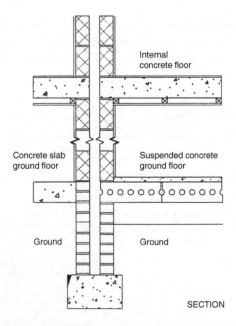

Figure 6.130 Wall type 2 – internal concrete floor and concrete ground floor.

The ground floor may be a solid slab, laid on the ground, or a suspended concrete floor.   E2.88

A concrete slab floor on the ground should not be continuous under a type 2 separating wall.   E2.88

A suspended concrete floor should not be continuous under a type 2 separating wall.   E2.89

A suspended concrete floor should be carried through to the cavity face of the leaf.   E2.89

The cavity should **not** be bridged.

6.6.2.30.2.6 JUNCTIONS WITH CEILING AND ROOF SPACE

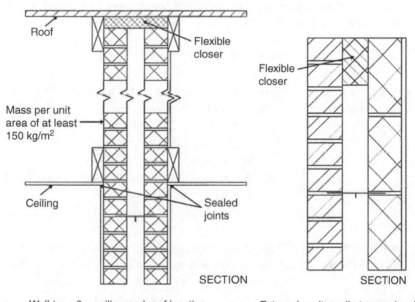

Wall type 2 – ceiling and roof junction   External cavity wall at eaves level

Figure 6.131 Junctions with ceiling and roof space.

A type 2 separating wall should be continuous to the underside of the roof.   E2.91

The junction between the separating wall and the roof should be filled with a flexible closer that is also suitable to act as a fire stop.   E2.92

If lightweight aggregate blocks (with a density less than 1200 kg/m³) are used above ceiling level, then one side should be sealed with cement paint or plaster skim.	E2.94
The cavity of an external cavity wall should be closed at eaves level with a suitable flexible material (e.g. mineral wool).	E2.95
A rigid connection between the inner and external wall leaves should be avoided.	E2.95
If a rigid material has to be used, then it should only be rigidly bonded to one leaf.	E2.95

Note: If the roof or loft space is not a habitable room (and there is a ceiling with a minimum mass per unit area of 10 kg/m² with sealed joints), then the mass per unit area of the separating wall above the ceiling may be reduced to 150 kg/m² – **but** it should still be a cavity wall.

6.6.2.30.2.7 GUIDANCE FOR OTHER WALL TYPE 2 JUNCTIONS

Junctions with internal masonry walls	Internal masonry walls that abut a type 2 separating wall should have a mass per unit area of at least 120 kg/m² excluding finish.	E2.81
	When there is a separating floor, the internal masonry walls should also have a mass per unit area of at least 120 kg/m² excluding finish.	
Junctions with internal framed walls	There are no restrictions on internal framed walls meeting a type 2 separating wall.	E2.80
Junctions with timber ground floors	If the floor joists are to be supported on a type 1 separating wall, they should be supported on hangers and should not be built in.	E2.49

6.6.2.30.2.8 JUNCTIONS WITH AN EXTERNAL SOLID MASONRY WALL

Currently there is no guidance available in the Approved Documents and it would be better to seek specialist advice if this is part of your building work.

6.6.2.31 Wall type 3 (masonry between independent panels)

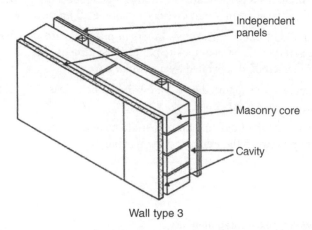

Wall type 3

Figure 6.132 Wall type 3 (masonry between independent panels).

Wall type 3 provides a high resistance to the transmission of both airborne sound and impact sound on the wall. As shown below, there are three different categories of wall type 3, which consist of either a solid or a cavity masonry core wall with independent panels on both sides. Their resistance to sound depends partly on the type (and mass) of the core and partly on the isolation and mass of the panels.

6.6.2.31.1 Wall type 3 – general requirements

Fill and seal all masonry joints with mortar.	E2.101a
Control flanking transmission from walls and floors connected to the separating wall.	E2.101b
The panels and any frame should not be in contact with the core wall.	E2.99
The panels and/or supporting frames should be fixed to the ceiling and floor only.	E101c
All joints should be taped and sealed.	E2.101d
Flue blocks shall not adversely affect the sound insulation.	E2.101e
A suitable finish is used over the flue blocks (see BS 1289-1:1986).	E2.101e
Free-standing panels and/or the frame should not be fixed, tied or connected to the masonry core.	E2.101
Wall ties in cavity masonry cores, used to connect the leaves of a cavity masonry core together, should be tie type A.	E2.102

The minimum mass per unit area of independent panels (excluding any supporting framework) should be 20 kg/m².	E2.104
Panels should be either at least two layers of plasterboard with staggered joints or a composite panel consisting of two sheets of plasterboard separated by a cellular core.	E2.104
Panels that are not supported on a frame should be at least 35 mm from the masonry core.	E2.104
Panels that are supported on a frame should have a gap of at least 10 mm between the frame and the masonry core.	E2.104

Table 6.48 Wall type 3 – categories

Wall type 3	
Category 3.1 Solid masonry core (dense aggregate concrete block), independent panels on both room faces	Minimum mass per unit area (including plaster) 300 kg/m²; independent panels on both room faces Minimum core width determined by structural requirements. For example: Size 140 mm core block Density 2200 kg/m³ Coursing 110 mm Independent panels – each panel of mass per unit area 20 kg/m², to be two sheets of plasterboard with joints staggered
Category 3.2 Solid masonry core (lightweight concrete block), independent panels on both room faces 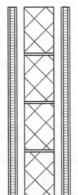 SECTION	Minimum mass per unit area (including plaster) 150 kg/m²; independent panels on both room faces Minimum core width determined by structural requirements For example: Size 140 mm core block Density 1400 kg/m³ 225 mm Coursing Independent panels – each panel of mass per unit area 20 kg/m², to be two sheets of plasterboard with joints staggered

Table 6.48 Wall type 3 – categories (Continued)

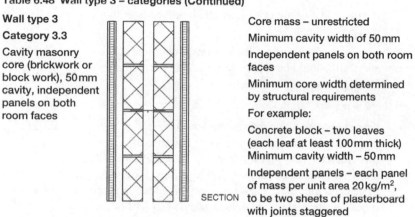

Wall type 3	Core mass – unrestricted
Category 3.3	Minimum cavity width of 50 mm
Cavity masonry core (brickwork or block work), 50 mm cavity, independent panels on both room faces	Independent panels on both room faces
	Minimum core width determined by structural requirements
	For example:
	Concrete block – two leaves (each leaf at least 100 mm thick) Minimum cavity width – 50 mm
SECTION	Independent panels – each panel of mass per unit area 20 kg/m^2, to be two sheets of plasterboard with joints staggered

6.6.2.31.2 Junction requirements for wall type 3

6.6.2.31.2.1 JUNCTIONS WITH AN EXTERNAL CAVITY WALL WITH MASONRY INNER LEAF

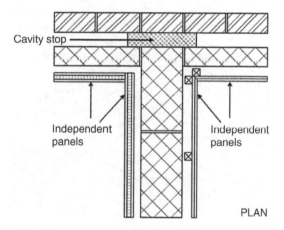

Figure 6.133 Wall type 3 – external cavity wall with masonry inner leaf.

If the external wall is a cavity wall:	E2.108
• the outer leaf of the wall may be of any construction;	
• the cavity should be stopped with a flexible closer.	
If the inner leaf of an external cavity wall is masonry:	E2.109
• the inner leaf of the external wall should be bonded or tied to the masonry core;	

- the inner leaf of the external wall should be lined with independent panels.

If there is a separating floor, the masonry inner leaf (of the external wall) should have a minimum mass per unit area of at least 120 kg/m² excluding finish.   E2.110

If there is no separating floor:

- the external wall may be finished with plaster or plasterboard of minimum mass per unit area 10 kg/m² (provided the masonry inner leaf of the external wall has a mass per unit area of at least 120 kg/m² excluding finish);   E2.111
- there is no minimum mass requirement on the masonry inner leaf (provided that the masonry inner leaf of the external wall is lined with independent panels in the same manner as the separating walls).   E2.112

6.6.2.31.2.2 JUNCTIONS WITH INTERNAL FRAMED WALLS

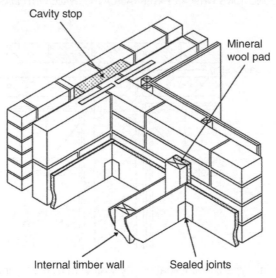

Figure 6.134 Wall type 3 – external cavity wall with internal timber wall.

Load-bearing (framed) internal walls should be fixed to the masonry core through a continuous pad of mineral wool.	E2.115
Non-load-bearing internal walls should be butted to the independent panels.	E2.116
All joints between internal walls and panels should be sealed with tape or caulked with sealant.	E2.117

6.6.2.31.2.3 JUNCTIONS WITH INTERNAL TIMBER FLOORS

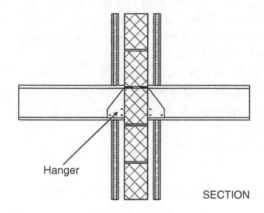

Hanger

SECTION

Figure 6.135 Wall type 3 – internal timber floor.

6.6.2.31.2.4 JUNCTIONS WITH INTERNAL MASONRY WALLS

If the floor joists are to be supported on the separating wall then they should be supported on hangers as opposed to being built in.	E2.119
Spaces between the floor joists should be sealed with full-depth timber blocking.	E2.120

6.6.2.31.2.5 JUNCTIONS WITH INTERNAL CONCRETE FLOORS

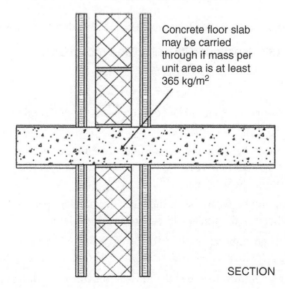

Concrete floor slab may be carried through if mass per unit area is at least 365 kg/m^2

SECTION

Figure 6.136 Wall types 3.1 and 3.2 – internal concrete floor.

For wall types 3.1 and 3.2 (i.e. those with solid masonry cores) internal concrete floor slabs may only be carried through a solid masonry core if the floor base has a mass per unit area of at least 365 kg/m^2.   E2.121

For wall type 3.3 (cavity masonry core):   E2.122

- internal concrete floors should generally be built into a cavity masonry core and carried through to the cavity face of the leaf;
- the cavity should not be bridged.

6.6.2.31.2.6 JUNCTIONS WITH CEILING AND ROOF SPACE

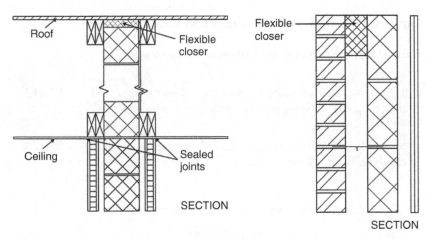

Wall types 3.1 and 3.2 – ceiling and roof junction   External cavity wall at eaves level

Figure 6.137 Junctions with ceiling and roof space.

The masonry core should be continuous to the underside of the roof.   E2.133

The junction between the separating wall and the roof should be filled with a flexible closer that is also suitable as a fire stop.   E2.134

The junction between the ceiling and independent panels should be sealed with tape or caulked with sealant.   E2.135

If there is an external cavity wall, the cavity should be closed at eaves level with a suitable flexible material (e.g. mineral wool).   E2.136

Rigid connections between the inner and external wall leaves should be avoided where possible.   E2.136

If a rigid material is used, then it should only be rigidly bonded to one leaf.   E2.136

For wall types 3.1 and 3.2 (solid masonry core):

- if the roof or loft space is not a habitable room (and there is a ceiling with a minimum mass per unit area of 10 kg/m^2 and it has sealed joints), the independent panels may be omitted in the roof space and the mass per unit area of the separating wall above the ceiling may be a minimum of 150 kg/m^2;   E2.137
- if lightweight aggregate blocks with a density less than 1200 kg/m^3 are used above ceiling level, one side should be sealed with cement paint or plaster skim.   E2.138

For wall type 3.3 (cavity masonry core) if the roof or loft space is not a habitable room (and there is a ceiling with a minimum mass per unit area of 10 kg/m^2 and it has sealed joints), the independent panels may be omitted in the roof space, but the cavity masonry core should be maintained to the underside of the roof.   E2.139

6.6.2.31.2.7 JUNCTIONS WITH INTERNAL MASONRY FLOORS

Internal walls that abut a type 2 separating wall should not be of masonry construction.

6.6.2.31.2.8 JUNCTIONS WITH TIMBER GROUND FLOORS

Floor joists supported on a separating wall should be supported on hangers as opposed to being built in.   E2.123

The spaces between floor joists should be sealed with full-depth timber blocking.   E2.124

6.6.2.31.2.9 JUNCTIONS WITH AN EXTERNAL CAVITY WALL WITH TIMBER FRAME INNER LEAF

 As no official guidance is currently available it is best to seek specialist advice.

6.6.2.31.2.10 JUNCTIONS WITH AN EXTERNAL SOLID MASONRY WALL

 As no official guidance is currently available it is best to seek specialist advice.

6.6.2.31.2.11 JUNCTIONS WITH CONCRETE GROUND FLOORS

The ground floor may be a solid slab, laid on the ground, or a   E2.126
suspended concrete floor.

For wall types 3.1 and 3.2 (solid masonry core):

*   a concrete slab floor on the ground may be continuous   E2.127
    under the solid masonry core of the separating wall;

*   a suspended concrete floor may only pass under the solid   E2.128
    masonry core if the floor has a mass per unit area of at
    least 365 kg/m^2;

*   hollow-core concrete plank (and concrete beams with   E2.129
    infilling block floors) should **not** be continuous under
    the solid masonry core of the separating wall.

For wall type 3.3 (cavity masonry core):

*   a concrete–masonry core of a slab floor on the ground   E2.130
    should **not** be continuous under the cavity masonry core
    of the separating wall;

*   a suspended concrete floor should **not** be continuous   E2.131
    under the cavity type 3.3 separating wall;

*   a suspended concrete floor **should** be carried through to   E2.132
    the cavity face of the leaf but the cavity should not be
    bridged.

6.6.2.31.2.12 JUNCTIONS WITH INTERNAL MASONRY WALLS

Internal walls that abut a type 3 separating wall should not   E2.118
be of masonry construction.

6.6.2.32 Wall type 4 (framed walls with absorbent material)

A wall type 4 consists of a timber frame with a plasterboard lining on the room
surface with an absorbent material between the frames. Its resistance to airborne sound depends on:

*   the mass per unit area of the leaves;
*   the isolation of the frames;
*   the absorption in the cavity between the frames.

6.6.2.32.1 General requirements

If a fire stop is required in the cavity between frames, then it should either be flexible or only be fixed to one frame.	E2.146
Layers of plasterboard should:	
• be independently fixed to the stud frame;	E2.146c
• not be chased.	E2.146b2
If two leaves have to be connected together for structural reasons, then:	
• the cross-section of the ties shall be less than 40 mm × 3 mm;	E2.146a2
• ties should be fixed to the studwork at or just below ceiling level;	E2.146a2
• ties should not be set closer than 1.2 m centres.	E2.146a2
Sockets should:	
• be positioned on opposite sides of a separating wall;	E2.146b
• not be connected back to back;	E2.146b2
• be staggered a minimum of 150 mm edge to edge.	E2.146b2
The flanking transmission from walls and floors connected to a separating wall should be controlled (see guidance on junctions).	E2.146d

6.6.2.32.2 Wall type 4.1 (double-leaf frames with absorbent material)

6.6.2.32.2.1 GENERAL REQUIREMENTS

The lining shall be two or more layers of plasterboard with a minimum sheet mass per unit area 10 kg/m2 and with staggered joints.	E2.147
If a masonry core is used for structural purposes, then the core should only be connected to one frame.	E2.147
The minimum distance between inside lining faces shall be 200 mm.	E2.147
Plywood sheathing may be used in the cavity if required for structural reasons.	E2.147
Absorbent material:	E2.147

- shall have a minimum density of 10 kg/m3;
- shall be unlaced mineral wool batts (or quilt);
- may be wire reinforced;
- shall have a minimum thickness of between 25 and 50 mm as shown in Figure 6.138.

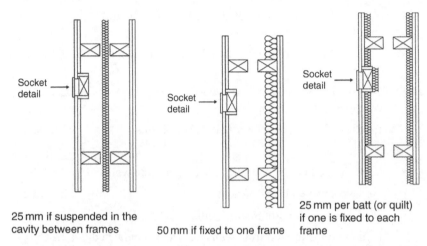

25 mm if suspended in the cavity between frames

50 mm if fixed to one frame

25 mm per batt (or quilt) if one is fixed to each frame

Figure 6.138 Wall type 4.1 – minimum thickness of absorbent material.

6.6.2.32.3 Junction requirements for wall type 4

6.6.2.32.3.1 JUNCTIONS WITH AN EXTERNAL CAVITY WALL WITH TIMBER-FRAME INNER LEAF

If the external wall is a cavity wall:

- the outer leaf of the wall may be of any construction;   E2.149

- the cavity should be stopped between the ends of the separating wall and the outer leaf with a flexible closer.   E2.149

The wall finish of the inner leaf of the external wall should be one layer of plasterboard (or two layers of plasterboard if there is a separating floor).   E2.150a E2.150b

Each sheet of plasterboard should be of minimum mass per unit area 10 kg/m².   E2.150c

All joints should be sealed with tape or caulked with sealant.   E2.150

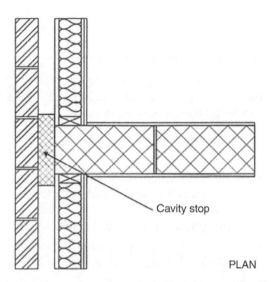

PLAN

Figure 6.139 Junctions with an external solid masonry wall.

6.6.2.32.3.2 JUNCTION WITH CEILING AND ROOF SPACE

The wall should preferably be continuous to the underside of the roof.	E2.160
The junction between the separating wall and the roof should be filled with a flexible closer.	E2.161
The junction between the ceiling and the wall linings should be sealed with tape or caulked with sealant.	E2.162
If the roof or loft space is not a habitable room (and there is a ceiling with a minimum mass per unit area of 10 kg/m^2 with sealed joints), then:	
• either the linings on each frame may be reduced to two layers of plasterboard, each sheet with a minimum mass per unit area of 10 kg/m^2; or	E2.162a
• the cavity may be closed at ceiling level without connecting the two frames rigidly together.	E2.162b

Note: In which case there need only be one frame in the roof space provided there is a lining of two layers of plasterboard, each sheet of minimum mass per unit area of 10 kg/m^2, on both sides of the frame.

> External wall cavities should be closed at eaves level with a   E2.163
> suitable material.

6.6.2.32.3.3 JUNCTIONS WITH TIMBER GROUND FLOORS

> Air paths through the wall into the cavity shall be blocked   E2.156
> using solid timber blockings, continuous ring beam or
> joists.

See also Approved Document C (Site preparation and resistance to moisture) and Approved Document L (Conservation of fuel and power).

6.6.2.32.3.4 JUNCTIONS WITH CONCRETE GROUND FLOORS

> If the ground floor is a concrete slab laid on the ground, it   E2.158
> may be continuous under a type 4 separating wall.
>
> If the ground floor is a suspended concrete floor, it may only
> pass under a wall type 4 if the floor has a mass per unit area
> of at least 365 kg/m^2.

See also Approved Document C (Site preparation and resistance to moisture) and Approved Document L (Conservation of fuel and power).

6.6.2.32.3.5 JUNCTIONS WITH INTERNAL TIMBER FLOORS

> Air paths through the wall into the cavity shall be blocked   E2.154
> using solid timber blockings, continuous ring beam or joists.

6.6.2.32.3.6 JUNCTIONS WITH INTERNAL CONCRETE FLOORS

As no official guidance is currently available it is best to seek specialist advice.

6.6.2.32.3.7 JUNCTIONS WITH INTERNAL FRAMED WALLS

There are no restrictions on internal framed walls meeting a type 4 separating wall.

6.6.2.32.3.8 JUNCTIONS WITH INTERNAL MASONRY WALLS

There are no restrictions on internal masonry walls meeting a type 4 separating wall.

6.6.2.32.3.9 JUNCTIONS WITH AN EXTERNAL SOLID MASONRY WALL

 As no official guidance is currently available it is best to seek specialist advice.

6.6.2.32.3.10 JUNCTIONS WITH AN EXTERNAL CAVITY WALL WITH MASONRY INNER LEAF

 As no official guidance is currently available it is best to seek specialist advice.

6.6.2.33 CONSERVATION OF ENERGY AND POWER

Following the publication of new Approved Documents L1A and L1B in November 2013, with effect from April 2014:

- all **new homes** (i.e. dwellings) need to achieve, or better, a fabric energy efficiency target in addition to the strengthened requirement to deliver **6 per cent carbon dioxide savings**; and
- all **new non-domestic buildings** need to make **9 per cent carbon dioxide savings**.

 Regulation 9 of the Building Regulations exempts some conservatory and porch extensions from the energy efficiency requirements. These consist of conservatories or porches:

- which are at ground level;
- where the floor area is less than $30\,m^2$;
- where the existing walls in the part of the dwelling which separates the conservatory are retained or, if removed, replaced by walls which meet the energy-efficiency requirements;
- where the heating system of the dwelling is not extended into the conservatory or porch;
- where the worst standards for fabric properties do not exceed the values given in Table 6.49.

Table 6.49 Limiting fabric parameters for conservatory and porch extensions

Element	Standard $(W/m^2\,K)$
Roof	0.20
Wall	0.30
Floor	0.25
Party wall	0.20
Windows, roof windows, glazed rooflights, curtain walling and pedestrian doors	2.00
Opaque elements	0.18

If a new conservatory or porch does not meet all these requirements then it is not exempt and must comply with the relevant energy-efficiency requirements of the 2014 revision of Approved Document L.

If a building extension is a conservatory or porch that is **not** exempt from the energy-efficiency requirements, then:	L1A L2B 4.8

- the effective thermal separation between the heated area in the existing building (i.e. the walls, doors and windows between the building and the extension) should be insulated and draught-proofed to at least the same extent as in the existing building; and
- independent temperature and on/off controls to any heating system should have been installed within the extension.

Removing and not replacing any of the thermal separation between the building and an existing exempt extension, or extending the building's heating system into the extension, mean that the extension ceases to be exempt!

6.6.2.33.1 Dwellings

Extensions to dwellings should use either newly constructed thermal elements (which meet the requirements for the conservation of fuel and power) or existing (or new) windows, roof windows and rooflights that meet the standards shown in Table 6.50.

If upgrades are proposed to the existing dwelling, such upgrades should be implemented to a standard that is no worse than that shown in column (b) of Table 6.50.	L1B 4.7

Table 6.50 Upgrading retained thermal elements

Element	(a) Threshold U-value (W/m² 1)	(b) Improved U-value (W/m² 1)
Wall – cavity insulation	0.70	0.55
Wall – external or cavity insulation	0.70	0.30

6.6.2.33.2 Building fabric

In accordance with the 2014 revision of Approved Document L and Regulation 7, reasonable provision shall be made for the conservation of fuel and power in buildings by limiting heat gains and losses through thermal elements and other parts of the building fabric.

The building fabric of all dwellings and buildings other than dwellings should be constructed to a reasonable standard so that:	L1A 3.2 L2A 3.2

- the insulation is reasonably continuous over the whole building envelope; and
- the air permeability is within reasonable limits.

Reductions in thermal performance can occur where the air barrier and the insulation layer are not adjacent and the cavity between them is subject to air movement. L2A 5.4

To avoid this problem, either the insulation layer should be adjacent to the air barrier at all points in the building envelope or the space between them should be filled with solid material such as in a masonry wall.

6.6.2.33.3 Limiting fabric standards

U-values shall be calculated using the methods and conventions set out in BR 443 (Conventions for U-value calculations), and should be based on the whole element or unit (e.g. in the case of a window it would be the combined performance of the glazing and the frame).

For windows, the U-value needs to be: L1A 2.34
L2A 2.40

- the smaller of the two standard windows defined in BS EN 14351-1; or
- the standard configuration set out in BR 443; or
- the specific size and configuration of the actual window.

For doors, the U-value needs to be: L1A 2.34
L2A 2.40

- the standard size as laid out in BS EN 14351-1; or
- the specific size and configuration of the actual door.

6.6.2.33.3.1 DOMESTIC BUILDINGS

Table 6.51 provides limiting standards for a domestic building's fabric.

Table 6.51 Limiting fabric parameters for domestic buildings

Roof	0.20 W/m² K
Wall	0.30 W/m² K
Floor	0.25 W/m² K
Party wall	0.20 W/m² K
Swimming pool basin	0.25 W/m² K
Windows, roof windows, glazed rooflights, curtain walling and pedestrian doors	2.00 W/m² K
Air permeability	10.0 m³/h m² at 50 Pa

Note: Where a swimming pool is constructed as part of a new building, reasonable provision should be made to limit the heat loss from the pool basin by achieving a U-value no worse than 0.25 W/m² K (see BS EN ISO 13370).

6.6.2.33.4 BUILDINGS OTHER THAN DOMESTIC BUILDINGS

Table 6.52 sets out the limiting standards for the properties of the fabric elements of the building.

Generally speaking, however, in order to achieve the target emissions rate (TER) you are likely to need a better fabric performance than that shown in Table 6.52.

Table 6.52 Limiting fabric parameters buildings (other than domestic buildings)

Roof	0.25 W/m² K
Wall	0.35 W/m² K
Floor	0.25 W/m² K
Swimming pool basin	0.25 W/m² K
Windows, roof windows, rooflights, curtain walling and pedestrian doors	2.2 W/m² K
Vehicle access and similar large doors	1.5 W/m² K
High-usage entrance doors	3.5 W/m² K
Roof ventilators (including smoke vents)	3.5 W/m² K
Air permeability	10.0 m³/h m² at 50 Pa

Note: Where a swimming pool is constructed as part of a new building, reasonable provision should be made to limit heat loss from the pool basin by achieving a U-value no worse than 0.25 W/m² K as calculated according to BS EN ISO 13370.

6.6.2.33.5 Thermal bridges

> The building fabric should be constructed so that there are   L1A 3.9
> no reasonably avoidable thermal bridges in the insulation
> layers caused by gaps within the various elements, at the
> joints between elements, and at the edges of elements, such
> as those around window and door openings

To meet this requirement, it is recommended that you use construction joint details that have been calculated following the guidance set out in BRE Report BR 497, *Conventions for Calculating Linear Thermal Transmittance and Temperature Factors*.

6.6.2.33.6 Party walls and other thermal bypasses

Party cavity walls, in many circumstance, may **not** be zero heat loss walls, as the air flow in the cavity provides a heat loss mechanism.

 Note: Where outside air flows into the party wall cavity, a cold zone is created which causes heat loss through the wall sections on either side.

> Heat loss can be reduced by fully filling the cavity and/or by   L1A 3.4
> effectively causing a seal around the perimeter and thereby
> restricting the amount of air movement through the cavity
> (see www.buildingcontrolalliance.org for further guidance).

 Note: Fully filling the cavity may affect sound transmission through party walls. Developers who plan to fill a party wall cavity must satisfy the BCB that the requirements of Part E (Sound) of Schedule 1 of the Building Regulations will be satisfied, either by adopting a full-fill detail accredited under the Robust Details scheme or through specific site testing.

When calculating the dwelling CO_2 emission rate (DER) and dwelling fabric energy-efficiency (DFEE) rate for a dwelling, a party wall U-value for the type of construction adopted, as set out in Table 6.53, should be applied.

Table 6.53 U-values for party walls

Party wall construction	U-value (W/m² K)
Solid	0.00
Unfilled cavity with no effective edge sealing	0.50
Unfilled cavity with effective sealing around all exposed edges and in line with insulation layers in abutting elements	0.20
A fully filled cavity with effective sealing at all exposed edges and in line with insulation layers in abutting elements	0.00

The party wall is a particular case of the more general thermal bypass problem that occurs if the air barrier and the insulation layer are not contiguous and the cavity between them is subject to air movement.

When considering heat losses via party walls, it is important to remember that wherever the wall penetrates an insulation layer, such as when the block-work of a masonry party wall penetrates insulation at ceiling level, a thermal bridge is likely to exist – even when the party wall U-value is zero.

To avoid the consequent reduction in thermal performance, either the insulation layer should be contiguous with the air barrier at all points in the building envelope, or the space between the air barrier and insulation layer should be filled with solid material, such as in a masonry wall.

6.7 Ceilings

6.7.1 Requirements

Fire protection
As a fire precaution, all materials used for internal linings of a building should have a low rate of surface flame spread and (in some cases) a low rate of heat release.

(Approved Document B2)

Airborne and impact sound
Dwellings shall be designed so that the noise from domestic activity in an adjoining dwelling (or other parts of the building) is kept to a level that:

- *does not affect the health of the occupants of the dwelling;*
- *will allow them to sleep, rest and engage in their normal activities in satisfactory conditions.*

(Approved Document E1)

Dwellings shall be designed so that any domestic noise that is generated internally does not interfere with the occupants' ability to sleep, rest and engage in their normal activities in satisfactory conditions.

(Approved Document E2)

Domestic buildings shall be designed and constructed so as to restrict the transmission of echoes.

(Approved Document E3)

Schools shall be designed and constructed so as to reduce the level of ambient noise (particularly echoing in corridors).

(Approved Document E4)

6.7.2 Meeting the requirements

6.7.2.1 General

6.7.2.2 Ceiling joists

Softwood timber used for roof construction or fixed in the roof space (including ceiling joists within the void spaces of the roof) should be adequately treated to prevent infestation by the house longhorn beetle (*Hylotrupes bajulus* L.), particularly in the following areas:

- the Borough of Bracknell Forest, in the parishes of   A1/2 2B2
  Sandhurst and Crowthorne;
- the Borough of Elmbridge;
- the District of Hart, in the parishes of Hawley and
  Yateley;
- the District of Runnymede;
- the Borough of Spelthorne;
- the Borough of Surrey Heath;
- the Borough of Rushmoor, in the area of the former
  district of Farnborough;
- the Borough of Woking.

Notes:

(1) Guidance on suitable preservative treatments is given within the *British Wood Preserving and Damp-Proofing Association's Manual* (2000 revision), available from 1 Gleneagles House, Vernongate, South Street, Derby DE1 1UP.
(2) Guidance on the sizing of certain members in floors and roofs is given in BS 5268: Part 2:2002 and Part 3:1998 as *Span Tables for Solid Timber Members in Floors, Ceilings and Roofs (Excluding Trussed Rafter Roofs) for Dwellings*, published by TRADA, available from Chiltern House, Stocking Lane, Hughenden Valley, High Wycombe, Bucks HP14 4ND.

6.7.2.2 Fire protection

6.7.2.2.1 Fire-resisting ceilings

The need for cavity barriers in some concealed floor   B2 3.6 (V1)
or roof spaces can be reduced by using a fire-resisting   B2 6.6 (V2)
ceiling below the cavity.

6.7.2.2.2 Suspended ceilings

A suspended, fire-resisting ceiling should meet the requirements given in Table 6.54.

Table 6.54 Limitations on fire-resisting suspended ceilings

Height of building or separated part	Type of floor	Provision for fire resistance of floor (minutes)
Less than 18	Not compartment	60 or less
	Compartment	Less than 60
18 or more	Any	60 or less
No limit	Any	More than 60

For further details see Approved Document B, Appendix A, Table A3.

6.7.2.2.3 Ceiling linings

To inhibit the spread of fire within the building, ceiling internal linings shall:

- adequately resist the spread of flame over their surfaces; and
- have, if ignited, a rate of heat release or a rate of fire growth that is reasonable in the circumstances.

 Note: Flame spread over wall or ceiling surfaces is controlled by ensuring that the lining materials or products meet given performance levels that are measured in terms of performance with reference to Tables A1 and A3 of Approved Document B.

For the purpose of this requirement, the content of ceilings is as described in Table 6.55.

Table 6.55 Content of ceiling linings

Include	Do not include
• The surface of glazing • Any part of a wall which slopes at an angle of 70° or less to the horizontal • The underside of a gallery • The underside of a roof exposed to the room below	• Trap doors and their frames • The frames of windows or rooflights and frames in which glazing is fitted • Architraves, cover moulds, picture rails • Exposed beams and similar narrow members

6.7.2.2.4 Classification of linings

In general terms (but see paragraphs 3.2 to 3.14 of Approved Document B V1 and 6.2 to 6.14 of Approved Document B V2 for more details), the surface linings for ceilings should meet the classifications given in Table 6.56.

Table 6.56 Classification of linings

Location	National class	European class
Small rooms less than 4 m² in residential accommodation, 30 m² in non-residential accommodation	3	D-s3, d2
Other rooms (including garages) and circulation spaces in dwellings	1	C-s3, d2
Other circulation spaces, including the common areas of blocks of flats	0	B-s3, d2

6.7.2.2.5 Heat alarms

Heat detectors and heat alarms should:	
• be designed and installed in accordance with BS 5839-6:2004;	B1 1.10
• be sited so that the sensor in ceiling-mounted devices is between 25 mm and 150 mm below the ceiling;	B1 1.15c
• be mains-operated and conform to BS 5446-2:2003; and	B1 1.4 (V1)
• have a standby power supply such as a rechargeable (or non-rechargeable) battery.	B1 1.5 (V2)

6.7.2.2.6 Thermoplastic materials

Thermoplastic materials may be used in rooflights and lighting diffusers in suspended ceilings.	B2 3.8 (V1) B2 6.10 (V2)
Flexible thermoplastic material may be used in panels to form a suspended ceiling.	B2 3.8 (V1) B2 6.10 (V2)
Flexible membranes covering a structure shall be in accordance with Appendix A of BS 7157:1989.	B2 6.8 (V2)

6.7.2.2.7 Rooflights

Rooflights should meet the relevant classification in Table 6.57.	B2 3.7 (V1) B2 6.7 (V2)
Rooflights may be constructed of a thermoplastic material if:	B2 3.10 (V1)
• the lower surface has a TPI(a) (rigid) or a TP(b) classification;	B2 6.12 (V2)
• the size and disposition of the rooflights accord with the limits in Table 6.57.	

Table 6.57 Limitations applied to thermoplastic rooflights and lighting diffusers in suspended ceilings and class 3 plastic rooflights

Minimum classification of lower surface	Use of space below the diffusers or rooflight	Maximum area of each diffuser panel or rooflight (m²)	Maximum total area of diffuser panels and rooflights as percentage of floor area of the space in which the ceiling is located (%)	Minimum separation distance between diffuser panels or rooflights (m)
TP(a)	Any (except a protected stairway)	No limit	No limit	No limit
D-s3, D2 or Class 3 or TP(b)	Rooms	1–5	50	3
	Circulation spaces (except protected stairways)	5	15	3

6.7.2.2.8 Lighting diffusers

Thermoplastic lighting diffusers (i.e. translucent or open-structured elements that allow light to pass through) should not be used in fire-protecting or fire-resisting ceilings, unless they have been satisfactorily tested as part of the ceiling system that is to be used to provide the appropriate fire protection.

B2 3.12 (V1)
B2 6.14 (V2)

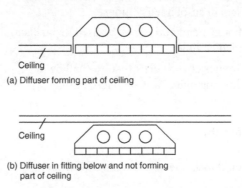

Ceiling
(a) Diffuser forming part of ceiling

Ceiling
(b) Diffuser in fitting below and not forming part of ceiling

Figure 6.140 Lighting diffuser in relation to ceiling.

6.7.2.2.9 Suspended or stretched-skin ceilings

The ceiling of a room may be constructed from panels of a thermoplastic material of the TP(d) flexible classification, provided that it is not part of a fire-resisting ceiling.

Each panel should not exceed 5 m² in area and should be supported on all of its sides.	B2 3.14 (V1) B2 6.16 (V2)

6.7.2.2.10 Smoke alarms

Smoke alarm systems (conforming to BS EN 14604 and/or BS EN 14604) should be ceiling-mounted and at least 300 mm from walls and light fittings.	B1 1.15b (V1) B1 1.14b (V2)
The sensor in ceiling-mounted devices shall be between 25 mm and 600 mm below the ceiling (25–150 mm in the case of heat detectors or heat alarms).	B1 1.15c (V1) B1 1.14c (V2)

6.7.2.2.11 Air-circulation systems

Transfer grilles of air-circulation systems should **not** be fitted in any ceiling enclosing a protected stairway.	B1 2.17 (V1) B1 2.18 (V2) B3 7.10 (V1) B3 10.2 (V2)

6.7.2.2.12 Venting of heat and smoke from basements

Smoke outlets should be sited at high level, either in the ceiling or in the wall of the space they serve.	B5 18.3 (V2) B5 18.5 (V2) B5 18.7 (V2)

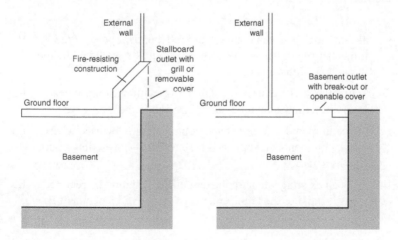

Figure 6.141 Fire-resisting construction for smoke outlet shafts.

6.7.2.3 Airborne and impact sound

6.7.2.3.1 Ceilings – general

The resistance to airborne and impact sound depends on the independence and isolation of the ceiling and the type of material used.

Three ceiling treatments (which are ranked in order of sound insulation) may be used:

- ceiling treatment A – independent ceiling with absorbent   E
  material;
- ceiling treatment B – plasterboard on proprietary resilient bars
  with absorbent material;
- ceiling treatment C – plasterboard on timber battens (or
  proprietary resilient channels) with absorbent material.

If the roof or loft space is not a habitable room (and provided that there is a ceiling with a minimum mass per unit area of 10 kg/m^2 with sealed joints and the cavity masonry core is maintained to the underside of the roof) then:

- the mass per unit area of the separating wall above the ceiling   E
  may be reduced to 150 kg/m^2;
- the independent panels may be omitted in the roof space;
- the linings on each frame may be reduced to two layers of
  plasterboard or the cavity may be closed at ceiling level
  without connecting the two frames rigidly together.

In addition:

- all junctions between ceilings and independent panels (and   E
  joints between casings and ceiling) should be sealed with tape
  or caulked with sealant;
- at junctions with external cavity walls (with masonry inner   E3
  leaf) the ceiling should be taken through to the masonry;
- the ceiling void and roof space detail can only be used where   E3
  the requirements of Approved Document B (Fire safety) can
  also be satisfied.

If there is an existing lath and plaster ceiling it should be retained   E3
as long as it satisfies the requirements of Approved Document B.

If the existing ceiling is not lath and plaster, it should be upgraded to provide:

- at least two layers of plasterboard with staggered joints;   E4
- a minimum total mass per unit area of 20 kg/m^2;
- an absorbent layer of mineral wool laid on the ceiling (minimum thickness 100 mm, minimum density 10 kg/m^3);
- plasterboard with joints staggered, total mass per unit area 20 kg/m^2.

Care should be taken at the design stage to ensure that adequate ceiling height is available in all rooms to be treated.

The ceiling should be supported by either: E2
- independent joists fixed only to the surrounding walls; or
- independent joists fixed to the surrounding walls with additional support provided by resilient hangers attached directly to the existing floor base.

 Note: A clearance of at least 25 mm should be left between the top of the independent ceiling joists and the underside of the existing floor construction.

6.7.2.3.2 Other considerations

Where a window head is near to the existing ceiling, the new independent ceiling may be raised to form a pelmet recess.	E4
A rigid or direct connection should not be created between an independent ceiling and the floor base.	E4
Where the roof or loft space is not a habitable room (and there is a ceiling with a minimum mass per unit area of 10 kg/m^2 with sealed joints) then the mass per unit area of the separating wall above the ceiling may be reduced to 150 kg/m^2.	E2.57
If lightweight aggregate blocks of density less than 1200 kg/m^3 are used above ceiling level, then one side should be sealed with cement paint or plaster skim.	E2.58 E2.94 E2.138
Where the external wall is a cavity wall with a masonry inner leaf (or a simple cavity masonry wall or masonry between independent panels), the ceiling should be taken through to the masonry.	E3.105 E3.125
Where a window head is near to the existing ceiling, the new independent ceiling may be raised to form a pelmet recess.	E4.29
A rigid or direct connection should not be created between the independent ceiling and the floor base.	

When considering heat losses via party walls, it is important to remember that, wherever the blockwork of a masonry party wall penetrates insulation at ceiling level, a thermal bridge is likely to exist – even when the party wall U-value is zero.

To avoid the consequent reduction in thermal performance, either the insulation layer should be contiguous with the air barrier at all points in the building envelope or the space between the air barrier and insulation layer should be filled with solid material, such as in a masonry wall.

6.8 Roofs

The roof of a brick-built house is normally an aitched (sloping) roof comprising rafters fixed to a ridge board, braced by purlins, struts and ties and fixed to wall plates bedded on top of the walls. They are then usually clad with slates or tiles to keep the rain out.

Timber-framed houses usually have trussed roofs – prefabricated triangulated frames that combine the rafters and ceiling joists – which are lifted into place and supported by the rails. The trusses are joined together with horizontal and diagonal ties. A ridge board is not fitted, nor are purlins required. Roofing-felt battens and tiling are applied in the usual way.

6.8.1 Requirements

Structure
The building shall be constructed so that the combined dead, imposed and wind loads are sustained and transmitted by it to the ground.

(Approved Document A1)

 Amendments to the 2010 version of Approved Document A came into force in 2013.

Precautions against moisture
The roof of the building shall be resistant to the penetration of moisture from rain or snow to the inside of the building.

All floors next to the ground, walls and roof shall not be damaged by moisture from the ground, rain or snow and shall not carry that moisture to any part of the building that it would damage.

(Approved Document C2)

 Amendments to the 2010 version of Part C came into force in 2013.

Rainwater drainage
Rainwater from roofs shall be carried away from the surface either by a drainage system or by other means.

The rainwater drainage system shall carry the flow of rainwater from the roof to an outfall (e.g. a soakaway, a watercourse, a surface water or combined sewer).

(Approved Document H3)

Fire precautions
As a fire precaution, all materials used for internal linings of a building should have a low rate of surface flame spread and (in some cases) a low rate of heat release.

(Approved Document B2)

External fire spread

• *The roof shall be constructed so that the risk of spread of flame and/or fire penetration from an external fire source is restricted.*
• *The risk of a fire spreading from the building to a building beyond the boundary, or vice versa shall be limited.*

(Approved Document B4)

Internal fire spread (structure)
Ideally the building should be sub-divided by elements of fire-resisting construction into compartments.

• *All openings in fire-separating elements shall be suitably protected in order to maintain the integrity of the continuity of the fire separation.*
• *Any hidden voids in the construction shall be sealed and sub-divided to inhibit the unseen spread of fire and products of combustion, in order to reduce the risk of structural failure, and the spread of fire.*

(Approved Document B3)

Ventilation
There shall be adequate means of ventilation provided for people in the building.

(Approved Document F)

Conservation of fuel and power
Reasonable provision shall be made for the conservation of fuel and power in buildings by:

(a) limiting heat gains and losses

   (i) through thermal elements and other parts of the building fabric; and
   (ii) from pipes, ducts and vessels used for space heating, space cooling and hot water services;

(b)   providing fixed building services which

    (i)   are energy efficient;
    (ii)   have effective controls; and are commissioned by testing and adjusting as necessary to ensure they use no more fuel and power than is reasonable in the circumstances.

(Approved Document L)

A new version of Approved Document L came into force on 6 April 2014.

Safety
Pedestrian guarding should be provided for any roof to which people have access.

(Approved Document K2)

A new version of Approved Document K came into force in 2013.

6.8.2 Meeting the requirements

6.8.2.1 General

> Roofs shall be constructed so that they: A1/2 1A2d
>
> •   provide local support to the walls;
> •   act as horizontal diaphragms capable of transferring the wind forces to buttressing elements of the building.

Note: A traditional cut timber roof (such as one using rafters, purlins and ceiling joists) generally has sufficient built-in resistance to instability and wind forces (e.g. from hipped ends, tiling battens, rigid sarking, etc.). However, the need for diagonal rafter bracing equivalent to that recommended in BS EN 1995-1-1:2004 with its UK National Annex should be considered, **especially** for single-hipped and non-hipped roofs of greater than 40° pitch to detached houses.

Roofs should:

> •   act to transfer lateral forces from walls to buttressing   A1/2 2C33a
>   walls, piers or chimneys;
> •   be secured to the supported wall.   A1/2 2C33b

The roof shall be braced (in accordance with BS 5268: Part 3):

- at rafter level; A1/2 2C38(i)h
- horizontally at eaves level;
- at the base of any gable by roof decking, rigid sarking or diagonal timber bracing (as appropriate).

Vertical strapping may be omitted if the roof:

- has a pitch of 15° or more; and A1/2 2C36a–d
- is tiled or slated; and
- is of a type known by local experience to be resistant to wind gusts; and
- has main timber members spanning on to the supported wall at not more than 1.2 m centres.

Gable walls should be strapped to roofs as shown in Figure 6.142(a) and (b) by tension straps.

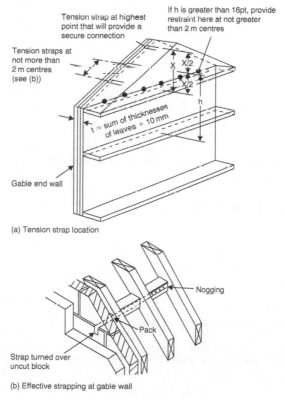

(a) Tension strap location

(b) Effective strapping at gable wall

Figure 6.142 Lateral support at roof level.

Walls shall be tied to the roof structure vertically and horizontally and have a horizontal lateral restraint at roof level.

Wall ties should also be provided, spaced not more than 300 mm apart vertically, within a distance of 225 mm from the vertical edges of all roof verges.	A1/2 2C8
Walls shall be tied to the roof structure vertically and horizontally and have a horizontal lateral restraint at roof level.	A1/2 2C38(i)i
Walls should be tied horizontally at no more than 2 m centres to the roof structure at eaves level, base of gables and along roof slopes (as shown in Figure 6.143) with straps.	A1/2 2C38(iv)

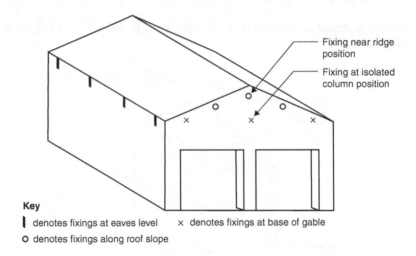

Key

| denotes fixings at eaves level   × denotes fixings at base of gable
O denotes fixings along roof slope

Figure 6.143 Lateral restraint at roof level.

Isolated columns should also be tied to the roof structure (see Figure 6.143).	A1/2 2C38(iv)
The roof structure of an annexe shall be secured to the structure of the main building at both rafter and eaves level.	A1/2 2C38(1)j
Access to the roof shall only be for the purposes of maintenance and repair.	A1/2 2C38(1)d

6.8.2.1.1 Timber

Figure 6.144 House longhorn beetle.

Figure 6.145 House longhorn larva.

Softwood timber used for roof construction or fixed in the   A1/2 2B2
roof space (including ceiling joists within the void spaces of
the roof), should be adequately treated to prevent infestation
by the house longhorn beetle (*Hylotrupes bajulus* L.),
particularly in the following areas:

- the Borough of Bracknell Forest, in the parishes of
  Sandhurst and Crowthorne;
- the Borough of Elmbridge;
- the District of Hart, in the parishes of Hawley and
  Yateley;
- the District of Runnymede;
- the Borough of Spelthorne;
- the Borough of Surrey Heath;

> • the Borough of Rushmoor, in the area of the former
>   district of Farnborough;
> • the Borough of Woking.

Notes:

(1) Guidance on suitable preservative treatments is given in the *British Wood Preserving and Damp-Proofing Association's Manual* (2000 revision), available from 1 Gleneagles House, Vernongate, South Street, Derby DE1 1UP.

(2) Guidance on the sizing of certain members in floors and roofs is given in *Span Tables for Solid Timber Members in Floors, Ceilings and Roofs (Excluding Trussed Rafter Roofs) for Dwellings*, published by TRADA, available from Chiltern House, Stocking Lane, Hughenden Valley, High Wycombe, Bucks HP14 4ND.

Note: Also see BS EN 1995-1-1:2004, *Design of Timber Structures*, and BS 8103-3:2009, *Structural Design of Low-Rise Buildings: Code of Practice for Timber Floors and Roofs for Housing*.

6.8.2.1.2 Rating of material and products

Table 6.58 Typical performance ratings of some generic materials and products

Rating	Material or product
Class 0	(1) Any non-combustible material or material of limited combustibility.
	(2) Brickwork, blockwork, concrete and ceramic tiles.
	(3) Plasterboard (painted or not with a PVC facing not more than 0.5 mm thick) with or without an air gap or fibrous or cellular insulating material behind.
	(4) Wood-wool cement slabs.
	(5) Mineral-fibre tiles or sheets with cement or resin binding.
Class 3	(6) Timber or plywood with a density more than 400 kg/ml, painted or unpainted.
	(7) Wood-particle board or hardboard, either untreated or painted.
	(8) Standard glass-reinforced polyesters.

6.8.2.1.3 Building height

> For residential buildings, the maximum height of the   A1/2 2C4i
> building measured from the lowest finished ground level
> adjoining the building to the highest point of any roof
> should not be greater than 15 m.

6.8.2.2 Openings

Where an opening in a roof for a stairway adjoins a supported wall and interrupts the continuity of lateral support:

- the maximum permitted length of the opening is 3 m,   A1/2 2C37a
  measured parallel to the supported wall;

- connections (if provided by means other than by   A1/2 2C37b
  anchor) should be throughout the length of each
  portion of the wall on each side of the opening;

- connections via mild steel anchors should be spaced   A1/2 2C37c
  closer than 2 m on each side of the opening to
  provide the same number of anchors as if there were
  no opening;

- there should be no other interruption of lateral   A1/2 2C37d
  support.

6.8.2.3 Pitched roofs covered with slates or tiles

Table 6.59 Pitched roofs – slates or tiles

Covering material	Supporting structure	Designation
(1) Natural slates (2) Fibre reinforced (3) Clay tiles (4) Concrete tiles	Timber rafters with or without underfelt, sarking, boarding, wood-wool slabs, compressed straw slabs, plywood, wood chipboard, or fibre cement slates insulating board.	AA

Although Table 6.59 does not include guidance for pitched roofs covered with bitumen felt, it should be noted that there is a wide range of materials on the market, and information on specific products is readily available from manufacturers.

6.8.2.4 Flat roofs covered with bitumen felt

A flat roof consisting of bitumen felt is (irrespective of the felt specification) deemed to be of designation AA if the felt is laid on a deck constructed of 6 mm plywood, 12.5 mm wood chipboard, 16 mm (finished) plain-edged timber boarding, compressed straw slab, screeded wood-wool slab, profiled fibre-reinforced cement or steel deck (single or double skin) with or without fibre insulating board overlay, profiled aluminium deck (single or double skin) with or without fibre insulating board overlay, or concrete or clay-pot slab (in situ or precast), and has a surface finish of:

- bitumen-bedded stone chippings covering the whole surface to a depth of at least 12.5 mm;
- bitumen-bedded tiles of a non-combustible material;
- sand and cement screed; or
- tarmacadam.

Table 6.60 Pitched roofs – self-supporting sheet

Covering material	Construction	Supporting structure	Designation
Profiled sheet of galvanized steel, aluminium, fibre-reinforced cement, or pre-painted (coil-coated) steel or aluminium with a PVC or PVF2 coating	Single skin without underlay, or with underlay or plasterboard, or wood-wool slab.	Structure of timber, steel or concrete.	AA
Profiled sheet of galvanized steel, aluminium, fibre-reinforced cement, or pre-painted (coil-coated) steel or aluminium with a PVC or PVF2 coating	Double skin without interlayer, or with interlayer of resin bonded or concrete glass fibre, mineral-wool slab, polystyrene or polyurethane.	Structure of timber, steel or concrete.	AA

6.8.2.5 Pitched or flat roofs covered with fully supported material

Table 6.61 Pitched roofs – fully supported material

Covering material	Supporting structure	Designation
(1) Aluminium sheet (2) Copper sheet (3) Zinc sheet (4) Lead sheet	Timber joists and tongued and grooved boarding, or plain-edged boarding.	AA
(5) Mastic asphalt (6) Vitreous enamelled steel (7) Lead/tin alloy-coated steel sheet (8) Zinc/aluminium alloy-coated steel sheet	Steel or timber joists with deck of wood-wool slabs, compressed-straw slab, wood chipboard, fibre insulating board or 9.5 mm plywood.	AA
(9) Pre-painted (coil-coated) steel sheet including liquid-applied PVC coatings	Concrete or clay-pot slab (in situ or precast) or non-combustible deck of steel, aluminium or fibre cement (with or without insulation).	AA

 Note: Lead sheet supported by timber joists and plain-edged boarding should be regarded as having a BA designation.

6.8.2.6 Wall type 1 – solid masonry

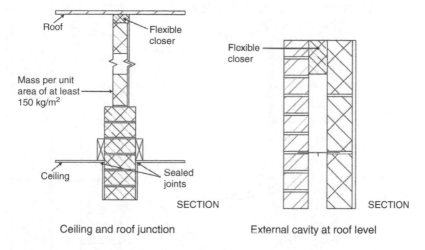

Ceiling and roof junction   External cavity at roof level

Figure 6.146 Wall type 1 – solid masonry.

If lightweight aggregate blocks of density less than 1200 kg/m³ are used above ceiling level, then one side should be sealed with cement, paint or plaster skim.	E2.58 E2.94 E2.138
Where the roof or loft space is not a habitable room (and there is a ceiling with a minimum mass per unit area of 10 kg/m² with sealed joints), the mass per unit area of the separating wall above the ceiling may be reduced to 150 kg/m².	E2.57
Where there is an external cavity wall, the cavity should be closed at eaves level with a suitable flexible material (e.g. mineral wool).	E2.59
A rigid connection between the inner and external wall leaves should be avoided.	Ep23
If a rigid material is used, then it should only be rigidly bonded to one leaf.	Ep23

6.8.2.7 Wall type 2 – cavity masonry

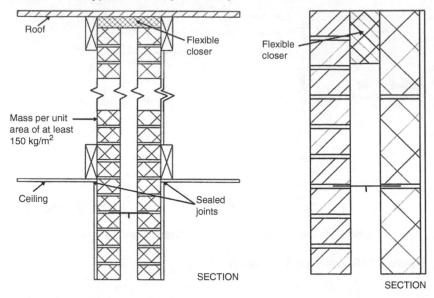

Wall type 2 – ceiling and roof junction   External cavity wall at eaves level

Figure 6.147 Wall type 2 – cavity masonry.

Note: Where a type 2 separating wall is used, it should be continuous to the underside of the roof.

The junction between the separating wall and the roof should be filled with a flexible closer that is also suitable as a fire stop.	E2.92
If lightweight aggregate blocks (with a density less than 1200 kg/m^3) are used above ceiling level, one side should be sealed with cement paint or plaster skim.	E2.94
The cavity of an external cavity wall should be closed at eaves level with a suitable flexible material (e.g. mineral wool).	E2.95
A rigid connection between the inner and external wall leaves should be avoided.	E2.95
If a rigid material has to be used, it should **only** be rigidly bonded to one leaf.	E2.95

If the roof or loft space is not a habitable room (and there is a ceiling with a minimum mass per unit area of 10 kg/m^2 with sealed joints), the mass per unit area of the separating wall above the ceiling may be reduced to 150 kg/m^2 – **but** it should still be a cavity wall.

6.8.2.8 Wall type 3 – masonry between independent panels

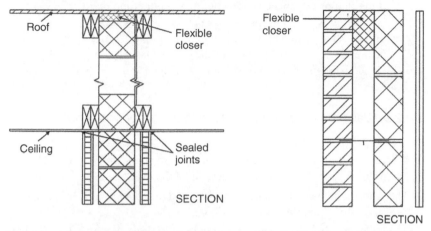

Wall types 3.1 and 3.2 – ceiling and roof junction   External cavity wall at eaves level

Figure 6.148 Wall type 3 – masonry between independent panels.

Where a type 3 separating wall is used, the masonry core should be continuous to the underside of the roof.	E2.133
The junction between the separating wall and the roof should be filled with a flexible closer that is also suitable as a fire stop.	E2.134
The junction between the ceiling and independent panels should be sealed with tape or caulked with sealant.	E2.135
If there is an external cavity wall, the cavity should be closed at eaves level with a suitable flexible material (e.g. mineral wool).	E2.136
Rigid connections between the inner and external wall leaves should be avoided where possible.	E2.136
If a rigid material is used, it should only be rigidly bonded to one leaf.	E2.136
For wall types 3.1 and 3.2 (solid masonry core):	
• if the roof or loft space is not a habitable room (and there is a ceiling with a minimum mass per unit area of 10 kg/m² and it has sealed joints), the independent panels may be omitted in the roof space and the mass per unit area of the separating wall above the ceiling may be a minimum of 150 kg/m²;	E2.137
• if lightweight aggregate blocks with a density less than 1200 kg/m³ are used above ceiling level, one side should be sealed with cement paint or plaster skim.	E2.138

For wall type 3.3 (cavity masonry core):

• if the roof or loft space is not a habitable room (and there is a ceiling with a minimum mass per unit area of 10 kg/m^2 and it has sealed joints), the independent panels may be omitted in the roof space, but the cavity masonry core should be maintained to the underside of the roof.	E2.139

6.8.2.9 Wall type 4 – framed walls with absorbent material

Where a type 4 separating wall is used, the wall should preferably be continuous to the underside of the roof.	E2.160
The junction between the separating wall and the roof should be filled with a flexible closer.	E2.161
The junction between the ceiling and the wall linings should be sealed with tape or caulked with sealant.	E2.162
If the roof or loft space is not a habitable room (and there is a ceiling with a minimum mass per unit area of 10 kg/m^2 with sealed joints), then:	E2.162a
• either the linings on each frame may be reduced to two layers of plasterboard, each sheet with a minimum mass per unit area of 10 kg/m^2; or	E2.162a
• the cavity may be closed at ceiling level without connecting the two frames rigidly together.	E2.162b
External wall cavities should be closed at eaves level with suitable material.	E2.163a

 Note: In which case there need only be one frame in the roof space provided there is a lining of two layers of plasterboard, each sheet of minimum mass per unit area 10 kg/m^2, on both sides of the frame.

6.8.2.10 Precipitation

Roofs should:

• resist the penetration of precipitation to the inside of the building;	C6.2a
• not be damaged by precipitation;	C6.2b
• not carry precipitation to any part of the building that would be damaged by it;	C6.2b
• be designed and constructed so that their structural and thermal performance is not adversely affected by interstitial condensation.	C6.2c

6.8.2.10.1 Resistance to moisture from the outside

Roofs should be designed so as to protect the building from precipitation either by holding the precipitation at the face of the roof or by stopping it from penetrating beyond the back of the roofing system.

Roofs that are jointless or have sealed joints should be impervious to moisture.	C6.4a
Roofs that have overlapping dry joints should be weather resistant and backed by a material (such as roofing felt) to direct any precipitation that does enter the roof towards the outer face.	C6.4b
Materials that can deteriorate rapidly without special care should only be used as the weather-resisting part of a roof.	C6.5
Weather-resisting parts of a roofing system shall not include paint or include any coating, surfacing or rendering that will not itself provide all the weather resistance.	C6.5
Roofing systems may be:	C6.6a
• impervious – such as metal, plastic and bituminous products;	
• weather-resistant – such as natural stone or slate, cement-based products, fired clay and wood;	C6.6b
• moisture-resisting – such as bituminous and plastic products lapped at the joints;	C6.6c
• jointless materials and sealed joints – which would allow for structural and thermal movement.	C6.6d
Dry joints between roofing sheets should be designed so that precipitation will not pass through them.	C6.7
Any precipitation that does enter a joint shall be drained away without penetrating beyond the back of the roofing system.	C6.7
Each sheet, tile and section of roof should be fixed in accordance with the guidance contained in BS 8000-6:1990.	C6.8

6.8.2.10.2 Resistance to damage from interstitial condensation

Roofs shall be designed and constructed in accordance with Clause 8.4 of BS 5250:2002 and BS EN ISO 13788:2001.

Cold deck roofs (i.e. those roofs where the moisture from the building can permeate the insulation) shall be ventilated.	C6.11
Any parts of a roof that have a pitch of 70° or more shall be insulated as though they were walls.	C6.11
All gaps and penetrations for pipes and electrical wiring should be filled and sealed to avoid excessive moisture transfer to roof voids.	C6.12
An effective draught seal should be provided to loft hatches to reduce inflow of warm air and moisture.	C6.12
Specialist advice should be sought when designing swimming pools and other buildings where interstitial condensation in the walls (caused by high internal temperatures and humidities) can cause high levels of moisture being generated.	C6.13

6.8.2.10.3 Resistance surface condensation and mould growth

Roofs shall be designed and constructed so that the:

- thermal transmittance (U-value) does not exceed 0.35 $W/m^2 K$ at any point;
- junctions between elements and the details of openings   C6.14b
  (such as windows) are in accordance with the recommendations in the report on robust construction details.

6.8.2.11 Fire

6.8.2.11.1 Means of escape

A flat roof being used as a means of escape should:

- be part of the same building from which escape is being   C1 made;
- lead to a storey exit or external escape route;
- provide 30 minutes of fire resistance.

 Where a balcony or flat roof is provided for escape purposes, guarding may be needed (see also Approved Document K (Protection from falling, collision and impact)).

6.8.2.11.2 Construction of escape stairs

If an escape route is over a flat roof:

- the roof should be part of the same building from which escape is being made;   B1 2.7 (V2)

- the route across the roof should lead to a storey exit or external escape route;   B1 5.35 (V2)

- the part of the roof forming the escape route and its supporting structure, together with any opening within 3 m of the escape route, should provide 30 minutes of fire resistance (see Appendix A, Table A1 of Approved Document B for fire resistance figures for elements of structure, etc.);

- the route should be adequately defined and guarded by walls and/or protective barriers (that meet the provisions of Approved Document K).   B1 2.11 (V1)

 Note: Approved Document K (Protection from falling, collision and impact) specifies a minimum guarding height of 800 mm, except in the case of a window in a roof, where the bottom of the opening may be 600 mm above the floor.

An escape over a flat roof is permissible if: B1 2.31 (V2)

- more than one escape route is available from a storey or part of a building;

- the route does not serve as an institutional building;

- part of a building is intended for use by members of the public;

- the roof is fire-resistant in accordance with Tables A1 and A2 of Appendix A to Approved Document B;   B1 5.3 (V2)

- the exit from a flat, etc. is remote from the main entrance door to that flat.   B1 2.17 (V2)

6.8.2.11.3 Provision of refuges

Refuges are relatively safe waiting areas for short periods. They are **not** areas where disabled people should be left alone indefinitely until rescued by the fire and rescue service or until the fire is extinguished.

> A refuge such as a flat roof (balcony, podium or similar   B1 4.8 (V2)
> compartment, protected lobby, protected corridor
> or protected stairway) should be provided for each
> protected stairway.

Note: The number of refuge spaces need not necessarily equal the number of wheelchair users who can be present in the building.

6.8.2.11.4 Compartmentation

To prevent the spread of fire within a building, whenever possible the building should be subdivided into compartments separated from one another by walls and/or floors of fire-resisting construction.

> Compartment walls in a top storey beneath a roof should   B3 5.8 (V1)
> be continued through the roof space.   B3 8.24 (V2)
>
> When a compartment wall meets the underside of the   B3 8.28 (V2)
> roof covering or deck, the wall/roof junction shall
> maintain continuity of fire resistance.
>
> Double-skinned insulated roof sheeting should   B3 8.29 (V2)
> incorporate a band of material of limited
> combustibility.

If a fire penetrates a roof near a compartment wall there is a risk that it will spread over the roof to the adjoining compartment. To reduce this risk either:

> • the wall should be extended up through the roof for   B3 8.29 (V2)
>   a height of at least 375 mm above the top surface of   B3 8.30 (V2)
>   the adjoining roof covering; or
> • a 1500 mm wide zone on either side of the wall
>   should have a suitable covering (Figure 6.149).

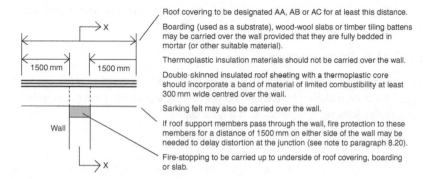

Roof covering to be designated AA, AB or AC for at least this distance.

Boarding (used as a substrate), wood-wool slabs or timber tiling battens may be carried over the wall provided that they are fully bedded in mortar (or other suitable material).

Thermoplastic insulation materials should not be carried over the wall.

Double-skinned insulated roof sheeting with a thermoplastic core should incorporate a band of material of limited combustibility at least 300 mm wide centred over the wall.

Sarking felt may also be carried over the wall.

If roof support members pass through the wall, fire protection to these members for a distance of 1500 mm on either side of the wall may be needed to delay distortion at the junction (see note to paragraph 8.20).

Fire-stopping to be carried up to underside of roof covering, boarding or slab.

Figure 6.149 Junction of compartment wall with roof.

6.8.2.11.5 Concealed spaces (cavities)

Concealed spaces or cavities in the construction of roofs will provide an easy route for smoke and flame to spread, which, because it is concealed, will present a greater danger than would a more obvious weakness in the fabric of the building. For this reason:

Cavity barriers should be provided at the edges of cavities (e.g. window and door openings) and particularly at any junction between an external cavity wall and a compartment wall which separates buildings (see Figure 6.150).	B3 6.3 (V1) B3 9.3 (V2)

Cavity barriers need not be provided between double-skinned corrugated or profiled insulated roof sheeting **if** the sheeting is a material of limited combustibility.

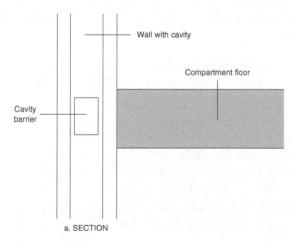

a. SECTION

Figure 6.150 Roof cavity barrier.

Note: Separate rules exist for bedrooms in institutional and other residential buildings (see Approved Document B3 9.7 (V2)).

6.8.2.11.6 Roof covering

Re-covering of roofs is commonly undertaken to extend the useful life of buildings; however, it should be remembered that roof structures may be required to carry underdrawing or insulation at a later date.

All materials used to cover roofs (including transparent or translucent materials, but excluding windows of glass in residential buildings with roof pitches of not less than 15°) shall be capable of safely withstanding the concentrated imposed loads upon roofs specified in BS EN 1991-1-1:2012.	A1/2 4.1

Note: Transparent or translucent covering materials for roofs not accessible except for normal maintenance and repair are excluded from the requirement to carry the concentrated imposed load upon roofs if they are non-fragile or are otherwise suitably protected against collapse.

Where the work involves a significant change in the applied loading, the structural integrity of the roof structure and the supporting structure should be checked to ensure that upon completion of the work the building is not less compliant.	A1/2 4.3

Note: To extend the useful life of a building, it is common for roofs to be re-covered. However, before work commences a check should be made to ensure that the existing roof structure is able to sustain any significant roof loading, and appropriate strengthening work or replacement of roofing members should be undertaken.

'A significant change in roof loading' is when the loading upon the roof is increased by more than 15 per cent. This is classified as *a material alteration*.

Where work will significantly decrease the roof dead loading, the roof structure and its anchorage to the supporting structure should be checked to ensure that an adequate factor of safety is maintained against uplift of the roof under imposed wind loading.	A1/2 4.7
Thatch and wood shingles should be regarded as having an AD, BD or CD designation (see Table 6.62).	B4 10.9 (V1) B4 14.9 (V2)

Table 6.62 Limitations on roof coverings

Designation of roof covering	Minimum distance from any point on relevant boundary			
	Less than 6 m	At least 6 m	At least 12 m	At least 20 m
AA, AB or AC	Acceptable	Acceptable	Acceptable	Acceptable
BA, BB or BC	Not acceptable	Acceptable	Acceptable	Acceptable
CA, CB or CC	Not acceptable	Acceptable	Acceptable	Acceptable
AD, BD or CD	Not acceptable	Acceptable	Acceptable	Acceptable
DA, DB, DC or DD	Not acceptable	Not acceptable	Not acceptable	Acceptable

Note: Separation distances (i.e. the minimum distance from the roof, or part of the roof, to the relevant or notional boundary) shall be in accordance with Approved Document B, Volume 1, Table 5 according to the type of roof covering and the size and use of the building.

6.8.2.11.7 Rooflights

Rooflights should meet the relevant classification in Table 6.63.	B2 3.7 (V1) B2 6.7 (V2)

Table 6.63 Classification of linings

Location	National class	European class
Small rooms less than 4 m²	3	D-s3, d2
Domestic garages less than 40 m²	3	D-s3, d2
Other rooms (including garages)	1	C-s3, d2
Circulation spaces within dwelling houses	1	C-s3, d2

When used in rooflights, unwired glass shall be at least 4 mm thick and shall be AA designated (see Table 6.62).	B4 10.8 (V1)

Note: Thermoplastic materials may be used in windows, rooflights and lighting diffusers in suspended ceilings.

Rooflights may be constructed of a thermoplastic material (made from polycarbonate or from unplasticized PVC) if:	B2 3.10 (V1) B2 6.12 (V2)

- the lower surface has a TPI(a) (rigid) or a TP(b) classification;
- the size and disposition of the rooflights accord with the limits in Table 6.64.

Table 6.64 Limitations applied to thermoplastic rooflights and lighting diffusers in suspended ceilings and class 3 plastic rooflights

Minimum classification of lower surface	Use of space below the diffusers or rooflight	Maximum area of each diffuser panel or rooflight	Maximum total area of diffuser panels and rooflights as percentage of floor area of the space in which the ceiling is located	Minimum separation distance between diffuser panels or rooflights
TP(a)	Any (except a protected stairway)	No limit	No limit	No limit
Class 3 or TP(b)	Rooms	5	50	3
	Circulation spaces (except protected stairways)	5	15	3

6.8.2.11.8 Thatched roofs

Thatched roofs can sometimes be vulnerable to spontaneous combustion caused by heat transferred from flues building up in thick layers of thatch in contact with the chimney. To reduce the risk it is recommended that rigid twin-walled insulated metal flue liners be used within a ventilated (top and bottom) masonry chimney void provided they are adequately supported and not in direct contact with the masonry. Non-metallic chimneys and cast in situ flue liners can also be used provided the heat transfer to the thatch is assessed in relation to the depth of thatch and risk of spontaneous combustion.

Spark arrestors are not generally recommended, as they can be difficult to maintain and may increase the risk of flue blockage and flue fires.

 Further information and recommendations are contained in HETAS Information Paper 1/007.

6.8.2.11.8.1 CHIMNEYS IN THATCHED PROPERTIES

In thatched roofs:

- the rafters should be at least 30 minutes fire-resistant;   B4 10.9 (V1)
- smoke alarms should be installed in the roof space.   B4 14.9 (V2)

6.8.2.11.9 Need for cavity barriers

The need for cavity barriers in some concealed roof spaces can be reduced by using a fire-resisting ceiling below the cavity.

Where the conversion of an existing roof space (such as a loft conversion to a two-storey house) means that a new storey is going to be added, then the stairway will need to be protected with fire-resisting doors and partitions.

6.8.2.12 Ventilation

6.8.2.12.1 Roof with a pitch of 15° or more

- Pitched roof spaces should have ventilation openings at least 10 mm wide at eaves level to promote cross-ventilation.
- A pitched roof that has a single slope and abuts a wall should have ventilation openings at eaves level at least 10 mm wide and at high level (i.e. at the junction of the roof and the wall) at least 5 mm wide.

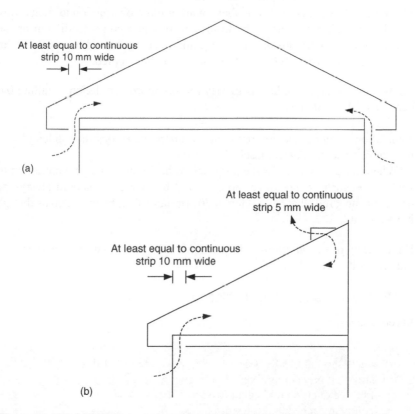

Figure 6.151 Ventilating roof voids. (a) Pitched roof. (b) Lean-to roof

6.8.2.12.2 Roof with a pitch of less than 15°

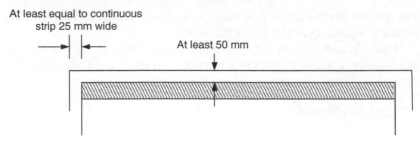

Figure 6.152 Ventilating roof void – flat roof.

- Roof spaces should have ventilation openings at least 25 mm wide in two opposite sides to promote cross-ventilation.
- The void should have a free air space of at least 50 mm between the roof deck and the insulation.
- Pitched roofs where the insulation follows the pitch of the roof need ventilation at the ridge at least 5 mm wide.
- Where the edges of the roof abut a wall or other obstruction in such a way that free air paths cannot be formed to promote cross-ventilation or the movement of air outside any ventilation openings would be restricted, an alternative form of roof construction should be adopted.

 Roofs with a span exceeding 10 m may require more ventilation, totalling 0.6 per cent of the roof area.

Ventilation openings may be continuous or distributed along the full length and may be fitted with a screen, fascia, baffle, etc.

Where necessary (i.e. for the purposes of health and safety), ventilation to small roofs such as those over porches and bay windows should always be provided and a roof that has a pitch of 70° or more shall be insulated as though it were a wall.

 If the ceiling of a room follows the pitch of the roof, ventilation should be provided as if it were a flat roof.

6.8.2.12.3 Passive stack ventilation

In roof spaces:

- ducts should, ideally, be secured to a wooden   F Table 5.2a to 5.2d strut that is securely fixed at both ends;
- flexible ducts should be allowed to curve gently at each end of the strut;
- ducts should be insulated with at least 25 mm of a material having a thermal conductivity of 0.04 W/m K.

For stability, rigid ducts should be used for any outside part of a passive stack ventilation (PSV) system that is above the roof slope and to provide stability and they should project down into the roof space far enough to allow firm support.	F Table 5.2a to 5.2d
If a duct penetrates the roof more than 0.5 m from the roof ridge, it must extend above the roof slope to at least the height of the roof ridge.	F Table 5.2a to 5.2d
If tile ventilators are used on the roof slope they must be positioned no more than 0.5 m from the roof ridge.	F Table 5.2a to 5.2d
If a duct extends above the roof level, then that section of the duct should be insulated or be fitted with a condensation trap just below roof level.	F Table 5.2a to 5.2d
Ducts should be securely fixed to the roof outlet terminal so that it cannot sag or become detached.	F Table 5.2a to 5.2d
Separate ducts shall be taken from the ceilings of wet rooms to separate terminals on the roof.	F Table 5.2a to 5.2d
Terminals should be designed such that any condensation forming inside them cannot run down into the dwelling but will run off on to the roof.	F Table 5.2a to 5.2d

Note: Placing the outlet terminal at the ridge of the roof is the preferred option, as it is not prone to wind gusts and/or certain wind directions.

6.8.2.12.4 Ceiling and roof junctions

Where a type 1 separating wall is used it should be continuous to the underside of the roof.

The junction between the separating wall and the roof should be filled with a flexible closer which is also suitable as a fire stop.	E2.56 E2.92 E2.134

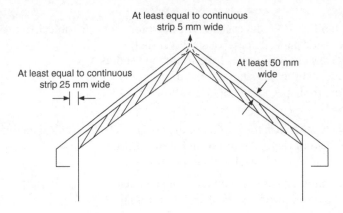

Figure 6.153 Ventilating roof void – ceiling following pitch of roof.

6.8.2.13 Conservation of fuel and power

Following the publication of new Approved Documents L1A and L1B in November 2013, with effect from April 2014:

*   all **new homes** (i.e. dwellings) need to achieve, or better, a fabric energy-efficiency target in addition to the strengthened requirement to deliver **6 per cent carbon dioxide savings**; and
*   all **new non-domestic buildings** now need to make **9 per cent carbon dioxide savings**.

Note: See Appendix B to this book (available at www.routledge.com/ 9780415721714) for further information concerning the requirements for the conservation of fuel and power.

6.8.2.13.1 Historic and traditional buildings

When undertaking work on any type of historic or traditional type of building, the aim should always be to improve energy-efficiency as far as is reasonably practicable, without prejudicing the character of the host building or increasing the risk of long-term deterioration of the building fabric or fittings.

In general, new extensions to historic or traditional buildings   L2B 3.11 should comply with the standards of energy-efficiency as set out in Approved Document L:2010. The only exception would be where there is a particular need to match the external appearance or character of the extension to that of the host building.

Building inspectors normally require the use of *sympathetic treatment* when restoring the historic character of a building that has been subject to previous inappropriate alteration (e.g. replacing a roof or a rooflight).

Particular issues that could warrant sympathetic treatment – and where advice from others would probably be beneficial – include:

> • restoring the historic character of a building that has   L2B 3.11
>   been subject to previous inappropriate alteration (e.g.
>   replacement roofs, rooflights, windows and doors,
>   etc.);
> • rebuilding a former historic building (e.g. following a
>   fire or filling a gap site in a terrace); and
> • renovating the fabric of historic buildings to enable
>   them to 'breathe' and to control any moisture and
>   potential long-term decay problems.

6.8.2.13.2 Extensions

Regulation 9 of the Building Regulations exempts some conservatory and porch extensions from the energy-efficiency requirements, **provided** that:

- the floor area of that extension does not exceed $30\,\text{m}^2$;
- glazing and fixed electrical installations comply with the appropriate sections of Building Regulations;
- the front entrance door between the existing house and the new porch remains in place;
- any conservatory is separated from the house by external-quality walls, doors or windows;
- the conservatory has an independent heating system with separate temperature and on/off controls;
- the porch does not adversely affect access to the house.

> If a building extension is a conservatory or porch that is   L2B 4.12
> **not** exempt from the energy-efficiency requirements, the
> effective thermal separation between the heated area in the
> existing building (i.e. the walls, doors and windows between
> the building and the extension) should be insulated and
> draught-proofed to at least the same extent as in the existing
> building.

6.8.2.13.3 Dwellings

> Where a dwelling is erected, it shall not exceed the target   L1A 2.8
> fabric energy-efficiency rate for the dwelling.

Extensions to dwellings should either use newly constructed thermal elements or make use of existing (or new) roof windows and rooflights, provided that they meet the requirements for the conservation of fuel and power.

If upgrades are proposed to the existing dwelling, such upgrades should be implemented to a standard that is no worse than that shown in column (b) of Table 6.65.

L1B 4.7

Table 6.65 Upgrading retained thermal elements

Element	(a) Threshold U-value (W/m² K)	(b) Improved U-value (W/m² K)
Pitched roof – insulation at ceiling level	0.35	0.16
Pitched roof – insulation between rafters	0.35	0.18
Flat roof or roof with integral insulation	0.35	0.18

6.8.2.13.4 Buildings other than dwellings

Where a building is erected, it shall not exceed the target CO_2 emission rate for the building.

L2A 2.7

It is recommended that the total area of roof windows in extensions to existing buildings does not exceed the sum of:

*   25 per cent of the floor area of the extension; plus
*   the total area of any roof windows, windows or doors which, as a result of the extension works, no longer exist – or are no longer exposed.

Fabric elements and fixed building services should satisfy minimum energy-efficiency standards.
Table 6.66 provides an indication of the worst acceptable standards for fabric properties.

L2A 4.29

Table 6.66 Limiting fabric parameters

Element	W/m² K
Roof	0.25
Roof windows, glazed rooflights, curtain walling and pedestrian doors	2.2
Roof ventilators (including smoke vents)	3.5
Air permeability	10.00 m³/h m² at 50 Pa

Replacement roof windows, lights, ventilators, etc. should have U-values that are no worse than those given in Table 6.67.

Table 6.67 Standards for controlled fittings

Fitting	Standard (W/m² K)
Windows, roof windows and glazed rooflights	1.8 for the whole unit
Plastic rooflight	1.8
Roof ventilators (including smoke extract ventilators)	3.5

Removing and **not** replacing any of the thermal separation between the building and an existing exempt extension, or extending the building's heating system into the extension, mean that the extension ceases to be exempt!

Table 6.68 Opening areas in the extension

Building type	Windows, rooflights and personnel doors as percentage of exposed wall	Rooflights as percentage of area of roof
Residential buildings where people temporarily or permanently reside	30	20
Places of assembly, offices and shops	40	20
Industrial and storage buildings	15	20

> The area of windows and rooflights in the extension should generally not exceed the percentage values given in Table 6.68. L2B 4.4

6.8.2.13.5 Building fabric

> The building fabric should be constructed to a reasonable standard so that:
>
> • the insulation is reasonably continuous over the whole building envelope; and L1A 3.2
>
> • the air permeability is within reasonable limits. L2A 3.2

Reductions in thermal performance can occur where the air barrier and the insulation layer are not contiguous and the cavity between them is subject to air movement. To avoid this problem, either:

• the insulation layer should be contiguous with the air barrier at all points in the building envelope; or
• the space between the insulation layer and air barrier should be filled with solid material such as in a masonry wall.

6.8.2.13.6 Thermal bridges

> The building fabric should be constructed so that there are no reasonably avoidable thermal bridges in the insulation layers caused by gaps within the various elements, at the joints between elements, and at the edges of elements, such as those around window and door openings. L1A 3.9

To meet this requirement, it is recommended that you use construction joint details that have been calculated following the guidance set out in BRE Report BR 497 (*Conventions for Calculating Linear Thermal Transmittance and Temperature Factors*).

6.9 Chimneys and fireplaces

Under the Clean Air Act 1993, local authorities may declare the whole or part of a district to be a smoke control area. It is an offence to emit smoke from a chimney of a building, from a furnace or from any fixed boiler if located in a designated smoke control area unless an authorized fuel is used. It is also an offence to acquire an *unauthorized fuel* for use within a smoke control area unless it is used in an *exempt* appliance (i.e. *exempted* from the controls which generally apply in the smoke control area).

Authorized fuels include inherently smokeless fuels, such as gas, electricity and anthracite, together with specified brands of manufactured solid smokeless fuels which have passed tests to confirm that they are capable of burning in an open fireplace without producing smoke.

Exempt appliances are appliances (ovens, wood burners, boilers and stoves) which have passed tests to confirm that they are capable of burning an unauthorized or inherently smoky solid fuel without emitting smoke.

Note: More information and details of authorized fuels and exempt appliances can be found at http://smokecontrol.defra.gov.uk.

6.9.1 Requirements

Structure
The building shall be constructed so that the combined dead, imposed and wind loads are sustained and transmitted by it to the ground:

*   *safely;*
*   *without causing such deflection or deformation of any part of the building (or such movement of the ground) as will impair the stability of any part of another building.*

(Approved Document A1)

Amendments to the 2010 version of Approved Document A came into force in 2013.

Fire precautions
Any hidden voids in the construction shall be sealed and subdivided to inhibit the unseen spread of fire and products of combustion, in order to reduce the risk of structural failure, and the spread of fire.

(Approved Document B3)

Combustion appliances and fluepipes shall be so installed, and fireplaces and chimneys shall be so constructed and installed, as to reduce to a reasonable level the risk of people suffering burns or the building catching fire in consequence of their use.

(Approved Document J3)

Installation
Combustion appliances shall be so installed that there is an adequate supply of
air to them for combustion, to prevent overheating and for the efficient working
of any flue.

(Approved Document J1)

Discharge of products of combustion
Combustion appliances shall have adequate provision for the discharge of
products of combustion to the outside air.

(Approved Document J2)

Warning of release of carbon monoxide
Where a fixed combustion appliance is provided, appropriate provision shall
be made to detect and give warning of the release of carbon monoxide.

(Approved Document J2A)

Provision of information
Where a hearth, fireplace, flue or chimney is provided or extended, a durable
notice containing information on the performance capabilities of the hearth,
fireplace, flue or chimney shall be affixed in a suitable place in the building for
the purpose of enabling combustion appliances to be safely installed.

(Approved Document J4)

6.9.2 Meeting the requirements

6.9.2.1 Smoke emissions

In accordance with the Clean Air Act 1993:

* it is an offence for dark smoke to be emitted from a chimney of any building;
* it is an offence for dark smoke to be emitted from a chimney which serves
  the furnace of any fixed boiler or industrial plant.

This prohibition does not apply to:

* emissions that are solely due to the lighting up of a furnace which was cold
  and that all practicable steps had been taken to prevent or minimize the
  emission of dark smoke;
* emissions that are solely due to some failure of a furnace, or of apparatus
  used in connection with a furnace, and that:

  ○ the failure could not reasonably have been foreseen or, if foreseen,
    could not reasonably have been provided against; and
  ○ the alleged emission could not reasonably have been prevented by
    action taken after the failure occurred;

* emissions that are solely due to the use of unsuitable fuel and:

  ○ suitable fuel was unobtainable and the least unsuitable fuel which was
    available was used; and
  ○ all practicable steps had been taken to prevent or minimize the emis-
    sion of dark smoke as the result of the use of that fuel.

A person found guilty of an offence under this section is liable, on summary conviction, to a fine in accordance with the 'Standard Scale' contained in the Criminal Justice Act (at the time of publication the maximum fine for such offences was £5000.00).

6.9.2.2 Construction

6.9.2.2.1 General

> Chimneys shall consist of a wall or walls enclosing one or more flues (Figure 6.154).   J 0.4–7

Note: Chimneys and flues should also provide satisfactory control of water condensation.

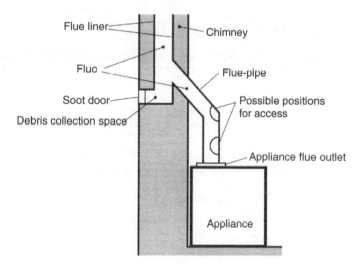

Figure 6.154 Chimneys and flues.

Down-draughts that could interfere with the combustion performance of an open-flued appliance (Figure 6.155) shall be minimized. This can be achieved by using a factory-made draught stabilizer (see Figure 6.155) to prevent excessive variations in the draught.

In the gas industry, the chimney for a gas appliance is commonly called the flue.

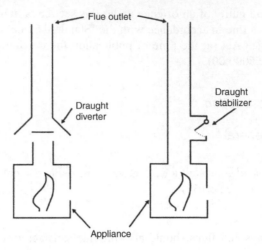

Figure 6.155 Draught diverters and draught stabilizers.

6.9.2.2.2 Open fire with a throat and gather

Permanently open air vents should have a total free area of at least 50 per cent of the throat opening area (see Figure 6.156).

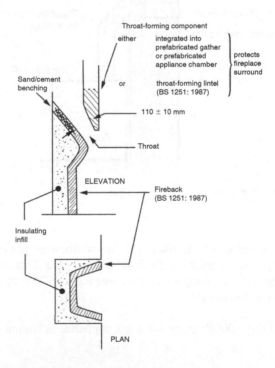

Figure 6.156 Open fireplaces – throat and fireplace components.

6.9.2.2.3 Open fire with no throat (e.g. a fire under a canopy)

Permanently open air vents should have a total free area of at least 50 per cent of the cross-sectional area of the flue (see Figure 6.157).

6.9.2.2.4 Appliances burning solid fuel (with a rated output up to 50 kW)

Any room or space containing an appliance burning solid fuel (with a rated output up to 50 kW) should have a permanent air vent opening of at least the size shown in Figure 6.157.   J 2.2

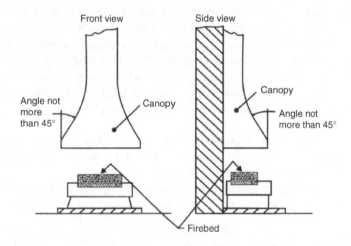

Figure 6.157 Canopy for an open solid-fuel fire.

6.9.2.2.5 Other appliances (such as a stove, cooker or boiler)

These should have permanently open air vent(s).

6.9.2.3 Fire precautions

Proprietary fire-stopping and sealing systems are available and may be used, as they have been shown by test to maintain the fire resistance of an element. Other fire-stopping materials include:

- cement mortar;
- gypsum-based plaster;
- cement or gypsum-based vermiculite/perlite mixes;
- glass fibre, crushed rock, blast furnace slag or ceramic-based products (with or without resin binders); and
- intumescent mastics.

Joints between fire-separating elements should be fire-stopped.	B3 7.12a (V1) B3 10.17a (V2)
To prevent displacement, materials used for fire-stopping should be reinforced with (or supported by) materials of limited combustibility.	B3 7.13 (V1) B3 10.18 (V2)

6.9.2.3.1 Carbon monoxide alarms

Note: Carbon monoxide (CO) is a colourless, odourless and tasteless gas that is slightly less dense than air. It is toxic to humans and animals when encountered in higher concentrations, although it is also produced in normal animal metabolism in low quantities, and is thought to have some normal biological functions. In the atmosphere it is spatially variable and short-lived, having a role in the formation of ground-level ozone. Carbon dioxide is normally expelled through the lungs, but if there is too much carbon dioxide in the blood stream then it can cause hypercapnia, which will trigger a reflex which increases breathing and access to oxygen, such as arousal and turning the head during sleep. A failure of this reflex can be fatal, as in sudden infant death syndrome.

Under current Building Regulations, in England and Wales it is now a **mandatory requirement** to fit carbon monoxide alarms where new or replacement fixed solid fuel appliances are installed in a home.

In Northern Ireland, it a **legal requirement** to fit CO alarms in **all** homes where a new or replacement appliance, not used solely for cooking, is installed.

Carbon monoxide alarms should comply with BS EN 50291:2001 and be powered by a battery designed to operate for the working life of the alarm.	J 2.35
Carbon monoxide alarms should incorporate a warning device to alert users when the working life of the alarm is due to pass.	J 2.35
The carbon monoxide alarm should be located in the same room as the appliance, either:	J 2.36

- on the ceiling at least 300 mm from any wall; or
- on a wall, as high up as possible (above any doors and windows) but not within 150 mm of the ceiling; and
- between 1 m and 3 m horizontally from the appliance.

6.9.2.4 Chimneys

6.9.2.4.1 End restraints

The ends of every wall (except single-leaf walls less than 2.5 m in storey height and length) in small single-storey non-residential buildings and annexes should be bonded or otherwise securely tied throughout their full height to a chimney.	A1/2 (2C25)
The buttressing chimney should provide support from the base to the full height of the wall.	A1/2 (2C25)
The sectional area, on plan, of chimneys (excluding openings for fireplaces and flues) should be not less than the area required for a pier in the same wall, and the overall thickness should not be less than twice the required thickness of the supported wall (Figure 6.158).	A1/2 (2C27b)

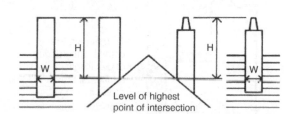

Figure 6.158 Proportions for masonry chimneys.

6.9.2.4.2 Masonry chimneys

Where a chimney is not adequately supported by ties or securely restrained in any way, its height H (measured from the highest point of any chimney pot or other flue terminal) should not exceed 4.5 times the width W (the least horizontal dimension of the chimney measured at the same point of intersection) – provided that the density of the masonry is greater than 1500 kg/m³ (see Figure 6.158).	A1/2 1D1
The foundation of piers, buttresses and chimneys should project as indicated in Figure 6.159 and the projection X should never be less than P.	A1/2 2E2f

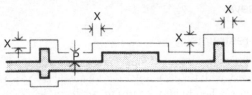

Projection X should not be less than P

Figure 6.159 Piers and chimneys.

6.9.2.4.3 Masonry chimneys (change of use)

Where a building is to be altered for a different use (e.g. it is   J 1.31
being converted into flats) the fire resistance of the walls of
existing masonry chimneys may need to be improved (see
Figure 6.160).

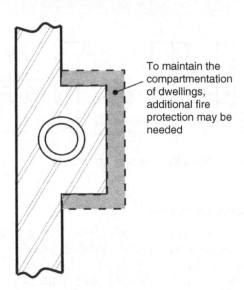

To maintain the
compartmentation
of dwellings,
additional fire
protection may be
needed

Figure 6.160 Fire protection of chimneys passing through other dwellings.

6.9.2.4.4 Flueblock chimneys

A flueblock chimney system is a factory-made set of components that are manufactured from precast concrete, clay or other masonry units and assembled on
site to provide a complete chimney.

There are two types, one for gas-burning appliances and the other (often
called a 'chimney block system') for solid-fuel-burning appliances.

Flueblock chimneys:

- should be constructed using factory-made components that   J 1.29
  are suitable for the intended application;
- should be installed with sealed joints;   J 1.30
- should only have bends and offsets that have been formed   J 1.30
  with matching factory-made components.   J 4.16

Flueblocks that are not intended to be bonded into surrounding   J 4.16
masonry should be supported and restrained in accordance with
the manufacturer's installation instructions.

Where a flue pipe or chimney penetrates a fire compartment   J 4.18
wall or floor, it must not breach the fire separation
requirements.

6.9.2.4.5 Masonry and flueblock chimneys

The thickness of the walls around the flues, excluding the   J 2.17
thickness of any flue liners, should be in accordance with
Figure 6.161.

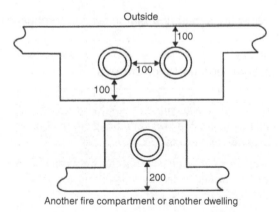

Figure 6.161 Wall thickness for masonry and flueblock chimneys. (Dimensions in mm.)

Combustible material should not be located where it could be   J 2.18
ignited by the heat dissipating through the walls of fireplaces or
flues.

Figure 6.162 shows a method of meeting this requirement so that combustible material is at least:

- 200 mm from the inside surface of a flue or fireplace recess; and/or
- 40 mm from the outer surface of a masonry chimney or fireplace recess.

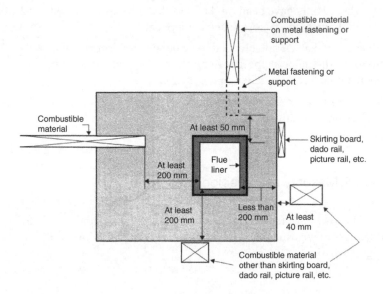

Figure 6.162 Minimum separation distances for combustible material in or near a chimney.

6.9.2.4.6 Factory-made metal chimneys

Where a factory-made metal chimney passes through a wall, sleeves should be provided to prevent damage to the flue or building through thermal expansion.	J 1.43
To allow gas-tightness to be checked, joints between chimney sections should not be concealed within ceiling joist spaces or within the thicknesses of walls.	J 1.43
Following installation of a factory-made metal chimney, it should be a simple measure to withdraw the appliance without having to dismantle the chimney.	J 1.44
Factory-made metal chimneys should **not** be installed near combustible materials.	J 1.45
Where a factory-made metal chimney passes through a cupboard, storage space or roof space, the chimney should be no closer than a distance X from combustible material, where X is defined in BS EN 1856-1:2003, as shown in Figure 6.163.	J 1.45

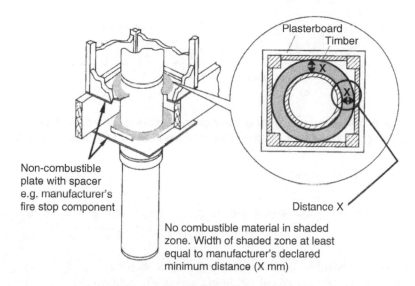

Plasterboard
Timber

Non-combustible
plate with spacer
e.g. manufacturer's
fire stop component

Distance X

No combustible material in shaded
zone. Width of shaded zone at least
equal to manufacturer's declared
minimum distance (X mm)

Figure 6.163 The separation of combustible material from a factory-made metal chimney meeting BS 4543: Part 1: 1990.

 Casing the chimney in non-combustible material is recommended.

6.9.2.4.7 Debris collection space

If a chimney cannot be cleaned directly through the appliance, a debris-collecting space should be provided at appropriate locations in the **chimney** for emptying and cleaning. J 2.16 J 3.38

6.9.2.5 *Flues*

6.9.2.5.1 General

Flue walls should have a fire resistance of at least one-half of that required for the compartment wall or floor and be of non-combustible construction. B3

If a flue (or duct containing flues and/or ventilation duct(s)) passes through a compartment wall, or is built into a compartment wall, each wall of the flue or duct should have a fire resistance of at least half that of the wall or floor (Figure 6.164). B3 7.11 (V1) B3 10.16 (V2)

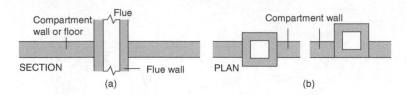

Figure 6.164 Flues penetrating compartment walls or floors. (a) Flue passing through compartment wall or floor. (b) Flue built into compartment wall.

6.9.2.5.2 Flue systems

Flue systems should offer least resistance to the passage of flue gases by minimizing changes in direction or horizontal length.	J 1.48
Wherever possible flues should be built so that they are straight and vertical **except** for the connections to combustion appliances with rear outlets where the horizontal section should not exceed 150 mm. Where bends are essential, they should be angled at no more than 45° to the vertical.	J 1.48
Provisions should be made to enable flues to be swept and inspected (Figure 6.165).	J 1.49

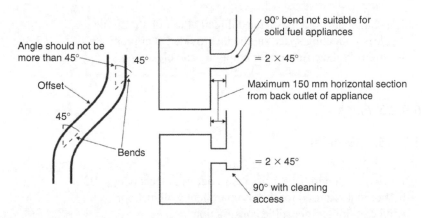

Figure 6.165 Bends in flues.

A flue should not have openings into more than one room or space except for the purposes of: • inspection and/or cleaning; or	J 1.50

- fitting an explosion door, draught break, draught stabilizer or draught diverter.

Openings for inspection and cleaning should have an access cover that has the same level of gas-tightness as the flue system and an equal level of thermal insulation.   J 1.51

After the appliance has been installed, it should be possible to sweep the **whole** flue.   J 1.50

6.9.2.5.3 Flues in chimneys

Table 6.69 Size of flues in chimneys (see Figure 6.166)

Type	Dimensions
Fireplace with an opening up to 500 mm × 550 mm	200 mm diameter or rectangular/square flues having the same cross-sectional dimension not less than 175 mm.
Fireplace with an opening in excess of 500 mm × 550 mm or a fireplace exposed on two or more sides	If rectangular/square flues are used, the minimum dimension should not be less than 200 mm.
Closed appliance up to 20 kW rated output which: • burns smokeless or low volatile fuel; or • is an appliance that meets the requirements of the Clean Air Act when burning an appropriate bituminous coal (these appliances are known as 'exempted fireplaces')	125 mm diameter or rectangular/square flues having the same cross-sectional area and a minimum dimension not less than 100 mm for straight flues or 125 mm for flues with bends or offsets.
Other closed appliance of up to 35 kW rated output burning any fuel	150 mm diameter or rectangular/square flues having the same cross-sectional area and a minimum dimension not less than 125 mm.
Closed appliance from 30 kW up to 50 kW rated output burning any fuel	175 mm diameter or rectangular/square flues having the same cross-sectional area and a minimum dimension not less than 150 mm.

For fireplaces with openings larger than 500 mm × 550 mm or fireplaces exposed on two or more sides (e.g. a fireplace under a canopy or open on both sides of a central chimney breast) a way of showing compliance would be to provide a flue with a cross-sectional area equal to 15 per cent of the total face area of the fireplace opening(s) using the formula:

Fireplace opening area (mm^2) × Total horizontal length of fireplace opening L (mm) × Height of fireplace opening H (mm)

Examples of L and H for large and unusual fireplace openings are shown in Figure 6.166.

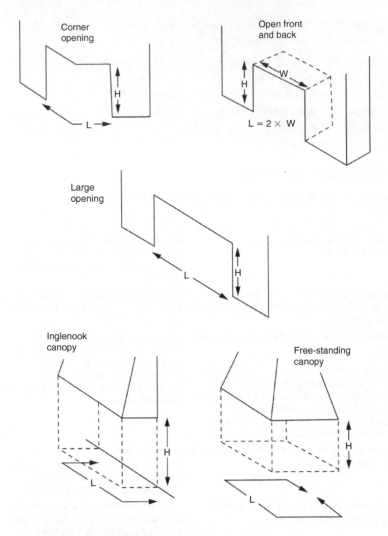

Figure 6.166 Large or unusual fireplace openings.

6.9.2.5.4 Height of flues

Flues should be high enough (normally 4.5 m is sufficient) to ensure there is always sufficient draught to clear the products of combustion.

J 2.8

The outlet from a flue should be above the roof of the building so that the products of combustion can discharge freely and will not present a fire hazard, whatever the wind conditions (Figure 6.167 and Table 6.70).

J 2.10

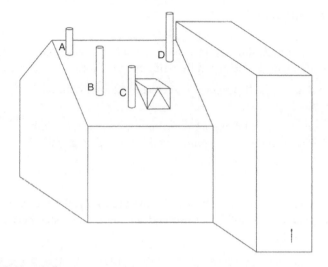

Figure 6.167 Flue outlet positions for solid fuel appliances.

Table 6.70 Flue outlet positions

Point where flue passes through weather surfaces (e.g. roof, tiles or external walls)	Clearance to flue outlet
A At or within 600 mm of the ridge	At least 600 mm above the ridge.
B Elsewhere on a roof (whether pitched or flat)	At least 2300 mm horizontally from the nearest point on the weather surface and: • at least 1000 mm above the highest point of intersection of the chimney and the weather surface; or • at least as high as the ridge.
C Below (on a pitched roof) or within 2300 mm horizontally to an openable rooflight, dormer window or other opening	At least 1000 mm above the top of the opening.
D Within 2300 mm of an adjoining or adjacent building	At least 600 mm above the adjacent building.

6.9.2.5.5 Size of flues

Flue pipes should have the same diameter or equivalent cross-sectional area as the appliance's flue outlet.	J 2.4
Flues should not be smaller than the appliance's flue outlet or that recommended by the appliance manufacturer.	J 2.5

6.9.2.5.6 Flue outlets

Outlets from flues should allow the dispersal of products of combustion and, if a balanced flue, the intake of air.	J 3.23
Flue outlets should be protected where flues are at significant risk of blockage.	J 3.24
Flues serving natural draught open-flued appliances should be fitted with outlet terminals if the flue diameter is no greater than 170 mm.	J 3.25

 In areas where nests of squirrels or jackdaws are likely, the fitting of a protective cage designed for solid fuel use and having a mesh size between 6 and 25 mm is advisable.

Flue outlets should be protected with a guard if persons could come into contact with them or if they could be damaged.	J 3.26

6.9.2.5.6.1 FLUES DISCHARGING AT LOW LEVEL NEAR BOUNDARIES

Flues that discharge at low level near boundaries should ensure safe flue gas dispersal (see Figure 6.168 for examples of achieving this).	J 1.52

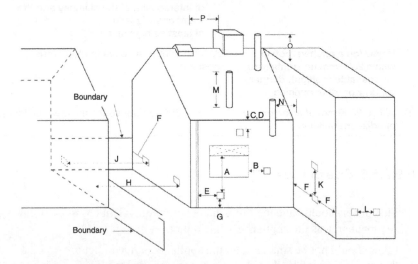

Figure 6.168 Location of outlets from flues serving oil-fired appliances (see Table 6.73 for minimum separation distances).

6.9.2.5.6.2 FLUE OUTLET CLEARANCES – THATCHED OR SHINGLED ROOF

The clearances to flue outlets that discharge on, or are in close J 2.12
proximity to, roofs with surfaces which are readily ignitable
(e.g. covered in thatch or shingles) should be increased to
those shown in Figure 6.169.

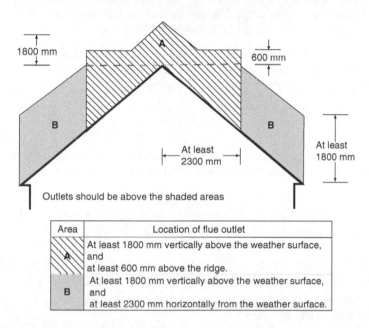

Area	Location of flue outlet
A	At least 1800 mm vertically above the weather surface, and at least 600 mm above the ridge.
B	At least 1800 mm vertically above the weather surface, and at least 2300 mm horizontally from the weather surface.

Figure 6.169 Flue outlet positions for solid fuel appliances – discharging near easily ignited roof coverings.

Thatched roofs can sometimes be vulnerable to spontaneous combustion caused by heat being transferred from flues and building up in thick layers of thatch that are in contact with the chimney. To reduce the risk it is recommended that rigid twin-walled insulated metal flue liners are used within a ventilated (top and bottom) masonry chimney void, provided they are adequately supported and not in direct contact with the masonry. Non-metallic chimneys and cast in situ flue liners can also be used, provided the heat transfer to the thatch is assessed in relation to the depth of thatch and risk of spontaneous combustion.

Spark arrestors are not generally recommended, as they can be difficult to maintain and may increase the risk of flue blockage and flue fires.

 Note: Further information and recommendations are contained in HETAS Information Paper 1/007, *Chimneys in Thatched Properties*.

6.9.2.5.7 Connecting flue pipes

Whenever possible, flue pipes should be manufactured from: J 1.32

- cast iron (BS41:1973 (1998));
- mild steel (BS1449, Part 1:1991, with a flue wall thickness of at least 3 mm);
- stainless steel (BS EN 10088-1:1995, grades 1.4401, 1.4404, 1.4432 or 1.4436 with a flue wall thickness of at least 1 mm);
- vitreous enamelled steel (BS 6999:1989 (1996)).

Flue pipes with spigot and socket joints should be fitted with   J 1.33
the socket facing upwards to contain moisture and other
condensates in the flue.

Joints should be made gas-tight.   J 1.33

Combustible materials in the building fabric should be protected from the heat
dissipation from flues so that they are not at risk of catching fire.

Connecting flue pipes should not pass through any roof space,   J 2.14
partition, internal wall or floor, unless they pass directly into
a chimney through either a wall of the chimney or a floor
supporting the chimney.

Connecting flue pipes should be guarded if they are likely   J 2.14
to be damaged or if the burn hazard they present to people is
not immediately apparent.

Connecting flue pipes should be located so as to avoid igniting   J 2.15
combustible material (see Figure 6.170).   and 1.45

6.9.2.5.8 Concealed flues

(1)   When a flue is routed within a void, access should be   J 1.47
       provided at various strategic points to allow the following
       aspects to be visually checked and confirmed whenever
       the appliance is first installed and subsequently when the
       appliance is serviced:

- that the flue is continuous throughout its length;
- that all joints have been correctly assembled and sealed;
- that the flue is adequately supported throughout its length;
- that required gradients of fallback to the boiler and any other required drain points have been provided.

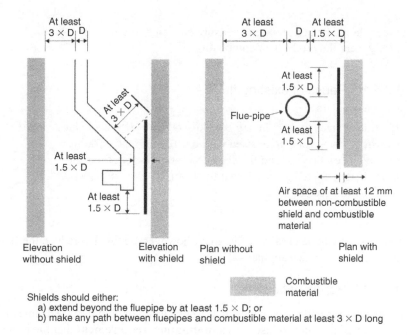

Figure 6.170 Protecting combustible material from uninsulated flue pipes for solid fuel appliances.

(2) Access for concealed flues shall be sufficiently sized and positioned so as to permit visual inspection of the flue (particularly at any joints in the flue).

(3) Concealed flues should not pass through another dwelling.

(4) Access hatches should be at least 300 mm × 300 mm or larger where necessary to allow sufficient access to the void to look along the length of the flue.

6.9.2.5.9 Ventilation

Any room or space intended to contain a decorative fuel effect (DFE) fire should have permanently open air vents.   J 3.11

Rooms or spaces intended to contain flueless appliances may need: permanent ventilation and purge ventilation (e.g. openable windows) to comply with Approved Document J; and adjustable ventilation and rapid ventilation to comply with Approved Document F.

> Flues should be high enough to ensure sufficient draught to safely clear the products of combustion.   J 3.21

6.9.2.5.10 Reuse of existing flues

> If a flue in an existing chimney is going to be brought back into use (or a flue is going to be used with a different type or rating of appliance), the flue **and** the chimney should be checked and, if necessary, altered to ensure that they satisfy the requirements for the proposed use   J 1.36

 Oversize flues can be unsafe. A flue may, however, be lined to reduce the flue area to suit the intended appliance.

6.9.2.5.11 Repair of flues

 If the installation and maintenance of a fireplace and/or chimney are deemed as being a material change of use, it is a **mandatory requirement** that the building is brought up to the standards required by Approved Documents J1 to J3.

If renovation, refurbishment or repair amounts to or involves the provision of a new or replacement flue liner, it is considered 'building work' within the meaning of Regulation 3 of the Building Regulations and must, therefore, **not** be undertaken without prior notification to the local authority. Examples of work that would need to be notified include:

> • relining work consisting of the creation of new flue walls by the insertion of new linings such as rigid or flexible prefabricated components;   J 1.34–1.35
> • a cast in situ liner that significantly alters the flue's internal dimensions.

 Note: If you are in any doubt about this requirement, you should consult the building control department of your local authority or an approved inspector.

6.9.2.5.12 Relining flues

 In certain circumstances, relining is considered *building work* within the meaning of Regulation 3 of the Building Regulations and must, therefore, **not** be undertaken without prior notification to the local authority. If you are in doubt you should consult the building control department of your local authority or an approved inspector.

If a chimney has been previously relined using a metal lining   J 1.39
system **and** the appliance is being replaced, then the metal liner
should also be replaced unless the metal liner has been recently
installed and is in good condition.

Flues should be swept to remove deposits before being relined.

 Flexible flue liners should **only** be used to reline a   J 1.40
chimney and should **not** be used as the primary liner of a
new chimney.

 Plastic flue pipe systems can be used in some cases   J 1.41
(e.g. with condensing boiler installations) provided
that they are supplied by or specified by the appliance
manufacturer.

6.9.2.6 Fireplaces

6.9.2.6.1 General

Fireplace recesses (sometimes called a 'builder's   J 0.4–17
opening') shall be formed in a wall or in a chimney
breast, from which a chimney leads and which has a
hearth at its base.

Simple recesses are suitable for closed appliances
such as room heaters, stoves, cookers or boilers.
They are **not** suitable for an open fire without a
canopy.

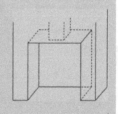

Fireplace recesses are used for accommodating open
fires and free-standing fire baskets.

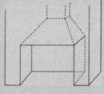

Fireplace recesses are often lined with firebacks to
accommodate inset open fires.

 Note: Lining components and decorative
treatments fitted around openings reduce
the opening area. It is the finished fireplace
opening area that determines the size of flue
required for an open fire in such a recess.

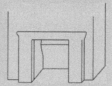

6.9.2.6.2 Fireplace recesses

Fireplaces for open fires should be constructed so that they   J 2.30
adequately protect the building fabric from catching fire.

Fireplace recesses can either: J 2.30a

- be from masonry or concrete, as shown in Figure 6.171; or
- be a prefabricated factory-made appliance chamber   J 2.30b
  made of insulating concrete having a density of between
  1200 kg/m^3 and 1700 kg/m^3 and with the minimum
  thickness as shown in Table 6.71.

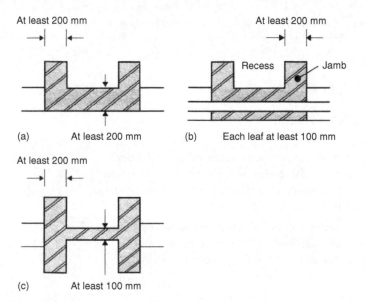

Figure 6.171 Fireplace recesses. (a) Solid wall. (b) Cavity wall. (c) Back-to-back (within the same dwelling).

Table 6.71 Flue outlet positions

Point where flue passes through weather surfaces (e.g. roof, tiles or external walls)	Clearance to flue outlet
A At or within 600 mm of the ridge	At least 600 mm above the ridge.
B Elsewhere on a roof (whether pitched or flat)	At least 2300 mm horizontally from the nearest point on the weather surface and: • at least 1000 mm above the highest point of intersection of the chimney and the weather surface; or • at least as high as the ridge.

Table 6.71 Flue outlet positions (Continued)

Point where flue passes through weather surfaces (e.g. roof, tiles or external walls)	Clearance to flue outlet
C Below (on a pitched roof) or within 2300mm horizontally to an openable rooflight, dormer window or other opening	At least 1000mm above the top of the opening.
D Within 2300mm of an adjoining or adjacent building	At least 600mm above the adjacent building.

6.9.2.6.3 Fireplace lining components

Fireplace recesses containing inset open fires need to be heat-protected and should be lined with either suitable firebricks or lining components as shown in Table 6.72.   J 2.31

Table 6.72 Prefabricated appliance chambers: minimum thickness

Component	Minimum thickness (mm)
Base	50
Side section, forming wall on either side of chamber	75
Back section, forming rear of chamber	100
Top slab, lintel or gather, forming top of chamber	100

6.9.2.6.4 Construction of fireplace gathers

To minimize resistance to the proper working of flues, tapered gathers should be provided in fireplaces for open fires or corbelling of masonry, as shown in Figure 6.172.

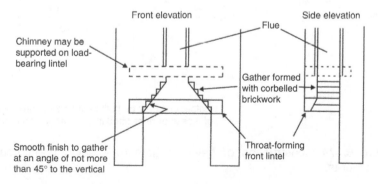

Figure 6.172 Construction of fireplace gathers – using masonry.

Alternatively a suitable canopy (as shown in Figure 6.173) or a prefabricated appliance chamber incorporating a gather may be used.

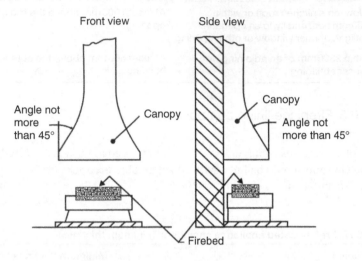

Figure 6.173 Canopy for an open fuel fire.

This can be achieved by using prefabricated gather components built into a fireplace recess, as shown in Figure 6.174.

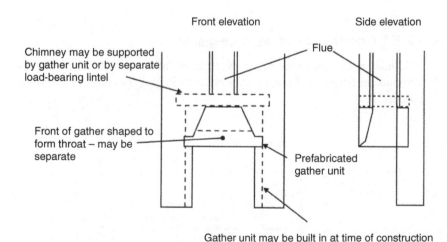

Figure 6.174 Construction of fireplace gathers – using prefabricated components.

6.9.2.6.4.1 Formation of gathers

Tapered gathers should be provided in fireplaces for open fires.   J 2.21

6.9.2.7 Hearths

All hearths shall safely isolate combustion appliances   J 0.4–27
from people, soft furnishings and combustible parts of the
building fabric.

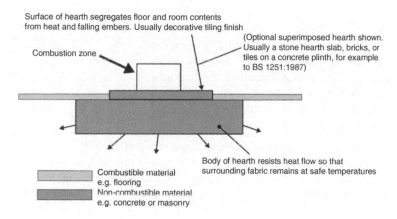

Surface of hearth segregates floor and room contents
from heat and falling embers. Usually decorative tiling finish

(Optional superimposed hearth shown.
Usually a stone hearth slab, bricks, or
tiles on a concrete plinth, for example
to BS 1251:1987)

Combustion zone

Combustible material
e.g. flooring
Non-combustible material
e.g. concrete or masonry

Body of hearth resists heat flow so that
surrounding fabric remains at safe temperatures

Figure 6.175 The functions of a hearth.

6.9.2.7.1 Construction of hearths

Hearths should be constructed so that, in normal use, they   J 2.22
prevent combustion appliances setting fire to the building
fabric and furnishings – as well as limiting the possibility of
people being accidentally burnt.

If the chimney is not independently supported, the hearth should   J 2.22
be able to support the weight of the appliance and its chimney.

Appliances should stand entirely above either: J 2.23

• hearths made of non-combustible board/sheet material; or
• tiles at least 12 mm thick; or
• constructional hearths.

Constructional hearths should have plan dimensions as shown   J 2.24a
in Figure 6.176.

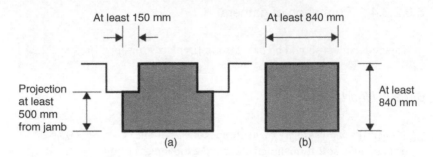

Figure 6.176 Constructional hearth suitable for solid fuel appliances (including open fires) – plan. (a) Fireplace recess. (b) Free-standing.

Constructional hearths should be made of solid, non-combustible material, such as concrete or masonry that is at least 125 mm thick, including the thickness of any non-combustible floor and/or decorative surface.   J 2.24b

Combustible material should **not** be placed beneath constructional hearths unless: J 2.25

- there is an air space of at least 50 mm between the underside of the hearth and the combustible material; or
- the combustible material is at least 250 mm below the top of the hearth (Figure 6.177).

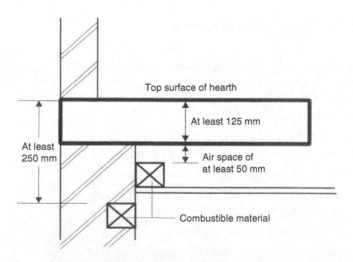

Figure 6.177 Constructional hearth suitable for solid fuel appliances (including open fires) – section.

An appliance should be located on a hearth so that: J 2.26

- it is surrounded by a surface that is free of combustible material (as shown in Figure 6.178); or
- the surface of a superimposed hearth is laid wholly or partly upon a constructional hearth.

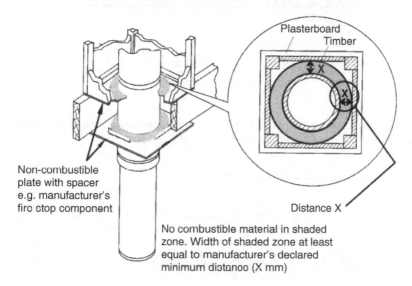

Plasterboard
Timber

Non-combustible plate with spacer e.g. manufacturer's fire stop component

Distance X

No combustible material in shaded zone. Width of shaded zone at least equal to manufacturer's declared minimum distance (X mm)

Figure 6.178 The separation of combustible material from a factory-made metal chimney meeting BS 4543: Part 1: 1990.

 Note: The edges of this surface should be marked (e.g. by a change in level) to provide a warning to the building occupants and to discourage combustible floor finishes such as carpet from being laid too close to the appliance.

Combustible material that is placed on or beside a constructional   J 2.28
hearth should neither extend under a superimposed hearth
by more than 25 mm **nor** be closer than 150 mm measured
horizontally to the appliance.

 Note: Some ways of making these provisions are shown in Diagram 27 of Approved Document J:2010.

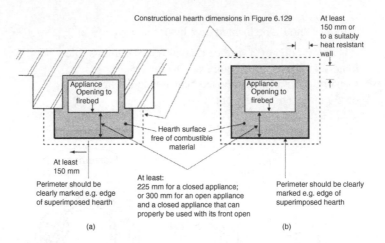

Figure 6.179 Non-combustible hearth surface surrounding a solid fuel appliance. (a) Fireplace recess. (b) Free-standing.

6.9.2.7.2 Walls adjacent to hearths

Walls that are not part of a fireplace recess or a prefabricated   J 2.32
appliance chamber but are adjacent to hearths or appliances
also need to protect the building from catching fire. A way of
achieving the requirement is shown in Figure 6.180.

6.9.2.7.3 Dry lining around fireplace openings

Gaps around decorative treatment, round a fireplace opening   J 1.53
(e.g. a fireplace surround, masonry cladding or dry lining),
should be sealed to prevent any leakage from the fireplace
opening into the void behind the decorative treatment.

The sealing material should be capable of remaining in   J 1.53
place despite any relative movement between the decorative
treatment and the fireplace recess.

6.9.2.8 Provision of information

On completion of work:

*   A report should be drawn up by the person carrying out the   J 1.54
    work, to show what materials and components have been
    used and to confirm that flues have passed appropriate tests.

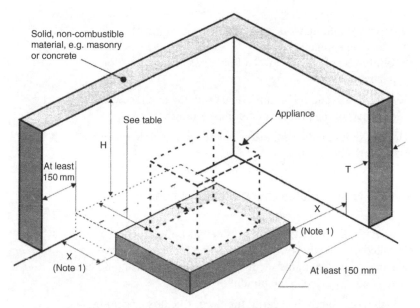

Note 1: There is no requirement for protection of the wall where X is more than 150 mm

Location of hearth or appliance	Solid, non-combustible material	
	Thickness (T)	Height (H)
Where the hearth abuts a wall and the appliance is not more than 50 mm from the wall	200 mm	At least 300 mm above the appliance and 1.2 m above the hearth
Where the hearth abuts a wall and the appliance is more than 50 mm but not more than 300 mm from the wall	75 mm	At least 300 mm above the appliance and 1.2 m above the hearth
Where the hearth does not abut a wall and is no more than 150 mm from the wall (see Note 1)	75 mm	At least 1.2 m above the hearth

Figure 6.180 Walls adjacent to hearths.

 Note: Guidance on testing is given at Appendix E to Approved Document J:2010.

> •   Flues should be checked to show that they are free from   J 1.55
>   obstructions, satisfactorily gas-tight and constructed with
>   materials and components of sizes which suit the intended
>   application.

 See Appendix A to Approved Document J:2010 for detailed checklists for checking and testing of hearths, fireplaces, flues and chimneys.

Where the building work includes the installation of a combustion appliance, tests should include flue pipes and the gas-tightness of joints between flue pipes and combustion appliance outlets.	J 1.55
A spillage test should be carried out with the appliance under fire, as part of the process of commissioning.	J 1.55
Hearths should indicate the area where combustible materials should not encroach.	J 1.56

6.9.2.8.1 Notice plates for hearths and flues

If a hearth, fireplace (including a flue box), flue or chimney is   J 1.57
installed or extended (e.g. as part of some refurbishment work),
a notice plate containing the following information should be
permanently posted in the building:

- the location of the hearth, fireplace, flue box or beginning of the flue;
- the category of the flue and generic types of appliances that can be safely accommodated;
- the type and size of the flue (or its liner if it has been relined) and the manufacturer's name;
- the installation date.

Notice plates should be securely fixed either:

- next to the electricity consumer unit; or
- next to the chimney or hearth described; or
- next to the water supply stopcock.

6.9.2.9 Maintenance

6.9.2.9.1 Access to combustion appliances for maintenance

Safe access to appliances for maintenance purposes should be provided.	J 1.60
Roof space installations of gas-fired appliances should comply with the requirements of BS 6798:2009.	J 1.60

6.9.2.10 Gas-burning devices

The Gas Safety (Installation and Use) Regulations require that:

(a) gas fittings, appliances and gas storage vessels must only be installed by a person with the required competence; and

(b) any person having control to any extent of gas work must ensure that the person carrying out that work has the required competence; and

(c) any person carrying out gas installation, whether an employee or self-employed, **must** be a member of a class of persons approved by the HSE and registered with the Gas Safety Register.

Important elements of the regulations include:

Precautions must be taken to ensure that all installation pipework, gas fittings, appliances and flues are installed safely.	J 3.5e
When any gas appliance is installed, checks are required for ensuring compliance with the regulations, including the effectiveness of the flue, the supply of combustion air, the operating pressure or heat input (or where necessary both), and the operation of the appliance to ensure its safe functioning.	J 3.5f
All flues must be installed in a safe position.	J 3.5g
No alteration is allowed to any premises in which a gas fitting or gas storage vessel is fitted that would adversely affect the safety of that fitting or vessel, causing it no longer to comply with the regulations.	J 3.5h

Note: Outlets from flues should be situated externally so as to allow the products of combustion to dispel and, if a balanced flue, the intake of air (Figure 6.181).

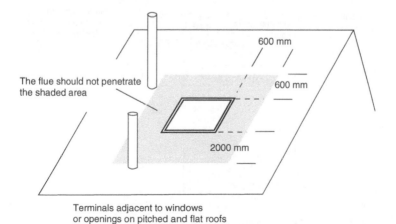

The flue should not penetrate the shaded area

600 mm

600 mm

2000 mm

Terminals adjacent to windows or openings on pitched and flat roofs

Figure 6.181 Location of outlets near roof windows from flues serving gas appliances.

6.9.2.10.1 Gas fires (other than flueless gas fires)

Gas-fired appliances should **only** be located where accidental contact is unlikely and where they can be surrounded by a non-combustible surface which provides adequate separation from combustible materials.

> Gas fires may be installed in fireplaces which have flues designed J 3.7
> to serve solid fuel appliances – provided it can be shown to be
> safe.

6.9.2.11 Back boilers

> Back boilers should adequately protect the fabric of the J 3.39
> building from heat (see the example in Figure 6.182).

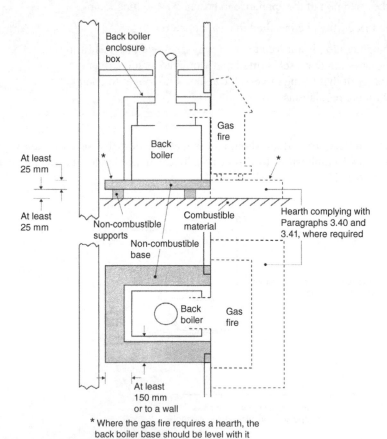

Figure 6.182 Bases for back boilers.

6.9.2.12 *Kerosene and gas-oil burning appliances*

Kerosene (class C2) and gas-oil (class D) appliances have the following, additional, requirements:

Open-fired oil appliances should not be installed in rooms such as bedrooms and bathrooms where there is an increased risk of carbon monoxide poisoning.	J 4.2
Flues should be sized to suit the intended appliance and to ensure sufficient discharge velocity to prevent flow reversal problems – without imposing excessive flow resistances.	J 4.4
The outlet from a flue should be so situated externally as to ensure:	J 4.6

* the correct operation of a natural draft flue;
* the intake of air if a balanced flue;
* dispersal of the products of combustion.

Note: Figure 6.183 (and Table 6.73) indicates typical positioning to meet this requirement.

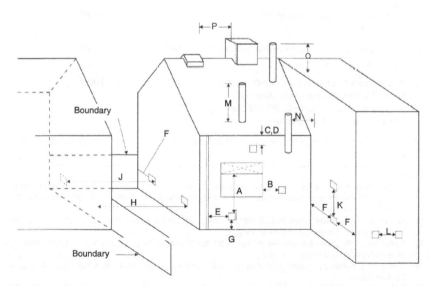

Figure 6.183 Location of outlets from flues serving kerosene and oil-fired appliances (see Table 6.73 for minimum separation distances).

Table 6.73 Location of outlets from flues serving oil-fired appliances

Location of outlet[1]	Minimum separation distances for terminals (mm)	
	Appliance with pressure-jet burner	Appliance with vaporizing burner
A Below an opening[2,3]	600	Should not be used
B Horizontally to an opening[2,3]	600	Should not be used
C Below a plastic/painted gutter, drainage pipe or eaves if combustible material protected[4]	75	Should not be used
D Below a balcony or a plastic/painted gutter, drainage pipe or eaves without protection to combustible material	600	Should not be used
E From vertical sanitary pipework	300	Should not be used
F From an external or internal corner or from a surface or boundary alongside the terminal	300	Should not be used
G Above ground or balcony level	300	Should not be used
H From a surface or boundary facing the terminal	600	Should not be used
J From a terminal facing the terminal	1200	Should not be used
K Vertically from a terminal on the same wall	1500	Should not be used
L Horizontally from a terminal on the same wall	750	Should not be used
M Above the highest point of an intersection with the roof	600[5]	1000[6]
N From a vertical structure to the side of the terminal	750[5]	2300
O Above a vertical structure which is less than 750 mm (pressure jet burner) or 2300 mm (vaporizing burner) horizontally from the side of the terminal	600[5]	1000[6]
P From a ridge terminal to a vertical structure on the roof	1500	Should not be used

Notes:
(1) Terminals should only be positioned on walls where appliances have been approved for such configurations when tested in accordance with BS EN 303-1:1999 or OFTEC standard OFS A100 or OFS A101.
(2) An opening means an openable element, such as an openable window, or a permanent opening such as a permanently open air vent.
(3) Notwithstanding the dimensions above, a terminal should be at least 300 mm from combustible material (e.g. a window frame).
(4) A way of providing protection of combustible material would be to fit a heat shield at least 750 mm wide.
(5) Outlets for vertical balanced flues in locations M, N and O should be in accordance with the manufacturer's instructions.
(6) Where a terminal is used with a vaporizing burner, the terminal should be at least 2300 mm horizontally from the roof.

Flexible metal flue liners should be installed in one complete   J.3.37
length without joints within the chimney.   J 4.22

Other than for sealing at the top and the bottom, the space   J 4.22
between the chimney and the liner should be left empty (unless
this is contrary to the manufacturer's instructions).

Flues that may be expected to serve appliances burning class D   J 4.23
oil (i.e. gas oil) should be made of materials that are resistant to
acids.

6.9.2.12.1 Relining chimney flues (for oil appliances)

In some circumstances, lining or relining flues **could** be considered as building
work, in which case the works must be brought up to the standards required by
Parts J2 to J4.

6.9.2.12.2 Hearths for oil appliances

Oil appliance hearths are needed to prevent the building   J 4.24
catching fire and, while it is not a health and safety provision, it
is customary to top them with a tray for collecting spilt fuel.

6.10 Stairs

6.10.1 Requirements

Structure
The building shall be constructed so that the combined dead, imposed and wind
loads are sustained and transmitted by it to the ground:

- *safely;*
- *without causing such deflection or deformation of any part of the building*
  (or such movement of the ground) as will impair the stability of any part of
  another building.

(Approved Document A1)

Amendments to the 2010 version of Approved Document A came into force
in 2013.

Fire protection
The building shall be designed and constructed so that there are appropriate
provisions for the early warning of fire, and appropriate means of escape in
case of fire from the building to a place of safety outside the building capable
of being safely and effectively used at all material times.

(Approved Document B)

For a typical one- or two-storey dwelling, the requirement is limited to the provision of smoke alarms and to the provision of openable windows for emergency exit (see B1.i).

Airborne and impact sound
Dwellings shall be designed so that the noise from domestic activity in an adjoining dwelling (or other parts of the building) is kept to a level that:

- *does not affect the health of the occupants of the dwelling;*
- *will allow them to sleep, rest and engage in their normal activities in satisfactory conditions.*

(Approved Document E1)

Dwellings shall be designed so that any domestic noise that is generated internally does not interfere with the occupants' ability to sleep, rest and engage in their normal activities in satisfactory conditions.

(Approved Document E2)

Domestic buildings shall be designed and constructed so as to restrict the transmission of echoes.

(Approved Document E3)

Schools shall be designed and constructed so as to reduce the level of ambient noise (particularly echoing in corridors).

(Approved Document E4)

Ventilation
There shall be adequate means of ventilation provided for people in the building.

(Approved Document F)

Stairs, ladders and ramps
Stairs, ladders and ramps shall be so designed, constructed and installed as to be safe for people moving between different levels in or about the building.

(Approved Document K1)

A new issue of Approved Document K came into force in 2013.

Note: In the Secretary of State's view, you can meet requirement K1 by ensuring that the steepness, rise and going, handrails, headroom, length and width of any stairs, ladders and ramps between levels are appropriate to afford reasonable safety to people gaining access to and moving about buildings.

This requirement only applies to stairs, ladders and ramps that form part of the building.

Any stairs, ramps, floors and balconies and any roof to which people have access, and any light well, basement area or similar sunken area connected to a building, shall be provided with barriers where it is necessary to protect people in or about the building from falling.

(Approved Document K2)

In the Secretary of State's view, you can meet requirement K2 as follows:

- in **dwellings**: provide pedestrian guarding that is capable of preventing people from being injured by falling from a height of more than 600 mm;
- in **buildings other than dwellings**: provide pedestrian guarding that is capable of preventing people from falling more than the height of two risers (or 380 mm, if not part of a stair).

Access to and use of buildings
Reasonable provisions shall be made for people to:

(a) gain access to; and
(b) use the building and its facilities.

(Approved Document M)

Amendments to the 2004 version of Approved Document M came into force in 2013.

In addition to the requirements of the Equality Act 2010, precautions need to be taken to ensure that:

- new non-domestic buildings and/or dwellings (e.g. houses and flats used for student living accommodation etc.);
- extensions to existing non-domestic buildings;
- non-domestic buildings that have been subject to a material change of use (e.g. so that they become a hotel, boarding house, institution, public building or shop)

are capable of allowing people, regardless of their disability, age or gender, to:

- gain access **to** buildings;
- gain access **within** buildings;
- be able to use the facilities of the buildings (both as visitors and as people who live or work in them);
- use sanitary conveniences in the principal storey of any new dwelling.

6.10.2 Meeting the requirements

As the upper surfaces of floors and stairs are not significantly involved in a fire until it is well developed, they do not play an important part in fire spread in the early stages of a fire.

6.10.2.1 Structure

6.10.2.1.1 Stairway openings

Where an opening in a floor or roof for a stairway or the like adjoins a supported wall and interrupts the continuity of lateral support:

- the maximum permitted length of the opening is to be 3 m, measured parallel to the supported wall; A1/2 2C37a
- connections (if provided by means other than by anchors) should be throughout the length of each portion of the wall situated on each side of the opening; A1/2 2C37b
- connections via mild steel anchors should be spaced closer than 2 m on each side of the opening to provide the same number of anchors as if there were no opening; A1/2 2C37c
- there should be no other interruption of lateral support. A1/2 2C37d

Stairs that separate a dwelling from another dwelling (or part of the same building) shall resist:

- the transmission of impact sound (e.g. footsteps and furniture moving); E2
- the flow of sound energy through walls and floors;
- the level of airborne sound;
- flanking transmission from stairs connected to the separating wall.

All new stairs constructed within a dwelling house (flat or room used for residential purposes) – whether purpose-built or formed by a material change of use – shall meet the laboratory sound insulation values set out in Table 6.74. E0.9

Table 6.74 Dwelling houses and flats – performance standards for separating floors and stairs that have a separating function

Type	Airborne sound insulation $D_{nT,w} + {}_{Ctr}$ (dB) (minimum values)	Impact sound insulation $L9_{nT,w}$ (dB) (maximum values)
Purpose-built rooms for residential purposes	45	62
Purpose-built dwelling houses and flats	45	62
Rooms for residential purposes formed by material change of use	43	64
Dwelling houses and flats formed by material change of use	43	64

Notes:

(1) The sound insulation values in Table 6.74 include a built-in allowance for 'measurement uncertainty', and so, if any of these test values are not met, that particular test will be considered as failed.
(2) Occasionally, a higher standard of sound insulation may be required between spaces used for normal domestic purposes and noise generated in and to an adjoining communal or non-domestic space. In these cases it would be best to seek specialist advice before committing yourself.
(3) If the stair is not enclosed, the potential sound insulation of the internal floor will not be achieved; nevertheless, the internal floor should still satisfy Requirement E2.
(4) In some cases it may be that an existing wall, floor or stair in a building will achieve these performance standards without the need for remedial work (e.g. if the existing construction was already compliant).

Figure 6.184 illustrates the relevant parts of the building that should be protected from airborne and impact sound in order to satisfy Requirement E2.

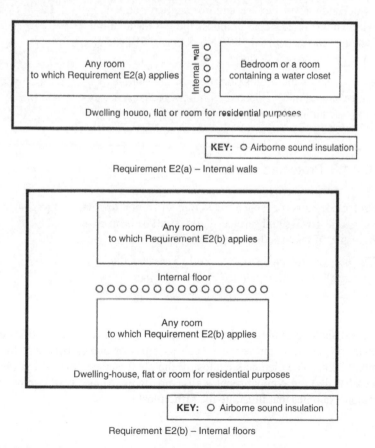

Figure 6.184 Airborne and impact sound requirements.

6.10.2.1.2 Protected stairways

Protected stairways should discharge:	B1 2.38 (V2)
• directly to a final exit; or • via a protected exit passageway to a final exit.	
Where two protected stairways (or exit passageways leading to different final exits) are adjacent, they should be separated by an imperforate enclosure.	B1 2.39 (V2)
A protected stairway should be relatively free of potential sources of fire.	B1 2.40 (V2)
In single-stair buildings, meters located within the stairway should be enclosed within a secure cupboard that is separated from the escape route with fire-resisting construction.	B1 2.40 (V2)
Gas service and installation pipes (together with their associated meters, etc.) should **not** be incorporated within a protected stairway unless the gas installation is in accordance with SI 1996 No. 825 and SI 1998 No. 2451.	B1 4.40 (V2) B1 2.42 (V2)
Refuse chutes and rooms provided for the storage of refuse should **not** be located within protected stairways or protected lobbies.	B1 5.55 and 5.56 (V2)

6.10.2.1.3 Protected shafts

Protected shafts (i.e. spaces that connect compartments, such as stairways and service shafts) shall be protected to restrict fire spread between the compartments.	B2 8.7 (V2)
The uses of protected shafts should be restricted to stairs, lifts, escalators, chutes, ducts and pipes.	B2 8.36 (V2)

Protected shafts provide for the movement of people (e.g. stairs, lifts) or for passage of goods, air or services such as pipes or cables between different compartments. The elements enclosing the shaft (unless formed by adjacent external walls) are compartment walls and floors. Figure 6.185 shows three common examples that illustrate the principles.

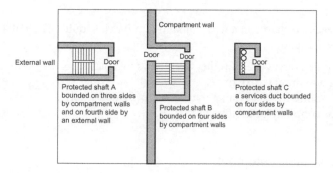

Figure 6.185 Protected shafts.

> Any stairway or other shaft passing directly from one compartment to another should be enclosed in a protected shaft so as to delay or prevent the spread of fire between compartments.
>
> B2 8.35 (V2)
>
> An uninsulated glazed screen may be incorporated in the enclosure to a protected shaft between a stair and a lobby or corridor which is entered from the stair provided that:
>
> B2 8.38 (V2)
>
> - the fire resistance for the stair enclosure is not more than 60 minutes; and
> - the glazed screen has at least 30 minutes' fire resistance; and
> - the lobby or corridor is enclosed to at least a 30-minute standard (Figure 6.186).

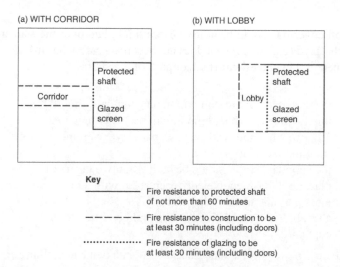

Figure 6.186 Uninsulated glazed screen separating protected shaft from lobby or corridor.

> If a protected shaft contains a stair and/or a lift, it should B2 8.40 (V2)
> **not** also contain:
>
> •   a pipe conveying oil (other than in the mechanism
>   of a hydraulic lift); or
> •   a ventilating duct (other than a duct provided solely
>   for ventilating the stairway).

6.10.2.1.4 Stair treatment

'Stair treatment' consists of a stair covering and independent ceiling with absorbent material.

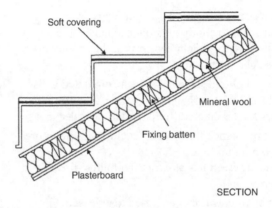

Soft covering

Mineral wool

Fixing batten

Plasterboard

SECTION

Figure 6.187 Stair covering and independent ceiling with absorbent material.

The soft covering should be at least 6 mm thick, laid over the stair treads and securely fixed (e.g. glued) so it does not become a safety hazard.

If there is a cupboard under all, or part, of the stair:

> •   the underside of the stair within the cupboard should be   E4.37
>   lined with plasterboard (minimum mass per unit area
>   10 kg/m^2) together with an absorbent layer of mineral wool
>   (minimum density 10 kg/m^3);
> •   the cupboard walls should be built from two layers of
>   plasterboard (or equivalent), each sheet with a minimum
>   mass per unit area of 10 kg/m^2;
> •   a small, heavy, well-fitted door should be fitted to the
>   cupboard.
>
> If there is no cupboard under the stair, an independent ceiling
> should be constructed below the stair (see floor treatment 1).

Where a staircase performs a separating function it shall conform to Approved Document B (Fire safety).

6.10.2.2 Fire protection

Except for kitchens, all habitable rooms in the upper storeys of a dwelling house that are served by only one stair should be provided with:	B1 2.4 (V1) B1 2.12 (V2)
• a window (or external door); or • direct access to a protected stairway.	B1 V (b)

Note: If direct escape to a place of safety is impracticable, it should be possible to reach a place of relative safety such as a protected stairway within a reasonable travel distance.

Table 6.75 Limitations on distance of travel in common areas of blocks of flats

Maximum distance of travel from flat entrance door to common stair or to stair lobby

Escape in one direction only	Escape in more than one direction
7.5 m	30 m

Escape routes should be planned so that people do not have to pass through one stairway enclosure to reach another.	B1 2.23 (V2)
Common corridors should be protected corridors.	B1 2.24 (V2)
The wall between each flat and the corridor should be a compartment wall.	B1 2.24 (V2)
Means of ventilating common corridors or lobbies (i.e. to control smoke and so protect the common stairs) should be available.	B1 2.25 (V2)
In large buildings, the corridor or lobby adjoining the stair should be provided with a vent that is located as high as practicable, and with its top edge at least as high as the top of the door to the stair.	B1 2.26 (V2)
There should also be a vent, with a free area of at least 1.0 m^2 from the top storey of the stairway to the outside.	B1 2.26 (V2)
In single-stair buildings the smoke vents on the fire floor and at the head of the stair should be actuated by means of smoke detectors in the common access space providing access to the flats.	B1 2.26 (V2)
In buildings with more than one stair, the smoke vents may be actuated manually.	B1 2.26 (V2)

Note: Self-closing fire doors should be positioned so that smoke will not affect access to more than one stairway.

6.10.2.2.1 Escape routes

Any storey which has more than one escape stair should be planned so that it is not necessary to pass through one stairway to reach another.	B1 3.13 (V2)
If an escape stair forms part of the only escape route from an upper storey of a large building it should **not** be continued down to serve any basement storey.	B1 2.44 (V2)

 The basement should be served by a separate stair.

If there is more than one escape stair from an upper storey of a building, only one of the stairs serving the upper storeys of the building need be terminated at ground level.	B1 2.45 (V2)

 Note: Other stairs may connect with the basement storey(s) if there is a protected lobby or a protected corridor between the stair(s) and accommodation at each basement level.

6.10.2.2.1.1 DOORS ON ESCAPE ROUTES

Unless escape stairways and corridors are protected by a pressurization system complying with BS EN 12101-6:2005, every dead-end corridor exceeding 4.5 m in length should be separated by self-closing fire doors (together with any necessary associated screens) from any part of the corridor that: • provides two directions of escape; • continues past one storey exit to another.	B1 3.27 (V2)
A door that opens towards a corridor or a stairway should be sufficiently recessed to prevent its swing from encroaching on the effective width of the stairway or corridor.	B1 5.16 (V2)
Vision panels shall be provided where doors on escape routes subdivide corridors, or where any doors are hung to swing both ways. (See also Approved Document M.)	B1 5.17 (V2)
Revolving doors, automatic doors and turnstiles should not be placed across escape routes.	B1 5.18 (V2)

6.10.2.2.2 Escape stairs

An external escape stair may be used, provided that:

• there is at least one internal escape stair from every part of each storey (excluding plant areas);	B1 2.49 (V2)
• in the case of an assembly and recreation building, the route is not intended for use by members of the public;	B1 4.44 (V2)
• in the case of an institutional building, the route serves only office or residential staff accommodation;	
• all doors giving access to the stair are fire-resisting and self-closing;	B1 2.15 (a, b and c) (V1)
• any part of the external envelope of the building within 1.8 m of (and 9 m vertically below) the flights and landings of an external escape stair is of fire-resisting construction (Figure 6.188);	B1 5.25 (V2)
• there is protection by fire-resisting construction for any part of the building within 1.8 m of the escape route from the stair to a place of safety;	
• glazing is fire-resistant and fixed shut.	

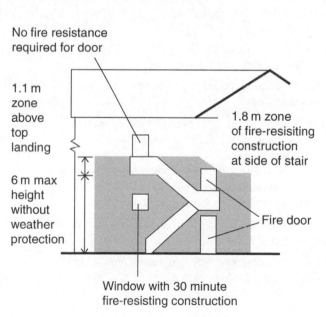

Figure 6.188 Fire resistance of areas adjacent to external stairs.

Escape stairs shall have a protected lobby or protected corridor at all levels (except the top storey, all basement levels and when the stair is a firefighting stair) if: — B1 4.34 (V2)

- the stair is the only one serving a building that has more than one storey above or below the ground storey;
- the stair serves a storey that is higher than 18 m;
- the building is designed for phased evacuation;
- the stairway is near (or potentially next to) a place of special fire hazard. — B1 4.35 (V2)

External escape stairs greater than 6 m in vertical extent shall be protected from the effects of adverse weather conditions. — B1 2.15d

If the building (or part of the building) is served by a single access stair, that stair may be external if it: — B1 2.48 (V2)

- serves a floor not more than 6 m above the ground level; and
- meets the provisions in paragraph B1 5.25 (V2).

6.10.2.2.2.1 PROTECTION OF ESCAPE STAIRS

Escape stairs need to have a satisfactory standard of fire protection. — B1 4.31 (V2)

Internal escape stairs should be a protected stairway within a fire-resisting enclosure. — B1 4.32 (V2)

Except for bars and restaurants, an escape stair may be open provided that: — B1 4.33 (V2)

- it does not connect more than two storeys and reaches the ground storey not more than 3 m from the final exit; and
- the storey is also served by a protected stairway; or
- it is a single stair in small premises with the floor area in any storey not exceeding 90 m².

A dwelling house with more than one floor 4.5 m above ground level may either: — B1 2.6 (V1)

- have a protected stairway that extends to the final exit (Figure 6.189(a)); or

- have a protected stairway that gives access to at least two escape routes at ground level, each delivering to final exits and separated from each other by fire-resisting construction and fire doors (Figure 6.189(b)); or
- have the top floor separated from the lower storeys by fire-resisting construction and have its own alternative escape route leading to its own final exit (Figure 6.189 (a and b)).

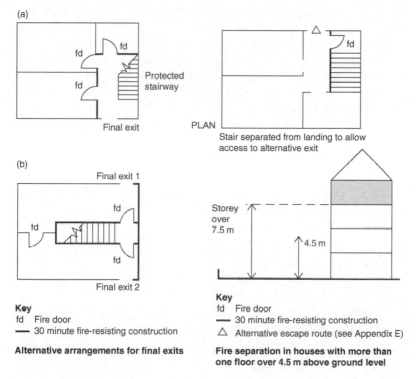

Figure 6.189 Alternative arrangements for final exits.

6.10.2.2.2.2 NUMBER OF ESCAPE STAIRS

The number of escape stairs required in a building (or part of a building) will be determined by:

B1 4.2 (V2)

- the constraints imposed by the design of horizontal escape routes;
- whether independent stairs are required in mixed occupancy buildings;

- whether a single stair is acceptable; and
- the width for escape (and the possibility that a stair cannot be used because of fire or smoke).

Provided that independent escape routes are not necessary from areas in different purpose groups, single escape stairs may be used from:　　B1 4.6 (V2)

- small premises (other than bars or restaurants);
- office buildings comprising not more than five storeys above the ground storey (provided that the travel distance from every point in each storey does not exceed 18 m – see Table 2 of Approved Document B2);
- factories comprising not more than one storey above the ground storey if the building is of normal risk (two storeys if the building is of low risk); or
- process plant buildings with an occupant capacity of not more than ten people.

6.10.2.2.2.3 CONSTRUCTION OF ESCAPE STAIRS

The flights and landings of every escape stair should be constructed using materials of limited combustibility, particularly if it is:　　B1 5.19 (V2)

- the only stair serving the building;
- within a basement storey;
- serving any storey having a floor level more than 18 m above ground or access level;
- external;
- a firefighting stair.

Single steps may **only** be used on an escape route if they are prominently marked.　　B1 5.21 (V2)

Helical and spiral stairs forming part of an escape route should be:　　B1 5.22a (V2)

- designed in accordance with BS 5395-2:1984;
- type B (public stair) if they are intended to serve members of the public.

Fixed ladders should not be used as a means of escape for members of the public.　　B1 5.22b (V2)

 Note: See Approved Document K for guidance on the design of helical and spiral stairs and fixed ladders.

If a protected stairway projects beyond, or is recessed from, or is within an internal angle adjoining an external wall of the building, the distance between any unprotected area in the external enclosures to the building and any unprotected area in the enclosure to the stairway should be at least 1800 mm (Figure 6.190).

B1 5.24 (V2)

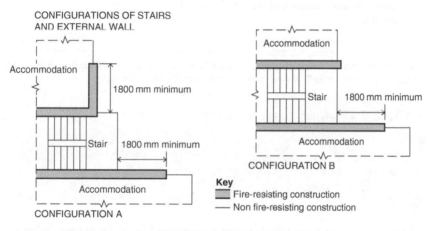

Figure 6.190 External protection to protected stairways.

The width of escape stairs should:

B1 4.15 (V2)

- not be less than the width of any exit(s);
- not be less than the minimum widths given in Table 6.76;
- not exceed 1.4 m if their vertical extent is more than 30 m, **unless** they are provided with a central handrail;
- not reduce in width at any point on the way to a final exit.

In public buildings, if the width of the stair is more than 1800 mm, the stair should have a central handrail.

B1 4.16 (V2)

Every escape stair should be wide enough to accommodate the number of persons needing to use it in an emergency.

B1 4.18 (V2)

 Note: For further guidance and worked examples see Approved Document B, Sections 4.18 (V2) to 4.25 (V2) and Appendix C of the same document.

Table 6.76 Minimum widths of escape stairs

Stair situation	Maximum number of people served	Minimum stair width
(1a) In an institutional building (unless the stair is only used by staff)	150	1000 mm
(1b) In an assembly building and serving an area used for assembly purposes (unless the area is less than 100 m²)	220	1100 mm
(1c) In any other building and serving an area with an occupancy of more than 50 people	Over 2200	1000–1800 mm*
(2) Any stair not described above	50	800 mm

* **Depending on whether the stairs are used for simultaneous evacuation or phased evacuation (see Table 7 of Approved Document B1 (Version 2)).**

6.10.2.2.2.4 LIGHTING TO ESCAPE STAIRS

Lighting to escape stairs should be on a separate   A1 5.36 (V2)
circuit from that supplying any other part of the
escape route.

The installation of an escape lighting system shall be in accordance with BS 5266-1:2005 and BS 7671 (latest edition). Full details are contained in the partner book to this publication, *Wiring Regulations in Brief* (3rd edition).

6.10.2.2.3 Basements

Because of their situation, basement stairways are more likely to be filled with smoke and heat than stairs located in ground and upper storeys. Special measures are, therefore, required in order to prevent a basement fire endangering upper storeys.

If an escape stair forms part of the only escape route   B1 4.42 (V2)
from an upper storey of a building, it should **not** be
continued down to serve a basement storey (i.e. the
basement should be served by a separate stair).

If there is more than one escape stair from an upper   B1 4.43 (V2)
storey of a building, **only** one of the stairs serving the
upper storeys of the building need be terminated at
ground level.

Owing to the possibility of a single stairway becoming blocked by smoke from a fire in the basement or ground storey:

basement storeys in a dwelling house that contain a habitable room **shall** be provided with either:

B1 2.13 (V1)

- a protected stairway leading from the basement to a final exit; or

B1 2.6 (V2)

- an external door or window suitable for egress from the basement.

6.10.2.2.4 Loft conversions

Where the conversion of an existing roof space (e.g. a loft conversion to a two-storey house) means that a new storey is going to be added, the stairway will need to be protected with fire-resisting doors and partitions.

B1 2.20b

Alternating tread stairs may be used, but only in a loft conversion and only when as shown in Figure 6.200, the stair is for access to only one habitable room and the construction of the alternate tread stair complies with Figure 6.191 and:

K1 (1.29)
K1 (1.30)

- alternating steps are uniform with parallel nosings;
- all treads have slip-resistant surfaces;
- tread sizes over the wider part of the step conform to Approved Document K1, Table 1.1; and
- a minimum clear headroom of 2 m is provided.

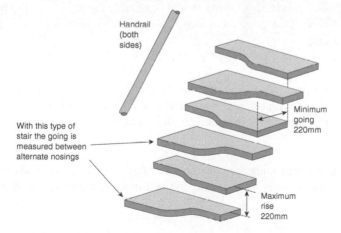

Figure 6.191 Alternating tread stair.

6.10.2.2.5 Wall cladding on escape routes

> Where wall cladding is required to function as pedestrian guarding to stairs, ramps or vertical drops of 600 mm or greater or as a vehicle barrier, account should be taken of the additional imposed loading as stipulated in Approved Document K.   A1/2 3.5
>
> Where wall cladding is required to safely withstand lateral pressures from crowds, an appropriate design loading is given in BS 6399: Part 1 and the *Guide to Safety at Sports Grounds* (5th edition, 2008) (colloquially known as the Green Guide).

6.10.2.3 Airborne and impact sound

6.10.2.3.1 Sound insulation testing

The person carrying out the building work should arrange for sound insulation testing to be carried out (by a test body with appropriate third-party accreditation) in accordance with the procedure described in Annex B of Approved Document E.

> Impact sound insulation tests should be carried out without a soft covering (e.g. carpet, foam-backed vinyl, etc.) on the stair floor.   E1.10
>
> Testing should not be carried out between living spaces, corridors, stairwells or hallways.   E1.8
>
> Test bodies conducting testing should preferably have UKAS accreditation (or a European equivalent) for field measurements.   E0.4

 Note: Some properties, for example loft apartments, may be sold before being fitted out with internal walls and other fixtures and fittings. In these cases sound insulation measurements should be made between the available spaces.

If stairs form a separating function they are subject to the same sound insulation requirements as floors. In this case, the resistance to airborne sound depends mainly on:

- the mass of the stair;
- the mass and isolation of any independent ceiling;
- the airtightness of any cupboard or enclosure under the stairs;
- the stair covering (which reduces impact sound at source).

6.10.2.3.2 Reverberation

Requirement E3 requires that *domestic buildings shall be designed and con-structed so as to restrict the transmission of echoes.* The guidance notes provided in Approved Document E cover two methods (Method A and Method B) which can be used to determine the amount of additional absorption to be used in corridors, hallways, stairwells and entrance halls that give access to flats and rooms for residential purposes. Method A is applicable to stairs and requires the following to be observed:

Cover the ceiling area with the additional absorption.	E7.10
Cover the underside of intermediate landings, the underside of the other landings, and the ceiling area on the top floor.	E7.11
The absorptive material should be equally distributed between all floor levels.	E7.12
For stairwells (or a stair enclosure), calculate the combined area of the stair treads, the upper surface of the intermediate landings, the upper surface of the landings (excluding ground floor) and the ceiling area on the top floor. Either cover an area equal to this calculated area with a class D absorber or cover an area equal to at least 50 per cent of this calculated area with a class C absorber or better.	E7.11

 Note: Method B can generally be satisfied by the use of proprietary acoustic ceilings.

6.10.2.3.3 Piped services

Piped services (excluding gas pipes) and ducts that pass through separating floors should be surrounded with sound-absorbent material for their full height and enclosed in a duct above and below the floor.

6.10.2.3.4 Junctions with floor penetrations (excluding gas pipes)

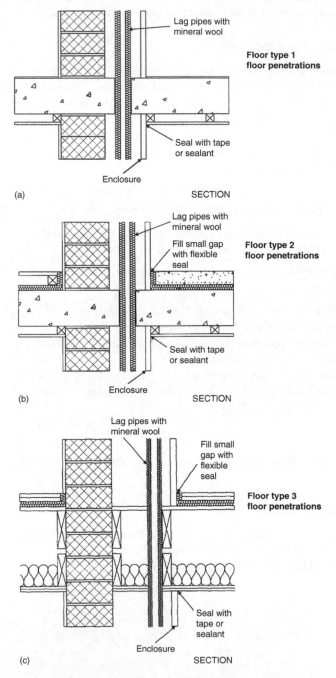

Figure 6.192 Junctions with floor penetrations (excluding gas pipes).

Pipes and ducts that penetrate a floor separating habitable rooms in different flats should be enclosed for their full height in each flat.	E3.41 E3.79 E3.117
The enclosure should be constructed of material having a mass per unit area of at least $15\,kg/m^2$.	E3.32 E3.80 E3.118
Either the enclosure should be lined or the duct (or pipe) within the enclosure should be wrapped with 25 mm unfaced mineral fibre.	E3.42 E3.80 E3.118
A small gap (sealed with sealant or neoprene) of about 5 mm should be left between the enclosure and the floating floor.	E3.81 E3.119
Where floating floor (a) or (b) is used the enclosure may go down to the floor base (provided that the enclosure is isolated from the floating layer).	E3.81 E3.119

6.10.2.3.5 Junctions with floor penetrations (including gas pipes)

Gas pipes may be contained in a separate (ventilated) duct or can remain unenclosed.	E3.43 E3.120
If a gas service is installed it shall comply with the Gas Safety (Installation and Use) Regulations 1998, SI 1998 No. 2451.	E3.43 E3.120

Note: In the Gas Safety Regulations there are requirements for ventilation of ducts at each floor where they contain gas pipes. Gas pipes may be contained in a separate ventilated duct or they can remain unducted.

6.10.2.4 Stairs and ladder

6.10.2.4.1 Steepness of stairs – rise and going

The rise and going should be in accordance with Figure 6.193.

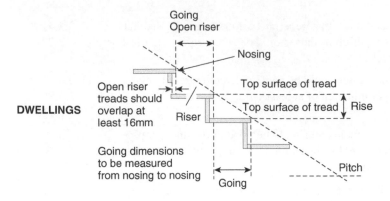

DWELLINGS

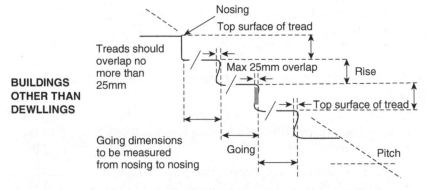

BUILDINGS OTHER THAN DEWLLINGS

Figure 6.193 Measuring rise and going.

The rise of a stair **shall** be between 150 mm and 220 mm, with   K1 1.2
any going between 220 mm and 400 mm and a maximum pitch
of 42° (see Table 1.1 of Approved Document K1).

 Note: The normal relationship between the dimensions of the rise and going is twice the rise plus the going (i.e. 2R + G).

6.10.2.4.2 Construction of steps

For all buildings:

steps should have level treads according to Table 1.1 of   K1 1.5
Approved Document F1.

For buildings other than dwellings:

> step risers should not be open, step nosings should be apparent   K1 1.6
> (without protruding over the tread below) and, if the soffit
> beneath a stair is less than 2m above floor level, the area
> beneath a stair should be protected with guarding or a barrier.

For dwellings:

> steps with open risers (especially those that are likely   K1 1.7 and 1.8
> to be used by children under five years old) should be
> constructed so that a 100 mm diameter sphere cannot
> pass through the open risers and overlap treads should
> be a minimum of 16 mm.

Steps which have more than 36 risers in consecutive flights should make at
least one change of direction, between flights, of at least 30° (see Approved
Document B2 for further details).
For common access areas in buildings that contain flats:

> step nosings should be apparent (with a suitable tread   K1 1.10
> nosing profile as shown in Figure 6.194 and risers should not
> be open.

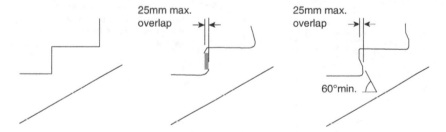

Figure 6.194 Examples of suitable tread nosings.

6.10.2.4.3 Headroom for stairs

For all buildings:

> on the access between levels, provide the minimum   K1 1.11
> headroom shown in Figure 6.195.

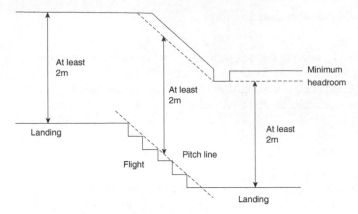

Figure 6.195 Minimum headroom.

For buildings other than dwellings:

> escape routes should have a minimum clear headroom of 2 m,   K1 1.12
> except in doorways.

For loft conversions in dwellings:

> Where there is not enough space to achieve the height shown   K1 1.13
> in Figure 6.195, provide the reduced headroom shown in
> Figure 6.196.

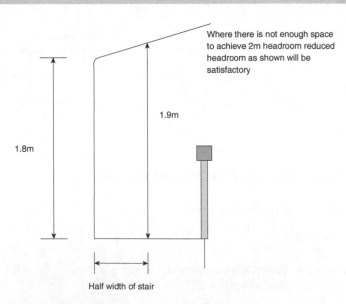

Figure 6.196 Reduced headroom for loft conversions.

6.10.2.4.4 Width of flights of stairs

For buildings other than dwellings:

> For flights of stairs which do not form part of the means of escape, provide all of the following:
>
> K1 1.15
>
> • a minimum stair width between enclosing walls, strings or upstands of 1200 mm;
> • a minimum width between handrails of 1000 mm.

 If the flight is more than 2 m wide, divide it into flights a minimum of 1000 mm wide, as shown in Figure 6.197.

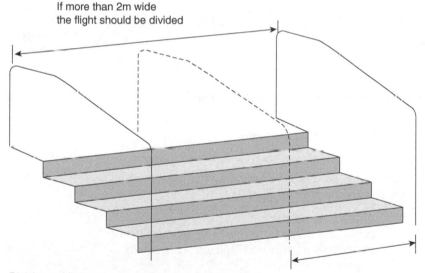

If more than 2m wide
the flight should be divided

Divisions of flights should not be more thand 2m wide,
measured between the inside edges of handrails

Figure 6.197 Dividing flights.

 Note: For stairs that form part of means of escape, refer to Approved Document B.

For dwellings:

> In exceptional circumstances where severely sloping plots are involved (and a stepped change of level within the entrance storey is unavoidable) the minimum stair width within the entrance storey of a dwelling is 900 mm.
>
> K1 1.16

6.10.2.4.5 Length of flight of stairs

For all buildings:

If stairs have more than 36 risers in consecutive flights, a minimum of one change of direction between flights shall be made as shown in Figure 6.198.	K1 1.17

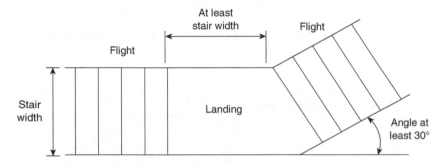

Figure 6.198 Change of direction in flights.

For buildings other than dwellings:

Single steps are disallowed and the number of risers for flights between landings should be: • utility stairs – 16 risers; • general access stairs – 12 risers (but no more than 16 in small premises where the plan area is restricted).	K1 1.18

6.10.2.4.6 Landings for stairs

For all buildings:

Landings shall be provided at the top and bottom of every flight (see Figure 6.198).	K1 1.20
A landing may include part of the floor of the building (but it should be kept clear of permanent obstructions) and it may have doors to cupboards and ducts that open over a landing at the top of a flight (see Figure 6.199), but only when they are kept locked or shut when in normal use.	K1 1.21

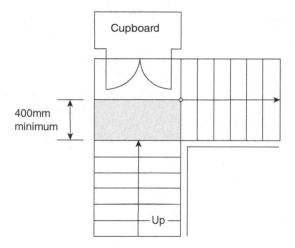

Figure 6.199 Cupboards opening on to a landing.

 Note: Also see Approved Document B.

> Landings must be level, unless the landing is at the top (or bottom) of a flight which is formed by the ground, in which case it may have a maximum gradient along the direction of travel of 1:60 provided that the surface is paved ground or permanently firm.   K1 1.22

For buildings other than dwellings:

> Landings shall have an unobstructed minimum length of 1200 mm on each landing and doors shall not swing across landings, except as stipulated in recommendation K1 1.22 (above).   K1 1.23

For dwellings:

> A door may swing across a landing at the bottom of flight, but only as shown in Figure 6.200.   K1 1.14

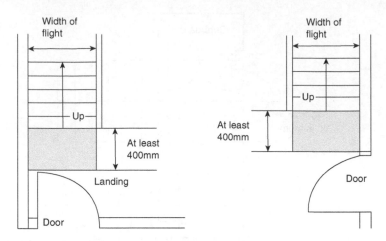

Figure 6.200 Landings next to doors in dwellings.

6.10.2.4.7 Tapered treads

The rise and going shall comply with the recommendations shown in compliance with Approved Document H (1.2 and 1.3). Measured tapered treads shall be as in Figure 6.201.	K1 1.25

Consecutive tapered treads shall use the same going.	K1 1.26
If a stair consists of straight and tapered treads, the going of the tapered treads shall not be less than the going of the straight treads.	K1 1.27

6.10.2.4.8 Spiral and helical stairs

The design of spiral and helical stairs shall be in accordance with BS 5395-2.	K1 1.28

6.10.2.4.9 Handrails for stairs

For all buildings:

If the stairs are 1000 mm or wider, there should be a handrail (whose top shall be 900 mm to 1100 mm from the pitch line or floor) on both sides and the handrail may form the top of a guarding if the heights are matched.	K1 1.34

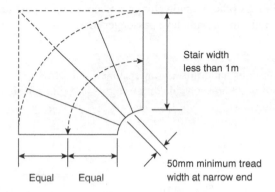

Measure going at centre of tread;
measure from curved stair line, even when tread
is in rectangular closure

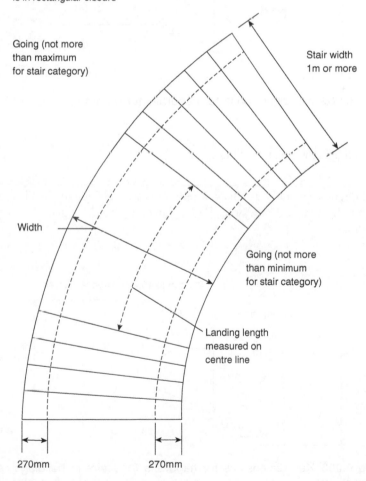

Figure 6.201 Measured tapered treads.

For buildings other than dwellings, and common access areas in buildings that contain flats and do not have passenger lifts:

> Continuous handrails, as dimensioned in Figure 6.202, shall be provided on each side of the flights and on each side of the landings.   K1 1.35

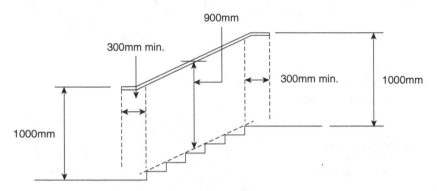

Figure 6.202 Key dimensions for handrails for common stairs in blocks of flats.

For buildings other than dwellings:

> Continuous handrails, as dimensioned in Figure 6.203, shall be provided on each side of the flights and on each side of the landings.   K1 1.35

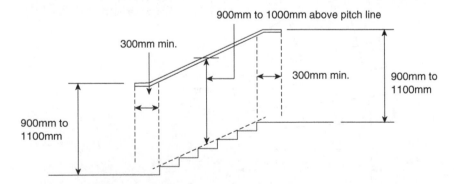

Figure 6.203 Key dimensions for handrails for stairs in buildings other than dwellings.

- A continuous handrail should be provided along the   K1 1.36
  flights and landings of a ramped or stepped flight.
- If a second (lower) handrail is provided, the vertical
  height from the pitch line of the steps (or the surface
  of the ramp) to the top of the second (lower) handrail
  should be 600 mm.
- Handrails shall not project into an access route.
- All handrails shall contrast visually with the background
  against which they are seen, without being highly
  reflective.
- The surface of the handrail must be slip-resistant and (in
  locations subject to extremely cold or hot temperatures)
  should not become excessively cold or hot to touch.
- The ends of the handrail shall be fashioned so as to
  reduce the risk of clothing being caught.

The desired handrail profile is shown in Figure 6.204.

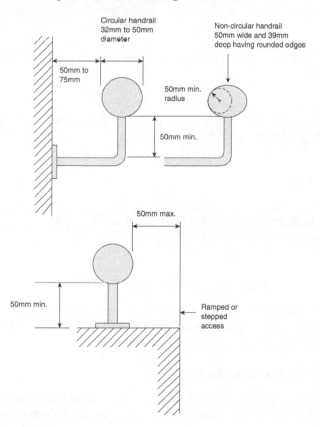

Figure 6.204 Handrail design.

In dwellings:

> Especially in exceptional circumstances where severely   K1 1.37
> sloping plots are involved (and a stepped change of
> level within the entrance storey is unavoidable), a flight
> comprising three or more risers may be provided with a
> suitable continuous handrail on each side of the flight and on
> each side of any intermediate landings.

6.10.2.4.10 Guarding of stairs

For all buildings:

> For stairs that are likely to be used by children under five   K1 1.39
> years old the construction of the guarding shall be such that
> a 100 mm sphere cannot pass through any openings in the
> guarding and children will not easily climb the guarding (see
> Figure 6.208).

For buildings other than dwellings and common access areas of buildings that contain flats:

> Guarding should be provided at the sides of flights and   K1 1.40
> landings when there are two or more risers.

In dwellings:

> Guarding should be provided at the sides of flights and   K1 1.41
> landings where there is a drop of more than 600 m.

6.10.2.4.11 Access for maintenance

Where the stairs or ladders will be used to access areas for maintenance they should comply with the provisions of BS 5395-3 and the Construction (Design and Management) Regulations 2007 as appropriate.

6.10.2.4.12 Fixed ladders

In dwellings, a fixed ladder may be used – with fixed handrails on both sides – only for access in a loft conversion that contains one habitable room.

 Retractable ladders may **not** be used as means of escape.

For industrial buildings, the design and construction of stairs, ladders and walkways shall be in accordance with BS 5395-3 or BS 4211.

6.10.2.5 Ramps

6.10.2.5.1 CONSTRUCTION OF RAMPS

For buildings other than dwellings:

• the surface of a ramp should be slip-resistant, with frictional characteristics similar to the landing and of a colour that will contrast visually with that of the landings;	K1 2.4 / K1 2.5
• in addition to guarding, on the open side of any ramp or landing a kerb should be provided that is a minimum of 100 mm high and which will contrast visually with the ramp or landing;	K1 2.6
• where the change of level is 300 mm or more, two or more clearly signed steps should be provided in addition to the ramp.	

Note: If the change in level is less than 300 mm, provide a ramp instead of a single step.

If the soffit beneath any ramp is less than 2 m above floor level, the area beneath the ramp should be protected with some sort of barrier (e.g. guarding, low-level cane detection, etc.).	K1 2.7

6.10.2.5.2 Design of ramps

For all buildings ramps and landings should be designed in accordance with Figure 6.205.

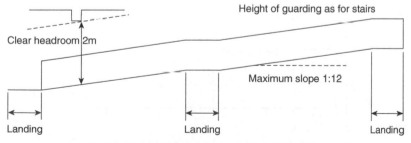

Clear headroom 2m

Height of guarding as for stairs

Maximum slope 1:12

Landing   Landing   Landing

Length of landings to be at least equal to the width of the ramp

Figure 6.205 Ramp design.

Note: For ramps that are likely to be used by children under five years old the construction of the guarding shall be such that a 100 mm sphere cannot pass through any openings in the guarding and children will not easily climb the guarding.

6.10.2.5.3 Appearance of ramps

For buildings other than dwellings, all ramps should be are readily apparent or clearly signposted.   K1 2.2

6.10.2.5.4 Obstruction of ramps

For all buildings, ramps should be kept clear of all permanent obstructions.

6.10.2.5.5 Steepness of ramps

For all buildings:

The relationship between the gradient of a ramp and its going between landings should be as shown in Figure 6.206.   K1 2.3

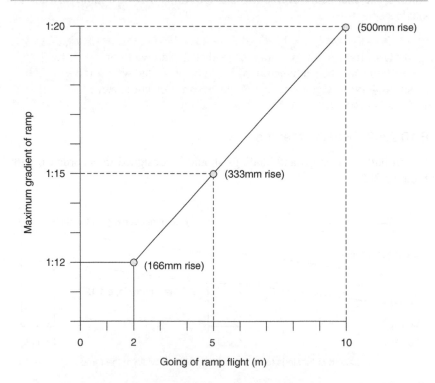

Figure 6.206 The relationship of ramp gradient and going of a flight.

6.10.2.5.6 Width of ramps

For buildings other than dwellings, for a ramp that provides access for people the width between walls, upstands and/or kerbs should be a minimum of 1500 mm.

6.10.2.5.7 Handrails for ramps

For buildings other than dwellings a handrail should be provided on both sides of the ramp.

In dwellings and common access areas in buildings that contain flats:

ramps that are less than 1000 mm wide should be provided   K1 2.12
with a handrail on one or both sides;
ramps that are 1000 mm or more wide should be provided
with a handrail on both sides;
ramps that are 600 mm or less in height do not need to be
provided with handrails;
the top of the handrails should be positioned 900 mm to
1000 mm above the surface of the ramp.

 Note: Handrails should give firm support and allow a firm grip and may form the top of the guarding.

6.10.2.5.8 Landings for ramps

For buildings other than dwellings:

- the foot and head of a ramp should be provided with   K1 2.13
  landings which are a minimum of 1200 mm long and
  are clear of any door swings or other obstructions;
- intermediate landings should be a minimum of
  1500 mm long and clear of any door swings or other
  obstructions;
- if a wheelchair user cannot see from one end of
  the ramp to the other or the ramp has three flights
  or more then 1800 mm × 1800 mm intermediate
  landings should be provided as passing places.

6.10.2.5.9 Protection from falling

Guarding should be provided whenever it is considered necessary from the point of view of safety to guard:

> - the edges of any part of a floor (including the edge   K2 3.1
>   below an opening window), gallery, balcony or roof
>   (including roof lights and other openings); and
> - any other place to which people have access; and
> - any light well, basement or similar sunken area next to
>   a building; and
> - in vehicle parks.

Note: Guarding is not required on ramps that are used only for vehicle access or in places such as loading bays where it would obstruct normal use.

> If a building is likely to be used by children under five years   K2 3.3
> old, the guarding should not have horizontal rails, it should
> stop children from easily climbing it, and the construction
> should prevent a 100 mm sphere being able to pass through
> any opening of that guarding.

6.10.2.5.9.1 DESIGN OF GUARDING

For all buildings:

> - guarding should be provided in accordance with the   K1 1.35
>   height shown in Figure 6.207;
> - any wall, parapet, balustrade or similar obstruction may
>   be used as guarding;
> - guarding must be capable of resisting, as a minimum, the
>   loads given in BS EN 1991-1-1;
> - where glazing is used as part the guarding, refer also to
>   Approved Document K5.

Building Category and location		Height (h)	
Single family dwellings	Stairs, landings, ramps, edges of internal floors	900mm for all elements	
	External balconies, including Juliette balconies and edges of roof	1100mm	
Factories and warehouses (light traffic)	Stairs, ramps	900mm	
	Landings and edges of floors	1100mm	
Residential, institutional, educational,office and public buildings	All locations	900mm for flights otherwise 1100mm	
Assembly	Within 530mm in front of fixed seating	800mm (h1)	
	All other locations	900mm for flights elsewhere 1100mm (h2)	
Retail	All locations	900mm for flights otherwise 1100mm	
Glazing in all buildings	At opening windows except roof windows in loft extensions	800mm	
	At glazing to changes of levels to provide containment	Below 800mm	

Figure 6.207 Guarding design.

Note: Typical locations for guarding are shown in Figure 6.208, and for further guidance on the design of barriers and infill panels refer to BS 6180.

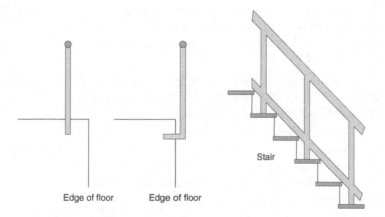

Edge of floor   Edge of floor

Stair

Figure 6.208 Typical locations for guarding.

6.10.2.5.10 Assembly buildings

To maintain sightlines for spectators, stairs and/or ramps that form part of the means of access within an assembly building such as a sports stadium, theatre or cinema should be in accordance with the following requirements:

- Ensure that the maximum pitch for gangways to seating   K1 1.4
  areas for spectators is 35°.
- Align the ends of all rows of seats/wheelchair spaces so
  that the width of the gangway remains the same.
- Provide transverse gangways that (in auditoria with tiered
  seating) do not cross.
- Ensure that stepped tiers are between 100 mm and 190 mm
  high.
- Where an exit is approached from a stepped gangway,
  place a landing the width of the exit and a minimum of
  1100 mm deep immediately in front of the exit doors.
- For stepped side gangways, provide a handrail.

 Note: Gangways should normally be 900 mm wide unless they are used by more than 50 persons, in which case they should be a minimum of 900 mm.

 Note: In areas where resistance to vandalism or low maintenance is a key factor, the use of metals with relatively low thermal conductivity is recommended.

 While there are no recommendations for minimum stair widths, designers should bear in mind the requirements of Approved Documents B (Means of escape) and M (Access and facilities for disabled people).

All common stairs should be situated within a fire-resisting enclosure (i.e. it should be a protected stairway), to reduce the risk of smoke and heat making use of the stair hazardous.	B1 2.36 (V2)
Where a common stair forms part of the only escape route from a flat, it should **not** also serve any covered car park, boiler room, fuel storage space or other ancillary accommodation of similar fire risk.	B1 2.46 (V2)
Common stairs which do not form part of the only escape route from a flat may also serve ancillary accommodation if they are separated from the ancillary accommodation by a protected lobby or a protected corridor.	B1 2.47 (V2)
If the stair serves an enclosed (i.e. non-open-sided) car park, or place of special fire hazard, the lobby or corridor should have not less than $0.4\,m^2$ permanent ventilation or be protected by a mechanical smoke-control system.	B1 2.47 (V2)

6.11 Windows

6.11.1 Requirements

Means of escape
In an emergency, the occupants of any part of the building shall be able to escape without any external assistance.
(Approved Document B1)

Ventilation
There shall be adequate means of ventilation provided for people in the building.
(Approved Document F)

Conservation of fuel and power
Reasonable provision shall be made for the conservation of fuel and power in buildings by:

(a) limiting heat gains and losses

   (i)   through thermal elements and other parts of the building fabric; and
   (ii) from pipes, ducts and vessels used for space heating, space cooling and hot water services;

(b) providing fixed building services which

   (i)   are energy efficient;
   (ii) have effective controls; and are commissioned by testing and adjusting as necessary to ensure they use no more fuel and power than is reasonable in the circumstances.
(Approved Document L)

A new version of Approved Document L came into force on 6 April 2014.

Protection against impact with glazing
Glazing, with which people are likely to come into contact whilst moving in or about the building, shall:

a.   *if broken on impact, break in a way which is unlikely to cause injury; or*
b.   *resist impact without breaking; or*
c.   *be shielded or protected from impact.*
(Approved Document K4)

A new issue of Approved Document K came into force in 2013.

Protection from collision with open windows
Provision shall be made to prevent people moving in or about the building from colliding with open windows, skylights or ventilators.
(Approved Document K5.1)

Transparent glazing
Transparent glazing, with which people are likely to come into contact while mov-
ing in or about the building, shall incorporate features which make it apparent.
(Approved Document K5.2)

Safe opening and closing of windows
Windows, skylights and ventilators which can be opened by people in or about
the building shall be so constructed or equipped that they may be opened,
closed or adjusted safely.
(Approved Document K5.3)

Safe access for cleaning windows etc.
Provision shall be made for any windows, skylights, or any transparent or
translucent walls, ceilings or roofs to be safely accessible for cleaning.
(Approved Document K5.4)

6.11.2 Meeting the requirements

6.11.2.1 Risks associated with glazing

The existence of large uninterrupted areas of transparent glazing represents
a significant risk of injury through collision. This risk is at its most severe
between areas of a building or its surroundings that are essentially at the same
level and where a person might reasonably assume direct access between locations that are separated by glazing.

The most likely places where people can sustain injuries are due to impacts
with doors or door side panels (especially between waist and shoulder level)
when initial impact can be followed by a fall through the glazing resulting in
additional injury to the face and body. Hands, wrists and arms are particularly
vulnerable. Walls and partitions are also a high-risk area, particularly where
children are concerned.

6.11.2.1.1 Protection against impact

Frequently people are injured by colliding with windows, skylights and ventilators that have been left open. In order to reduce the likelihood of this occurring:

- windows, skylights and ventilators should be installed so that projecting parts cannot come into contact with people moving in and around the
  building; and/or
- features which guide people moving in or around the building away from
  any open window, skylight or ventilator should be incorporated.

6.11.2.1.2 Projecting parts

If any part of a window, skylight and/or ventilator projects inside or outside a
building, this should be indicated as shown in Figures 6.209 and 6.210.

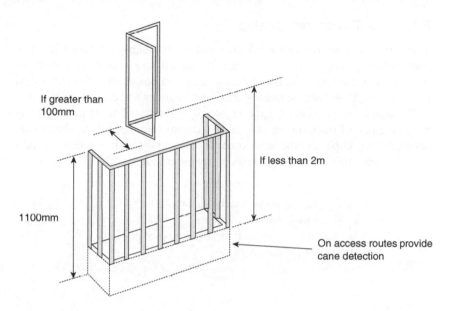

Figure 6.209 Marking by a barrier.

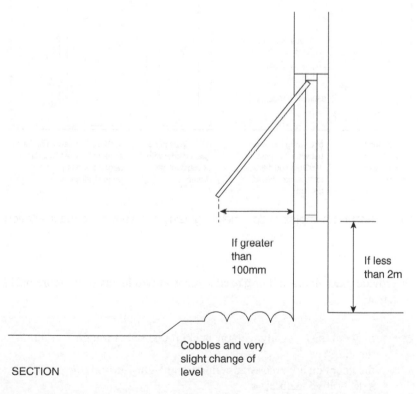

Figure 6.210 Marking by a surface.

6.11.2.1.3 Transparent glazing

Transparent glazing is another risk area and is particularly relevant when two parts of the building (or the building and its immediate surroundings) are at the same level, but separated by transparent glazing and people moving in or around a building might not see the glazing in critical locations and can collide with it!

To avoid this possibility, large uninterrupted areas of transparent glazing which are part of (or form) an internal or external wall, shop door, showroom, office, factory, public or other non-domestic building should be protected or indicated in such a manner as to avoid risk of people colliding with them, either:

> • by make the glazing immediately apparent; or   K5.2 (7.3)
> • using an alternative method of glazing, such as
>   mullions, transoms, door framing or large pull or
>   push handles (see Figure 6.211).

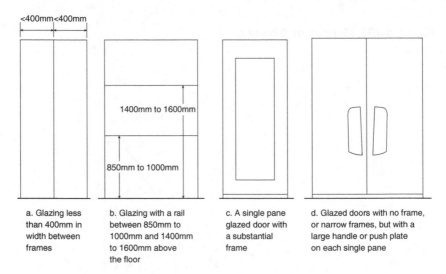

a. Glazing less than 400mm in width between frames

b. Glazing with a rail between 850mm to 1000mm and 1400mm to 1600mm above the floor

c. A single pane glazed door with a substantial frame

d. Glazed doors with no frame, or narrow frames, but with a large handle or push plate on each single pane

Figure 6.211 Examples of door-height glazing not warranting manifestation.

Or provide glass doors and/or glazed screens at two levels (see Figure 6.212) such that:

> • they contrast visually with the background seen through   K4 (7.4)
>   the glass; or
> • they contain a logo or sign (minimum height 150 mm) or
>   a decorative feature; or

- where glazed doors are beside or part of a glazed screen, they are clearly marked with a high-contrast strip at the top and on both sides; or
- where glass doors may be held open, they are protected with guarding.

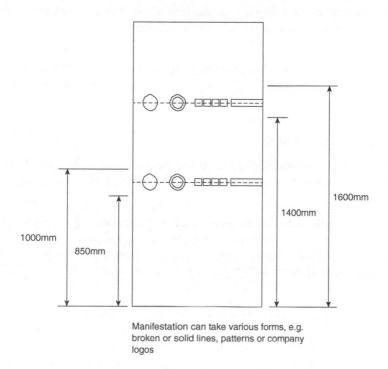

Manifestation can take various forms, e.g. broken or solid lines, patterns or company logos

Figure 6.212 Height of manifestation for glass doors and glazed screens.

6.11.2.2 Design of windows

Part of the regulations regarding the design of openable windows revolves around the actual percentage that the window needs to be opened. For example:

- if the window is designed to open **more** than 30°, then the height times the width of the opening part of a hinged or pivot window and/or sliding sash windows should be at least 1/20 of the floor area of the room.

F Tables 5.2a to 5.2d

> • if the window is designed to open **less**   F Tables 5.2a to 5.2d
> than 30°, then the height times width
> of the opening part of a hinged or pivot
> window should be at least 1/10 of the
> floor area of the room.

If a room contains more than one openable window, the areas of **all** the opening parts may be added together to achieve the required floor area.

> Cavity barriers should also be provided around   B3 6.3 (V1)
> window openings.   B3 9.3 (V2)

6.11.2.2.1 Robustness

 Some glazing materials (such as annealed glass) gain strength through thickness; others such as polycarbonates or glass blocks are inherently strong.

> The maximum dimensions for annealed glass of different   K4 (5.5)
> thicknesses for use in large areas forming fronts to shops,
> showrooms, offices, factories and public buildings with four
> edges supported are shown in Figure 6.213 to a minimum height.

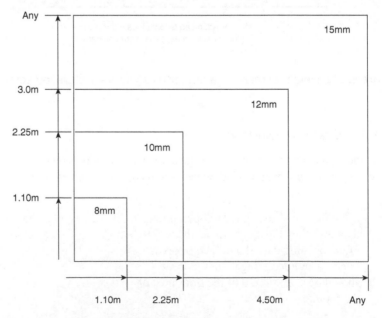

Figure 6.213 Annealed glass thickness and dimension limits.

In an impact test, a breakage is considered 'safe' if it creates a small clear opening only, with detached particles no larger than the specified maximum size (as defined in BS EN 12600 section 4 and BS 6206 clause 5.3), and disintegrates into small detached particles that are not sharp or pointed.

K4 (5.3)

6.11.2.2.2 Solar gain

Reasonable provision should be made to limit solar gains.

L1A and L2A

Solar gains are certainly beneficial in winter, as they reduce the demand for heating. However, they can cause overheating in the summer.

To overcome this problem (and whenever possible) during the summer, the effects of solar gain should be limited by a suitable combination of window size and orientation, shading, solar control measures, ventilation (day and night) and high thermal capacity.

If ventilation is provided using a balanced mechanical system, consideration should be given to providing a summer bypass function for use during warm weather (or allow the dwelling to operate via natural ventilation) so that the ventilation is more effective in reducing overheating.

The building fabric should also be constructed so that there are no reasonably avoidable thermal bridges in the insulation layers caused by gaps within the various elements, at the joints between elements, and at the edges of elements around windows and door openings. Provision should also be made to reduce unwanted air leakage through the new envelope parts.

 Although many buildings are equipped with air-conditioning for comfort, they are high energy users and their use should, whenever possible, be limited. This can be achieved by either reducing the need for air-conditioning or reducing the installed capacity of any air-conditioning system that is installed.

6.11.2.2.3 Thermal bridges

The building fabric should be constructed so that there are no reasonably avoidable thermal bridges in the insulation layers caused by gaps within the various elements, at the joints between elements, and at the edges of elements, such as those around window and door openings.

L1A 3.9

To meet this requirement, it is recommended that you use construction joint details that have been calculated following the guidance set out in BRE Report BR 497, *Conventions for Calculating Linear Thermal Transmittance and Temperature Factors*.

Use a conservative default U-value of 0.15 W/m² K, rather than linear transmittance values for each construction joint, in the dwelling CO_2 emission rate (DER) and dwelling fabric energy-efficiency (DFEE) rate calculation for dwellings.

6.11.2.2.4 Thermoplastic materials

Thermoplastic materials may be used in windows, rooflights and lighting diffusers in suspended ceilings.	B2 3.8 (V1) B2 6.10 (V2)
External windows to rooms (other than to circulation spaces) may be glazed with thermoplastic materials.	B2 3.9 (V1) B2 6.11 (V2)

6.11.2.2.5 Replacement windows

Where there is a change in the building's energy status, such work, if any, shall be carried out as to ensure that the building complies with the applicable requirements for the conservation of fuel and power.

Subsequent to a material change of use, the following will affect the building's energy status, for example where a replacement unit is provided (e.g. for an existing window, roof window, rooflight or door separating a conditioned space from an unconditioned space or external environment) which has a U-value worse than 3.3 W/m² K.

If the area of the openings in the newly created dwelling is more than 25 per cent of the total floor area, either the area of openings should be reduced or the larger area should be compensated for in some other way using the procedure described in SAP 2009.	L1B 4.15

Although the erection of a new dwelling is **not** a material change of use, the requirements for the conservation of fuel and power still apply where a dwelling is being created in an existing building as the result of a material change of use of all, or part, of the building.

Where windows are to be replaced, the replacement work should comply with the requirements of Parts L and K.	B1 2.19

> If a window is currently located where, in the case of   B1 2.19
> a new dwelling house, an escape window would be
> necessary, the replacement window opening should be
> sized to provide at least the same potential for escape as
> the window it replaces.

Note: If the original window is larger than necessary (i.e. for escape purposes) the window opening **could** be reduced to a minimum of 450 mm high and 450 mm wide, provided that it has an unobstructed openable area of at least 0.33 m^2 and the openable area is not more than 1100 mm above the floor.

Although the installation of replacement windows or glazing (e.g. by way of repair) is not considered as building work under Regulation 3 of the Building Regulations, glazing that:

* is installed in a location where there was none previously;
* is installed as part of an erection;
* is installed as part of an extension or material alteration of a building

is nevertheless subject to the requirements of these regulations.

6.11.2.2.6 Ventilation

Habitable rooms **without** openable windows may be ventilated through another habitable room (e.g. an internal room) **provided** that the other room has:

> * purge ventilation; and   F 5.14
> * an 8000 mm^2 background ventilator; and
> * there is a permanent opening between the two rooms.

Habitable rooms **without** openable windows may also be ventilated through a conservatory provided that the conservatory has:

> * purge ventilation;   F 5.15
> * an 8000 m^2 background ventilator; and
>
> there is a closable opening between the room and the
> conservatory that is equipped with:
>
> * purge ventilation; and
> * an 8000 mm^2 background ventilator.

Windows with night latches should **not** be used, as they are more liable to draughts as well as being a potential security risk. F 4.19

If a fan is installed in an internal room or an office **without** an openable window, the fan should have a 15-minute overrun. F Table 5.2c

6.11.2.3 Safety requirements

6.11.2.3.1 Prevention of falls

Where a person might fall through a window that is above ground-floor level, opening limiters should be provided to restrain the window sufficiently to prevent such falls, or guarding should be used. K5.3 (8.2)

6.11.2.3.2 Safe opening and closing of windows

Windows, skylights and ventilators that can be opened by people should be capable of being opened, closed or adjusted safely by ensuring that:

the controls for operating the windows, etc. are positioned as shown in Figure 6.214. K5.3 (8.1)

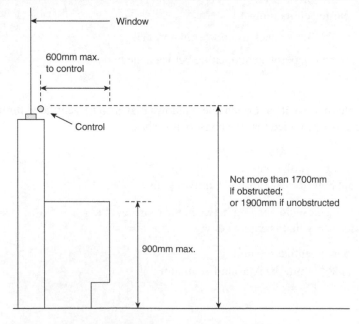

Figure 6.214 Height of controls.

Note: Where this proves to be impossible or impracticable, a safe (manual or electrical) opening system may be used. (Also see Approved Document M for further advice.)

6.11.2.3.3 Safe access for cleaning

All windows and skylights (or any transparent or translucent walls, ceilings or roofs) of a building should be safely accessible for cleaning and, where a person may fall from a window, safe means of access for cleaning **both** sides of the glass should be provided.

Where a person standing on a permanent stable surface cannot safely clean a glazed surface:

windows should be designed so that people can safely clean the outside safely from inside the building (see Figure 6.215).	K5.4 (9.1a)

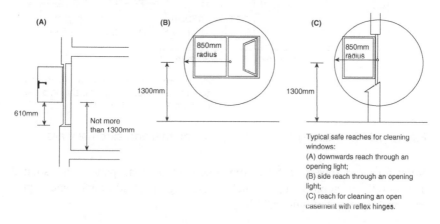

Figure 6.215 Safe reaches for cleaning.

In buildings where this is not possible, then one of the following should be provided:

• ladders; or	K5.4 (9.1b)
• access equipment (such as suspended cradles or travelling ladders, with attachments for safety harnesses); or	K5.4 (9.1c) K5.4 (9.1d) K5.4 (9.1e)
• suitable anchorage points for safety harnesses or abseiling hooks; or	
• walkways (at least 400 mm wide and with guarding or with anchorages for sliding safety harnesses.	

If none of the above alternatives is possible, then scaffolding towers (portable or fixed) will have to be used.

6.11.2.3.4 Glazing in critical locations

In critical locations (see Figure 6.216) all buildings should:

> • permanently protect glazing;   K4 (5.2)
> • choose glazing that is in small panes and is robust;
> • ensure that glazing, if it breaks, will break safely.

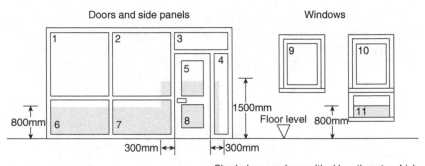

Shaded areas show critical locations to which requirement K4 applies (i.e. glazing in areas numbered 2, 4, 5, 6, 7, 8, 11)

Figure 6.216 Critical glazing locations in internal and external walls.

Note: for the purposes of this recommendation, a 'small pane' is an isolated pane or one of a number of panes held in glazing bars, traditional leaded lights or copper lights (see Figure 6.217).

If glazing in a critical location is protected by a permanent screen (see Figure 6.218) then the glazing itself does not need to comply with requirement K4. The permanent screen should, however, comply with all of the following:

> • prevent a sphere of 75 mm from coming into contact   K4 (5.8)
>   with the glazing;
> • be robust; and
> • if it protects glazing installed to help prevent people
>   from falling, be difficult to climb (e.g. no horizontal
>   rails).

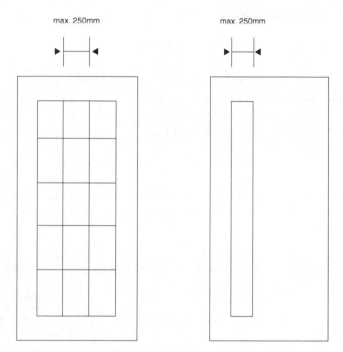

max. 250mm   max. 250mm

Maximum area of single pane not to exceed 0.5m², small panes of annealed glass should not be less than 6mm thick

Figure 6.217 Dimensions and areas of small panes.

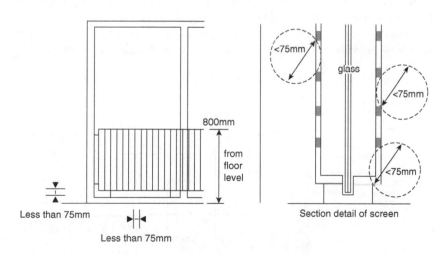

800mm
from floor level

Less than 75mm

Less than 75mm

<75mm

glass

<75mm

<75mm

Section detail of screen

Figure 6.218 Permanent screen protection.

6.11.2.4 Means of escape

6.11.2.4.1 Emergency egress windows

Except for kitchens, all habitable rooms on the ground floor (and the upper storey(s) of a dwelling house that is served by only one stair) should be provided with an emergency egress window.	B1 2.3 and 2.4 (V1) B1 2.11 and 2.12 (V2)

Note: There are some other alternatives if this is not possible, such as having access to a protected route, so it is best to see the regulations if you need confirmation of this point.

An emergency egress window should be at least 450 mm high and 450 mm wide and have an unobstructed openable area of at least 0.33 m².	B1 2.8 (V1)
The bottom of the openable area should be not more than 1100 mm above the floor.	B1 2.9 (V2)
The window should enable the person escaping to reach a place free from danger of fire (e.g. a courtyard or back garden which is at least as deep as the dwelling house is high (Figure 6.219).	

Approved Document K (Protection from falling, collision and impact) specifies a minimum guarding height of 800 mm.

Locks (with or without removable keys) and stays may be fitted to egress windows, **provided** that the stay is fitted with a child-resistant release catch.

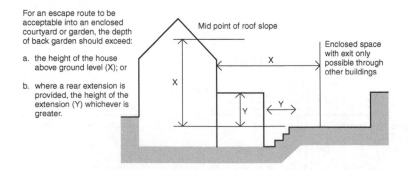

For an escape route to be acceptable into an enclosed courtyard or garden, the depth of back garden should exceed:

a. the height of the house above ground level (X); or

b. where a rear extension is provided, the height of the extension (Y) whichever is greater.

Mid point of roof slope

Enclosed space with exit only possible through other buildings

Figure 6.219 Ground or basement storey exit into an enclosed space.

Emergency egress windows should be designed so that they remain in the open position without needing to be held open by the person making an escape.

Any inner room that is a kitchen, laundry or utility room, dressing room, bathroom, WC or shower room situated not more than 4.5 m above ground level and whose only escape route is through another room shall be provided with an emergency egress window.	B1 2.9 (V1) B1 2.5 (V2)
All galleries shall be provided with an alternative exit or, where the gallery floor is not more than 4.5 m above ground level, an emergency egress window.	B1 2.12 (V1) B1 2.8 (V2)
All basement storeys in a dwelling house that contain a habitable room shall be provided with either an external door or a window suitable for egress from the basement.	B1 2.13 (V1) B1 2.6 (V2)

Note: There are certain alternatives. See the regulations if you require further confirmation of these points.

6.11.2.5 Conservation of fuel and power

Following the publication of new Approved Documents L1A and L1B in November 2013, with effect from April 2014:

* all **new homes** (i.e. dwellings) need to achieve, or better, a fabric energy-efficiency target in addition to the strengthened requirement to deliver **6 per cent carbon dioxide savings**; and
* all **new non-domestic buildings** need to make **9 per cent carbon dioxide savings**.

Note: Regulation 258 (Nearly zero-energy requirements for new buildings) will **not** come into force until 2019 at the earliest.

There have also been changes in the legal requirements for the conservation of fuel and power. Regulation 9 of the Building Regulations exempts some conservatory and porch extensions from the energy-efficiency requirements as long as:

* they are at ground level and under $30\,m^2$;
* glazing and any fixed electrical installations comply with the appropriate sections of Building Regulations;
* the conservatory is separated from the house by external-quality walls, doors or windows;
* there is an independent heating system with separate temperature and on/off controls;

- the front entrance door between the existing house and the new porch remains in place.

It is recommended that the total area of windows, roof windows and doors in extensions to existing buildings does not exceed the sum of:

- 25 per cent of the floor area of the extension; and
- the total area of any windows or doors which, as a result of the extension works, no longer exist or are no longer exposed.

6.11.2.5.1 Energy performance certificates

Regulation 29 of the Building Regulations requires that, when a dwelling or a new building is erected, the person carrying out the work must give an energy performance certificate to the owner of the building and a notice to the building control body (BCB) that a certificate has been given, including the reference number under which the certificate has been registered.

See www.legislation.gov.uk and www.planningportal.gov.uk for further guidance.

Note: Responsibility for achieving compliance with the requirements of Approved Document L rests with the person carrying out the work. That person may be, for example, a developer, a main (or sub-) contractor, or a specialist firm directly engaged by a private client.

The person responsible for achieving compliance should either provide a certificate him- or herself or obtain a certificate from the subcontractor that commissioning has been successfully carried out. The certificate should be made available to the client and the building control body.

6.11.2.6 Building fabric

The building fabric should be constructed to a reasonable standard so that:	L1A 3.2 L2A 3.2
<ul><li>the insulation is reasonably continuous over the whole building envelope; and</li><li>the air permeability is within reasonable limits.</li></ul>	

Reductions in thermal performance can occur where the air barrier and the insulation layer are not contiguous and the cavity between them is subject to air movement. To avoid this problem, either:

- the insulation layer should be contiguous with the air barrier at all points in the building envelope; or

- the space between the insulation layer and air barrier should be filled with solid material such as in a masonry wall.

Note: The method for demonstrating compliance with the requirements of Approved Document L are fully explained in Appendix B, which is available at www.routledge.com/9780415721714.

Further guidance may also be obtained through:

- the Planning Portal website: www.planningportal.gov.uk;
- your local authority building control service or an approved inspector (if you are the person undertaking the building work);
- the scheme operator (if you are registered with a competent person scheme);
- a specialist or an industry technical body for the relevant subject (if your query is of a highly technical nature).

6.11.2.6.1 Controlled fittings

In the context of conservation of fuel and power, the application of the term *controlled fitting* to a window, roof window, rooflight or door refers to a whole unit, i.e. **including** the frame. Consequently, replacing the glazing whilst retaining an existing frame is not providing a controlled fitting, and so such work is **not** notifiable and does not have to meet the Approved Document L standards – although, where practical, it would naturally be sensible to do so! Similar arguments apply to a new door in an existing frame.

6.11.2.6.2 Dwellings

Where a dwelling is erected, it shall not exceed the target fabric energy-efficiency rate for the dwelling.	L1A 2.8
The dwelling CO_2 emission rate (DER) and the dwelling fabric energy-efficiency (DFEE) rate must be no worse than the target emissions rate (TER) and target fabric energy-efficiency (TFEE) rate.	L1A 2.9
Dwellings should be constructed and equipped so that performance is consistent with the calculated DER and DFEE rate.	L1A 3.1
The final DER and DFEE rate must be based on the building as constructed, incorporating: • any changes to the list of specifications that have been made during construction; and • the assessed air permeability.	L1A 2.9
All windows, roof windows, rooflights and/or doors should be provided with draught-proofed units, the performance of which is no worse than as given in Table 6.77.	L1B 4.19

Note: The final DER and DFEE rate must be based on the building as constructed, incorporating:

- any changes to the list of specifications that have been made during construction; and
- the assessed air permeability.

Table 6.77 Standards for control fittings

Fitting	Standard (W/m² K)
Windows, roof windows or rooflights	1.60
Doors with >50% of internal face glazed	1.80
Other doors	1.80

Where replacement windows are unable to meet the requirements of Table 6.77 (e.g. because of the need to maintain the external appearance of the façade and/ or the character of the building) replacement windows should meet a centre pane U-value of 1.2 W/m² K, or single glazing should be supplemented with low-e secondary glazing. In the latter case, the weather stripping should be on the secondary glazing to minimize condensation risk between the primary and secondary glazing.

U-values shall be calculated using the methods and conventions set out in BR 443 (*Conventions for U-value Calculations*), and should be based on the whole element or unit (e.g. in the case of a window, it would be the combined performance of the glazing and the frame).

For windows, the U-value needs to be: L1A 2.34
                                                               L1B 4.19
- the smaller of the two standard windows defined in BS EN 14351-1; or
- the standard configuration set out in BR 443; or
- the specific size and configuration of the actual window.

Note: Table 6.78 sets out the worst acceptable standards for fabric properties, but, whilst the stated value represents the area-weighted average value for **all** elements of that type in general, it is likely to require significantly better fabric performance than is set out in Table 6.78.

Table 6.78 Limiting fabric parameters

Fitting	W/m² K
Roof	0.20
Wall	0.30
Floor	0.25
Party wall	0.20
Windows, roof windows, glazed rooflights, curtain walling and pedestrian doors	2.00
Air permeability	10.00 m³/h m² at 50 Pa

If a window is enlarged or a new one created, the area of windows, roof windows, rooflights and doors should not exceed 25 per cent of the total floor area of the **dwelling** unless compensating measures are included elsewhere in the work

L1B 4.23

Note: Each fixed building service should be at least as efficient as the minimum acceptable value for the particular type of service, as set out in DCLG's *Domestic Building Services Compliance Guide*.

6.11.2.6.3 Buildings other than dwellings

Where a building is erected, it shall not exceed the target CO_2 emission rate for the building.

L2A 2.7

Buildings should be constructed and equipped so that performance is consistent with the calculated building efficiency rate (BER).

L2A 3.1

This calculated BER shall be submitted to the BCB after completion to take account of any changes in performance between design and construction.

The energy-efficiency requirements given in Table 6.79 are relevant to buildings other than dwellings.

Table 6.79 Standards for controlled fittings in conservatories and porches

Fitting	Standard (W/m² K)
Windows, roof windows and glazed rooflights	1.8 for the whole unit
Alternative option for windows in buildings that are essentially domestic in character	A window energy rating of band C
Plastic rooflight	1.8
Curtain walling	<1.8
Pedestrian doors where the door has more than 50% of its internal face area glazed	1.8 for the whole unit
Other doors	1.8
Roof ventilators (including smoke extract ventilators)	3.5

6.11.2.6.4

U-values shall be calculated using the methods and conventions set out in BR 443 (*Conventions for U-value Calculations*) (http://www.bre.co.uk/filelibrary/pdf/rpts/BR_443_(2006_Edition).pdf), and should be based on the whole element or unit (e.g. in the case of a window, it would be the combined performance of the glazing and the frame).

> For windows, the U-value needs to be: L1B 4.19
>
> • the smaller of the two standard windows defined in BS EN 14351-1; or
> • the standard configuration set out in BR 443; or
> • the specific size and configuration of the actual window.

 Note: Table 6.80 sets out the worst acceptable standards for fabric properties.

 The stated value represents the area-weighted average value for **all** elements of that type.

Table 6.80 Limiting fabric parameters (non-domestic buildings)

Fitting	W/m² K
Roof	0.25
Wall	0.35
Floor	0.25
Windows, roof windows, glazed rooflights, curtain walling and pedestrian doors	2.20
Air permeability	10.00 m³/h m² at 50 Pa

Where windows, roof windows, rooflights or doors are
to be provided, reasonable provision would be to install:

L2B 4.24

- draught-proofed units whose area-weighted average
  performance is no worse than that given in Table
  6.81; and

- insulated cavity closers where appropriate.

Where the replacement windows are unable to meet the requirements of Table
6.81 because of the need to maintain the external appearance of the façade or
the character of the building, replacement windows should meet a centre pane
U-value of 1.2 W/m^2 K, or single glazing should be supplemented with low-e
secondary glazing.

Table 6.81 Standards for controlled fittings

Fitting	Standard (W/m^2 K)
Windows, roof windows and glazed rooflights	1.8 for the whole unit
Alternative option for windows in buildings that are essentially domestic in character	A window energy rating of band C
Plastic rooflight	1.8
Curtain walling	<1.8
Pedestrian doors where the door has more than 50% of its internal face area glazed	1.8 for the whole unit
High-usage entrance doors for people	3.5
Vehicle access and similar large doors	1.5
Other doors	1.8
Roof ventilators (including smoke extract ventilators)	3.5

If a window, pedestrian door or rooflight is enlarged (or
a new one is created), the area of windows and pedestrian
doors and of rooflights expressed as a percentage of the
total floor area of the building should not exceed the relevant value from Table 6.82, or should be compensated for
in some other way.

L2B 4.24

Table 6.82 Opening areas in the extension

Building type	Windows and personnel doors as percentage of exposed wall	Rooflights as percentage of area of roof
Residential buildings where people temporarily or permanently reside	30	20
Places of assembly, offices and shops	40	20
Industrial and storage buildings	15	20
Vehicle access doors and display windows and similar glazing	As required	n/a
Smoke vents	n/a	As required

Note: In certain classes of building with high internal gains, a less demanding U-value for glazing may be an appropriate way of reducing overall CO_2 emissions, in which case the average U-value for windows, doors and rooflights can be relaxed from the values given in Table 6.81, but the value should not exceed 2.7 W/m^2 K.

6.11.2.6.5 Historic and traditional buildings

Building inspectors normally require the use of 'sympathetic treatment' when restoring the historic character of a building that has been subject to previous inappropriate alteration (e.g. replacement doors).

Particular issues that could warrant sympathetic treatment – and where advice from others would probably be beneficial – include:

> • restoring the historic character of a building that has been subject to previous inappropriate alteration (e.g. replacement windows, doors and rooflights).　　　　L2B 3.11

6.11.2.6.6 Conservatories and porches

If a conservatory or porch is installed at the same time as the construction of a new dwelling and:

> • adequate thermal separation (see Tables 6.83 and 6.84) is provided between the dwelling and the conservatory or porch, and the dwelling's heating system is not extended into the conservatory or porch, then follow the guidance in Approved Document L1B;　　　　L1A 2.29

- no, or inadequate, thermal separation is included   L1A 2.29
  between the dwelling and the conservatory or porch,
  or the dwelling's heating system is extended into
  the conservatory or porch, follow the guidance from
  Approved Document L1B, including a TER/DER and
  TFEE/DFEE rate calculation.

Regulation 9 of the Building Regulations exempts some conservatory and porch extensions from the energy-efficiency requirements.

This exemption applies only to conservatories or porches:

- which are at ground level;
- where the floor area is less than $30\,m^2$;
- where the existing walls, doors and windows which separate the conservatory from the building are retained or, if removed, are replaced by walls, windows and doors which meet the energy-efficiency requirements; and
- where the heating system of the building is not extended into the conservatory or porch.

If a new conservatory or porch does not meet all these requirements it is **not** exempt and must comply with the relevant energy-efficiency requirements of Approved Document L.

In accordance with Approved Document L, conservatories or porches should provide:

- effective thermal separation between the heated area in the existing dwelling (i.e. the walls, doors and windows between the dwelling and the extension should be insulated and draught-proofed to at least the same extent as in the existing dwelling);
- glazed elements that meet the standards set out in Table 6.83 and opaque elements that meet the standards set out in Table 6.84.

Removing, and not replacing, any or all of the thermal separation between the dwelling and an existing exempt extension, or extending the dwelling's heating system into the extension, means the extension ceases to be exempt and should be treated as a conventional extension.

Table 6.83 Standards for control fittings in conservatories and porches (dwellings)

Fitting	Standard (W/m^2 K)
Window, roof window or rooflight	1.6
Doors with >50% of internal face glazed	1.8
Other doors	1.8

Table 6.84 Standards for new thermal elements in conservatories and porches (dwellings)

Element	Standard (W/m² K)
Wall	0.28
Pitched roof – insulation at ceiling level	0.16
Pitched roof – insulation at rafter level	0.18
Flat roof or roof with integral insulation	0.18
Floor	0.22
Swimming pool basin	0.25

6.11.2.6.7 Building extensions

In a dwelling, the total area of windows, roof windows and doors in extensions shall not exceed the sum of:

•   25 per cent of the floor area of the extension; and
•   the total area of any windows or doors.

> In non-domestic buildings, the area of windows and roof-   L2B 4.4
> lights in the extension should not exceed the values given in
> Table 6.85.
>
> However, where a greater proportion of glazing is present in
> the part of the building to which the extension is attached,
> reasonable provision would be to limit the proportion of glaz-
> ing in the extension so that it is no greater than the proportion
> that exists in the part of the building to which it is attached.

Table 6.85 Opening areas in the extension of a non-domestic building

Building type	Windows and personnel doors as percentage of exposed wall	Rooflights as percentage of roof area
Residential buildings where people temporarily or permanently reside	30	20
Places of assembly, offices and shops	40	20
Industrial and storage buildings	15	20
Vehicle access doors and display windows and similar glazing	As required	n/a
Smoke vents	n/a	As required

If the additional room is connected to an existing habitable   F 7.8ai
room that now has no windows that open to the outside (or
if it still has windows opening to outside, but with a total
background ventilator equivalent area less than 5000 mm^2),
the ventilation opening (or openings) shall be greater than
8000 mm^2 equivalent area.

If the additional room is connected to an existing habitable   F 7.8aiii
room that still has windows opening to the outside (but
with a total background ventilator equivalent area of at least
5000 mm^2 equivalent area), there should be:

* background ventilators of at least 8000 m^2 equivalent area
  between the two rooms; and
* background ventilators of at least 8000 mm^2 equivalent
  area between the additional room and the outside.

6.11.2.6.7.1 FABRIC STANDARDS

Reasonable provision would be for the proposed extension to   L2B 4.3
include the following (**provided** that they meet the standards
set by the Building Regulations):

* doors, windows, roof windows, rooflights and smoke vents;
* newly constructed thermal elements; and
* existing opaque fabric which becomes a thermal element
  (and which has been upgraded).

6.11.2.6.7.2 OPENING AREAS

The area of windows and rooflights in the extension should   L2B 4.4
generally not exceed the values given in Table 6.86.

Table 6.86 Opening areas in the extension

Building type	Windows and personnel doors as percentage of exposed wall	Rooflights as percentage of area of roof
Residential buildings where people temporarily or permanently reside	30	20
Places of assembly, offices and shops	40	20
Industrial and storage buildings	15	20
Vehicle access doors and display windows and similar glazing	As required	n/a
Smoke vents	n/a	As required

Note: As well as satisfying the energy-efficiency requirements in respect of the material change of use or change in energy status, such building work may be one of the triggers for consequential improvements (see section 6.11.2.6.8).

6.11.2.6.8 Consequential improvements

If a building has a total useful floor area greater than $1000\,\text{m}^2$ and the proposed building work includes:

- an extension;
- the initial provision of any fixed building services; or
- an increase to the installed capacity of any fixed building services

then, in addition to the proposed building work (i.e. the principal works), consequential improvements will have to be completed provided that they are technically, functionally and economically feasible.

For example, replacing existing windows, roof windows or rooflights (but excluding display windows) or doors (but excluding high-usage entrance doors) which have a U-value worse than $3.3\ \text{W/m}^2\ \text{K}$ will achieve a simple payback within 15 years provided that they are economically feasible and there are no unusual circumstances.

Note: *Consequential improvements* means those energy-efficiency improvements required by Regulation 28.

Where the installed capacity per unit area of a **heating system** is increased:	L2B 6.10a
• the thermal elements within the area served which have U-values worse than those set out in column (a) of Table 6.87 should be upgraded; and	
• existing windows, roof windows or rooflights (but excluding display windows) or doors (but excluding high-usage entrance doors) within the area served and which have U-values worse than $3.3\ \text{W/m}^2\ \text{K}$ should be replaced.	L2B 6.10b
Where the installed capacity per unit area of a **cooling system** is increased:	L2B 6.11a
• thermal elements within heated areas which have U-values worse than those set out in column (a) of Table 6.87 should be upgraded; and	

- if the area of windows and roof windows (but   L2B 6.11b
  excluding display windows) within the area served
  exceeds 40 per cent of the façade area (or the area of
  rooflights exceeds 20 per cent of the area of the roof)
  and the design solar load exceeds 25 W/m², the solar
  control provisions should be upgraded such that at
  least one of the following four criteria is met:

  (1) the solar gain per unit floor area averaged over
      the period 06:30 to 16:30 GMT is not greater
      than 25 W/m² when the building is subject to
      solar irradiances for July as given in the table of
      design irradiancies in *CIBSE Design Guide A*;

  (2) the design solar load is reduced by at least 20 per
      cent;

  (3) the effective g-value is no worse than 0.3;

  (4) the zone (or zones) satisfies the criterion 3
      check in Approved Document L2A – based on
      calculations by an approved software tool.

  Note: Any general lighting system within the area
  served by the relevant fixed building service which has
  an average lamp efficacy of less than 45 lamp-lumens per
  circuit-watt should be upgraded with new luminaires and/
  or controls following the guidance in the *Non-Domestic
  Building Services Compliance Guide*.

Table 6.87 Upgrading retained thermal elements

Element	(a) Threshold U-value W/m² K	(b) Improved U-value W/m² K
Wall – cavity insulation	0.70	0.55
Wall – external or cavity insulation	0.70	0.30
Floor	0.70	0.25
Pitched roof – insulation at ceiling level	0.35	0.16
Pitched roof – insulation between rafters	0.35	0.18
Flat roof or roof with integral insulation	0.35	0.18

All existing windows (less display windows), roof   L2B 6.10b
windows, rooflights or doors (excluding high-usage
entrance doors) that are within the area served by the fixed
building service and which have a U-value worse than 3.3
W/m² K should be replaced.

Where the original windows were fitted with trickle   F 7.3
ventilators, the replacement windows should include them
and they should be sized as set out in Table 6.88.

Table 6.88 Equivalent areas for replacement windows – dwellings

Type of room	Equivalent area (mm²)
Habitable rooms	2500
Kitchen	2500
Utility room	2500
Bathroom (without WC)	2500

Where the original windows were not fitted with trickle ventilators and the
room is not ventilated adequately by other installed provisions, it would be
good practice to fit trickle ventilators (or an equivalent means of ventilation) to
help with control of condensation and improve indoor air quality. Ventilation
devices should be fitted with accessible controls.

Where there was no previous ventilation opening, or where   F 7.6
the size of the original ventilation opening is not known, the
replacement window(s) **shall** be greater than the minimum
requirements shown in Tables 6.88 and 6.89.

Table 6.89 Equivalent areas for replacement windows – buildings other than dwellings

Type of room	Equivalent area (mm²)
Occupiable rooms with a floor area 10 m²	2500
Occupiable rooms with a floor area 10 m²	250 per m² of floor area
Kitchens (domestic type)	2500
Bathrooms and shower rooms	2500 per bath or shower
Sanitary accommodation (and/or washing facilities)	2500 per WC

6.12 Doors

6.12.1 Requirements

Ventilation
There shall be adequate means of ventilation provided for people in the building.

(Approved Document F)

Conservation of fuel and power
Reasonable provision shall be made for the conservation of fuel and power in buildings by:

(a) *limiting heat gains and losses*

    (i) *through thermal elements and other parts of the building fabric; and*

    (ii) *from pipes, ducts and vessels used for space heating, space cooling and hot water services;*

(b) *providing fixed building services which*

    (i) *are energy efficient;*

    (ii) *have effective controls; and are commissioned by testing and adjusting as necessary to ensure they use no more fuel and power than is reasonable in the circumstances.*

(Approved Document L)

A new version of Approved Document L came into force on 6 April 2014.

Fire safety

•   *There shall be sufficient escape routes that are suitably located to enable persons to evacuate the building in the event of a fire.*

•   *Safety routes shall be protected from the effects of fire.*

•   *In an emergency, the occupants of any part of the building shall be able to escape without any external assistance.*

(Approved Document B1)

Access to and use of buildings
Reasonable provisions shall be made for people to gain access to and use the building and its facilities.

(Approved Document M)

Amendments to the 2004 version of Approved Document M came into force in 2013.

In addition to the requirements of the Equality Act 2010, precautions need to be taken to ensure that:

- new non-domestic buildings and/or dwellings (e.g. houses and flats used for student living accommodation, etc.);
- extensions to existing non-domestic buildings;
- non-domestic buildings that have been subject to a material change of use (e.g. so that they become a hotel, boarding house, institution, public building or shop)

are capable of allowing people, regardless of their disability, age or gender, to:

- gain access to buildings;
- gain access within buildings;
- be able to use the facilities of the buildings (both as visitors and as people who live or work in them);
- use sanitary conveniences in the principal storey of any new dwelling.

Transparent glazing
Transparent glazing, with which people are likely to come into contact while moving in or about the building, shall incorporate features which make it apparent.
(Approved Document K5.2)

A new issue of Approved Document K came into force in 2013.

Protection against impact from and trapping by doors
Provision shall be made:

a.   *to prevent any door or gate:*

- *which slides or opens upwards, from falling onto any person; and*
- *which is powered, from trapping any person.*

b.   *for powered doors and gates to be opened in the event of a power failure.*
c.   *to ensure a clear view of the space on either side of a swing door or gate.*
(Approved Document K6)

Note: Requirement K6 does not apply to dwellings, or any door or gate which is part of a lift.

6.12.2 Meeting the requirements

6.12.2.1 Ventilation

To ensure ventilation throughout the dwelling, there shall be an undercut of 7600 mm² (minimum) in all internal doors above the floor finish (equivalent to an undercut of 10 mm for a standard 760 mm width door).	F Table 5.2a
The height times the width of an external door (including patio doors) should be at least 1/20 of the floor area of the room.	F App B

> **Note:** If a room contains more than one external door (or a combination of at least one external door and at least one openable window), the areas of **all** the opening parts may be added together to achieve at least 1/20 of the floor area of the room.

6.12.2.2 Conservation of fuel and power

 Following the publication of new Approved Documents L1A and L1B in November 2013, with effect from April 2014:

- all **new homes** (i.e. dwellings) need to achieve, or better, a fabric energy-efficiency target in addition to the strengthened requirement to deliver **6 per cent carbon dioxide savings**; and
- all **new non-domestic buildings** need to make **9 per cent carbon dioxide savings**.

Energy-efficiency measures must, therefore, be provided that limit the heat loss through the doors, etc. by suitable means of insulation.

Where a dwelling is erected, it shall not exceed the target fabric energy-efficiency rate for the dwelling.	L1A 2.8
Where a building is erected, it shall not exceed the target CO_2 emission rate for the building.	L2A 2.7
Each fixed building service should be at least as efficient as the minimum acceptable value for the particular type of service, as set out in DCLG's *Domestic Building Services Compliance Guide*.	L1A 2.36

Responsibility for achieving compliance with the requirements of Approved Document L for the conservation of fuel and power rests with the main (or sub-) contractor, or a specialist firm directly engaged by a private client.

The person responsible for achieving compliance should either provide a certificate him- or herself or obtain a certificate from the subcontractor that commissioning has been successfully carried out. The certificate should be made available to the client and the building control body.

6.12.2.2.1 Extensions

It is recommended that the total area of windows, roof windows and doors in extensions to existing buildings does not exceed the sum of:

- 25 per cent of the floor area of the extension; and
- the total area of any windows or doors which, as a result of the extension works, no longer exist or are no longer exposed.

Note: If the total floor area of the proposed extension exceeds these limits, the work should be regarded as a **new building** and the requirements of Approved Documents L2A and L2B should be used.

The building fabric should be constructed so that there are no reasonably avoidable thermal bridges in the insulation layers caused by gaps within the various elements, at the joints between elements, and at the edges of elements, such as those around door openings	L1A 3.9

In addition, doors between the building and an extension should be insulated and weather-stripped to at least the same extent as in the existing building.

To meet this requirement, it is recommended that you use construction joint details that have been calculated following the guidance set out in BRE Report BR 497, *Conventions for Calculating Linear Thermal Transmittance and Temperature Factors*.

6.12.2.2.2 U-values

U-values shall be calculated using the methods and conventions set out in BR 443 (*Conventions for U-value Calculations*), and should be based on the whole element or unit (e.g. in the case of a window, it would be the combined performance of the glazing and the frame).

For doors, the U-value needs to be: • the standard size as laid out in BS EN 14351-1; or • the specific size and configuration of the actual door.	L1A 2.34 L2A 2.40

Table 6.90 Limiting U-value standards

Element or system	Values (W/m^2 K)
Opening areas (windows and doors)	Same as actual dwelling up to a maximum proportion of 25% of total floor area
Glazed doors	1.4
Opaque doors	1.0
Semi-glazed doors	1.2
Pedestrian doors	2.00
Vehicle access and similar large doors	1.5
High-usage entrance doors	3.5

6.12.2.3 Consequential improvements (non-domestic buildings)

If a building has a total useful floor area greater than $1000\,m^2$ and the proposed building work includes:

- an extension; or
- the initial provision of any fixed building services; or
- an increase to the installed capacity of any fixed building services

then consequential improvements should be made to improve the energy-efficiency of the whole building by replacing:

> all existing doors (excluding high-usage entrance doors) within the area served by the fixed building service that have a U-value worse than $3.3\,W/m^2\,K$ with doors whose U-value is less than $2.2\,W/m^2\,K$.   L2B 18-7

6.12.2.4 Work on controlled services or fittings (domestic buildings)

When working on a controlled service or fitting (i.e. where the service or fitting is subject to the requirements of Approved Document G, H, J, L or P of Schedule 1), and where windows, roof windows, rooflights and/or doors are to be provided:

> - doors should be provided with draught-proofed units;   L1B 32
> - the area-weighted average performance of draught-proofed units for new and replacement fittings that are provided as part of the construction of an extension shall be no worse than given in Table 6.91.   L1B 32

Table 6.91 Standards for thermal elements for new fittings in an extension

Element	New fittings in an extension ($W/m^2\,K$)	Replacement fittings in an existing dwelling ($W/m^2\,K$)
Doors	1.8	2.0 (whole unit)
		1.2 (centre pane)
Doors with more than 50% of their internal face glazed	2.2	2.2
Other doors	3.0	3.0

Dwellings should be constructed and equipped so that there   L2A 68
are no reasonably avoidable thermal bridges in the insulation
layers caused by gaps within the various elements, at the
joints between elements, and at the edges of door openings.

6.12.2.5 Safety features

People moving in or around a building might not see door glazing in critical
locations and can collide with it, and in order to avoid this happening either:

- make the glazing immediately apparent; or   K4 7.3
- use an alternative method of glazing, such as mullions,
  transoms, door framing or large pull or push handles
  (see Figure 6.220).

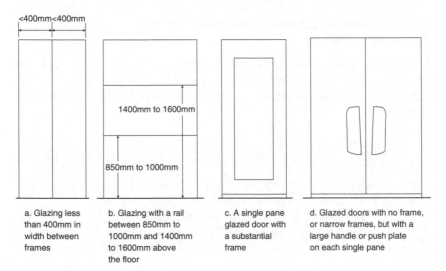

a. Glazing less than 400mm in width between frames

b. Glazing with a rail between 850mm to 1000mm and 1400mm to 1600mm above the floor

c. A single pane glazed door with a substantial frame

d. Glazed doors with no frame, or narrow frames, but with a large handle or push plate on each single pane

Figure 6.220 Examples of door-height glazing not warranting manifestation.

Or provide glass doors and/or glazed screens at two levels (see Figure 6.221)
such that:

- they contrast visually with the background seen   K4 7.4
  through the glass; or
- they contain a logo or sign (minimum height 150mm)
  or a decorative feature; or

- where glazed doors are beside or part of a glazed screen, they are clearly marked with a high-contrast strip at the top and on both sides; or
- where glass doors may be held open, they are protected with guarding.

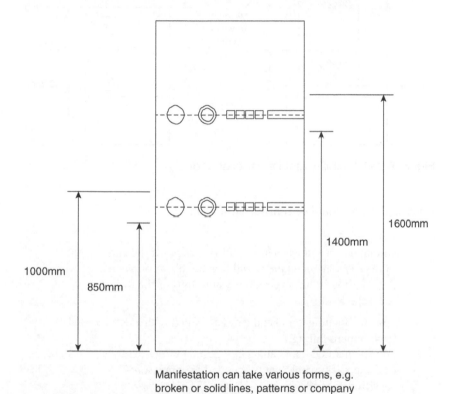

Manifestation can take various forms, e.g. broken or solid lines, patterns or company logos

Figure 6.221 Height of manifestation for glass doors and glazed screens.

Doors and gates should be constructed so that:

door leaves and side panels wider than 450 mm are designed to include vision panels towards the leading edge of the door that provide, as a minimum, the zone(s) of visibility shown in Figure 6.222.   K4 10.1a

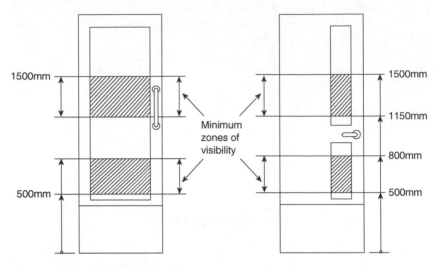

Figure 6.222 Visibility requirements of doors.

Doors and gates should be constructed so that:

• sliding doors and gates are unable to come off the end of the track and a retaining rail is available to prevent them falling if the suspension system fails or the rollers leave the track;	K4 10.1b
• upward-opening doors and gates are fitted with a device to stop them falling;	K4 10.1c
• power-operated doors and gates include safety features (e.g. a power switch that is operated by a pressure-sensitive door edge);	K4 10.1d
• the stop switch is readily identifiable and accessible;	K4 10.1d
• in the event of a power failure, the facility for manual or automatic opening is retained.	K4 10.1d

6.12.2.5.1 Internal doors – opening force

The opening force for a manually operated door should not exceed 30N.	M 3.10a
The effective clear width through a single-leaf door (or one leaf of a double-leaf door) should be in accordance with Table 6.92 and Figure 6.223.	M 3.10b

Table 6.92 Minimum effective clear widths of doors

Direction and width of approach	New buildings (mm)	Existing buildings (mm)
Straight on (without a turn or oblique approach)	800	750
At right angles to an access route at least 1500 mm wide	800	750
At right angles to an access route at least 1200 mm wide	825	775
External doors to buildings used by the general public	1000	775

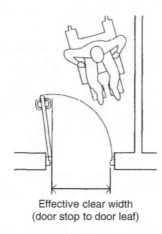

Effective clear width
(door stop to door leaf)

Figure 6.223 Effective clear width and visibility requirements of doors.

There should be an unobstructed space of at least 300 mm on the pull side of the door between the leading edge of the door and any return wall (unless the door is a powered entrance door, see Figure 6.223). M 3.10c

A space alongside the leading edge of a door should be provided to enable a wheelchair user to reach and grip the door handle, which should:

- be easy to operate by people with limited manual dexterity; M 3.10d
- be capable of being operated with one hand using a closed fist (e.g. a lever handle); M 3.10d
- contrast visually with the surface of the door. M 3.10e

Door frames should contrast visually with the surrounding wall. M 3.10f

The surface of the leading edge of a non-self-closing door should contrast visually with the other door surfaces and its surroundings. M 3.10g

Doors and gates should be constructed so that:

• door leaves and side panels wider than 450 mm are designed to include vision panels towards the leading edge of the door that provide, as a minimum, the zone(s) of visibility shown in Figure 6.222;	K4 10.1a
• sliding doors and gates are unable to come off the end of the track and a retaining rail is available to prevent them falling if the suspension system fails or the rollers leave the track;	K4 10.1b
• upward-opening doors and gates are fitted with a device to stop them falling;	K4 10.1c
• power-operated doors and gates include safety features (e.g. a power switch that is operated by a pressure-sensitive door edge);	K4 10.1d
• the stop switch is readily identifiable and accessible;	K4 10.1d
• in the event of a power failure, the facility for manual or automatic opening is retained.	K4 10.1d

 Note: It should be possible to tell between a fully glazed door and any adjacent glazed wall/partition by providing a high-contrast strip at the top and on both sides.

Fire doors (**particularly those in corridors**) should be held open with an electromagnetic device that is capable of self-closing when:

• the power supply fails;	M 3.10k
• activated by smoke detectors;	
• activated by a hand-operated switch.	

Fire doors (**particularly to individual rooms**) should be fitted with swing-free devices that close when:

• activated by smoke detectors;	M 3.10l
• the building's fire alarm system is activated;	
• the power supply fails.	

 Note: Low-energy powered door systems may be used in locations not subject to frequent use or heavy traffic.

Low-energy powered swing-door systems should be capable of being operated:	M 3.10m

- in manual mode;
- in powered mode; or
- in power-assisted mode.

Note: The use of self-closing devices should be minimized, as they disadvantage many people (e.g. those pushing prams or carrying heavy objects).

If closing devices are needed for fire control: M 3.7

- they should be electrically powered hold-open devices or swing-free closing devices;
- their closing mechanism should only be activated in case of emergency.

The presence of doors, whether open or closed, should be apparent to visually impaired people.

Note: Also see BS 8300 for guidance on electrically powered hold-open devices, swing-free systems and low-energy powered door systems.

6.12.2.6 Means of escape

Common corridors that connect two or more storey exits should be subdivided by a self-closing fire door with, if necessary, an associated fire-resisting screen.	B1 2.28 (V2)
Every corridor more than 12 m long and which connects two or more storey exits should be subdivided by self-closing fire doors positioned approximately midway between the two storey exits.	B1 3.26 (V2)
Except for doorways, all escape routes should have a clear headroom of not less than 2 m.	B1 3.17 (V2)
Except for kitchens, all habitable rooms shall be provided with suitable means for emergency egress from each storey via doors or windows.	B1 2.1–2.4 (V1) B1 2.11–2.12 (V2)
If the only escape route from an inner room is through another room, there should be a vision panel not less than 0.1 m^2 located in the door or walls of the inner room.	B1 3.10 (V2)
Unless escape stairways and corridors are protected by a pressurization system complying with BS EN 121016:2005, every dead-end corridor exceeding 4.5 m in length should be separated by self-closing fire doors.	B1 3.27 (V2)

> In residential care homes, bedrooms should be enclosed in fire-resisting construction with fire-resisting doors, and every corridor serving bedrooms should be a protected corridor.   B1 3.48 (V2)
>
> The dead-end portion of any common corridor should be separated from the rest of the corridor by a self-closing fire door.   B1 2.29 (V2)

 In residential care homes for the elderly, it is generally reasonable to assume that at least a proportion of the residents will need some assistance to evacuate.

6.12.2.6.1 Hazards on access routes

> If, during normal use, any door (other than a fire escape door) swings out by more than 100 mm towards an access route, it is protected as shown in Figure 6.224.   K4 (10.2)

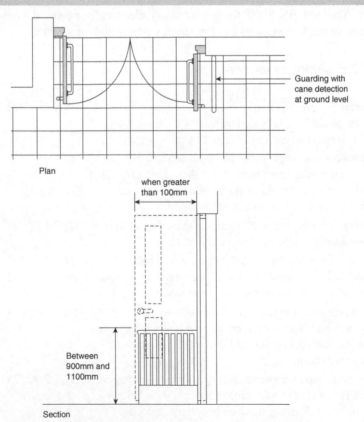

Figure 6.224 Avoiding doors on access routes.

6.12.2.6.2 Landings

For dwellings, a door may swing across a landing at the bottom of a flight, but only as shown in Figure 6.225.

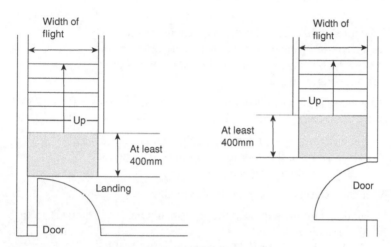

Figure 6.225 Landings next to doors in dwellings.

For buildings other than dwellings:

landings shall have an unobstructed minimum length of 1200 mm on each landing and doors shall not swing across landings.	K1 1.23

Note: This is unless the landing is at the top (or bottom) of a flight which is formed by the ground, in which case it may have a maximum gradient along the direction of travel of 1:60, provided that the surface is paved ground or otherwise made permanently firm.

6.12.2.7 Fire safety

Note: When fitted, manual call points for fire alarm systems shall be adjacent to exit doors.

6.12.2.7.1 Cavity barriers

Cavity barriers should be provided for all door openings.	B3 6.3 (V1) B3 9.3 (V2)
Openings in a cavity barrier should be limited to those for doors that have at least 30 minutes' fire resistance.	B3 6.8 (V1) B3 9.13 (V2)

Note: Detailed guidance on door openings and fire doors is given in Appendix B of Approved Document B.

6.12.2.72 Emergency egress doors

The door should enable the person escaping to reach a place free from danger of fire (e.g. a courtyard or back garden which is at least as deep as the dwelling house is high – see Figure 6.226).

In a gallery, if the floor is not provided with an alternative exit or escape window:

B1 2.12 (V1)
B1 2.8 (V2)

* the distance between the foot of the access stair to the gallery and the door to the room containing the gallery should not exceed 3 m.

Where an external escape stair is provided:

* all doors giving access to the stair should be fire-resisting;
* doors within 1800 mm of the escape route shall be protected by fire-resisting construction.

B1 2.15 (a, b and c)

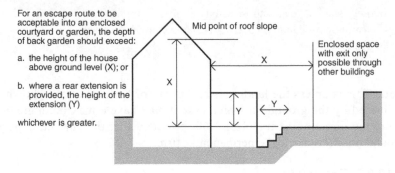

For an escape route to be acceptable into an enclosed courtyard or garden, the depth of back garden should exceed:

a. the height of the house above ground level (X); or

b. where a rear extension is provided, the height of the extension (Y)

whichever is greater.

Mid point of roof slope

Enclosed space with exit only possible through other buildings

Figure 6.226 Ground and basement storey exit into an enclosed space.

 Note: Glazing in any fire-resisting construction should be fire-resisting and fixed shut.

6.12.2.7.3 Fire doors

Doors that need to be fire-resisting should meet the requirements given in Table B1 of Appendix B to Approved Document B.

Doors on escape routes should be hung to open not less than 90°.	B1 5.15 (V2)
Doors giving access to an external escape should be fire-resisting and self-closing.	B1 5.25 (V2)
Door-closing devices for fire doors:	B1 3.51 (V2)
• shall be in accordance with BS EN 1155:1997; • should take account of the needs of residents; • should have free-swing door closers in bedrooms; • should have hold-open devices in circulation spaces.	
Doors on escape routes (both within and from the building) should be readily openable.	B1 5.10 (V2)
Doors on escape routes (whether or not the doors are fire doors), should either not be fitted with lock, latch or bolt fastenings or only be fitted with simple fastenings that can be readily operated from the side approached by people making an escape, without the use of a key and without having to manipulate more than one mechanism.	B1 5.11 (V2)

 Note: Doors that open towards a corridor or a stairway should be sufficiently recessed to prevent the swing from encroaching on the effective width of the stairway or corridor.

Fire doors should be capable of performing in accordance with Table 6.75 of Appendix B (to Volume 1 of Approved Document B).	App B 1 (V1)
Fire doors should be fitted with a self-closing device except for:	App B 2 (V2)
• fire doors to cupboards and to service ducts that are normally kept locked shut; and • fire doors within flats.	

Fire doors should be marked with the appropriate fire safety sign complying with BS 5499-5:2002 according to whether the door is:

• to be kept closed when not in use (**Fire door – keep shut**); • to be kept locked when not in use (**Fire door – keep locked shut**); or	App B 8 (V2)

- held open by an automatic-release mechanism or free-swing device (**Automatic fire door – keep clear**).

Fire doors serving an attached or integral garage should   App B 2 (V1)
be fitted with a self-closing device.

Electrically powered locks should return to the unlocked position:

- on operation of the fire alarm system;   B1 5.11 (V2)
- on loss of power or system error;
- on activation of a manual door-release unit.

 Note: In assembly places, shops and commercial buildings, doors on escape routes from rooms with an occupant capacity of more than 60 should either not be fitted with lock, latch or bolt fastenings or be fitted with panic fastenings in accordance with BS EN 1125: 1997.

No more than 25 per cent of the length of a   App B 5 (V2)
compartment wall should consist of door
openings.

Revolving doors, automatic doors and turnstiles should   B1 5.18 (V2)
not be placed across escape routes.

Secure doors that need to be operated by a code,   B1 5.11 (V2)c
combination, swipe or proximity card, biometric data
or similar means should also be capable of being
overridden from the side approached by people making
their escape.

Self-closing fire doors may be held open by: App B 3 (V2)

- a fusible link; or
- an automatic-release mechanism actuated by an
  automatic fire detection and alarm system; or
- a door-closer delay device.

The door of any doorway or exit should be hung to   B1 5.14 (V2)
open in the direction of escape.

The essential components of any hinge on which a fire door is hung should be made entirely from materials having a melting point of at least 800° C.	App B 3 (V1) App B 7 (V2)
Two fire doors may be fitted in the same opening so that the total fire resistance is the sum of their individual fire resistances, provided that each door is capable of closing the opening.	App B 4 (V2)
Vision panels shall be provided where doors on escape routes subdivide corridors, or where any doors are hung to swing both ways.	B1 5.17 (V2)
Fire doors (except those to cupboards and to service ducts) should be marked on both sides.	B1

 Note: See also Approved Documents M and N.

6.12.2.7.4 Fire doors in dwellings

The minimum fire resistance for doors in buildings other than dwellings is as given in Table B1 (page 134) of Approved Document B, Volume 2.

BS 8214:1990 gives recommendations for the specification, design, construction, installation and maintenance of fire doors constructed with non-metallic door leaves.

Guidance on timber fire-resisting doorsets may be found in *Timber Fire-Resisting Doorsets: Maintaining Performance under the New European Test Standard*, published by TRADA.

In flats where the habitable rooms do not have direct access to the entrance hall, bedrooms should be separated from the living accommodation by fire-resisting construction and fire doors (B1 2.14 (V2)).

Self-closing fire doors should be positioned so that smoke will not affect access to more than one stairway.

6.12.2.7.5 Fire-protected stairways

Fire-protected stairways shall (as far as is reasonably possible) consist of fire-resistant material and fire-resistant doors, and have an appropriate form of smoke-control system.	B1 1.viii

6.12.2.7.6 Basements

Owing to the risk that a single stairway may be blocked by smoke from a fire in the basement or ground storey:

Basement storeys in a dwelling house that contain a habitable room shall be provided with either an external door or a window suitable for egress from the basement.	B1 2.13 (V1) B1 2.6 (V2)

6.12.2.7.7 Loft conversions

Where a loft conversion means that a new storey is going to be added, then the stairway will need to be protected with fire-resisting doors and partitions.	B1 2.20b

6.12.2.7.8 Garages

Fire doors now only need to be provided with self-closing devices if they are between a dwelling house and an integral garage.

If a door is provided between a dwelling house and the garage: <ul><li>the floor of the garage should be laid to allow fuel spills to flow away from the door to the outside; or</li><li>the door opening should be positioned at least 100 mm above garage floor level.</li></ul>	B3 5.5 (V1)

6.12.2.7.9 Fire service (access and facilities)

Doors should be provided such that there is no more than 60 m between each door and/or the end of that elevation (e.g. a 150 m elevation would need at least two doors).	B5 16.5 (V2)
Every elevation to which vehicle access is provided should have a suitable door or doors, not less than 750 mm wide, giving access to the interior of the building.	B5 11.3 (V2) B5 16.5 (V2)

6.12.2.7.10 Sprinkler systems

Where a sprinkler system is provided, fire doors to bedrooms need not be fitted with self-closing devices.	B1 3.52 (V2)

6.12.2.7.11 Protected stairways

Dwelling houses with one floor more than 4.5 m above ground level should have a protected stairway separated by fire doors.	B1 2.6 (V1)
Transfer grilles belonging to air-circulation systems shall not be fitted in any door leading to a protected stairway.	B1 2.17

6.13 Access routes to, from and within buildings

6.13.1 Requirements

Stairs, ladders and ramps
Stairs, ladders and ramps shall be so designed, constructed and installed as to be safe for people moving between different levels in or about the building

(Approved Document K1)

 A new issue of Approved Document K came into force in 2013.

Protection from falling
Any stairs, ramps, floors and balconies and any roof to which people have access, and any light well, basement area or similar sunken area connected to a building shall be provided with barriers where it is necessary to protect people in or about the building from falling.

(Approved Document K2)

Vehicle barriers and loading bays
Vehicle ramps and any levels in a building to which vehicles have access, shall be provided with barriers where it is necessary to protect people in or about the building.

(Approved Document K3)

Vehicle loading bays shall be constructed in such a way, or be provided with such features, as may be necessary to protect people in them from collision with vehicles.

(Approved Document K3)

Safe access for cleaning windows etc.
Provision shall be made for any windows, skylights, or any transparent or translucent walls, ceilings or roofs to be safely accessible for cleaning.

(Approved Document K5.4)

Disabled access
In addition to the requirements of the Equality Act 2010, precautions need to be taken to ensure that:

- *new non-domestic buildings and/or dwellings (e.g. houses and flats used for student living accommodation etc.);*
- *extensions to existing non-domestic buildings;*
- *non-domestic buildings that have been subject to a material change of use (e.g. so that they become a hotel, boarding house, institution, public building or shop)*

are capable of allowing people, regardless of their disability, age or gender, to:

- *gain access to buildings;*
- *gain access within buildings;*
- *be able to use the facilities of the buildings (both as visitors and as people who live or work in them);*
- *use sanitary conveniences in the principal storey of any new dwelling.*

(Approved Document M)

Amendments to the 2004 version of Approved Document M came into force in 2013.

Fire safety
- *There shall be sufficient escape routes that are suitably located to enable persons to evacuate the building in the event of a fire.*
- *Safety routes shall be protected from the effects of fire.*
- *In an emergency, the occupants of any part of the building shall be able to escape without any external assistance.*

(Approved Document B1)

6.13.2 Meeting the requirements

6.13.2.1 Access routes – hazards

> If, during normal use, any door (other than a fire escape   K4 10.2
> door) swings out by more than 100 mm towards an access
> route, it shall be protected as shown in Figure 6.227.

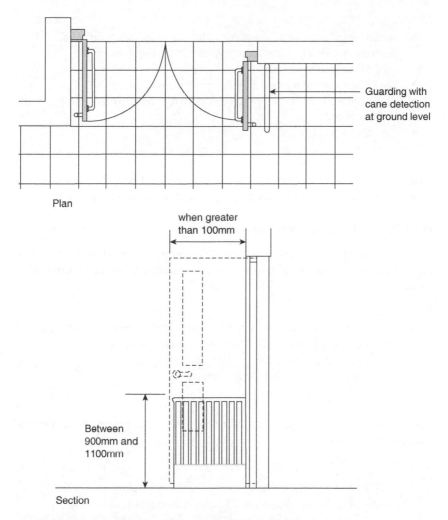

Plan

when greater
than 100mm

Between
900mm and
1100mm

Guarding with
cane detection
at ground level

Section

Figure 6.227 Avoiding doors on access routes.

6.13.2.2 Access routes – guarding

For buildings other than dwellings and common access areas of buildings that
contain flats:

• guarding should be provided at the sides of flights and landings when there are two or more risers.	K1 1.40

6.13.2.2.1 Pedestrian guarding

Guarding should be provided whenever it is considered necessary from the
point of view of safety to guard:

> • the edges of any part of a floor (including the edge below an   K2 3.1
>   opening window), gallery, balcony, roof (including roof lights
>   and other openings); and
> • any other place to which people have access; and
> • any light well, basement or similar sunken area next to a
>   building; and
> • in vehicle parks.

Note: Guarding is not required on ramps that are used only for vehicle access or in places such as loading bays where it would obstruct normal use.

6.13.2.2.2 Balcony guarding

If a balcony or flat roof is provided for escape purposes (or whenever it is considered necessary from a safety point of view) guarding should be provided to guard:

> • the edges of any part of a floor (including the edge below   K2 3.1
>   an opening window) of a gallery or balcony (and any
>   other openings); and
> • any other place to which people have access.

Note: Guarding is not required in places such as loading bays where it would obstruct normal use.

> 💡 If a building is likely to be used by children under   K2 3.3
> five years old, the guarding should not have horizontal rails
> and should stop children from easily climbing it, and the
> construction should prevent a 100 mm sphere being able to
> pass through any opening of that guarding.

6.13.2.3 Access routes – handrails for stairs

> Handrails shall not project into an access route.   K1 1.1.36

The desired handrail profile is shown in Figure 6.228.

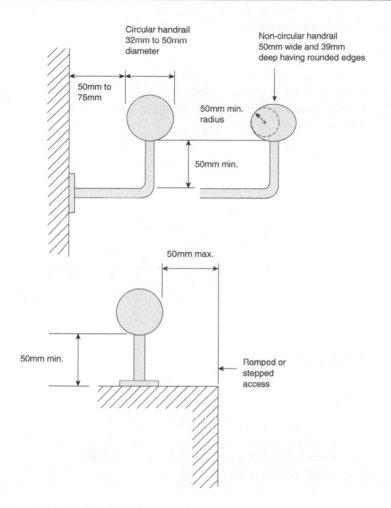

Figure 6.228 Handrail design.

6.13.2.4 Access routes – width of ramps

For buildings other than dwellings, for a ramp that provides access for people, the width between walls, upstands and/or kerbs should be a minimum of 1500 mm.

In dwellings and common access areas in buildings that contain flats:

> •   ramps that are less than 1000 mm wide should be   K1 2.12
>   provided with a handrail on one or both sides.

 Note: Handrails should give firm support and allow a firm grip and may form the top of the guarding.

6.13.2.5 Access – vehicles

For all buildings:

> • barriers should be provided at any edges which are level   K3 4.1
>   with (or above) the floor, ground or any other route used
>   for vehicles (see Figure 6.229).

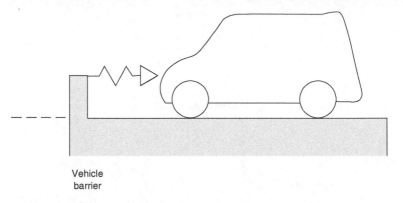

Vehicle
barrier

Figure 6.229 Barrier siting.

For all buildings:

> • barriers (which can be a wall, parapet, balustrade or   K3 4.1
>   similar obstacle) should be constructed in accordance
>   with Figure 6.230 to a minimum height.

As shown
in **BS EN
1991-1-1**
with its UK
National   Barrier   At least
Annex   375mm
          Floor or
          roof level

As shown
in **BS EN
1991-1-1**
with its UK   At least
National   Barrier   610mm
Annex   Ramp level

(a) **Any building:**   (b) **Any building:**
floor or roof edge   ramp edge

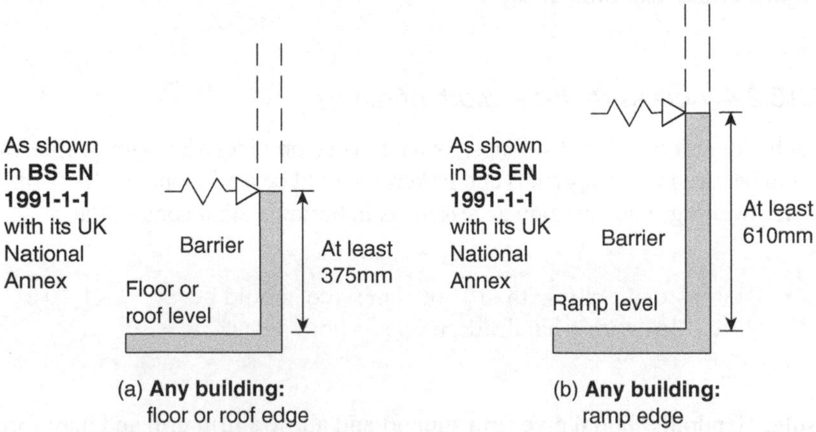

Figure 6.230 Barrier design.

Note: Also see BS EN 1991-1-1.

6.13.2.6 Access – for maintenance

Where the stairs or ladders will be used to access areas for maintenance they should comply with the provisions of BS 5395-3 and the Construction (Design and Management) Regulations 2007 as appropriate.

6.13.2.6.1 Access – for cleaning windows, etc.

Where a person standing on a permanent stable surface cannot safely clean a glazed surface:

• windows should be designed so that people can safely clean the outside safely from inside the building (see Figure 6.231).	K4 9.1a

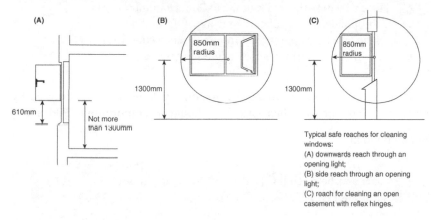

(A)

610mm

Not more than 1300mm

(B)

850mm radius

1300mm

(C)

850mm radius

1300mm

Typical safe reaches for cleaning windows:
(A) downwards reach through an opening light;
(B) side reach through an opening light;
(C) reach for cleaning an open casement with reflex hinges.

Figure 6.231 Safe reaches for cleaning.

In buildings where this is not possible, then one of the following should be provided:

• ladders; or	K4 9.1b
• access equipment (such as suspended cradles or travelling ladders, with attachments for safety harnesses); or	K4 9.1c
• suitable anchorage points for safety harnesses or abseiling hooks; or	K4 9.1d
• walkways (at least 400 mm wide and with guarding or with anchorages for sliding safety harnesses).	K4 9.1e

If none of the above alternatives is possible, then scaffolding towers (portable or fixed) will have to be used.

6.13.2.7 Access – assembly buildings

To maintain sightlines for spectators, stairs and/or ramps that form part of the means of access within an assembly building such as a sports stadium, theatre or cinema should be in accordance with the following requirements:

> • Ensure that the maximum pitch for gangways to seating areas for spectators is 35°.   K1 1.4
> • Align the ends of all rows of seats/wheelchair spaces so that the width of the gangway remains the same.
> • Provide transverse gangways that (in auditoria with tiered seating) do not cross.
> • Ensure that stepped tiers are between 100 mm and 190 mm high.
> • Have a landing the width of the exit and a minimum of 1100 mm deep immediately in front of the exit doors where an exit is approached from a stepped gangway.
> • Provide a handrail for stepped side gangways.

Note: Gangways should normally be 900 mm wide unless they are used by more than 50 persons, in which case they should be a **minimum** of 900 mm.

For all buildings:

> • for access between levels, the minimum headroom shown in Figure 6.232 shall be provided.   K1 1.11

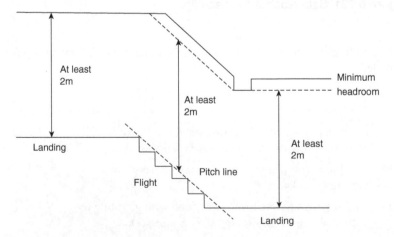

Figure 6.232 Minimum headroom.

Grandstands and other structures that are erected in places of public assembly may need to sustain the synchronous or rhythmic movement of numbers of people. To avoid the structure being damaged (or people using the structure becoming alarmed by the movement) it is important to ensure that the design of the structure takes these factors into account so as

Note: See *Dynamic Performance Requirements for Permanent Grandstands Subject to Crowd Action: Recommendations for Management, Design and Assessment* (published by the Institution of Structural Engineers) for further guidance.

6.13.2.8 Means of escape

In dwellings, a fixed ladder may be used – with fixed handrails on both sides – for access in a loft conversion that contains one habitable room.

Retractable ladders may **not** be used as means of escape.

6.13.2.9 General requirements for lifting devices

There should be an unobstructed manoeuvring space of 1500 mm × 1500 mm, or a straight access route 900 mm wide, in front of each lifting device.	M 3.28a
The landing call buttons should be located between 900 mm and 1100 mm from the floor and at least 500 mm from any return wall.	M 3.28b
Landing call buttons and lifting-device control button symbols: • should contrast visually with the surrounding face plate; • should be raised to facilitate tactile reading; • should be accessible by wheelchair users.	M 3.28c and d M 3.27
The floor of the lifting device should not be of a dark colour.	M 3.28e
A handrail (at 900 mm nominal) should be provided on at least one wall of the lifting device.	M 3.28f
A suitable emergency communication system should be fitted.	M 3.28g

Note: See also:

- Lift Regulations 1997, SI 1997/831;
- Lifting Operations and Lifting Equipment Regulations 1998, SI 1998/2307;
- Provision and Use of Work Equipment Regulations 1998, SI 1998/2306;
- Management of Health and Safety at Work Regulations 1999, SI 1999/3242;
- BS 8300 (*Design of Buildings and Their Approaches to Meet the Needs of Disabled People: Code of Practice*).

6.13.2.10 Lifts

Lifts should **not** be used when there is a fire in the building – unless it is a firefighting lift.

Lifts (except suitably designed and installed evacuation lifts) are **not** considered acceptable as a means of escape.	B1 vi
Lift entrances should be separated from the floor area on every storey by a protected lobby.	B1 5.42 (V2)
Lift shafts should not be continued down to serve any basement storey if they are: • in a building served by only one escape; • within the enclosure to an escape stair that is terminated at ground level.	B1 5.44 (V2)
If a protected shaft contains a stair and/or a lift, it should **not** also contain: • a pipe conveying oil (other than in the mechanism of a hydraulic lift); or • a ventilating duct (other than a duct provided solely for ventilating the stairway).	B2 8.40 (V2)
Lift wells should be either: • contained within the enclosures of a protected stairway; or • enclosed throughout their height with fire-resisting construction.	B1 5.42 (V2)
A lift well connecting different compartments should form a protected shaft.	B1 5.42 (V2)
Lift machine rooms should be sited **over** the lift well whenever possible.	B1 5.45 (V2)
In basements and enclosed (i.e. non-open-sided) car parks, the lift should be approached only by a protected lobby or protected corridor (unless it is within the enclosure of a protected stairway).	B1 5.43 (V2)
Evacuation lifts and refuges should be clearly identified by appropriate fire safety signs.	B1 4.10 (V2)

 Note: If a refuge is in a lobby or stairway the sign should be accompanied by a blue mandatory sign worded **Refuge – keep clear**.

6.13.2.10.1 Passenger lifts

For all buildings, a passenger lift is considered the most suitable form of access for people moving from one storey to another.

Wherever possible a lifting device (e.g. a passenger lift or a lifting platform) serving all storeys should be provided in: • new developments; • existing buildings.	M 3.17, 3.24a–d

 Note: In exceptional circumstances (e.g. a listed building, or an infill site in a historic town centre) where a passenger lift cannot be accommodated, a wheelchair-platform stairlift serving an intermediate level or a single storey may be used.

The location of lifting devices that are accessible by mobility-impaired people should be clearly visible from the building entrance.	M 3.18
Signs should be available at each landing to identify the floor reached by the lifting device.	M 3.18
In addition to the lifting device, internal stairs (designed to suit ambulant disabled people and those with impaired sight) should always be provided.	M 3.19

Lift sizes should be chosen to suit the anticipated density of use of the building and the needs of disabled people.

Passenger lifts should conform to the requirements of:

• the Lift Regulations 1997, SI 1997/831; • the relevant British Standards and the EN 81 series of standards; • BS EN 81-70:2003 (*Safety Rules for the Construction and Installation of Lifts*).	M 3.34a

Passenger lifts should:

- be accessible from the remainder of the storey;   M 3.34b

- have power-operated horizontal sliding doors   M 3.34e
  which provide an effective clear width of at
  least 800 mm (nominal);

- have doors fitted with timing devices (and   M 3.34f
  reopening activators) to allow enough time
  for people and any assistance dogs to enter or
  leave;

- have the car controls located between 900 mm   M 3.34g
  and 1200 mm (preferably 1100 mm) from the
  car floor and at least 400 mm from any return
  wall;

- have all landing call buttons located between   M 3.34h
  900 mm and 1100 mm from the floor of the
  landing and at least 500 mm from any return
  wall;

- be fitted with lift landing and car doors that are   M 3.34i
  visually distinguishable from adjoining walls;

- include (in the lift car and the lift lobby)   M 3.31 and 3.34j
  audible and visual information to tell
  passengers that a lift has arrived, which floor
  it has reached and where in a bank of lifts it is
  located;

- allow all glass areas to be easily identified by   M 3.34k
  people with impaired vision;

- conform to BS 5588-8 if the lift is to be used to   M 3.34l
  evacuate disabled people in an emergency.

If the lift is not large enough to allow a wheelchair user   M 3.34d
to turn around within the lift car, then a mirror should be
provided that enables the wheelchair user to see behind the
wheelchair.

The minimum dimensions of the lift cars should be 1100 mm   M 3.34c
wide and 1400 mm deep (Figure 6.233).

 Note: Visually and acoustically reflective wall surfaces should not be used.

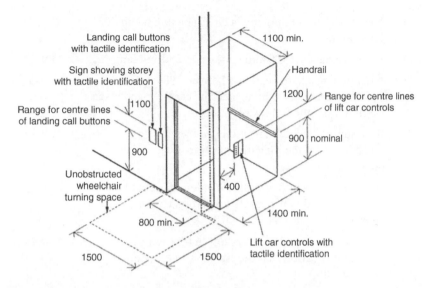

Figure 6.233 Key dimensions associated with passenger lifts (mm).

Where possible, lift cars (used for access between two levels only) may be provided with opposing doors to allow a wheelchair user to leave without having to reverse out.	M 3.33

Passenger lifts in dwelling houses (which serve any floor more than 4.5 m above ground level) should be located in either the enclosure to a protected stairway or a fire-resisting lift shaft.	B1 2.18

6.13.2.10.2 Passenger lifts in blocks of flats

If a passenger lift is installed, it should:

• be suitable for an unaccompanied wheelchair user;	M 9.4
• have a minimum load capacity of 400 kg;	M 9.6
• have a clear landing at least 1500 mm wide and 1500 mm long in front of its entrance;	M 9.7a

• have a door (or doors) with a clear opening width of at least 800 mm;	M 9.7b
• have a car at least 900 mm wide and 1250 mm long;	M 9.7c
• have landing and car controls between 900 mm and 1200 mm above the landing and the car floor and at least 400 mm from the front wall;	M 9.7d
• have suitable tactile indication (on the landing and adjacent to the lift call button) to identify the storey;	M 9.7e
• have suitable tactile indication (on or adjacent to lift buttons within the car) to confirm the floor selected;	M 9.7f
• incorporate a signalling system that provides visual notification that the lift is answering a landing call;	M 9.7g
• have a dwell time of five seconds before its doors begin to close after they are fully open;	M 9.7g
• provide a visual and audible indication of the floor reached (when the lift serves more than three storeys).	M 9.7h

6.13.2.10.3 Firefighting lifts

Buildings with a floor at more than 18 m above (or with a basement at more than 10 m below) the fire and rescue service vehicle access level should be provided with at least two firefighting shafts containing firefighting lifts.	B5 17.2 (V2) B5 17.8 (V2)
Other than in blocks of flats, all firefighting lifts should be approached from the accommodation through a firefighting lobby.	B5 17.11 (V2)

6.13.2.11 Lifting platforms

A lifting platform should only be provided to transfer wheelchair users, people with impaired mobility and their companions vertically between levels or storeys.

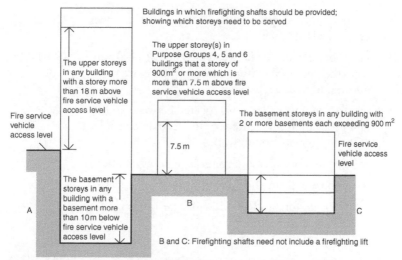

Buildings in which firefighting shafts should be provided; showing which storeys need to be served

The upper storey(s) in Purpose Groups 4, 5 and 6 buildings that a storey of 900 m² or more which is more than 7.5 m above fire service vehicle access level

The upper storeys in any building with a storey more than 18 m above fire service vehicle access level

Fire service vehicle access level

The basement storeys in any building with 2 or more basements each exceeding 900 m²

Fire service vehicle access level

7.5 m

The basement storeys in any building with a basement more than 10 m below fire service vehicle access level

A

B

C

B and C: Firefighting shafts need not include a firefighting lift

A. Firefighting shafts should include firefighting lift(s)

Note: Height excludes any top storey(s) consisting exclusively of plant rooms.

Figure 6.234 Provision of firefighting shafts.

All users including wheelchair users should be able to reach and use the controls that summon and direct the lifting platform.	M 3.36
Where possible, lifting platforms (used for access between two levels only) may be provided with opposing doors to allow a wheelchair user to leave without reversing out.	M 3.41
Visually and acoustically reflective wall surfaces should **not** be used.	M 3.42

Lifting platforms should:

- conform to the requirements of the Supply of Machinery (Safety) Regulations 1992, SI 1992/3073, the relevant British Standards and the EN81 series of standards;   M 3.43a

- restrict the vertical travel distance to: M 3.43b
    - not more than 2 m if there is no liftway enclosure and/or floor penetration;
    - more than 2 m if there is a liftway enclosure;

- restrict the rated speed of the platform so that it does not exceed 0.15 m/s;   M 3.43c
- have controls located between 80 mm and 1100 mm from the floor of the lifting platform and at least 400 mm from any return wall;   M 3.43d
- have all landing call buttons located between 900 mm and 1100 mm from the floor of the landing and at least 500 mm from any return wall;   M 3.43f
- have continuous pressure controls (e.g. push buttons);   M 3.43e
- have doors with an effective clear width of at least: M 3.43h
  - 900 mm for an 1100 mm wide and 1400 mm deep lifting platform;
  - 800 mm in other cases;
- be fitted with clear instructions for use;   M 3.43i
- have their entrances accessible from the remainder of the storey;   M 3.43j
- have doors visually distinguishable from adjoining walls;   M 3.43k
- have an audible and visual announcement of platform arrival and level reached;   M 3.43l
- have areas of glass that are identifiable by people with impaired vision.   M 3.43m

The minimum clear dimensions of the platform should be:

- 800 mm wide by 1250 mm deep (where the lifting platform is not enclosed and provision has been made for an unaccompanied wheelchair user);   M 3.43g
- 900 mm wide by 1400 mm deep (where the lifting platform is enclosed and provision has been made for an unaccompanied wheelchair user);
- 1100 mm wide by 1400 mm deep (where two doors are located at 90° relative to each other, the lifting platform is enclosed or provision is being made for an accompanied wheelchair user).

6.13.2.12 Wheelchair platform stairlifts

Wheelchair platform stairlifts are only intended for the transportation of wheelchair users and should only be considered for conversions and alterations where it is not practicable to install a conventional passenger lift or a lifting platform.

Wheelchair platform stairlifts should:

• be fitted with clear instructions for use;	M 3.49f
• provide an effective clear width of at least 800 mm;	M 3.49g
• not be installed where their operation restricts the safe use of the stair by other people;	M 3.44
• not exceed a speed of 0.15 m/s;	M 3.49c
• be provided with continuous pressure controls (e.g. joysticks);	M 3.47 and 3.49d
• have minimum **clear** dimensions of 800 mm wide and 1250 mm deep.	M 3.49e

Wheelchair platform stairlifts should conform to the requirements of the Supply of Machinery (Safety) Regulations 2008 (SI 2008/ 1597), the relevant British Standards and the EN81 series of standards.

6.14 Corridors and passageways

6.14.1 Requirements

Fire precautions
There shall be sufficient escape routes that are suitably located to enable persons to evacuate the building in the event of a fire; and

In an emergency, the occupants of any part of the building shall be able to escape without any external assistance.

All safety routes shall be protected from the effects of fire.

(Approved Document B1)

Access and use
Reasonable provision shall be made for people to gain access to and be able to use the building and its facilities.

(Approved Document M1)

Amendments to the 2004 version of Approved Document M came into force in 2013.

6.14.2 Meeting the requirements

6.14.2.1 General requirements for corridors

Every corridor more than 12 m long that connects two or more storey exits should be subdivided by self-closing fire doors positioned approximately midway between the two storey exits.	B1 3.26 (V2)
Unless escape stairways and corridors are protected by a pressurization system all dead-end corridors exceeding 4.5 m in length should be separated by self-closing fire doors (together with any necessary associated screens) from any part of the corridor that: • provides two directions of escape; • continues past one storey exit to another.	B1 3.27 (V2)
In large buildings, the corridor adjoining the stair should be provided with a vent that is located as high as is practicable and with its top edge at least as high as the top of the door to the stair.	B1 2.26 (V2)
In buildings containing flats, the wall between each flat and the corridor should be a compartment wall.	B1 2.24 (V2)
Stores and other ancillary accommodation should **not** form part of the only common escape route from a flat on the same storey as that ancillary accommodation.	B1 2.30 (V2)

6.14.2.2 Protected corridors

Protected corridors shall be installed for:

• corridors serving bedrooms; • dead-end corridors (excluding recesses not exceeding 2 m deep); • any corridor that is common to more than one different occupancy.	B1 3.24 (V2)

Escape stairs shall have a protected lobby or protected corridor at all levels (except the top storey, all basement levels and when the stair is a firefighting stair) if:

• the stair is the only one serving a building that has more than one storey above or below the ground storey;	B1 4.34 (V2)

- the stair serves any storey at a height greater than 18 m; or

- the building is designed for phased evacuation.

 Corridors serving bedrooms in residential care homes should be protected corridors.

6.14.2.3 Common corridors

Common corridors:

- should be protected corridors;   B1 2.24 (V2)

- should have available means of ventilation;   B1 2.25 (V2)

- that connect two or more storey exits should be subdivided by a self-closing fire door with, if necessary, an associated fire-resisting screen (see Figure 6.235).   B1 2.28 (V2)

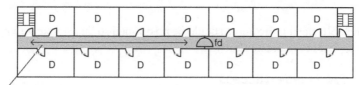

30 m max.

(a) Corridor access without dead ends

Note:
The arrangements shown also apply to the top storey.

Key

D   Dwelling
fd   Fire door
▨   Shaded area indicates zone where ventilation should be provided in accordance with paragraph 2:26
(An external wall vent or smoke shaft located anywhere in the shaded area)

Figure 6.235 Flats served by more than one common stair – corridor access without dead ends.

The dead-end portion of any common corridor should be separated from the rest of the corridor by a self-closing fire door (see Figure 6.236).   B1 2.29 (V2)

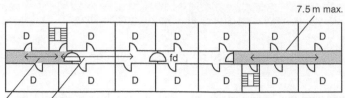

(b) Corridor access with dead ends
The central door may be omitted if maximum travel distance is not
more than 15 m

Note:
The arrangements shown also apply to the top storey.

Key

D   Dwelling
fd   Fire door
      Shaded area indicates zone where ventilation
      should be provided in accordance with
      paragraph 2:26
      (An external wall vent or smoke shaft located
      anywhere in the shaded area)

Figure 6.236 Flats served by more than one common stair – corridor access with dead ends.

Note: Self-closing fire doors should be positioned so that smoke will not affect access to more than one stairway.

6.14.2.4 Disabled access

Corridors and passageways should:

• be wide enough to allow wheelchair users, people with buggies, people carrying cases and/or people on crutches to pass others on the access route;	M 3.11
• **not** be hindered by projecting elements such as columns, radiators and fire hoses;	M 3.14a
• have an unobstructed width of at least 1200 mm.	M 3.14b

Note: For school buildings, the preferred corridor width is 2700 mm where there are lockers within the corridor.

Corridors and passageways should:

• have passing places at least 1800 mm long and at least 1800 mm wide at corridor junctions;	M 3.14c

- have a floor level no steeper than 1:60;   M 3.14d
- have an internal ramp in accordance with Table 6.93   M 3.14d
  and Figure 6.237 for floors with a gradient of 1:20 or
  steeper.

Table 6.93 Limits for ramp gradients

Going of a flight (m)	Maximum gradient	Maximum rise (mm)
10	1:20	500
5	1:15	333
2	1:12	16

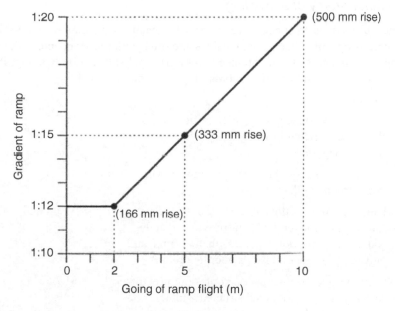

Figure 6.237 Relationship of ramp gradients.

Corridors should be at least 1800 mm wide where doors from a unisex wheelchair-accessible toilet project open into that corridor.		M 3.14h
If a section of the floor has a gradient steeper than 1:60 (but less than 1:20) it should rise no more than 500 mm without a level rest area at least 1500 mm long.		M 3.14e
Sloping sections should extend the full width of the corridor or have the exposed edge protected by guarding.		M 3.14f

Doors opening towards a corridor that is a major access route or an escape route should be recessed.	M 3.14g
On a major access route (or an escape route) the wider leaves of double doors should be on the same side of the corridor.	M 3.14i
Floor finishes should be slip-resistant.	M 3.14k
Glazed screens alongside a corridor should be clearly marked (i.e. with a logo or sign).	M 3.14l
The acoustic design should be neither too reverberant nor too absorbent.	M 3.13

6.14.2.5 Corridors, passageways and internal doors within the entrance storey of a dwelling

The objective is to make it easy for wheelchair users, ambulant disabled people, people of either sex with babies and small children, or people with luggage to gain access to an entrance and/or principal storey of a dwelling, into habitable rooms and/or a room containing a WC on that level.

Corridors and passageways in the entrance storey should be sufficiently wide for a wheelchair user to circumnavigate.	M 7.2
Internal doors should be wide enough for wheelchairs to go through with ease.	M 7.4
Permanent obstructions in a corridor (e.g. a radiator) should be no longer than 2 m (provided that the width of the corridor is not less than 750 mm and the obstruction is not opposite a door to a room).	M 7.2 and 7.5b
Corridors and/or other access routes in the entrance storey should have an unobstructed width in accordance with Table 6.94 and Figure 6.238.	M 7.5a

Table 6.94 Minimum widths of corridors and passageways for a range of doorway widths

Doorway clear opening width (mm)	Corridor/passageway width (mm)
750 or wider	900 (when approached head on)
750	1200 (when not approached head on)
775	1050 (when not approached head on)
800	900 (when not approached head on)

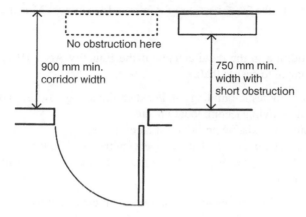

No obstruction here

900 mm min.
corridor width

750 mm min.
width with
short obstruction

Figure 6.238 Corridors, passages and internal doors.

Doors to habitable rooms and/or rooms containing a WC   M 7.5c
should have the minimum clear opening widths shown in
Table 6.94 and Figure 6.238.

6.15 Facilities in buildings other than dwellings

6.15.1 Requirements

Fire safety

- *There shall be sufficient escape routes that are suitably located to enable persons to evacuate the building in the event of a fire.*
- *Safety routes shall be protected from the effects of fire.*
- *In an emergency, the occupants of any part of the building shall be able to escape without any external assistance.*
- *There shall be an early-warning fire alarm system for persons in the building.*

(Approved Document B1)

Access to and use of buildings
Reasonable provisions shall be made for people to gain access to, and use the building and its facilities.

(Approved Document M)

Amendments to the 2004 version of Approved Document M came into force in 2013.

Note: In addition to the requirements of the Equality Act 2010, precautions need to be taken to ensure that:

- new non-domestic buildings and/or dwellings (e.g. houses and flats used for student living accommodation, etc.);
- extensions to existing non-domestic buildings;
- non-domestic buildings that have been subject to a material change of use (e.g. so that they become a hotel, boarding house, institution, public building or shop)

are capable of allowing people, regardless of their disability, age or gender, to:

- gain access to and within buildings;
- be able to use the facilities of the buildings (both as visitors and as people who live or work in them);
- use sanitary conveniences in the principal storey of any new dwelling.

Sanitary conveniences
*All buildings containing one or more rooms for residential purposes will be provided with a bathroom which contains either a fixed bath or a shower **and** a washbasin.*

(Approved Document G5)

Ventilation
There shall be adequate means of ventilation provided for people in the building.

(Approved Document F)

Stairs, ladders and ramps
Stairs, ladders and ramps shall be so designed, constructed and installed as to be safe for people moving between different levels in or about the building.

(Approved Document K1)

A new issue of Approved Document K came into force in 2013.

Protection against impact with glazing
Glazing with which people are likely to come into contact whilst moving in or about the building shall:

a.   if broken on impact, break in a way which is unlikely to cause injury; or
b.   resist impact without breaking; or
c.   be shielded or protected from impact.

(Approved Document K4)

Conservation of fuel and power
Reasonable provision shall be made for the conservation of fuel and power in buildings by:

(a)   *limiting heat gains and losses*

(i)   *through thermal elements and other parts of the building fabric; and*
(ii)   *from pipes, ducts and vessels used for space heating, space cooling and hot water services;*

(b)   *providing fixed building services which*

(i)   *are energy efficient;*
(ii)   *have effective controls; and are commissioned by testing and adjusting as necessary to ensure they use no more fuel and power than is reasonable in the circumstances.*

 A new version of Approved Document L came into force on 6 April 2014.

6.15.2 Meeting the requirements

6.15.2.1 Conservation of energy and power

Following the publication of new Approved Document L2B in November 2013, with effect from April 2014 all new non-domestic buildings now need to make 9 per cent carbon dioxide savings.

> Where a building is erected, it shall not exceed the target CO_2 L2A 2.7
> emission rate for that particular building.

6.15.2.1.1 Target emission rate

For a **new** building (other than a few exemptions – see section 6.15.2.1.1.1) the target emission rate (TER) is established by using approved software to calculate the CO_2 emission rate from a notional building of the same size and shape as the actual building, but with specified properties as set out in the NCM Modelling Guide. The key components of the notional building specification are shown in Table 6.95.

6.15.2.1.1.1 EXEMPTIONS

In the case of a new building, this is subject to the following exemptions:

*   new buildings other than dwellings which are roofed constructions having walls and which use energy to condition the indoor climate;
*   buildings which are used primarily or solely as places of worship;
*   temporary buildings with a planned time of use of two years or less;
*   industrial sites;

Table 6.95 Summary of concurrent notional building specification

Element	Side lit or unlit (where HVAC specification is heating only)	Side lit or unlit (where HVAC specification includes cooling)	Top lit
Roof U-value (W/m^2 K)	0.18	0.18	0.18
Wall U-value (W/m^2 K)	0.26	0.26	0.26
Floor U-value (W/m^2 K)	0.22	0.22	0.22
Window U-value (W/m^2 K)	11.6 (105 FF)	11.6 (105 FF)	n/a
G-value (%)	40	40	n/a
Light transmittance (%)	71	71	n/a
Rooflight U-value (W/m^2 K)	n/a	n/a	1.8 (15% FF)
G-value (%)	n/a	n/a	55
Light transmittance (%)	n/a	n/a	60
Air permeability (m^3/m^2/hour) (gross internal area less than or equal to 250 m^2)	5	5	7
Air permeability (m^3/m^2/hour) (gross internal area greater than 250 m^2 and less than 3500 m^2)	3	3	7
Air permeability (m^3/m^2/hour) (gross internal area greater than or equal to 3500 m^2 and less than 10,000 m^2)	3	3	5
Air permeability (m^3/m^2/hour) (gross internal area greater than or equal to 10,000 m^2)	3	3	3
Lighting luminaire (lm/circuit-watt)	60	60	60
Occupancy control (Yes/No)	Yes	Yes	Yes
Daylight control (Yes/No)	Yes	Yes	Yes
Maintenance factor	0.8	0.8	0.8
Constant illuminance control	No	No	No
Heating efficiency (heating and hot water) (%)	91	91	91
Central ventilation SFP (W/l/s)	1.8	1.8	1.8
Terminal unit SFP (W/l/s)	0.3	0.3	0.3
Cooling (air-conditioned) (SEER/SSEER)	n/a	4.5/3.6	4.5/3.6
Cooling (mixed mode) (SSEER)	n/a	2.7	2.7
Heat recovery efficiency (%)	70	70	70
Variable speed control of fans and pumps, controlled via multiple sensors	Yes	Yes	Yes
Demand control (mechanical ventilation only). Variable speed control of fans via	Yes	Yes	Yes

- workshops;
- non-residential agricultural buildings with low energy demand;
- stand-alone buildings with a total useful floor area of less than 50 m^2;
- 'some' conservatories and porches. (If this affects you, we would suggest that you always seek further advice from your planning officer before proceeding.)

6.15.2.2 Building fabric

The building fabric should be constructed to a reasonable standard so that: L2A 3.2

- the insulation is reasonably continuous over the whole building envelope; and
- the air permeability is within reasonable limits.

Reductions in thermal performance can occur where the air barrier and the insulation layer are not contiguous and the cavity between them is subject to air movement. To avoid this problem, either:

- the insulation layer should be contiguous with the air barrier at all points in the building envelope; or
- the space between the insulation layer and air barrier should be filled with solid material such as in a masonry wall.

6.15.2.3 Doors

If a door is enlarged or a new one created in a building other than a dwelling, the area of the pedestrian door, expressed as a percentage of the total floor area of the building, should not exceed the relevant value from Table 6.96, or should be compensated for in some other way.   L2B 4.24

Table 6.96 Opening areas in the extension

Building type	Personnel doors as percentage of exposed wall	Rooflights as percentage of area of roof
Places of assembly, offices and shops	40	20
Industrial and storage buildings	15	20
Vehicle access doors	As required	N/A

6.15.2.3.1 Limiting fabric standards for doors

For doors, the U-value needs to be: L2A 2.40

- the standard size as laid out in BS EN 14351-1; or
- the specific size and configuration of the actual door.

6.15.2.4 Windows

The area of windows and rooflights in the extension should   L2B 4.4
generally not exceed the values given in Table 6.97.

Table 6.97 Opening areas in the extensions to buildings other than dwellings

Building type	Windows and personnel doors as percentage of exposed wall	Rooflights as percentage of area of roof
Places of assembly, offices and shops	40	20
Industrial and storage buildings	15	20
Vehicle access doors and display windows and similar glazing	As required	N/A
Smoke vents	N/A	As required

6.15.2.4.1 Limiting fabric standards for windows

U-values shall be calculated using the methods and conventions set out in BR 443 (Conventions for U-value calculations), and should be based on the whole element or unit (e.g. in the case of a window, this would be the combined performance of the glazing and the frame).

For windows, the U-value needs to be: L2A 2.40

- the smaller of the two standard windows defined in BS EN 14351-1; or
- the standard configuration set out in BR 443; or
- the specific size and configuration of the actual window.

6.15.2.5 Glazing

Some glazing materials (such as annealed glass) gain strength through thickness; others such as polycarbonates or glass blocks are inherently strong.

The maximum dimensions for annealed glass of different   K4 5.5
thicknesses for use in large areas forming fronts to shops,
showrooms, offices, factories and public buildings with four edges
supported are shown in Figure 6.239 to a minimum height.

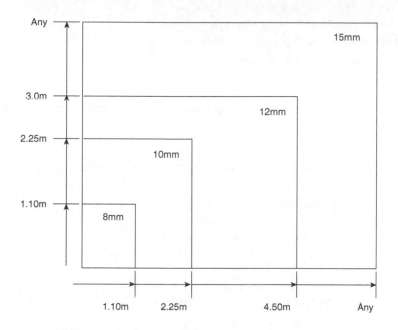

Figure 6.239 Annealed glass thickness and dimension limits.

6.15.2.6 Ventilation

Fresh air supplies should be protected from contaminants that   F 6.3
would be injurious to health.

6.15.2.6.1 Offices

All office sanitary accommodation, washrooms and food and beverage preparation areas shall be provided with intermittent air extract ventilation capable
of meeting the requirements of Table 6.98 and the following requirements of
Approved Document F:

Extract fans that are located in an internal room   F Tables 5.2a to 5.2d
which does not have an openable window
should have a 15-minute overrun.

Extract ventilators should be located as high as practicable and preferably not less than 400 mm below the ceiling level.	F Tables 5.2a to 5.2d
Passive stack ventilators (PSVs) (which can be used as an alternative to a mechanical extract fan for office sanitary accommodation and washrooms and food preparation areas) should be located in the ceiling of the room.	F Tables 5.2a to 5.2d
Purge ventilation shall be provided in each office.	F 6.12

 Purged air should be taken directly outside and **not** be recirculated to any other part of the building.

PSV controls and extract fans can be either manual or automatic.	F Tables 5.2a to 5.2d
Printers and photocopiers that are being used in large numbers and which are in almost constant use (i.e. greater than 30 minutes per hour) shall:	F 6.12

- be located in a separate room;
- have extract facilities capable of provid-ing an extract rate greater than 20 l/s per machine during use (Table 6.98).

 Note: The air flow rates given in Table 6.98 can mainly be provided by natural ventilation.

Table 6.98 Extract ventilation rate

Room	Air extract rate
Rooms containing printers and photocopiers in substantial use (greater than 30 minutes per hour)	20 l/s per machine during use
Office sanitary accommodation and washrooms	15 l/s per shower/bath 6 l/s per WC/urinal
Food and beverage preparation areas (not commercial kitchens)	15 l/s with microwave and beverages only 30 l/s adjacent to the hob with cooker(s) 60 l/s elsewhere with cooker(s)
Specialist buildings and spaces (e.g. commercial kitchens, fitness rooms)	See Table 2.3 of Approved Document F

The outdoor air supply rates shown in Table 6.98 for offices are based on controlling body odours with low levels of other pollutants.

> The whole-building ventilation rate for the supply of air to   F 6.12
> the offices should be greater than 10 l/s per person.

6.15.2.6.2 Other types of building

The ventilation requirements for other buildings (e.g. assembly halls, broadcasting studios, computer rooms, factories, hospitals, hotels, museums, schools, sports centres, warehouses, etc.) are listed in Table 2.3 of Approved Document F, which also provides a link to the relevant controlling Acts of Parliament, statutory instruments, British Standards, and Chartered Institution of Building Services Engineers (CIBSE) and Health and Safety Executive (HSE) standards, practices and recommendations.

6.15.2.6.3 Underground car parks

In underground car parks or car parks that are not open-sided, ventilation is an important factor, as heat and smoke cannot be dissipated readily, and non-combustible materials should always be used in their construction.

For more guidance see Section 11 of Approved Document B3 (Volume 2).

> Underground car parks, enclosed car parks and   F 6.18
> multi-storey car parks should be designed to limit the
> concentration of carbon monoxide to not more than 30
> parts per million, averaged over an eight-hour period
> and peak concentrations.
>
> Naturally ventilated car parks shall have openings at   F 6.20a
> each car-parking level:
>
> • at least 1/20 of the floor area at that level;
> • with a minimum of 25 per cent on each of two
>   opposing walls.
>
> Ramps and exits shall not go above 90 parts per million   F 6.18
> for periods not exceeding 15 minutes.
>
> Car parks should have a separate and independent   B1 5.50 (V2)
> extraction system and the extracted air should **not** be
> recirculated.
>
> Mechanically ventilated car parks can have either   F 6.20b
> natural ventilation openings (that are not less than 1/40
> of the floor area) or a mechanical ventilation system
> capable of at least three air changes per hour (ach).

Mechanically ventilated basement car parks shall be capable of at least six air changes per hour (ach).	F 6.20b
Mechanically ventilated exits and ramps (i.e. where cars queue inside the building with engines running) shall be capable of at least ten air changes per hour (ach).	F 6.20b
Where a common stair forms part of the only escape route from a flat or dwelling, it should **not** also serve any covered car park.	B1 2.46 (V2)
In enclosed (i.e. non-open-sided) car parks, lifts should be approached by a protected lobby or protected corridor (unless it is within the enclosure of a protected stairway).	B1 5.43 (V2)

Car parks are **not** normally expected to be fitted with sprinklers.

6.15.2.7 Fire safety

6.15.2.7.1 Hospitals

All hospitals and similar healthcare premises should be designed in accordance with HTM 05, Firecode.

6.15.2.7.2 Small offices

In small premises:

• floor areas should be undivided (except for ancillary offices and stores) to ensure that exits are clearly visible from all parts of the floor areas; • clear-glazed areas should be provided in partitions separating an office from the open floor area to enable any person within the office to have early, visual warning of an outbreak of fire.	B1 3.36 (V2)

6.15.2.8 Assembly buildings

To maintain sightlines for spectators, stairs and/or ramps that form part of the means of access within an assembly building such as a sports stadium, theatre or cinema should be in accordance with the following requirements:

- Ensure that the maximum pitch for gangways to seating   K1 1.4
  areas for spectators is 35°.
- Align the ends of all rows of seats/wheelchair spaces
  so that the width of the gangway remains the same.
- Provide transverse gangways that (in auditoria with
  tiered seating) do not cross.
- Ensure that stepped tiers are between 100 mm and
  190 mm high.
- Where an exit is approached from a stepped gangway,
  place a landing the width of the exit and a minimum of
  1100 mm deep immediately in front of the exit doors.
- For stepped side gangways, provide a handrail.

 Note: Gangways should normally be 900 mm wide unless they are used by more than 50 persons, in which case they should be a minimum of 900 mm.

6.15.2.9 Disabled facilities

In accordance with the Equality Act 2010, the overall aim should be that **all** people can have access to (and be able to use) **all** of the facilities provided within a building, and all floor areas (even when located at different levels) should be accessible.

6.15.2.9.1 Hotels, motels and student accommodation

In hotels, motels and student accommodation: M 4.4

- a proportion of the sleeping accommodation should
  be designed for wheelchair users;
- the remainder should include facilities suitable
  for people with sensory, dexterity or learning
  difficulties.

If there is a reception point: M 3.2 to 3.5

- it should be easily accessible and convenient to use;
- information about the building should be clearly
  available from noticeboards and signs;
- the floor surface should be slip-resistant.

6.15.2.9.1.1 SLEEPING ACCOMMODATION

Sleeping accommodation in hotels, motels and student   M 4.17
accommodation should be convenient for all types of people.

A proportion of rooms should be available for wheelchair users.	M 4.17
Wheelchair users should be able to:	
• reach all the facilities available within the building;	M 4.18
• manoeuvre around and use the facilities in the room, and operate switches and controls.	M 4.19
En suite sanitary facilities are the preferred option for wheelchair-accessible bedrooms.	M 4.19

 There should be at least as many en suite shower rooms as en suite bathrooms – some mobility-impaired people may find it easier to use a shower than a bath.

In all bedrooms, built-in wardrobes and shelving should be accessible and convenient to use.	M 4.20
Bedrooms that are **not** designed for independent use by a person in a wheelchair should have an outer door wide enough to be accessible to a wheelchair user.	M 4.21
For all bedrooms:	
• built-in wardrobe swing doors should open through 180°;	M 4.24b
• handles on hinged and sliding doors:	M 4.24c
◦ should be easy to grip and operate;	
◦ should contrast visually with the surface of the door;	
• windows and window controls should be:	M 4.24d
◦ located between 800 and 1000 mm above the floor;	
◦ easy to operate without using both hands simultaneously;	
• a visual fire alarm signal should be provided in addition to the requirements of Approved Document B;	M 4.24e
• room numbers should be embossed characters;	M 4.24f
• the effective clear width of the door from the access corridor should comply with Table 6.99.	M 4.24a

At least one wheelchair-accessible bedroom should be provided for every 20 bedrooms.

Table 6.99 Minimum effective clear widths of doors

Direction and width of approach	New buildings (mm)	Existing buildings (mm)
Straight on (without a turn or oblique approach)	800	750
At right angles to an access route at least 1500mm wide	800	750
At right angles to an access route at least 1200mm wide	825	775
External doors to buildings used by the general public	1000	775

Wheelchair-accessible bedrooms should:

• be located on accessible routes;	M 4.24h
• be designed to provide a choice of location;	M 4.24i
• have a standard of amenity equivalent to that of other bedrooms;	M 4.24j
• have (i.e. for en suite bathroom and shower room doors) an effective clear width complying with Table 6.99;	M 4.24k
• be large enough to enable a wheelchair user to manoeuvre with ease (Figure 6.240).	M 4.24l

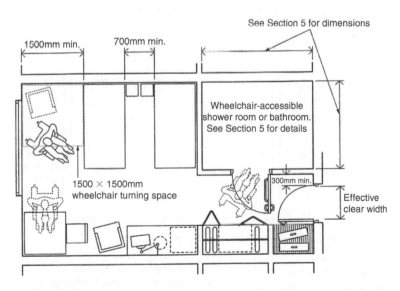

Figure 6.240 Example of a wheelchair-accessible hotel bedroom. (Dimensions in mm.)

In wheelchair-accessible bedrooms:

- if wide-angle viewers are provided in the entrance door, they should be located at 1050 mm and 1500 mm above floor level;   M 4.24n

- if a balcony is provided, it should: M 4.24o
  - comply with Table 6.99;
  - have a level threshold;

- have no horizontal transoms between 900 mm and 1200 mm above the floor;

- there should be no permanent obstructions in a zone 1500 mm back from any balcony doors;   M 4.24p

- emergency assistance alarms should be provided.   M 4.24q

The door from the access corridor to a wheelchair-accessible bedroom should: M 4.24j

- not require more than 20 N opening force;

- have an effective clear width through a single-leaf door (or one leaf of a double-leaf door) in accordance with Table 6.99 and Figure 6.241;

- have an unobstructed space of at least 300 mm on the pull side of the door (Figure 6.241).

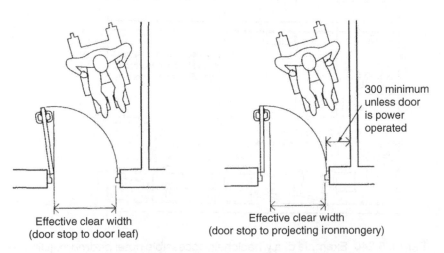

300 minimum unless door is power operated

Effective clear width
(door stop to door leaf)

Effective clear width
(door stop to projecting ironmongery)

Figure 6.241 Effective clear width of doors. (Dimensions in mm.)

Sanitary facilities, en suite to a wheelchair-accessible bedroom, should comply with the provisions of M 5.15 to M5.21 for wheelchair-accessible bathrooms or wheelchair-accessible shower facilities.

6.15.2.9.1.2 SWITCHES, OUTLETS AND CONTROLS

The aim should be to ensure that all switches, outlets and controls are easy to operate, visible and free from obstruction.

Light switches should:	M 4.30h and l
• have large push pads; • align horizontally with door handles; • be within 900 mm and 1100 mm from the entrance door opening.	
Switches and controls should be located between 750 mm and 1200 mm above the floor.	M 4.30c and d
The operation of all switches, outlets and controls should not require the simultaneous use of both hands (unless necessary for safety reasons).	M 4.30j
Switched socket outlets should indicate whether they are 'ON'.	M 4.30k
Mains and circuit isolator switches should clearly indicate whether they are 'ON' or 'OFF'.	M 4.30l
Individual switches on panels and on multiple-socket outlets should be well separated.	M 4.29
All socket outlets should be wall mounted.	M 4.30a and b
All telephone points and TV sockets should be located between 400 mm and 1000 mm above the floor (or 400 mm and 1200 mm above the floor for permanently wired appliances).	M 4.30a and b
Socket outlets should be located no nearer than 350 mm from room corners.	M 4.30g
Controls that need close vision (e.g. thermostats) should be located between 1200 mm and 1400 mm above the floor.	M 4.30f

Emergency alarm pull cords should:

• be coloured red; • be located as close to a wall as possible;	M 4.30e

> •   have two red 50 mm diameter bangles.
>
> Front plates should contrast visually with their backgrounds.   M 4.30m

 The colours red and green should **not** be used in combination as indicators of 'ON' and 'OFF' for switches and controls.

6.15.2.9.2 Audience and spectator facilities

Audience and spectator facilities fall primarily into three categories:

• lecture/conference facilities;
• entertainment facilities (e.g. theatres/cinemas);
• sports facilities (e.g. stadiums).

Wheelchair users (as well as those with mobility and/or sensory problems) may need to see or listen from a particular side, or sit at the front to lip-read or read sign interpreters. For this reason, a selection of spaces should be provided into which they can manoeuvre easily and that offer them a clear view of an event – taking particular care that these do **not** become segregated into 'special areas'.

> The route to wheelchair spaces should be accessible to users.   M 4.12a
>
> Stepped access routes to audience seating should be provided with fixed handrails.   M 4.12b
>
> Handrails to external stepped or ramped access should be positioned as shown in Figure 6.242.   M 4.12a

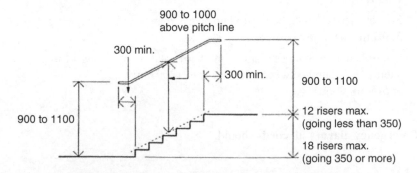

Figure 6.242 Handrails to external stepped and ramped access – key dimensions (mm).

Handrails to external stepped or ramped access should: M 1.37

- be continuous across flights and landings;
- extend at least 300 mm horizontally beyond the top and bottom of a ramped access;
- not project into an access route;
- contrast visually with the background;
- have a slip-resistant surface which is not cold to the touch;
- terminate in such a way as to reduce the risk of clothing being caught;
- be either circular (with a diameter of between 32 mm and 50 mm) or non-circular (50 mm wide and 39 mm deep) having rounded edges with a minimum radius of 15 mm (Figure 6.243).

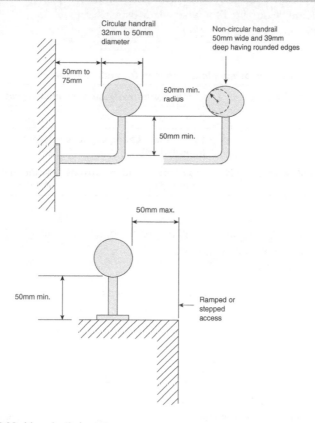

Figure 6.243 Handrail design.

 Note: In areas where resistance to vandalism or low maintenance is a key factor, the use of metals with relatively low thermal conductivity is recommended.

Handrails to external stepped or ramped access should: M 4.12b

- not protrude more than 100 mm into the surface width of the ramped or stepped access;
- have a clearance of between 60 mm and 75 mm between the handrail and any adjacent wall surface;
- have a clearance of at least 50 mm between a cranked support and the underside of the handrail;
- have their inner face located no more than 50 mm beyond the surface width of the ramped or stepped access;
- be spaced away from the wall and rigidly supported in a way that avoids impeding finger grip;
- be set at heights that are convenient for all users of the building.

The minimum number of permanent and removable spaces   M 4.12c
that should be provided for wheelchair users is as given in
Table 6.100.

Table 6.100 Provision of wheelchair spaces for audience seating

Seating capacity	Minimum provision of spaces for wheelchairs	
	Permanent	Removable
Up to 600	1% of total seating capacity (rounded up)	Remainder to make a total of 6
Over 600 but less than 10,000	1% of total seating capacity (rounded up)	Additional provision if desired

Some wheelchair spaces should be provided in pairs, with   M 4.12d
standard seating on at least one side (Figure 6.244).

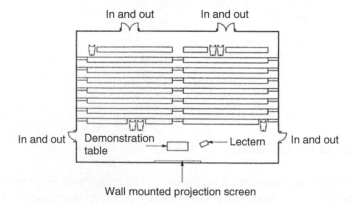

Figure 6.244 An example of wheelchair spaces in a lecture theatre.

If more than two wheelchair spaces are provided, they should be located so as to give a range of views of the event.

The minimum clear space for:	
• access to wheelchair spaces should be 900 mm;	M 4.12f
• wheelchair spaces in a parked position should be 900 mm wide by 1400 mm deep.	M 4.12g
The floor of each wheelchair space should be horizontal.	M 4.12h
Seats at the ends of rows and next to wheelchair spaces should have detachable (or lift-up) arms.	M 4.12j
Wheelchair spaces at the back of a stepped terraced floor should be in accordance with Figure 6.245.	M 4.12k

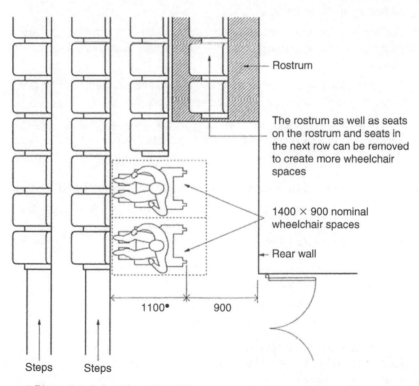

• Dimension derived from BS 8300

Figure 6.245 Possible location of wheelchair spaces in front of a rear aisle. (Dimensions in mm.)

6.15.2.9.3 Lecture/conference facilities

Everyone (no matter their disability) should be able to fully participate in lecture/conference facilities as well as be able to enjoy entertainment, leisure and social venues – not just as spectators, but also as participants and/or staff. To achieve these aims:

Disabled people should be able:	M 4.2
• to have a choice of seating location at spectator events; • to have a clear view of the activity taking place (whilst not obstructing the view of others); • to use e-presentation facilities.	
Bars and counters in refreshment areas should be at a suitable level for wheelchair users.	M 4.3
The design of the acoustic environment should ensure that audible information can be heard clearly.	M 4.9
Artificial lighting should be designed to give good colour rendering of all surfaces.	M 4.9
Glare and reflections from shiny surfaces (and large repeating patterns) should be avoided in spaces where visual accuracy is critical.	M 4.9
Uplighters mounted at low or floor level should be avoided, as they can disorientate some visually impaired people.	M 4.9
Where a podium or stage is provided, wheelchair users should have access to it by means of a ramp or lifting platform.	M 4.12

 A clearly audible public address system should be supplemented by visual information.

Hearing enhancement systems should be installed:	M 4.36
• in rooms and spaces designed for meetings, lectures, classes, performances, spectator sports or films; • at service or reception counters (especially when situated in noisy areas or behind glazed screens).	
The availability of an induction loop or infrared hearing enhancement system should be indicated by the standard symbol (see Figure 6.248).	
Telephones suitable for hearing aid users should:	
• be clearly indicated by the standard ear symbol;	

- incorporate an inductive coupler and volume control.

Text telephones for deaf and hard-of-hearing people should be clearly indicated by the standard symbol.

Artificial lighting should be designed to be compatible with other electronic and radio frequency installations.

6.15.2.9.4 Entertainment, leisure and social facilities

In all entertainment, leisure and social facilities, it is important that there are adequate signs available to indicate facilities (see Figure 6.246).

Toilets are available that have been adapted for people with mobility impairments

Changing rooms are available for people with mobility impairments

Assistance dogs are welcome on the premises

Figure 6.246 Examples of facility signs.

In theatres and cinemas (where seating is normally closely packed together) special care should be given to the design and location of wheelchair spaces.   M 4.10

6.15.2.9.5 Refreshment facilities

Restaurants and bars should be designed so that they can be reached and used by all people independently or with companions.

All people should have access to:

- all parts of the facility;   M 4.13

- staff areas;

- public areas (e.g. lavatory accommodation, public   M 4.14
  telephones and external terraces);

- self-service facilities (when provided).

Changes of floor level are permitted, provided that all the different levels are accessible and raised thresholds are avoided.

Part of the working surface of a bar or serving counter should: M 4.16b

- be permanently accessible to wheelchair users;
- be at a level of not more than 850 mm above the floor.

 Note: If a raised threshold is unavoidable, the total height should not be more than 15 mm, with a minimum number of upstands and slopes and with any upstands higher than 5 mm chamfered or rounded.

In addition: M 4.16c

- the worktop of a shared refreshment facility (e.g. for tea making) should be 850 mm above the floor with a clear space beneath at least 700 mm above the floor (see Figure 6.247);
- basin taps should either be controlled automatically or be capable of being operated using a closed fist (e.g. by lever action).

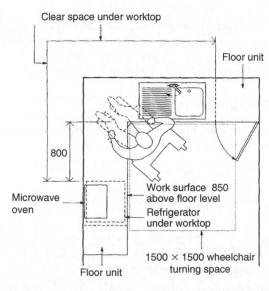

Figure 6.247 An example of a typical shared refreshment facility. (Dimensions in mm.)

6.15.2.9.6 Aids to communication

The design of the acoustic environment should ensure that audible information can be heard clearly.	M 4.33
A clearly audible public address system should be supplemented by visual information.	

Note: Assisting people with impaired hearing to fully participate in public discussions (meetings and performances) may require an advanced sound level system (e.g. induction loop, infrared system or radio system) to be installed.

A hearing system is available   Mini-com and/or text   Staff have received disability
in certain locations   phone facility available   awareness training

Figure 6.248 Examples of aids to communication.

Hearing enhancement systems should be installed:	M 4.36b
• in all rooms and spaces designed for meetings, lectures, classes, performances, spectator sport or films; • at service or reception counters (especially when they are situated in noisy areas or they are behind glazed screens).	
The availability of an induction loop or infrared hearing enhancement system should be indicated by the standard symbol.	M 4.36c
Telephones suitable for hearing aid users should:	M 4.36d
• be clearly indicated by the standard ear and 'T' symbol; • incorporate an inductive coupler and volume control.	
Text telephones for deaf and hard-of-hearing people should be clearly indicated by the standard symbol.	M 4.36e

6.16 Sanitary accommodation, bathrooms and showers

6.16.1 Requirements

Sanitary conveniences

(1) Adequate and suitable sanitary conveniences must be provided in rooms provided to accommodate them or in bathrooms.
(2) Adequate stand washing facilities must be provided in rooms containing sanitary conveniences, or in rooms or spaces adjacent to rooms containing sanitary conveniences.
(3) Any room containing a sanitary convenience, a bidet, or any facility for washing hands provided in accordance with paragraph (2), must be separated from any kitchen or any area where food is prepared.

(Approved Document G4)

There must be a suitable installation for the provision of water (of suitable quality) to any sanitary convenience fitted with a flushing device.

(Approved Document G1)

*All dwellings (and buildings containing one or more rooms for residential purposes) will be provided with a bathroom which contains either a fixed bath or a shower **and** a washbasin.*

(Approved Document G5)

Ventilation
There shall be adequate means of ventilation provided for people in the building.

(Approved Document F)

Access to and use of buildings
Reasonable provisions shall be made for people to gain access to, and use the building and its facilities.

(Approved Document M)

 Amendments to the 2004 version of Approved Document M came into force in 2013.

6.16.1.1 The Building Act 1984 Sections 64–68

Under existing regulations, all buildings (except factories and buildings used as workplaces) shall be provided with sufficient closet (or privy) accommodation according to the intended use of that building and the amount of people using that building. The only exceptions are if the building (in the view of the local authority) has an insufficient water supply and a sewer is not available.

If a building already has a sufficient water supply and sewer available, the local authority has the authority to insist that the owner of the property replaces any other closet (e.g. an earth closet) with a water closet. In these cases the owner is entitled to claim 50 per cent of the expense of doing this from the local authority.

Notes:

(1) If you propose using an earth closet, the local authority **cannot** reject the plans unless it considers that there is insufficient water supply to that earth closet.
(2) If the local authority completes the work, it is entitled to claim 50 per cent back from the owner.
(3) In the Greater London area, a 'water closet' can also be taken to mean a 'urinal'.

6.16.2 Meeting the requirements

Although the above are the minimum requirements for meeting sanitary convenience and washing facility regulations, other local authority regulations may apply, and it is worth seeking the advice of the local planning officer before proceeding with any new work.

6.16.2.1 Sanitary conveniences

Sanitary conveniences of an appropriate type for the sex and age of the persons using the building shall be provided in sufficient numbers. Hand-washing facilities shall be provided in, or adjacent to, rooms containing sanitary conveniences.

Water intended for sanitary conveniences shall:	G1

- be reliable;
- be either wholesome, softened wholesome or of suitable quality;
- possess a pressure and flow rate sufficient for the operation of the sanitary appliances;
- convey water to sanitary appliances and locations without waste, misuse, undue consumption or contamination of wholesome water.

It may be provided from the following alternative sources:	G1-6

- water abstracted from wells, springs, boreholes or watercourses;
- harvested rainwater;
- reclaimed greywater; or
- reclaimed industrial-process water.

> Measures should be incorporated to minimize the impact on   G1-6
> water quality from:
> - failure of any components;
> - failure due to lack of maintenance;
> - power failure; and
> - any other measures identified in a risk assessment.

Notes:

(1) Any system/unit used to supply dwellings with water from alternative sources should be subject to a risk assessment by the system designer and the manufacturer.
(2) A record of all sanitary appliances used in the water consumption calculation and installed in the dwelling shall be maintained.

6.16.2.1.1 Sanitary fittings

> The surface finish of sanitary fittings and grab bars should   M 5.4k
> contrast visually with background wall and floor finishes.
>
> Taps and WC cubicle doors should be operable by people   M 5.3
> with limited strength and/or manual dexterity.
>
> Bath and washbasin taps should either be controlled   M 5.4a
> automatically or be capable of being operated using a closed
> fist (e.g. by lever action).

6.16.2.1.2 Protection of openings for pipes

Pipes that pass through a compartment wall or a compartment floor (unless the pipe is in a protected shaft), or through a cavity barrier, should:

> - contain proprietary seals (any pipe diameter)   B3 11.5–11.6
>   that maintain the fire resistance of the wall,
>   floor or cavity barrier; or
> - use fire-stopping around the pipe, keeping the   B3 11.5 and 11.7
>   opening as small as possible; or
> - use some form of sleeving.   B3 11.5 and 11.8

Note: A lead, aluminium, aluminium alloy, fibre-cement or UPVC pipe, with a maximum nominal internal diameter of 160 mm, may be used with a sleeving of non-combustible pipe as shown in Figure 6.249, ensuring that the opening in the structure is as small as possible and provides fire-stopping between pipe and structure.

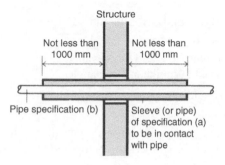

Figure 6.249 Pipes penetrating a structure.

Joints between fire-separating elements should be fire-stopped and all openings for pipes that pass through any part of a fire-separating element should also be fire-stopped, as small as practicable and limited in number.   B3 7.12 (V1)
B3 10.17 (V2)

6.16.2.1.3 Outlets, controls and switches

The aim is to ensure that all controls and switches should be easy to operate, visible and free from obstruction.

- They should be located between 750 mm and 1200 mm above the floor.   M 5.4i and 4.30
- They should not require the simultaneous use of both hands (unless necessary for safety reasons) to operate.
- Light switches should:
  - have large push pads;
  - align horizontally with door handles;
  - be within 900 mm to 1100 mm from the entrance door opening.
- Switched socket outlets should indicate whether they are 'ON'.
- Mains and circuit isolator switches should clearly indicate whether they are 'ON' or 'OFF'.
- Individual switches on panels and on multiple socket outlets should be well separated.
- Controls that need close vision (e.g. thermostats) should be located between 1200 mm and 1400 mm above the floor.

- **Emergency alarm pull cords** should be:
    - coloured **red**;
    - located as close to a wall as possible;
    - have two **red** 50 mm diameter bangles.
- Front plates should contrast visually with their backgrounds.

Heat emitters should either be screened or have their exposed surfaces kept at a temperature below 43° C.   M 5.4j

Where possible, light switches with large push pads should be used in preference to pull cords.   M 5.3

The colours **red** and **green** should not be used in combination as indicators of 'ON' and 'OFF' for switches and controls.   M 5.3

6.16.2.1.4 Scale of provision and layout in dwellings

Any dwelling (house or flat) should have at least **one** 'sanitary convenience' and associated hand-washing facility located in the principal/entrance storey of the dwelling.   G4.7

Where additional sanitary conveniences are provided, each should have an associated hand-washing facility.   G4.8

Hand-washing facilities should be located in: G4.9

- the room containing the sanitary convenience; or
- an adjacent room or place that provides the sole means of access to the room containing the sanitary convenience.

Provided that it is **not** used for the preparation of food, a place containing a sanitary convenience and/or associated hand-washing facilities should be separated by a door from any place used for the preparation of food – including a kitchen. (See Figures 6.250 and 6.251.)

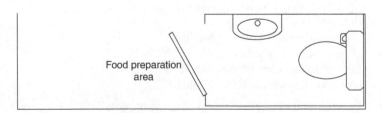

Food preparation area

Figure 6.250 Separation between hand washbasin/WC and food preparation area – single room.

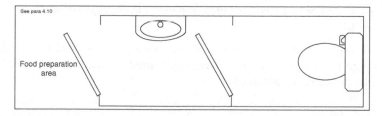

Figure 6.251 Separation between hand washbasin/WC and food preparation area – two rooms.

6.16.2.1.4.1 WC PROVISION IN THE ENTRANCE STOREY OF THE DWELLING

Whenever possible, a WC should be provided in the entrance storey of the dwelling so that there is no need to negotiate a stair to reach it from the habitable rooms in that storey.

 If there is a bathroom in the principal storey, then a WC may be collocated with it.

The door to the WC compartment should:	M 10.3b

- open outwards;
- be positioned so as to allow wheelchair users access to it;
- have a clear opening width in accordance with Table 6.101.

The WC compartment should: M 10.3c

- provide a clear space for wheelchair users to access the WC;
- position the washbasin so that it does not impede access.

Table 6.101 Minimum widths of corridors and passageways for a range of doorway widths

Doorway clear opening width (mm)	Corridor/passageway width (mm)
750 or wider	900 (when approached head on)
750	1200 (when not approached head on)
775	1050 (when not approached head on)
800	900 (when not approached head on)

 For further information see the website of the Equality and Human Rights Commission (http://www.equalityhumanrights.com).

6.16.2.2 Scale of provision and layout in buildings other than dwellings

The Workplace (Health, Safety and Welfare) Regulations 1992 require that a minimum number of sanitary conveniences **must** be provided in workplaces and that these may be provided in:

• a self-contained room which also contains hand-washing facilities;	G4.16
• a cubicle with shared hand-washing facilities located in a room containing a number of cubicles; or	
• a self-contained room with hand-washing facilities provided in an adjacent room.	

In addition, these regulations require that:

• a place containing a sanitary convenience and/or associated hand-washing facilities should be separated by a door from any place used for the preparation of food (including a kitchen).	G4.17

 Urinals, WC cubicles and hand-washing facilities may be in the same room, **but**, where hot and cold taps are provided on a sanitary appliance, the hot tap shall be on the **left**!

6.16.2.2.1 Provision of toilet accommodation in a building

Where sanitary facilities are provided in a building, at least one wheelchair-accessible unisex toilet should be available.	M 5.7b
In large building complexes (e.g. retail parks and large sports centres) there should be one wheelchair-accessible unisex toilet capable of including an adult changing table.	M 5.17
If there is only space for one toilet in a building, then it should be a wheelchair-accessible unisex type.	
In separate-sex toilet accommodation, at least one WC cubicle should be provided for ambulant disabled people.	M 5.7c
In separate-sex toilet accommodation having four or more WC cubicles, at least one should be an enlarged cubicle for use by people who need extra space.	M 5.7d
Wheelchair-accessible unisex toilets should always be provided in addition to any wheelchair-accessible accommodation in separate-sex toilet washrooms.	M 5.5

Note: Facilities incorporating adult changing tables are more commonly known as changing places toilets.

6.16.2.3 Chemical and composting toilets

Chemical toilets or composting toilets may be used where:

- suitable arrangements can be made for the disposal of   G4.19
  the waste either on or off the site; and
- the waste can be removed from the premises without
  carrying it through any living space or food preparation
  areas (including a kitchen); and
- no part of the installation is installed in a place where
  it might be rendered ineffective by the entry of flood
  water.

> Composting toilets should **not** be connected to an   G4.21
> energy source other than for purposes of ventilation
> or sustaining the composting process.

6.16.2.3.1 Discharges to drains

Any WC fitted with flushing apparatus should discharge to an   G4.22
approved drainage system.

A urinal fitted with flushing apparatus should discharge   G4.23
through a grating (trap or mechanical seal and a branch pipe) to
a discharge stack or a drain.

A WC fitted with a macerator and pump may be connected to a   G4.24
small-bore drainage system that discharges to a discharge stack
if:

- there is also access to a WC discharging directly to a
  gravity system; and
- the macerator and pump meet the requirements of BS EN
  12050-1:2001 or BS EN 12050-3:2001.

Sanitary appliances used for personal washing: G5.9

- should discharge through a grating, a trap and a branch
  discharge pipe to an adequate system of drainage;
- that are fitted with a macerator and pump may be
  connected to a small-bore drainage system discharging to a
  discharge stack if there is also access to washing facilities
  discharging directly to a gravity system.

6.16.2.4 Public conveniences

6.16.2.4.1 Erection of public conveniences

You are **not** allowed to erect a public sanitary convenience in (or on) any location that is accessible from a street without the consent of the local authority. Any person who contravenes this requirement is liable to a fine and can be made (at his or her own expense) to remove or permanently close the convenience. This requirement does not apply to sanitary conveniences erected by a railway company within its railway station, yard or approaches, or erected by dock undertakers on land belonging to them.

6.16.2.4.2 Loan of temporary sanitary conveniences

If the local authority is maintaining, improving or repairing drainage systems and this requires the disconnection of existing buildings from these sanitary conveniences, then, on request from the occupier of the building, the local authority is required to supply (on temporary loan and at no charge) sanitary conveniences:

*   if the disconnection is caused by a defect in a public sewer;
*   if the local authority has ordered the replacement of earth closets (see section 6.16.1.1);
*   for the first seven days of any disconnection.

6.16.2.5 Workplace conveniences

If the proposed building is going to be used as a workplace or a factory in which persons of both sexes are going to be employed, separate closet accommodation **must** be provided unless the local authority approves otherwise.

 There may be some additional requirements regarding the number, types and siting of appliances in business premises. If this applies to you, then you will need to look at:

*   the Offices, Shops and Railway Premises Act 1963;
*   the Factories Act 1961; or
*   the Food Hygiene (General Hygiene) Regulations 1995.

6.16.2.6 "Disabled" access

The aim of Approved Document M is that suitable sanitary accommodation should be available to *everybody*, including sanitary accommodation specifically designed for wheelchair users, ambulant disabled people, people of either sex with babies and small children, and/or people with luggage.

6.16.2.6.1 Wheelchair-accessible unisex toilets

A wheelchair-accessible compartment (where provided) shall have the same layout and fittings as the unisex toilet.

 Where sanitary facilities are provided in a building, at least one wheelchair-accessible unisex toilet should be available. (This is a comparatively new requirement and many builders, and indeed architects, have been caught out on this one!)

Wheelchair-accessible unisex toilets should **not** be used for baby changing.	M 5.5
Wheelchair-accessible unisex toilets should:	
• be located as close as possible to the entrance and/ or waiting area of the building;	M 5.10a
• not be located in a way that compromises the privacy of users;	M 5.10b
• be located in a similar position on each floor of a multi-storey building;	M 5.10c
• allow for right- and left-hand transfer on alternate floors;	M 5.10c and d
• be located on accessible routes that are direct and obstruction free;	M 5.10f
• always be provided in addition to any wheelchair-accessible accommodation in separate-sex toilet washrooms;	M 5.5
The minimum overall dimensions and arrangement of fittings within a wheelchair-accessible unisex toilet should comply with Figure 6.252.	M 5.10i

6.16.2.6.1.1 DOORS

Doors to wheelchair-accessible unisex toilets should:

• (ideally) open outwards;	M 5.3
• be operable by people with limited strength or manual dexterity;	
• be capable of being opened if a person has collapsed against them while inside the cubicle.	

Doors to wheelchair-accessible unisex toilets, changing rooms or shower rooms should:

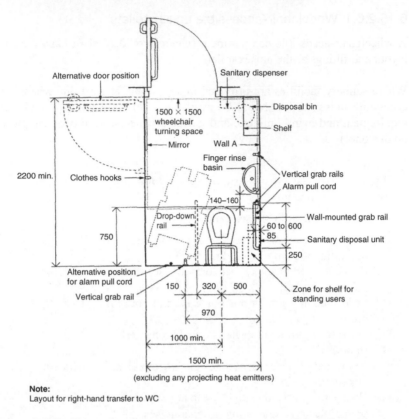

Note:
Layout for right-hand transfer to WC

Figure 6.252 Example of a unisex wheelchair-accessible toilet with a corner WC. (Dimensions in mm.)

• be fitted with light-action privacy bolts;	M 5.4d
• be capable of being opened using a force no greater than 30 N;	M 5.4d
• have an emergency release mechanism so that they can be opened outwards (from the outside) in case of emergency.	M 5.4e

Door-opening furniture should:

• be easy to operate by people with limited manual dexterity;	M 5.4c
• be easy to operate with one hand using a closed fist (e.g. a lever handle);	
• contrast visually with the surface of the door.	

Doors, when open, should **not** obstruct emergency escape   M 5.4f
routes.

6.16.2.6.2 Approach to toilets

Wheelchair users should:

- not have to travel more than 40 m on the same   M 5.9 and 5.10h
  floor to reach a unisex toilet;
- not have to travel more than the combined
  horizontal distance where the unisex toilet
  accommodation is on another floor of the
  building (accessible by passenger lift);
- be able to approach, transfer to and use   M 5.8
  the sanitary facilities provided within a
  building.

6.16.2.6.3 Doors

Doors to compartments for ambulant disabled   M 5.14d and 5.10g
people should:

- preferably open outwards;
- be fitted with a horizontal closing bar fixed
  to the inside face.

The swing of any inward-opening doors to   M 5.14a
standard WC compartments should enable a
450 mm diameter manoeuvring space to be
maintained between the swing of the door,
the WC pan and the side wall of the
compartment.

6.16.2.6.4 Heights and arrangements

The heights and arrangement of fittings in a wheelchair-   M 5.10
accessible unisex toilet should comply with Figure 6.252
and (as appropriate) Figure 6.253.

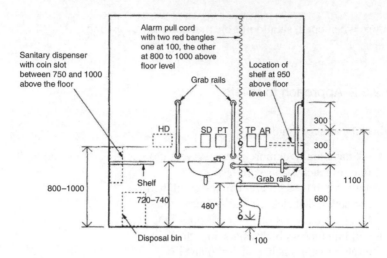

* Height subject to manufacturing tolerance of WC pan

HD: Possible position for automatic hand dryer
SD: Soap dispenser
PT: Paper-towel dispenser
AR: Alarm reset button
TP: Toilet-paper dispenser

* Height of drop-down rails to be the same as the other horizontal grab rails

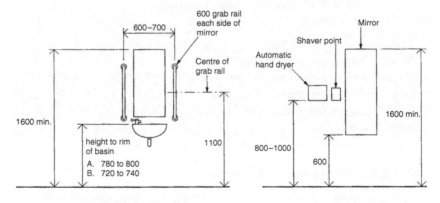

Height of independent wash basin and location of associated fittings, for wheelchair users and standing people

A. For people standing
B. For use from WC

Mirror located away from wash basin suitable for seated and standing people (Mirror and associated fittings used within a WC compartment or serving a range of compartments)

Figure 6.253 Typical heights of arrangement of fittings in a unisex wheel-chair-accessible toilet. (Dimensions in mm.)

The space provided for manoeuvring should enable wheelchair users to adopt various transfer techniques that allow independent or assisted use.   M 5.8

The transfer space alongside the WC should be kept clear to the back wall.   M 5.8

The relationship of the WC to the finger-rinse basin and other accessories should allow a person to wash and dry hands while seated on the WC.   M 5.8

Heat emitters (if located) should not restrict:

- the minimum clear wheelchair manoeuvring space;   M 5.10p
- the space beside the WC used for transfer from the wheelchair to the WC.

6.16.2.6.5 Height and arrangement – ambulant disabled people

The minimum dimensions of compartments for ambulant disabled people should comply with Figure 6.254.   M (M 5.14b

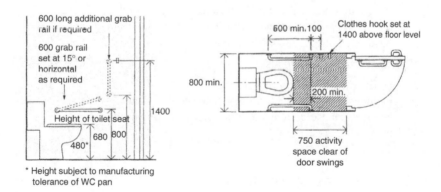

Figure 6.254 Example of a WC cubicle for an ambulant disabled person. (Dimensions in mm.)

Compartments used for ambulant disabled people should:

- include a horizontal grab bar adjacent to the WC;   M 5.14d
- include a vertical grab bar on the rear wall;
- include space for a shelf and fold-down changing table.

6.16.2.6.6 Support rails

The rail on the open side can be a drop-down rail, but on the wall side it can be either a wall-mounted grab rail or, alternatively, a second drop-down rail in addition to the wall-mounted grab rail.	M 5.8
If the horizontal support rail (on the wall adjacent to the WC) is set at the minimum spacing from the wall, an additional drop-down rail should be provided on the wall side, 320 mm from the centreline of the WC.	M 5.10j
If the horizontal support rail (on the wall adjacent to the WC) is set so that its centreline is 400 mm from the centreline of the WC, there is no additional drop-down rail.	M 5.10k

6.16.2.6.7 Wheelchair-accessible washroom

A wheelchair-accessible washroom (where provided) shall have:

• at least one washbasin with its rim set at 720 mm to 740 mm above the floor; • for men, at least one urinal with its rim set at 380 mm above the floor, with two 600 mm long vertical grab bars with their centrelines at 1100 mm above the floor, positioned either side of the urinal.	M 5.14g
The compartment should also:	M 5.11
• be fitted with support rails; • include a minimum activity space to accommodate people who use crutches or otherwise have impaired leg movements.	

6.16.2.6.8 Emergency assistance

An emergency assistance alarm system should be provided that has:

• an outside emergency assistance call signal that can be easily seen and heard by those able to give assistance;	M 5.10n
• visual and audible indicators to confirm that an emergency call has been received;	M 5.10m
• a reset control reachable from a wheelchair, WC or shower/changing seat; • a signal that is distinguishable visually and audibly from the fire alarms provided.	

Emergency assistance pull cords should:

> • be easily identifiable;   M 5.10o
> • be reachable from the WC and from the floor, close to
>   the WC;
> • be coloured;
> • be located as close to a wall as possible;
> • have two **red** 50 mm diameter bangles.

6.16.2.6.9 Toilets in separate-sex washrooms

There should be at least the same number of WCs (for women) as urinals (for men).

> Ambulant disabled people should have the opportunity to use   M 5.11
> a WC compartment within any separate-sex toilet washroom.
>
> Where a separate-sex toilet washroom can be accessed by   M 5.13
> wheelchair users, it should be possible for them to use both a
> urinal (where appropriate) and a washbasin at a lower height
> than is provided for other users.
>
> Separate-sex toilet washrooms above a certain size should   M 5.12
> include an enlarged WC cubicle for use by people who need
> extra space (e.g. parents with children and babies, people
> carrying luggage, and ambulant disabled people).
>
> Consideration should be given to providing a low-level urinal   M 5.13
> for children in male washrooms.

6.16.2.6.10 WC pans

WC pans should:

> • conform to BS EN 997:2012;   M 5.10q and 5.14e
> • be able to accept a variable-height toilet seat
>   riser;
> • have a flushing mechanism positioned   M 5.10r
>   on the open or transfer side of the space
>   – irrespective of handing.

See BS 8300 for more detailed guidance on the various techniques used to transfer from a wheelchair to a WC, as well as appropriate sanitary and other fittings.

6.16.2.6.11 Changing and shower facilities

A choice of shower layout, together with correctly located shower controls and fittings, will enable disabled people independently to make use of the facilities or to be assisted by others where necessary.

The dimensions of the self-contained compartment should allow sufficient space for a helper.	M 5.16
A combined facility should be divided into distinct 'wet' and 'dry' areas.	M 5.16
A choice of layouts suitable for left-hand and right-hand transfer should be provided when more than one individual changing compartment or shower compartment is available.	M 5.18a
Wall-mounted drop-down support rails and wall-mounted slip-resistant tip-up seats (not spring-loaded) should be provided.	M 5.18b
Subdivisions (with the same configuration of space and equipment as for self-contained facilities) should be provided for communal shower facilities and changing facilities.	M 5.18c
In addition to communal separate-sex facilities, individual self-contained shower and changing facilities should be available in sports amenities.	M 5.18d

An emergency assistance alarm system should be provided and should have:

• visible and audible indicators to confirm that an emergency call has been received; • a reset control reachable from a wheelchair, WC or shower/changing seat; • a signal that is distinguishable visually and audibly from the fire alarm.	M 5.18f

An **emergency assistance pull cord** should be provided that should:

• be easily identifiable and reachable from the wall-mounted tip-up seat (or from the floor); • be located as close to a wall as possible;	M 5.18e

- be coloured;
- have two **red** 50 mm diameter bangles.

Facilities for limb storage should be included for the benefit of amputees.   M 5.18g

6.16.2.6.11.1 CHANGING FACILITIES

The floor of a changing area should be level and slip-resistant when dry or when wet, particularly when associated with shower facilities.   M 5.18i

There should be a manoeuvring space 1500 mm deep in front of lockers in self-contained and/or communal changing areas.   M 5.18j

The minimum overall dimensions of (and the arrangement of equipment and controls within) individual self-contained changing facilities should comply with Figure 6.255.   M 5.18h

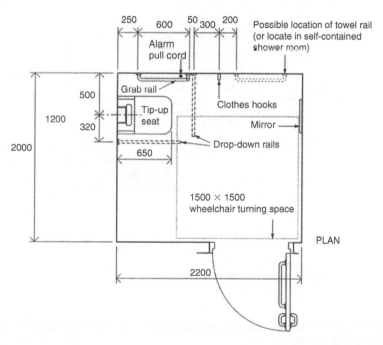

Figure 6.255 Example of a self-contained changing room for individual use. (Dimensions in mm.)

6.16.2.6.11.2 SHOWER FACILITIES

A shower curtain (enclosing the seat and rails and which can be operated from the shower seat) should be provided.	M 5.18m
A shelf (that can be reached from the shower seat and/or from the wheelchair) should be provided for toiletries.	M 5.18n
The shower controls should be positioned between 750 mm and 1000 mm above the floor in all communal area wheelchair-accessible shower facilities.	M 5.18q
If showers are provided in commercial developments for the benefit of staff, at least one wheelchair-accessible shower compartment (complying with Figure 6.256) should be made available.	M 5.18l
Individual self-contained shower facilities should comply with Figure 6.256.	M 5.18k

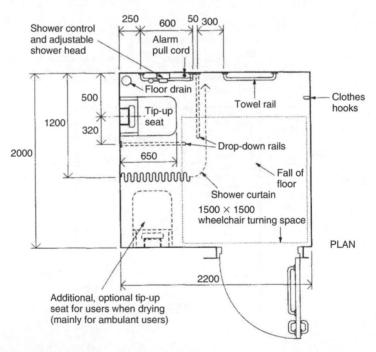

Figure 6.256 Example of a self-contained shower room for individual use. (Dimensions in mm.)

For shower facilities incorporating a WC:

> A choice of left-hand and right-hand transfer layouts should be available when more than one shower area includes a corner WC.   M 5.18s
>
> The minimum overall dimensions of (and the arrangement of fittings within) an individual self-contained shower area incorporating a corner WC (e.g. in a sports building) should comply with Figure 6.257.   M 5.18r

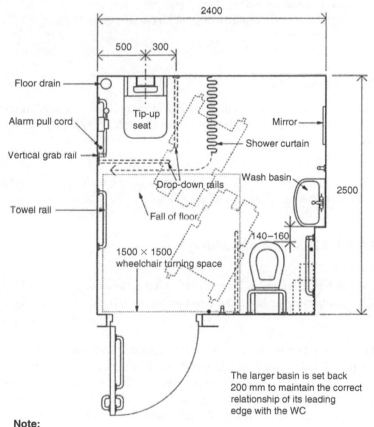

Note:
Layout shown for right-hand transfer to shower seat and WC

Figure 6.257 Example of a shower room incorporating a corner WC for individual use. (Dimensions in mm.)

 Note: More detailed guidance on appropriate sanitary and other fittings is given in BS 8300.

6.16.2.6.12 Fire safety

There shall be an early warning fire alarm system for all persons in the building.

Fire alarms should emit an audio and visual signal to warn occupants with hearing or visual impairments.	M 5.4g
Emergency assistance alarm systems should have:	M 5.4h

- visual and audible indicators to confirm that an emergency call has been received;
- a reset control reachable from a wheelchair, WC or shower/changing seat;
- a signal that is distinguishable visually and audibly from the fire alarm.

6.16.2.7 Bathrooms

All dwellings (and buildings containing one or more rooms for residential purposes) will be provided with a bathroom which contains either a fixed bath or a shower **and** a washbasin.

Any dwelling (house or flat) must have at least one bathroom with a fixed bath or shower, and a washbasin.	G5.6

6.16.2.7.1 Prevention of scalding

Where hot and cold taps are provided on a sanitary appliance, **the hot tap should be on the left**.	G4.6

Note: We think that the reasoning for this comparatively new requirement is to stop people (particularly children and poorly sighted people) inadvertently scalding themselves. To be on the safe side, therefore, it is recommended that, even though the requirement only applies to new buildings, you always ensure the hot water taps are located on the left-hand side of all of your basins and baths.

The hot water supply temperature to a bath should be limited to a maximum of 48° C by use of an in-line blending valve or other appropriate temperature control device, with a maximum temperature stop and a suitable arrangement of pipework.	G3.65

In-line blending valves and composite thermostatic mixing valves should be compatible with the hot and cold water sources that serve them.	G3.67
The length of supply pipes between in-line blending valves and outlets should be kept to a minimum in order to prevent the colonization of waterborne pathogens.	G3.68

6.16.2.7.2 Extract ventilation

Extract ventilation concerns the removal of air directly from a space or spaces to outside. Extract ventilation may be by natural means such as passive stack ventilation (PSV) or by mechanical means (e.g. by an extract fan or central system).

The main requirements are that:

- all sanitary accommodation shall be provided with extract ventilation to the outside that is capable of operating either intermittently or continuously with a minimum extract rate of 6 l/s;
- ventilation devices designed to work continuously shall not have automatic controls such as a humidity control when used for sanitary accommodation;
- as odour is the main pollutant, humidity controls should not be used for intermittent extract in sanitary accommodation;
- common outlet terminals and/or branched ducts shall not be used for wet rooms such as WCs;
- PSV can be used as an alternative to a mechanical extract fan for office sanitary accommodation and washrooms.

Extract to outside is required in **all** office sanitary accommodation and washrooms.

Where there was no previous ventilation opening, or where the size of the original ventilation opening is not known, replacement windows shall have an equivalent area greater than 2500 mm² per WC.	F 7.6
All bathrooms shall be provided with extract ventilation to the outside which is capable of operating either: • intermittently at a minimum extract rate of 15 l/s; or • continuously with a minimum extract rate of 8 l/s.	F.6

In bathrooms (without a WC) where there was no previous ventilation opening, or where the size of the original ventilation opening is not known, replacement windows shall have an equivalent area greater than 2500 mm².	F 7.6
Common outlet terminals and/or branched ducts shall **not** be used for a wet room such as a bathroom.	F Tables 5.2a to 5.2d
Wall- or ceiling-mounted centrifugal fans (which are fitted with a 100 mm diameter flexible duct or rectangular duct and which are designed to achieve 15 l/s for kitchens) should not be ducted further than 6 m and should have no more than one 90° bend.	F Tables 5.2a to 5.2d

6.16.2.7.3 Appliances fitted in bathrooms

Any appliance installed in a room that is used (or intended to be used) as a bathroom must be of the room-sealed type.

Open-flued oil-fired appliances should **not** be installed in bathrooms, where there is an increased risk of carbon monoxide poisoning.	J 4.2

If locating combustion appliances in such rooms cannot be avoided, a way of meeting the requirements is to provide room-sealed appliances.

Bathrooms (and for that matter WCs or showers) that are situated less than 4.5 m above ground level and whose only escape route is through another room shall be provided with an emergency egress window.

Smoke alarms should **not** be fixed in bathrooms.

6.16.2.7.4 Wheelchair-accessible bathrooms

Wheelchair users and ambulant disabled people (in hotels, motels, relatives' accommodation in hospitals, student accommodation and sports facilities) should be able to wash or bathe either independently or with assistance from others.

The minimum overall dimensions of (and the arrangement of fittings within) a bathroom for individual use incorporating a corner WC should comply with Figure 6.258.	M 5.21a

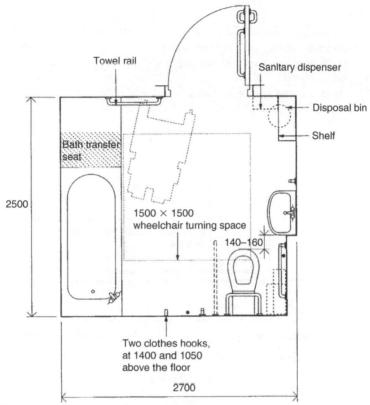

Towel rail

Sanitary dispenser

Disposal bin

Shelf

Bath transfer seat

2500

1500 × 1500 wheelchair turning space

140–160

Two clothes hooks, at 1400 and 1050 above the floor

2700

Note:
Layout shown for right-hand transfer to bath and WC.

Figure 6.258 Example of a bathroom containing a WC. (Dimensions in mm.)

A choice of layouts suitable for left-hand and right-hand transfer should be provided when more than one bathroom is available.	M 5.21b
The floor of a bathroom should be slip-resistant when dry or when wet.	M 5.21c
The bath should be provided with a transfer seat that is 400 mm deep and equal to the width of the bath.	M 5.21d
Outward-opening doors, fitted with a horizontal closing bar fixed to the inside face, should be provided.	M 5.21e

An **emergency assistance pull cord** should be provided which should:

> - be easily identifiable and reachable from the wall-mounted tip-up seat (or from the floor);   M 5.21f
> - be located as close to a wall as possible;
> - be coloured;
> - have two **red** 50 mm diameter bangles.

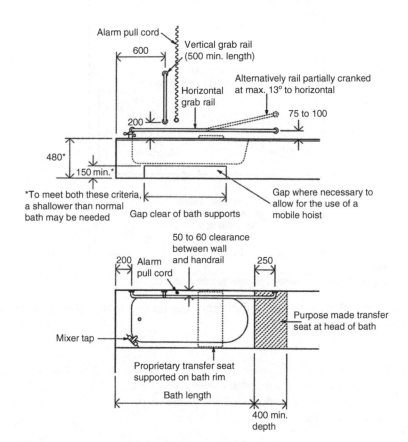

Figure 6.259 Grab rails and fitting associated with a bath.

 Note: More detailed guidance on appropriate sanitary and other fittings, including facilities for the use of mobile and fixed hoists, is given in BS 8300.

6.17 Electrical safety – dwellings

 All electrical work shall be carried out in accordance with Amendment No. 1 (2011) to BS 7671:2008 (*Requirements for Electrical Installations*, commonly referred to as the *IEE Wiring Regulations*, 17th edition).

The previous 2006 version of Approved Document P had duplicated much of that contained in BS 7671. This standard had developed from the Institution of Engineering Technology (IET)'s Wiring Regulations, which was first implemented in 1882 *For the prevention of Fire Risks from Electric Lighting*. With the upgrade of this standard in 2008 and further amendments made in 2011, it was decided that, rather than having two documents regulating the same thing, in future all electrical requirements and recommendations for the built environment would use BS 7671:2008 (incorporating Amendment No. 1:2011) and Approved Document P would only apply to electrical installations that are intended to operate at low or extra-low voltage and are:

- in or attached to a dwelling;
- in the common parts of a building serving one or more dwellings, but excluding power supplies to lifts;
- in a building that receives its electricity from a source located within or shared with a dwelling; or
- in a garden or in or on land associated with a building where the electricity is from a source located within or shared with a dwelling

and which are designed and installed so that:

- they afford appropriate protection against mechanical and thermal damage; and
- they do not present electric shock and fire hazards to people.

6.17.1 Main changes to the 2013 version of Approved Document P

6.17.1.1 Changes in the legal requirements

- The range of electrical installation work that is notifiable has been reduced.
- An installer who is not a registered competent person may use a registered third party to certify notifiable electrical installation work as an alternative to using a building control body.
- All electrical installation work **must** now fully comply with the requirements of BS 7671:2008 (incorporating Amendment No. 1:2011), usually referred to as the *IET Wiring Regulations* or the 17th edition.

Note: The latest edition of this standard has now grown to a massive 464-page document that defines the way in which all electrical installation work must be carried out. It does not matter whether the work is carried out by a professional electrician or an unqualified DIY enthusiast, but the installation **must** still totally comply with the *Wiring Regulations*.

These regulations form the basis of safe working practice throughout the electrical industry, but although BS 7671 is well structured and has separate sections for all the main topics (e.g. safety protection, selection and erection of equipment and so on) it is not the easiest of standards to get to grips with for a particular situation. Occasionally it can be very confusing and requires the reader to constantly flick backwards and forwards through the book to find what it is all about.

For example, Regulation 411.4.7 states that:

Where a circuit-breaker is used to satisfy the requirements of Regulation 411.3.2.2 or Regulation 411.3.2.3, the maximum value of earth fault loop impedance (Zs) shall be determined by the formula in Regulation 411.4.5. Alternatively, for a nominal voltage (Uo) of 230 V and a disconnection time of 0.4 s in accordance with Regulation 411.3.2.2 or 5.s in accordance with Regulation 411.3.2.3, the values specified in Table 41.3 for the types and ratings of overcurrent devices listed may be used instead of calculation!

For your assistance one of Ray Tricker's other 'In Brief' books (*Wiring Regulations in Brief*, 3rd edition) has recently been published and is available from the publisher (http://www.taylorandfrancis.com/books/details/9780415526876) or other booksellers.

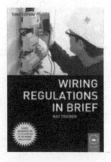

The intention of this book (similar to that of Ray Tricker's other 'In Brief' books) is to peel away some of this confusion and '*officialese*' and provide the reader with an on-the-job reference source that can be quickly used without having to delve backwards and forwards though the standard.

6.17.2 Interaction with other parts of the Building Regulations

Although Approved Document P covers the safety of electrical installation work, it does not cover system functionality. This is covered by other parts of the Building Regulations and/or legislation which deal with electrically powered products such as fire alarm systems, fans and pumps, which are contained in (but are not limited to) the following:

- Approved Document B (Fire safety): fire safety of certain electrical installations; provision of fire alarm and fire detection systems; fire resistance of service penetrations through floors, walls and ceilings;
- Approved Document G (Sanitation, hot water and water efficiency);
- Approved Document L (Conservation of fuel and power): energy-efficient lighting;
- Approved Document M (Access to and use of buildings): height of socket outlets and switches.

For completeness, we have also included system-specific requirements from these Approved Documents.

6.17.3 Requirements

Electrical safety – dwellings
Reasonable provision shall be made in the design and installation of electrical installations in order to protect persons operating, maintaining or altering the installations from fire or injury.

(Approved Document P1)

 A new issue of Approved Document P came into force in 2013.

Fire safety
The building shall be designed and constructed so that there are appropriate provisions for the early warning of fire, and appropriate means of escape in case of fire from the building to a place of safety outside the building capable of being safely and effectively used at all material times.

(Approved Document B)

Hot water supplies and systems

(1) *There must be a suitable installation for the provision of heated wholesome water or heated softened wholesome water to:*

- *any washbasin or bidet provided in or adjacent to a room containing a sanitary convenience;*
- *any washbasin, bidet, fixed bath and shower in a bathroom; and*
- *any sink provided in any area where food is prepared.*

(2) A hot water system, including any cistern or other vessel that supplies water to or receives expansion water from a hot water system, shall be designed, constructed and installed so as to resist the effects of temperature and pressure that may occur either in normal use or in the event of such malfunctions as may reasonably be anticipated, and must be adequately supported.

(3) A hot water system that has a hot water storage vessel shall incorporate precautions to:

- *prevent the temperature of the water stored in the vessel at any time exceeding 100° C; and*
- *ensure that any discharge from safety devices is safely conveyed to where it is visible but will not cause a danger to persons in or about the building.*

(Approved Document G3)

Access to and use of buildings
Reasonable provisions shall be made for people to gain access to and use the building and its facilities.

(Approved Document M)

Amendments to the 2004 version of Approved Document M came into force in 2013.

6.17.4 Meeting the requirements

6.17.4.1 General

All electrical installations **shall** be designed and installed in accordance with BS 7671:2008 incorporating Amendment No. 1:2011.

6.17.4.2 Provision of information

Sufficient information should be provided to ensure that people can operate, maintain and/or alter an electrical installation with reasonable safety.

This information should comprise items listed in BS 7671, including:

- electrical installation certificates or reports describing the installation and giving details of the work carried out;
- permanent labels, for example on earth connections and bonds, and on items of electrical equipment such as consumer units and residual current devices (RCDs);

- operating instructions and logbooks; and for unusually large or complex installations
- detailed plans.

6.17.4.3 New dwellings

Wall-mounted socket outlets, switches and consumer units in new dwellings should be easy to reach, in accordance with Approved Document M of the Building Regulations (Access to and use of buildings).	P1.4

6.17.4.3.1 New dwellings formed by a change of use

Where a material change of use creates a new dwelling, or changes the number of dwellings in a building, all necessary work should be carried out to ensure that the building complies with the requirements of Part	P1.5

Although in some cases the existing electrical installation will need to be upgraded to meet current standards, **if** the existing cables are adequate it is not necessary to replace them, even if they do use old colour codes!

6.17.4.4 Additions and alterations to existing electrical installations

In accordance with the Building Regulations, when building work is complete, *the building should be no more unsatisfactory in terms of complying with the applicable requirements of the Regulations than before the building work was started.* Thus, when extending or altering an electrical installation, **only the new work** must meet current standards. There is no obligation to upgrade the existing installation unless either of the following applies:

• the new work adversely affects the safety of the existing installation; • the state of the existing installation is such that the new work cannot be operated safely.	P1.6

All new work should be carried out in accordance with BS 7671.

6.17.4.5 Scope of Approved Document P

Approved Document P applies to electrical installations in a dwelling house or flat, and to parts of the installation that are:

- outside the dwelling, for example fixed lighting and air-conditioning units attached to outside walls, photovoltaic panels on roofs, and fixed lighting and pond pumps in gardens;
- in outbuildings such as sheds, detached garages and domestic greenhouses;
- in the common access areas of blocks of flats such as corridors and staircases;
- in shared amenities of blocks of flats such as laundries, kitchens and gymnasiums;
- in business premises (other than agricultural buildings) connected to the same meter as the electrical installation in a dwelling, for example shops and public houses below flats.

Approved Document P does **not** apply to electrical installations:

- in business premises in the same building as a dwelling but with separate metering;
- that supply the power for lifts in blocks of flats (but Approved Document P does apply to lift installations in single dwellings).

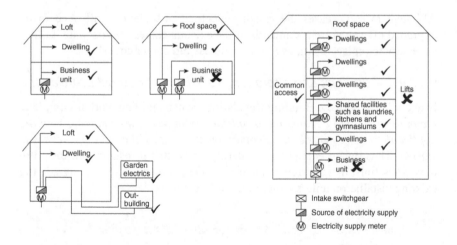

Figure 6.260 Scope of Approved Document P.

6.17.4.6 Notifiable work

Electrical installation work that is notifiable is set out in Regulation 12(6A).

Notifiable work involves activities (usually electrical) which should be 'notified' by the installer to the local authority building control. If you are not sure whether the work you want to undertake is notifiable, you should contact your local authority building control department for advice. A list of non-notifiable work is at section 6.17.4.8.

A person intending to carry out building work in relation to which Approved Document P of Schedule 1 imposes a requirement is required to give a building notice or deposit full plans where the work consists of:

* the installation of a new circuit;
* the replacement of a consumer unit; or
* any addition or alteration to existing circuits in a 'special location'.

6.17.4.6.1 Special locations

In this regulation 'special location' means:

(a) within a room containing a bath or shower, the space surrounding a bath tap or shower head, where the space extends:

* vertically from the finished floor level to:

  o a height of 2.25 m; or
  o the position of the shower head where it is attached to a wall or ceiling at a point higher than 2.25 m from that level; and

* horizontally:

  o where there is a bath tub or shower tray, from the edge of the bath tub or shower tray to a distance of 0.6 m; or
  o where there is no bath tub or shower tray, from the centre point of the shower head where it is attached to the wall or ceiling to a distance of 1.2 m; or

(b) a room containing a swimming pool or sauna heater.

Figure 6.261 illustrates the space around a bath tub or shower tray (a special location) within which minor additions and alterations to existing circuits, as well as the installation of new circuits, are notifiable.

6.17.4.7 Certifications, inspection and testing

For notifiable electrical installation work, one of the following three procedures must be used to certify that the work complies with the requirements set out in the Building Regulations and to verify that the design and installation of electrical work is adequate and that installations will be safe to use, maintain and alter. The electrical work should be inspected and tested in accordance with the procedures in BS 7671:

(1) self-certification by a registered competent person;
(2) third-party certification by a registered third-party certifier;
(3) certification by a building control body.

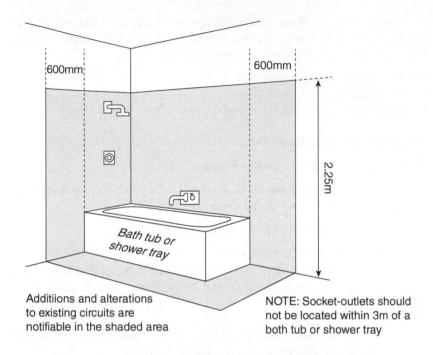

600mm

600mm

2.25m

Bath tub or
shower tray

Additiions and alterations
to existing circuits are
notifiable in the shaded area

NOTE: Socket-outlets should
not be located within 3m of a
both tub or shower tray

Figure 6.261 Notifiable work in rooms containing a bath or a shower.

6.17.4.7.1 Self-certification by a registered competent person

A *registered competent person* is defined as being a competent person reg-istered with an Approved Document P competent person self-certification scheme.

Registered competent persons should complete a BS 7671 electrical instal-lation certificate for every job they undertake and provide the certificate to the person ordering the work.

> An electrical installation certificate containing details of   WR 631.1
> the installation together with a record of the inspections
> made and the test results shall be provided for new
> installations and changes to existing installations.

6.17.4.7.2 Certification by a registered third party

If an installer is not registered as a competent person, then he or she must use a *registered third-party certifier* (i.e. a skilled person registered with an Approved Document P competent person third-party certification scheme) to ensure that his or her electrical installation meets the requirements of BS 7671.

Form 2   Form No /2

ELECTRICAL INSTALLATION CERTIFICATE (notes 1 and 2)

(REQUIREMENTS FOR ELECTRICAL INSTALLATIONS - BS 7671 [IEE WIRING REGULATIONS])

DETAILS OF THE CLIENT (note 1) ....
....
....

INSTALLATION ADDRESS
....
....
....Postcode..

DESCRIPTION AND EXTENT OF THE INSTALLATION Tick boxes as appropriate (note 1)

Description of installation:	New installation ☐
Extent of Installation covered by this Certificate:	Addition to an existing installation ☐
....	Alteration to an existing installation ☐

FOR DESIGN

I/We being the person(s) responsible for the design of the electrical installation (as indicated by my/our signatures below), particulars of which are described above, having exercised reasonable skill and care when carrying out the design, hereby CERTIFY that the design work for which I/we have been responsible is to the best of my/our knowledge and belief in accordance with BS 7671: ...., amended to.......... (date) except for the departures, if any, detailed as follows:

> Details of departures from BS 7671 (Regulations 120-01-03, 120-02):

The extent of liability of the signatory or the signatories is limited to the work described above as the subject of this Certificate. For the DESIGN of the installation.   **(Where there is mutual responsibility for the design)

Signature: .... Date .... Name (BLOCK LETTERS): .... Designer No. 1
Signature: .... Date .... Name (BLOCK LETTERS): .... Designer No. 2**

FOR CONSTRUCTION

I/We being the person(s) responsible for the construction of the electrical installation (as indicated by my/our signatures below), particulars of which are described above, having exercised reasonable skill and care when carrying out the construction, hereby CERTIFY that the construction work for which I/we have been responsible is to the best of my/our knowledge and belief in accordance with BS 7671: ....amended to ....(date) except for the departures, if any, detailed as follows:

> Details of departures from BS 7671 (Regulations 120-01-03, 120-02):

The extent of liability of the signatory is limited to the work described above as the subject of this Certificate. For CONSTRUCTION of the installation:

Signature: .... Date ....

Name (BLOCK LETTERS): .... Constructor

FOR INSPECTION & TESTING

I/We being the person(s) responsible for the inspection & testing of the electrical installation (as indicated by my/our signatures below), particulars of which are described above, having exercised reasonable skill and care when carrying out the inspection & testing, hereby CERTIFY that the work for which I/we have been responsible is to the best of my/our knowledge and belief in accordance with BS 7671:........., amended to .... (date) except for the departures, if any, detailed as follows:

> Details of departures from BS 7671 (Regulations 120-01-03, 120-02):

The extent of liability of the signatory is limited to the work described above as the subject of this Certificate. For INSPECTION & TEST of the installation: **(Where there is mutual responsibility for the design)

Signature: .... Date ....

Name (BLOCK LETTERS): .... Inspector

NEXT INSPECTION (notes 4 and 7)

I/We the designer(s) recommend that this installation is further inspected and tested after an interval of not more than............ years/months

Figure 6.262 Electrical installation certificate.

PARTICULARS OF THE SIGNATORIES TO THE ELECTRICAL INSTALLATION CERTIFICATE (note 3)

Designer (No. 1)
Name: .... Company: ....

Address: ....

.... Postcode: .... Tel No.: ....

Designer (No. 2)
(if applicable) Name: .... Company: ....

Address: ....

.... Postcode: .... Tel No.: ....

Constructor
Name: .... Company: ....

Address: ....

.... Postcode: .... Tel No.: ....

Inspector
Name: .... Company: ....

Address: ....

.... Postcode: .... Tel No.: ....

SUPPLY CHARACTERISTICS AND EARTHING ARRANGEMENTS Tick boxes and enter details, as appropriate

Earthing arrangements	Number and Type of Live Conductors	Nature of Supply Parameters	Supply Protective Device Characteristics
TN-C ☐	a.c. ☐ d.c. ☐	Nominal voltage, $U/U_o^{(1)}$ V	
TN-S ☐	1-line, 2-wire ☐ 2-pole ☐	Nominal frequency, $f^{(1)}$ .... Hz	Type: ....
TN-C-S ☐	1-line, 3-wire ☐ 3-pole ☐	Prospective fault current, $I_{pf}^{(2)}$ (note 6)....... kA	....
TT ☐	2-line, 3-wire ☐ other ☐	External loop impedance, $Z_e^{(2)}$ ....Ω	
IT ☐	3-line, 3-wire ☐	(Note: (1) by enquiry, (2) by enquiry or by measurement)	Nominal current rating
	3-line, 4-wire ☐		...............................A
Alternative source ☐ of supply (to be detailed on attached schedules)			

PARTICULARS OF INSTALLATION REFERRED TO IN THE CERTIFICATE Tick boxes and enter details, as appropriate

Means of Earthing	Maximum Demand
Distributor's facility ☐	Maximum demand (load) ....Amps per phase

Details of Installation Earth Electrode (where applicable)

	Type	Location	Electrode resistance to earth
Installation ☐ earth electrode	(e.g. rod(s), tape etc.)		
	....	....	......................................Ω

Main Protective Conductors

Earthing conductor: material .... csa ....mm^2 connection verified ☐

Main equipotential
bonding conductors: material .... csa ....mm^2 connection verified ☐

To incoming water and/or gas service ☐   To other elements ....

Main Switch or Circuit-breaker

BS. Type .... No. of poles .... Current rating ....A   Voltage rating V

Location ....   Fuse rating or setting ....A

Rated residual operating current $I_{\Delta n}$ = ... mA, and operating time of ....ms (at $I_{\Delta n}$)
(applicable only where an RCD is suitable and is used as a main circuit-breaker)

COMMENTS ON EXISTING INSTALLATION: (In the case of an alteration or addition see Section 743)

....

....

....

....

SCHEDULES (note 2)
The attached Schedules are part of this document and this Certificate is valid only when they are attached to it.
.... Schedules of Inspections and .... Schedules of Test Results are attached. (Enter quantities of schedules attached)

Figure 6.262 *(Continued)*

- Before work begins, the installer should appoint a registered   P3.5
  third-party certifier to inspect and test the work as
  necessary.

- Within five days of completing the work, the installer   P3.6
  must notify the registered third-party certifier, who,
  subject to the results of the inspection and testing
  being satisfactory, should then complete an electrical
  installation condition report and give it to the person
  ordering the work.

Note: The electrical installation condition report should be the model BS 7671
form or one developed specifically for Approved Document P purposes.

An electrical installation condition report shall contain   WR 631.2
details of the extent of the installation and limitation of
the inspection and testing covered by the report; together
with records of inspection, the results of testing and a
recommendation when the next periodic inspection should
occur shall be provided.

6.17.4.7.3 Certification by a building control body

If an installer is not a registered competent person and has not appointed a registered third-party certifier, then before work begins the installer must notify a
building control body, which:

- will determine the extent of inspection and testing needed   P3.9
  for it to establish that the work is safe, based on the nature of
  the electrical work and the competence of the installer;
- may choose to carry out any necessary inspection and
  testing itself, or it may contract a specialist to carry out some
  or all of the work and furnish it with an electrical installation
  condition report.

Once the building control body has decided that, as far as can be ascertained,
the work meets all Building Regulations requirements, it will issue to the occupier a Building Regulations completion certificate (if a local authority) or a
final certificate (if an approved inspector).

MINOR ELECTRICAL INSTALLATION WORKS CERTIFICATE
(REQUIREMENTS FOR ELECTRICAL INSTALLATIONS - BS 7671 [IEE WIRING REGULATIONS])
To be used only for minor electrical work which does not include the provision of a new circuit

---

PART 1: Description of minor works

1.   Description of the minor works: ....

2.   Location/Address: ....

3.   Date minor works completed: ....

4.   Details of departures, if any, from BS 7671

      ....

      ....

      ....

PART 2: Installation details

1.   System earthing arrangement: TN-C-S ☐   TN-S ☐   TT ☐

2.   Method of protection against indirect contact: ....

3.   Protective device for the modified circuit: Type BS .... Rating ....A

4.   Comments on existing installation, including adequacy of earthing and bonding arrangements:
      (see Regulation 130-07) ....

      ....

      ....

PART 3: Essential Tests

1.   Earth continuity: satisfactory   ☐

2.   Insulation resistance:

                              Phase/neutral ....MΩ

                              Phase/earth ....MΩ

                              Neutral/earth ....MΩ

3.   Earth fault loop impedance: ....Ω

4.   Polarity: satisfactory   ☐

5.   RCD operation (if applicable): Rated residual operating current $I_{\Delta n}$ ....mA and operating time of ....ms (at $I_{\Delta n}$)

PART 4: Declaration

1.   I/We CERTIFY that the said works do not impair the safety of the existing installation, that the said works have been designed, constructed, inspected and tested in accordance with BS 7671: .... (IEE Wiring Regulations), amended to .... and that the said works, to the best of my/our knowledge and belief, at the time of my/our inspection, complied with BS 7671 except as detailed in Part 1.

2.   Name: ....   3.   Signature: ....

      For and on behalf of: ....   Position: ....

      Address: ....

      ....   Date: ....

      ....

      ....Postcode: ....

Figure 6.263 Minor electrical installation certificate.

6.17.4.8 Non-notifiable work

Additions and alterations to existing electrical installations outside special locations together with replacements, repairs and maintenance anywhere are considered as non-notifiable (i.e. work that does not need to be notified to the local authority building control body). The work can be completed by a DIY enthusiast (family member or friend) but it will **still** need to be installed in accordance with the manufacturer's instructions and done in such a way that it does not present a safety hazard. Examples of non-notifiable work are:

* replacing any electrical fitting (for example socket outlets, light fittings, control switches);
* adding a fused spur (which is a socket that has a fuse and a switch that is connected to an appliance, e.g. heater) to an existing circuit (but not in a kitchen, bathroom or outdoors);
* any repair or maintenance work;
* installing or upgrading main or supplementary equipotential bonding;
* installing cabling at extra-low voltage for signalling, cabling or communication purposes (for example telephone cabling or cabling for fire alarm, burglar alarm or heating control systems).

6.17.4.8.1 Inspection and testing of non-notifiable work

Similar to notifiable work, non-notifiable electrical installation work should be designed and installed, and inspected, tested and certificated in accordance with BS 7671.

 If local authorities find that non-notifiable work is unsafe and non-compliant, they can take enforcement action.

6.17.4.9 Material alterations

New habitable rooms that are the result of a material alteration and which are above ground-floor level (or at ground-floor level where no final exit has been provided) shall be equipped with: • a fire detection and fire alarm system; • smoke alarms in the circulation spaces.	B1 1.8 (V1) B1 1.6 (V2)

6.17.4.10 Fire safety

Where it is critical for electrical circuits to be able to continue to function during a fire, protected circuits (meeting the requirements of BS EN 50200:2006) shall be installed. Electrically powered locks should return to the unlocked position.	B1 5.38 (V2)

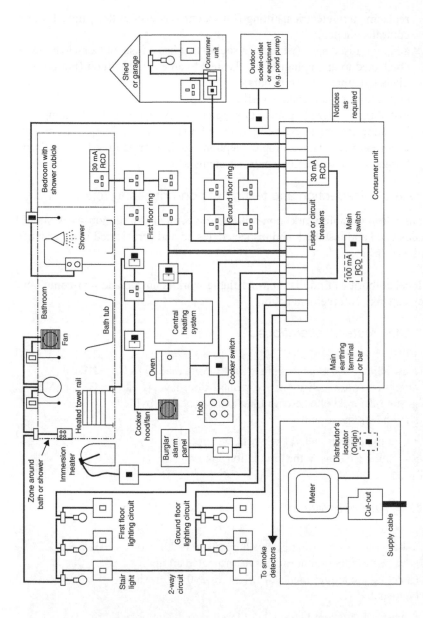

Figure 6.264 Typical fixed installations that might be encountered in new or upgraded existing dwellings.

6.17.4.10.1 Fire alarms

A visual and audible fire alarm signal should be provided in buildings where it is anticipated that one or more persons with impaired hearing may be in relative isolation (e.g. hotel bedrooms and sanitary accommodation).	B1 1.34 (V2)
Fire alarms should emit an audio and visual signal to warn occupants with hearing or visual impairments.	M 5.4 g

 All fire detection and fire-warning systems shall be properly designed, installed and maintained.

All buildings should have arrangements for detecting fire.	B1 1.27 (V2)
All buildings should have the means of raising an alarm in case of fire (e.g. rotary gongs, handbells or shouting 'fire') or be fitted with a suitable electrically operated fire-warning system (in compliance with BS 5839).	B1 1.28 (V2)
The fire-warning signal should be distinct from other signals that may be in general use.	B1 1.32 (V2)
In premises used by the general public (e.g. large shops and places of assembly), a staff alarm system (complying with BS 5839) may be used.	B1 1.33 (V2)
In small buildings, raising the alarm may be a comparatively simple matter, but, when this does not apply, the building should be provided with an electrically operated fire-warning system with manual call points adjacent to exit doors which shall comply with BS 5839-1:2002.	B1 1.29 and 1.30 (V2)
Call points for electrical alarm systems should be installed in accordance with BS 5839-1 and comply with either BS 5839-2:1983 or type A of BS EN 54-11:2001.	B1 1.31 (V2)

 Note: Type B call points should **only** be used with the approval of the building control body.

• Where it is critical for electrical circuits to be able to continue to function during a fire, protected circuits (meeting the requirements of BS EN 50200:2006) shall be installed.	B1 5.38 (V2)

6.17.4.10.2 Smoke alarms

With the introduction of Approved Document B 2007:

- smoke alarms need to be installed in accordance with BS EN 14604:2005 or BS 5446-2:2003;
- all smoke alarms should have a standby power supply;
- the provision of smoke alarms shall be based on an assessment of the risk to the occupants in the event of fire.

 Note: If a dwelling house is extended, smoke alarms should be provided in all circulation spaces.

Smoke alarms should:	B1 1.4
• be mains-operated and conform to BS 5446-1:2000;	
• have a standby power supply such as a rechargeable (or non-rechargeable) battery;	
• be designed and installed in accordance with BS 5839-6:2004;	B1 1.10 (V1) B1 1.9 (V2)
• be positioned in the circulation spaces between sleeping spaces and places where fires are most like to start (e.g. in kitchens and living rooms);	B1 1.11 (V1) B1 1.10 (V2)
• be installed on every storey of a dwelling house.	B1 1.12 (V1)
Where the kitchen area is not separated from the stairway or circulation space by a door, there should be a compatible interlinked heat detector or heat alarm in the kitchen, in addition to whatever smoke alarms are needed in the circulation space(s).	B1 1.11 (V2) B1 1.13 (V1) B1 1.12 (V2)
If more than one alarm is installed, the alarms should be linked so that the detection of smoke or heat by one unit operates the alarm signal in all of them.	B1 1.14 (V1) B1 1.13 (V2)
Smoke alarms/detectors should be sited so that:	B1 1.15a (V1)
• there is a smoke alarm in the circulation space within 7.5 m of the door to every habitable room;	B1 1.14a (V2)
• they are ceiling-mounted and at least 300 mm from walls and light fittings;	B1 1.15b (V1) B1 1.14b (V2)
• the sensor in ceiling-mounted devices is between 25 mm and 600 mm below the ceiling (25–150 mm in the case of heat detectors or heat alarms).	B1 1.15c (V1) B1 1.14c (V2)
Smoke alarms should **not** be fixed:	B1 1.16 (V1)

- over a stair or any other opening between floors;   B1 1.15 (V2)
- next to or directly above heaters or air-   B1 1.17 (V1)
  conditioning outlets;   B1 1.16 (V2)
- in bathrooms, showers, cooking areas or garages;   B1 1.17 (V1)
                                                                        B1 1.16 (V2)
- in any place where steam, condensation or fumes   B1 1.17 (V1)
  could give false alarms;   B1 1.16 (V2)
- in places that get very hot (such as a boiler room);   B1 1.18 (V1)
                                                                        B1 1.17 (V2)
- in places that get very cold (such as an unheated   B1 1.18 (V1)
  porch);   B1 1.17 (V2)
- to surfaces which are normally much warmer or   B1 1.18 (V1)
  colder than the rest of the space.   B1 1.17 (V2)

6.17.4.10.2.1 SMOKE ALARMS IN THATCHED ROOFS

In thatched roofs: B4 10.9 (V1)

- the rafters should be overdrawn with construction   B4 14.9 (V2)
  having not less than 30 minutes' fire resistance;
- a smoke alarm should be installed in the roof space.

6.17.4.10.2.2 SMOKE DETECTORS – POWER SUPPLIES

The power supply for a smoke alarm system should: B1 1.19

- be derived from the dwelling house's mains electricity
  supply;
- comprise a single independent circuit at the dwelling's
  main distribution board (consumer unit or a single
  regularly used local lighting circuit).

It should be possible to isolate the power to the smoke alarms   B1 1.19
without isolating the lighting.

The electrical installation should comply with Approved   B1 1.20
Document P (Electrical safety).

Any cable suitable for domestic wiring may be used for the   B1 1.21
power supply and interconnection to smoke alarm systems
(except in large buildings, where the cable needs to be fire-
resistant (see BS 5839-6:2004)).

Conductors used to interconnect alarms (e.g. signalling)   B1 1.21
should be colour-coded so as to distinguish them from those
supplying mains power.

> Mains-powered smoke alarms may be interconnected using   B1 1.21
> radio-links, provided that this does not reduce the lifetime or
> duration of any standby power supply below 72 hours.

Where the vertical distance between the floor of the entrance storey and the floors
above and below it does not exceed 7.5 m, multi-storey flats are required to:

> • provide a protected stairway plus additional   B1 216c (V2)
>   smoke alarms in all habitable rooms and a heat
>   alarm in any kitchen; or
> • provide a protected stairway plus a sprinkler   B1 2.16d (V2)
>   system and smoke alarms.

6.17.4.11 Emergency alarms

> Emergency assistance alarm systems should have: M 5.4h
>
> • visual and audible indicators to confirm that an
>   emergency call has been received;
> • a reset control reachable from a wheelchair, WC or
>   shower/changing seat;
> • a signal that is distinguishable visually and audibly
>   from the fire alarm.
>
> **Emergency alarm pull cords** should be: M 4.30e
>
> • coloured **red**;
> • located as close to a wall as possible;
> • have two **red** 50 mm diameter bangles.
>
> Front plates should contrast visually with their backgrounds.   M 4.30 m
>
> The colours **red** and **green** should **not** be used in combination   M 4.28
> as indicators of 'ON' and 'OFF' for switches and controls.

6.17.4.12 Outlets, controls and switches

> All controls and switches should be easy to operate,   M 5.4i and 4.30
> visible and free from obstruction and should:
>
> • be located between 750 mm and 1200 mm above
>   the floor;
> • not require the simultaneous use of both hands
>   (unless necessary for safety reasons) to operate;

- have switched socket outlets that indicate whether they are 'ON';
- have mains and circuit isolator switches that clearly indicate whether they are 'ON' or 'OFF';
- for individual switches on panels and on multiple socket outlets, be well separated;
- for controls that need close vision (e.g. thermostats), be located between 1200 mm and 1400 mm above the floor;
- have front plates that contrast visually with their backgrounds.

The operation of all switches, outlets and controls should not require the simultaneous use of both hands (unless necessary for safety reasons).	M 4.30j
Where possible, light switches with large push pads should be used in preference to pull cords.	M 5.3
The colours **red** and **green** should **not** be used in combination as indicators of 'ON' and 'OFF' for switches and controls.	M 5.3

6.17.4.12.1 Socket outlets

Socket outlets should be wall-mounted.	M 4.30a and b
Socket outlets should be located no nearer than 350 mm from room corners.	M 4.30g
Switched socket outlets should indicate whether they are 'ON'.	M 4.30k
Front plates should contrast visually with their backgrounds.	M 4.30m
Mains and circuit isolator switches should clearly indicate whether they are 'ON' or 'OFF'.	M 4.30l

 Older types of socket outlet designed for non-fused plugs must not be connected to a ring circuit.

The colours **red** and **green** should **not** be used in combination as indicators of 'ON' and 'OFF' for switches and controls.	M 4.28

6.17.4.12.2 Wall-mounted switches and socket outlets

Individual switches on panels and on multiple socket outlets should be well separated. M 4.29

Switches and socket outlets for lighting and other equipment should be located so that they are easily reachable. M 8.2

Switches and socket outlets for lighting and other equipment in habitable rooms should be located between 450 mm and 1200 mm from finished floor level (see Figure 6.265). M 8.3

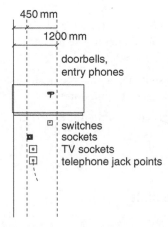

Figure 6.265 Heights of switches and sockets, etc.

Note: The aim is to help people with limited reach (e.g. seated in a wheelchair) to access a dwelling's wall-mounted switches and socket outlets.

6.17.4.12.3 Light switches

Light switches should: M 4.30h and l

- have large push pads;
- align horizontally with door handles;
- be within 900 mm and 1100 mm from the entrance door opening.

Switches and controls should be located between 750 mm and 1200 mm above the floor. M 4.30c and d

Where possible, light switches with large push pads should be used in preference to pull cords. M 5.3

> The colours **red** and **green** should **not** be used in   M 5.3
> combination as indicators of 'ON' and 'OFF' for
> switches and controls.

6.17.4.13 *Fixed internal lighting (domestic buildings)*

In order that dwelling occupiers may benefit from the installation of efficient electric lighting, whenever:

- a dwelling is extended; or
- a new dwelling is created from a material change of use; or
- an existing lighting system has been replaced as part of rewiring works

the rewiring works **must** comply with BS 7671.

Lighting fittings (including lamp, control gear, housing, reflector, shade, diffuser or other device for controlling the output light) should only take lamps with a luminous efficiency greater than 40 lumens per circuit-watt.

Note: Light fittings in less-frequented areas such as cupboards and other storage areas do not count.

6.17.4.14 *Fixed external lighting*

Fixed external lighting (i.e. lighting that is fixed to an external surface of the dwelling and which is powered from the dwelling's electrical system) should either:

- have a lamp capacity not exceeding 150 W per light fitting that automatically switches off:
  - when there is enough daylight; and
  - when it is not required at night; or
- include sockets that can only be used with lamps which have an efficiency greater than 40 lumens per circuit-watt.

Fixed energy-efficient light fittings (one per 25 m^2 dwelling floor area (excluding garages) and one per four fixed light fittings) should be installed in the most frequented locations in the dwelling.

External lighting (including lighting in porches, but not lighting in garages and car ports) should:

- automatically extinguish when there is enough daylight, and when not required at night;
- have sockets that can only be used with lamps having an efficacy greater than 40 lumens per circuit-watt (such as fluorescent or compact fluorescent lamp types, and **not** GLS tungsten lamps with bayonet cap or Edison screw bases).

6.17.4.15 Cables to outside buildings

Cables to an outside building (e.g. garage or shed):

> • if run underground, should be routed and positioned so   BS 7671
>   as to give protection against electric shock and fire as a
>   result of mechanical damage to a cable;
> • if concealed in floors and walls, are (in certain
>   circumstances) required to have an earthed metal
>   covering, be enclosed in steel conduit or have additional
>   mechanical.

 See BS 7671 or *Wiring Regulations in Brief* (3rd edition) by Ray Tricker for more information.

6.17.4.16 Electrically powered locks

> Electrically powered locks should return to the unlocked   B1 5.11 (V2)
> position:
> • on operation of the fire alarm system;
> • on loss of power or system error;
> • on activation of a manual door release unit.

6.17.4.17 Power-operated doors

Power-operated doors and gates should:

> • be provided with a manual or automatic opening device   K5 5.2d
>   in the event of a power failure where and when necessary
>   for health or safety;
> • have safety features to prevent injury to people who are
>   struck or trapped (e.g. a pressure-sensitive door edge
>   which operates the power switch);
> • have a readily identifiable and accessible stop switch.

> Doors to accessible entrances shall be provided with a   M 2.13a
> power-operated door opening and closing system if a force
> greater than 30 N is required to open or shut a door.

Once open, all doors to accessible entrances should be wide enough to allow unrestricted passage for a variety of users, including wheelchair users, people carrying luggage, people with assistance dogs, and those with pushchairs and small children.

The effective clear width through a single-leaf door (or one leaf of a double-leaf door) should be in accordance with Table 6.102.	M 2.13b

Table 6.102 Minimum effective clear widths of doors

Direction and width of approach	New buildings (mm)	Existing buildings (mm)
Straight on (without a turn or oblique approach)	800	750
At right angles to an access route at least 1500 mm wide	800	750
At right angles to an access route at least 1200 mm wide	825	775
External doors to buildings used by the general public	1000	775

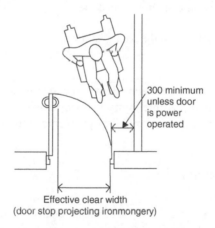

300 minimum
unless door
is power
operated

Effective clear width
(door stop projecting ironmongery)

Figure 6.266 Effective clear width and visibility requirements of doors.

Power-operated entrance doors should have a sliding, swinging or folding action controlled manually (by a push pad, card swipe, coded entry or remote control) or automatically controlled by a motion sensor or proximity sensor such as a contact mat.

Power-operated entrance doors should:

•	open towards people approaching the doors;	M 2.21a
•	provide visual and audible warnings that they are operating (or about to operate);	M 2.21c
•	incorporate automatic sensors to ensure that they open early enough (and stay open long enough) to permit safe entry and exit;	M 2.21c
•	incorporate a safety stop that is activated if the doors begin to close when a person is passing through;	M 2.21b
•	revert to manual control (or fail-safe) in the open position in the event of a power failure;	M 2.21d
•	when open, should **not** project into any adjacent access route;	M 2.21e
•	ensure that their manual controls:	M 2.21f

  ○ are located between 750 mm and 1000 mm above floor level;
  ○ are operable with a closed fist;

•	be set back 1400 mm from the leading edge of the door when fully open;	M 2.21g
•	be clearly distinguishable against the background;	M 2.21g
•	contrast visually with the background.	M 2.19 and 2.21g

 Note: Revolving doors are **not** considered 'accessible', as they create particular difficulties for (and possible injury to) people who are visually impaired, people with assistance dogs or mobility problems, and people with children and/or pushchairs.

6.17.4.18 Heat emitters

In toilets and bathrooms designed for disabled people, heat   M 5.10
emitters (if located) should not restrict:

• the minimum clear wheelchair manoeuvring space;
• the space beside a WC used to transfer from the
  wheelchair to the WC.

Heat emitters should either be screened or have their exposed   M 5.4j
surfaces kept at a temperature below 43° C.

6.17.4.19 *Telephone points and TV sockets*

All telephone points and TV sockets should be located   M 4.30a and b
between 400 mm and 1000 mm above the floor (or
400 mm and 1200 mm above the floor for permanently
wired appliances).

6.17.4.20 *Thermostats*

Controls that need close vision (e.g. thermostats) should   M 4.30f
be located between 1200 mm and 1400 mm above the
floor.

6.17.4.21 *Water safety devices*

Water safety devices normally consist of:

* non-self-resetting energy cut-outs;
* temperature and pressure relief devices.

6.17.4.21.1 Non-self-resetting energy cut-outs

Non-self-resetting energy cut-outs may only be used where   G3.28
they would have the effect of instantly disconnecting the
supply of energy to the storage vessel.

Where an electrical device, such as a relay or motorized valve,   G3.31
is connected to the energy cut-out, the device should operate to
interrupt the supply of energy if the electrical power supply is
disconnected.

Where there is more than one energy cut-out, each   G3.32
non-self-resetting energy cut-out should be independent
and have a separate motorized valve and a separate
temperature sensor.

Where an energy cut-out is fitted, each heat source should have   G3.33
a separate non-self-resetting energy cut-out.

6.17.4.21.2 Temperature and pressure relief devices

Where relevant, appropriate pressure-, temperature-, or temperature- and pressure-activated safety devices should be fitted in addition to a safety device such as energy cut-out.	G3.34
Temperature relief valves and combined temperature and pressure relief valves should not be used in systems which do not automatically replenish the stored water (e.g. unvented primary thermal storage vessels).	G3.35 G3.36

In such cases there should be a second non-self-resetting energy cut-out independent of the one provided.

Temperature relief valves or combined temperature and pressure relief valves should:	G3.37 G3.38
• give a discharge rating at least equal to the total power input to the hot water storage system; • be located directly on the storage vessel, to ensure the stored water does not exceed 100° C.	
In hot water storage system units and packages, the temperature relief valve(s):	G3.39
• should be factory fitted; • should not be disconnected other than for replacement; and • should not be relocated in any other device or fitting installed.	

Fixed building services, including controls, should be commissioned by testing and adjusting as necessary to ensure that they use no more fuel and power than is reasonable in the circumstances.

6.17.4.21.3 Electric water heating

Electric fixed immersion heaters should comply with BS EN 60335-2-73:2003.	G3.43
Electric instantaneous water heaters should comply with BS EN 60335-2-35:2002.	G3.44

| Electric storage water heaters should comply with BS EN 60335-2-21:2003. | G3.45 |

6.17.4.21.4 Solar water heating

Factory-made solar water heating systems should comply with BS EN 12976-1:2006.	G3.46
Other solar water heating systems should comply with prEN/TS 12977-1:2008 or BS 5918:1989.	G3.47
Where solar water heating systems are used, an additional heat source should be available in order to maintain the water temperature and thus to restrict microbial growth.	G3.48
As some solar hot water systems operate at high temperatures and pressures, all components should be rated to the appropriate temperatures and pressures.	

6.17.4.22 Cellars or basements

| LPG storage vessels and LPG-fired appliances fitted with automatic ignition devices or pilot lights must not be installed in cellars or basements. | J 3.5i |

6.17.4.23 Where can I get more information?

Further guidance concerning the requirements for electrical safety is available from:

- the Institution of Engineering and Technology (IET) at http://www.theiet.org;
- the National Inspection Council for Electrical Installation Contracting (NICEIC) at http://www.niceic.org.uk;
- the Electrical Contractors' Association (ECA) at http://www.niceic.org.uk or http://www.eca.co.uk.

To download pdf copies of Approved Document P (and other Approved Documents), go to http://www.planningportal.gov.uk. For details of fixed wire colour changes, go to http://www.niceic.org.uk.

6.18 Combustion appliances

6.18.1 Requirements

Air supply
Combustion appliances shall be so installed that there is an adequate supply of air to them for combustion, to prevent overheating and for the efficient working of any flue.

(Approved Document J1)

Discharge of products of combustion
Combustion appliances shall have adequate provision for the discharge of products of combustion to the outside air.

(Approved Document J2)

Warning of release of carbon monoxide
Where a fixed combustion appliance is provided, appropriate provision shall be made to detect and give warning of the release of carbon monoxide.

(Approved Document J2)

Protection of building
Combustion appliances and fluepipes shall be so installed, and fireplaces and chimneys shall be so constructed and installed, as to reduce to a reasonable level the risk of people suffering burns or the building catching fire in consequence of their use.

(Approved Document J3)

Provision of information
Where a hearth, fireplace, flue or chimney is provided or extended, a durable notice containing information on the performance capabilities of the hearth, fireplace, flue or chimney shall be affixed in a suitable place in the building for the purpose of enabling combustion appliances to be safely installed.

(Approved Document J4)

Protection of liquid fuel storage systems
Liquid fuel storage systems and the pipes connecting them to combustion appliances shall be so constructed, and separated from buildings and the boundary of the premises as to reduce to a reasonable level the risk of the fuel igniting in the event of fire in adjacent buildings or premises.

(Approved Document J5)

Protection against pollution
Oil storage tanks and the pipes connecting them to combustion appliances shall:

• *be so constructed and protected as to reduce to a reasonable level the risk of the oil escaping and causing pollution; and*

- *have affixed in a prominent position a durable notice containing informa-tion on how to respond to an oil escape so as to reduce to a reasonable level the risk of pollution.*

(Approved Document J6)

Conservation of fuel and power
Reasonable provision shall be made for the conservation of fuel and power in buildings by:

(a) limiting heat gains and losses

   (i)   through thermal elements and other parts of the building fabric; and
   (ii) from pipes, ducts and vessels used for space heating, space cooling and hot water services;

(b) providing fixed building services which

   (i)   are energy efficient;
   (ii) have effective controls; and are commissioned by testing and adjust-ing as necessary to ensure they use no more fuel and power than is reasonable in the circumstances.

(Approved Document L)

 A new version of Approved Document L came into force on 6 April 2014.

6.18.2 Meeting the requirements

Combustion appliances require ventilation to supply them with air for combustion. Ventilation is also required to ensure the proper operation of flues or, in the case of flueless appliances, to ensure that the products of combustion are safely dispersed to the outside air.

6.18.2.1 Air supplies for combustion installations

A room containing an open-flued appliance may need permanently open air vents (Figure 6.267(a) and (c)).	J 1.4
Appliance compartments that enclose open-flued combustion appliances should be provided with vents large enough to admit all the air required by the appliance for combustion and proper flue operation, whether the compartment draws its air from a room directly from outside or not (Figure 6.267(b) and (c)).	J 1.5
Where appliances require cooling air, appliance compartments should be large enough to enable air to circulate, and high- and low-level vents should be provided (Figure 6.268).	J 1.6

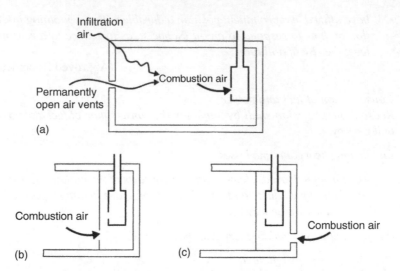

Figure 6.267 Air for combustion and operation of the flue (open flued). (a) Appliance in room. (b) Appliance in appliance compartment with internal vent. (c) Appliance in appliance compartment with external vent.

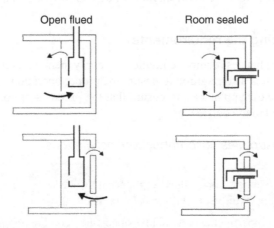

Figure 6.268 Combustion and operation requiring air cooling.

 Where appliances are to be installed within balanced compartments, special provisions will be necessary.

 Note: In a flueless situation, air for combustion (and to carry away its products) can be achieved as shown in Figure 6.269.

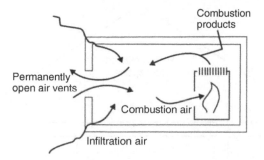

Figure 6.269 Air for combustion and operation of the flue (flueless).

If an appliance is room-sealed but takes its combustion air from another space in the building (e.g. the roof void), or if a flue has a permanent opening to another space in the building (e.g. where it feeds a secondary flue in the roof void), that space should have ventilation openings directly to outside.

Where flued appliances are supplied with combustion air through air vents which open into adjoining rooms or spaces, the adjoining rooms or spaces should have air vent openings of at least the same size direct to the outside.	J 1.9

 Note: Air vents for flueless appliances, however, should open directly to the outside air.

Any hidden voids in the construction shall be sealed and subdivided to inhibit the unseen spread of fire and products of combustion.	B3

6.18.2.2 Air vents

Permanently open air vents should be non-adjustable, sized to admit sufficient air for the purpose intended and positioned where they are unlikely to become blocked.

Air vents should be sufficient for the appliances to be installed (taking account where necessary of obstructions such as grilles and anti-vermin mesh).	J 1.11
Air vents should be sited outside fireplace recesses and beyond the hearths of open fires so that dust or ash will not be disturbed by draughts.	J 1.11a

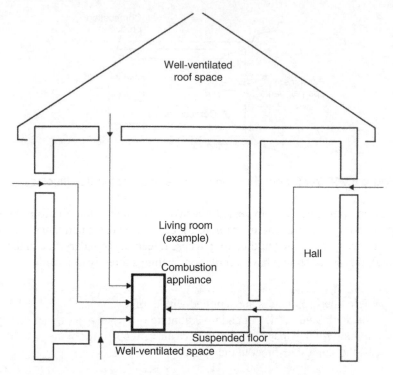

Figure 6.270 Locating permanent air vent openings (examples).

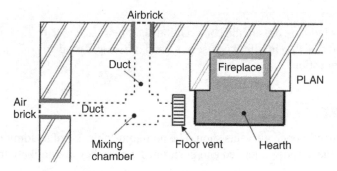

Figure 6.271 Air vent openings in a solid floor.

Air vents should be sited in a location unlikely to cause discomfort from cold draughts.

Grilles or meshes protecting air vents from the entry of animals   J 1.15
or birds should have aperture dimensions no smaller than 5 mm.

In noisy areas, it may be necessary to install proprietary noise-attenuated ventilators to limit the entry of noise into the building.

In buildings where it is intended to install open-flued   J 1.20
combustion appliances and extract fans, the combustion
appliances should be able to operate safely whether or not the
fans are running.

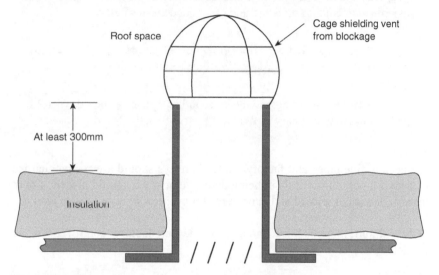

Roof space

Cage shielding vent
from blockage

At least 300mm

Insulation

Figure 6.272 Ventilator used in a roof space (e.g. a loft).

For gas appliances where a kitchen contains an open-flued   J 1.20a
appliance, the extract rate of the kitchen extract fan should not
exceed 20 l/s (72 m3/h).

When installing ventilation for solid fuel appliances, avoid   J 1.20c
installing extract fans in the same room.

Note: Discomfort from cold draughts can be avoided by placing vents close to appliances (for instance, by using floor vents), by drawing air from intermediate spaces (such as from a hallway) or by ensuring good mixing of incoming cold air by placing air vents close to ceilings.

6.18.2.3 Flues

Appliances other than flueless appliances should incorporate or be connected to suitable flues which discharge to the outside air.

New chimneys should be constructed with flue liners (clay, concrete or pre-manufactured) and masonry (bricks, medium-weight concrete blocks or stone) suitable for the intended application.	J 1.27
Liners should be selected to form the flue without cutting and joints should be kept to a minimum.	J 1.28
Liners need to be placed with the sockets or rebate ends uppermost to contain moisture and other condensates in the flue.	J 1.28
Joints should be sealed with fire cement or refractory mortar or installed in accordance with their manufacturer's instructions.	J 1.28
Spaces between the lining and the surrounding masonry should not be filled with ordinary mortar.	J 1.28

Connecting flue pipes and factory-made chimneys should always be guarded if there is a possibility of them being damaged or if they could present a burn hazard (that is not immediately apparent) for people.

Note: Chimneys and flues should provide satisfactory control of water condensation.

If a flue passes through a compartment wall or compartment floor, or is built into a compartment wall, each wall of the flue should have a fire resistance of at least half that of the wall or floor (see Figure 6.273).	B1 7.11 (V1) B3 10.16 (V2)

(a) FLUE PASSING THROUGH COMPARTMENT WALL OR FLOOR

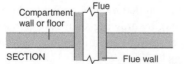

Flue walls should have a fire resistance of at least one half of that required for the compartment wall or floor, and be of non-combustible construction.

(b) FLUE BUILT INTO COMPARTMENT WALL

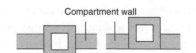

In each case flue walls should have a fire resistance at least one half of that required for the compartment wall and be of non-combustible construction.

Figure 6.273 Flues penetrating compartment walls or floors.

6.18.2.4 *Conservation of heat and energy*

6.18.2.4.1 Secondary heating

In a dwelling:

> • if a secondary heating appliance is fitted, the efficiency   L1A 2.14
>   of the actual appliance with its appropriate fuel must be
>   used in the calculation of the dwelling efficiency rate
>   (DER);
> • if a chimney or flue is provided (but no actual appli-
>   ance has been installed) then the DER calculation must
>   assume that:
>
>   ○ if a gas point is available adjacent to the hearth,
>   then the possible effect of a decorative fuel-effect
>   gas fire open to the chimney or flue with an effi-
>   ciency of 20 per cent must be assumed;
>   ○ if there is no gas point available adjacent to the
>   hearth then the possible effect of an open fire
>   in a grate burning multi-fuel with an efficiency
>   of 37 per cent must be assumed (unless the
>   dwelling is in a smoke control area, when the
>   fuel should be taken as smokeless solid mineral
>   fuel).

6.18.2.4.2 Modular and portable buildings with a planned service life of more than two years

Portable buildings with a planned service life of more than two years but with an intended time of use in a given location of less than two years are often 'distress purchases' (e.g. following a fire), and the buildings must be up and operational in a matter of days. In these circumstances:

> • where the planned time of use in a given location is less   L2B 2.32
>   than two years, the only practical heating technology
>   is electric resistance heating. In such cases, reasonable
>   provision would be to provide energy-efficiency
>   measures that are 15 per cent better than if using
>   conventional fossil fuel heating.

6.18.2.4.3 Energy meters

Energy metering systems should be capable of enabling and ensuring that:

- at least 90 per cent of the estimated annual energy   L2A 2.47
  consumption of each fuel is assigned to the various
  end-use categories such as heating, lighting etc.
  (see CIBSE TM39, 'Building energy metering');
  and
- the output of any renewable system is separately
  monitored; and
- in buildings with a total useful floor area greater than
  1000m2, automatic meter reading and data collection
  facilities are available.

6.18.2.4.4 Centralized switching of appliances

Wherever possible, centralized switches should be available   L2A 2.49
to the facilities manager to switch off appliances when they
are not needed (e.g. overnight and at weekends).

Note: Where appropriate, these should be automated (with manual override) so
that energy savings are maximized.

6.19 Hot water storage

6.19.1 Requirements

Hot water supplies and systems

*(1) There must be a suitable installation for the provision of heated whole-
some water or heated softened wholesome water to:*

   *(a) any washbasin or bidet provided in or adjacent to a room containing
       a sanitary convenience;*
   (b) any washbasin, bidet, fixed bath and shower in a bathroom; and
   (c) any sink provided in any area where food is prepared.

*(2) A hot water system, including any cistern or other vessel that supplies
    water to or receives expansion water from a hot water system, shall be
    designed, constructed and installed so as to resist the effects of tempera-
    ture and pressure that may occur either in normal use or in the event
    of such malfunctions as may reasonably be anticipated, and must be
    adequately supported.*

*(3) A hot water system that has a hot water storage vessel shall incorporate
    precautions to:*

(a) *prevent the temperature of the water stored in the vessel at any time exceeding 100° C; and*

(b) *ensure that any discharge from safety devices is safely conveyed to where it is visible but will not cause a danger to persons in or about the building.*

(Approved Document G3)

Note: Requirement G3(3) does not apply to a system which heats or stores water for the purposes only of an industrial process.

(4) *The hot water supply to any fixed bath must be so designed and installed as to incorporate measures to ensure that the temperature of the water that can be delivered to that bath does not exceed 48° C.*

(Approved Document G3)

Note: Requirement G3(4) applies only when a dwelling is erected or formed by a material change of use.

Hot water storage
The hot water system shall:

*   *be installed by a competent person;*
*   *not exceed 100° C;*
*   *discharge safely;*
*   *not cause danger to persons in or about the building.*

(Approved Document G3)

Conservation of fuel and power
Reasonable provision shall be made for the conservation of fuel and power in buildings by:

(a) *limiting heat gains and losses*

   (i) *through thermal elements and other parts of the building fabric; and*

   (ii) *from pipes, ducts and vessels used for space heating, space cooling and hot water services;*

(b) *providing fixed building services which*

   (i) *are energy efficient;*

   (ii) *have effective controls; and are commissioned by testing and adjusting as necessary to ensure they use no more fuel and power than is reasonable in the circumstances.*

(Approved Document L)

A new version of Approved Document L came into force on 6 April 2014.

6.21.2 Meeting the requirements

6.19.1.1 Hot water supply and systems

 All electrical work associated with hot water systems shall be carried out in accordance with Amendment No. 1 (2011) to BS 7671:2008 (*Requirements for Electrical Installations*, commonly referred to as the *IEE Wiring Regulations*, 17th edition).

All fixed building services, including controls, should be commissioned to ensure that they use no more fuel and power than is considered reasonable in the circumstances.

Hot water (heated wholesome water or heated softened water) shall be supplied to the sanitary appliances and locations specified in the requirement without waste, misuse or undue consumption of water.	G3
The hot water outlet temperature device used to limit the maximum temperature that can be supplied at the outlet should not be capable of being easily altered by building users.	G3(4)

All components of the hot water system (including any cistern that supplies water to, or receives expansion water from, the hot water system) shall continue to safely contain the hot water:

• during normal operation of the hot water system; • following failure of any thermostat used to control temperature; and • during operation of any of the safety devices.	G3(2)

6.19.1.2 Hot water storage systems

Pipework should be designed and installed in such a way as to minimize the transfer time between the hot water storage system and hot water outlets.

Hot water storage systems should be designed and installed in accordance with BS 6700:2006 or BS EN 12897:2006. Specifically:	G3.10
• the temperature relief valve(s) in hot water storage system units and packages should:	G3.39

- o   be factory fitted;
- o   not be disconnected other than for replacement; and
- o   not be relocated in any other device or fitting
        installed;

- all components of the hot water system (including any   G3(2)
  cistern that supplies water to, or receives expansion
  water from, the hot water system) shall continue to safely
  contain the hot water:

  - o   during normal operation of the hot water system;
  - o   following failure of any thermostat used to control
          temperature; and
  - o   during operation of any of the safety devices.

6.19.1.3 Vented hot water storage systems

Hot water storage vessels should conform to BS 853-1:1,   G3.11
BS 1566-1:2002 or BS 3198:1981, or other relevant national
standards as appropriate.

All vented hot water storage systems should have a vent pipe connected to the
top of the hot water storage system and above the level of the water in the cold
water storage cistern; and the system should incorporate either:

- a non-self-resetting energy cut-out in the event of the storage system
  overheating; and

- an overheat cut-out to prevent the stored water exceeding 100° C; or

- a safety device (such as a temperature relief valve or a combined
  temperature and pressure relief valve) that will discharge the water
  in the event of overheating, either directly or by way of a manifold
  via a short length of metal pipe to a tundish.

6.19.1.4 Unvented storage system

The installation of an unvented system is notifiable building work which
must be reported to the building control body before work commences, unless
the installer is registered with a competent person scheme, in which case the
installer may self-certify that the work complies with all relevant requirements
in the Building Regulations and provide the building owner or occupier with a
Building Regulations certificate of compliance.

A hot water storage system that has an unvented storage vessel shall have:

> • at least two independent safety devices that release   G3(3)
>   pressure and, in doing so, prevent the temperature of the
>   stored water at any time exceeding 100° C;
> • pipework that is capable of discharging hot water from safety   G3(3)
>   devices so that it is visible at some point and safely conveys
>   it to an appropriate place open to the atmosphere where it
>   will cause no danger to persons in or about the building.

Note: In addition to any thermostat provided to control the desired tempera-ture of the stored water, unvented hot water storage systems should incorpo-rate a minimum of two independent safety devices such as a non-self-resetting energy cut-out, temperature relief valve, or combined temperature and pressure relief valve.

> Any unvented hot water storage system unit or package should   G3.23
> be indelibly marked (see Figure 6.274) with the following
> information:
> • the manufacturer's name and contact details;
> • the model reference;
> • the rated storage capacity;
> • the operating pressure of the system;
> • the operating pressure of the expansion valve;
> • relevant operating data on each of the safety devices fitted;
>   and
> • the maximum primary circuit pressure and flow
>   temperature of indirect hot water storage system units or
>   packages.

6.19.1.5 Commissioning heating and hot water systems

Reasonable provision should be made to limit heat losses from pipes as set out in DCLG's *Domestic Building Services Compliance Guide.* This includes insulating primary circulation pipes for domestic hot water services throughout their length.

> All fixed building services shall be commissioned by   L1A 3.23
> testing (and adjustments made where considered necessary)   L2A 3.15
> to ensure that they use no more fuel and power than is
> reasonable in the circumstances.

WARNING TO USER

a. Do not remove or adjust any component part of this unvented water heater–contact the Installer.

b. If this unvented water heater develops a fault, such as a flow of hot water from the discharge pipe, switch the heater off and contact the Installer.

WARNING TO INSTALLER

a. This Installation is subject to the Building Regulations.

b. Use only appropriate components for Installation or maintenance. Installed by:

Name...

Address...

Tel. No ....

Completion date

Figure 6.274 Hot water warning sign.

A commissioning plan, identifying the systems that need   L1A 3.24
to be tested and the tests that will be carried out, should be   L2A 3.16
prepared during the design stage as part of the builders rate
calculation (i.e. TER/DER and TFEE/DFEE in the case of a
dwelling and TER/BER for a building) so that the building
control body can check that the commissioning is being
done as the work proceeds.

Obviously not all fixed building services will need to be commissioned, as the only controls are possibly 'on' and 'off' switches (e.g. a mechanical extraction system or single fixed electrical heater).

Fixed building services which do not require commissioning   L1A 3.25
should be identified in the commissioning plan, along with
the reason for not requiring commissioning.

When commissioning a heating and/or hot water system, the following should be included in the test and inspection schedule:

- that any hot water outlet temperature device being used   G3(4)
  to limit the maximum temperature supplied at the outlet
  shall not be capable of being easily altered by building
  users;

- that the installation shall convey hot water to the sanitary   G3
  appliances and locations specified in the requirement
  without waste, misuse or undue consumption of water;
  and the water supplied shall be heated wholesome water
  or heated softened water;

- that pipework should be designed and installed in such   G3.7
  a way as to minimize the transfer time between the hot
  water storage system and hot water outlets.

In addition, when commissioning heating and hot water systems, the person carrying out the commission should ensure that:

- the performance of the building fabric and the heating and hot water systems is no worse than the design limits;
- the systems and their controls have been left in working order and are capable of operating efficiently for the purposes of the conservation of fuel and power;
- independent temperature and on/off controls to all heating appliances have been provided;
- the heating system uses heat-raising appliances that have an efficiency not less than that recommended in the *Domestic Heating Compliance Guide DCLG*;
- if both heating and cooling are provided, they are capable of being controlled so as not to operate simultaneously;
- energy meters have been included so as to allow building occupants to assign at least 90 per cent of the estimated annual energy consumption of each fuel used for heating, lighting, etc.;
- meters have been provided to enable installed low or zero carbon (LZC) systems to be separately monitored;
- automatic meter reading and data collection have been provided in all buildings with a total useful floor area that is greater than $1000\,m^2$.

The person carrying out the work shall provide the local authority with a notice confirming that all fixed building services have been properly commissioned in accordance with a procedure approved by the Secretary of State.

In non-domestic buildings:

- new heating, ventilation and cooling (HVAC) systems should be provided with controls that are capable of achieving a reasonable standard of energy-efficiency;

- separate control zones should be capable of independent switching and control set-point;
- if both heating and cooling are provided, then they should not operate simultaneously;
- the central plant serving zone-based systems should:

    ○ only operate as and when required;
    ○ have a default condition that is off.

Any heating system more than 15 years old should either be replaced or be equipped with improved controls.

6.19.1.6 Extensions

In an extension to a building, where the installed capacity per unit area of a heating system is increased, existing doors (but excluding high-usage entrance doors) within the area served which have U-values worse than 3.3 W/m² K should be replaced.	L2B 6.10 L2B 6.10

Replacing these sorts of doors will normally achieve a simple payback within 15 years.

6.19.1.7 Conservatories and porches

Regulation 9 of the Building Regulations exempts some conservatory and porch extensions from the energy-efficiency requirements, if doors and windows which separate the conservatory from the building are retained or, if removed, are replaced by walls, windows and doors which meet the energy-efficiency requirements, and where the heating system of the building is not extended into the conservatory or porch.

This exemption applies to conservatories as long as:

- they are at ground level and under 30 m²;
- glazing and any fixed electrical installations comply with the appropriate sections of the Building Regulations;
- the conservatory is separated from the house by external-quality walls, doors or windows;
- there is an independent heating system with separate temperature and on/off controls; and
- the front entrance door between the existing house and the new porch remains in place.

Note: In May 2013 Parliament agreed to increase the size of single-storey rear extensions which can be built under permitted development and brought into force the associated neighbour consultation scheme. For a period of three years, between 30 May 2013 and 30 May 2016, householders will be able to build larger single-storey rear extensions under permitted development. In essence the permitted development size limits for extensions have doubled from $4\,m^2$ to $8\,m^2$ for detached houses, and from $3\,m^2$ to $6\,m^2$ for all other houses.

On the downside:

(1) under Regulation 28 of the Building Regulations, the construction of an extension may trigger the requirement for consequential improvements;
(2) if a new conservatory or porch does not meet all these requirements, it is not exempt and **must** comply with the relevant energy-efficiency requirements of Approved Document L, and particular attention should be paid to independent temperature and on/off controls to any heating system installed within the extension;
(3) removing and not replacing any of the thermal separation between the building and an existing exempt extension, or extending the building's heating system into the extension, mean that the extension ceases to be exempt!

6.19.1.8 Consequential improvements (non-domestic buildings)

If work is being undertaken on an existing building which has a total useful floor area of over $1000\,m^2$, then, in addition to the principal works (which **must** still comply with the energy-efficiency requirements detailed in Approved Document L in the normal way), consequential improvements (where technically, functionally and economically feasible) will have to be completed provided that they are:

where the installed capacity per unit area of a heating system is increased, existing doors (but excluding high-usage entrance doors) within the area served and which have U-values worse than 3.3 W/m^2 K should be replaced.	L2B 6.10 L2B 6.10

Replacing these sorts of doors will normally achieve a simple payback within 15 years.

6.19.1.8.1 Insulation of pipes, ducts and vessels

For buildings other than dwellings:

• insulation should not be less than that shown in the *Non-Domestic Heating, Cooling and Ventilation Compliance Guide*;

- hot and chilled water pipework, storage vessels, refrigerant pipework and ventilation ductwork should be insulated so as to conserve energy and to maintain the temperature of the heating or cooling service;
- fans that are rated at more than 1100 W and which form part of the environmental control system should be equipped with variable-speed drives.

6.19.1.8.2 Ducted warm air heating systems

A room thermostat for a ducted warm air heating system should be mounted in the living room, at a height between 1370 mm and 1830 mm, and its maximum setting should not exceed 27° C.	B1 2.17 (V1) B1 2.18 (V2) B3 7.10 (V1) B3 10.2 (V2)

6.19.1.9 Controls

Systems should be provided with appropriate controls to enable reasonable standards of energy-efficiency whilst in use by: • subdividing the systems into separate control zones corresponding to each area of the building that has a significantly different solar exposure or pattern or type of use; • ensuring that each separate control zone is capable of independent timing and temperature control and, where appropriate, ventilation and air recirculation rate; and • providing an appropriate service according to the requirements of the space it serves.	L2A 2.43

 Note: If both heating and cooling are provided, they should be controlled so as not to operate simultaneously; and central plant should operate.

6.20 Liquid fuel

6.20.1 Requirements

Protection of liquid fuel storage systems
Liquid fuel storage systems and the pipes connecting them to combustion appliances shall be so constructed, and separated from buildings and the boundary of the premises as to reduce to a reasonable level the risk of the fuel igniting in the event of fire in adjacent buildings or premises.

(Approved Document J5)

Fire precautions
- *Buildings shall be sub-divided by elements of fire-resisting construction into compartments.*
- *Openings in fire-separating elements shall be suitably protected in order to maintain the integrity of the element (i.e. the continuity of the fire separation).*

(Approved Document B3)

6.20.2 Meeting the requirements

6.20.2.1 Storage and supply

Oil and liquid petroleum gas (LPG) fuel storage installations (including the pipework connecting them to the combustion appliances in the buildings they serve) shall:

• be located and constructed so that they are reasonably protected from fires that may occur in buildings or beyond boundaries;	J 5.1a
• be reasonably resistant to physical damage and corrosion;	J 5.1a
• be designed and installed so as to minimize the risk of oil escaping during the filling or maintenance of the tank;	J 5.1bi
• incorporate secondary containment when there is a significant risk of pollution;	J 5.1bii
• contain labelled information on how to respond to a leak.	J 5.1biii

6.20.2.2 Oil pollution

The Control of Pollution (Oil Storage) (England) Regulations 2001 (SI 2001/2954) came into force on 1 March 2002. They apply to a wide range of oil storage installations in England, but they do not apply to the storage of oil on any premises used wholly or mainly as one or more private dwellings, **if** the capacity of the tank is 3500 litres or less.

6.20.2.2.1 Provisions where there is a risk of oil pollution

The main problems with regard to leakage concerns inland freshwater streams, rivers, reservoirs and lakes, as well as ditches and ground drainage (e.g. perforated drainage pipes) that feed into them.

When secondary containment is considered necessary, a way of meeting the requirement would be to:	J 5.10

- provide an integrally bunded prefabricated tank; or
- construct a bund from masonry or concrete.

Note: Bunds (whether part of a prefabricated tank system or constructed on site) should have a capacity of at least 110 per cent of the largest tank they contain.

An oil storage installation should carry a label in a prominent position giving advice on what to do if an oil spill occurs, and the telephone number of the Environment Agency's Emergency Hotline (see Appendix F to Part J:2010 for further details).	J 5.12

Table 6.103 Limits on design flexibility for mechanical ventilation

System type	Performance
Specific fan power (SFP) for continuous supply only and continuous extract only	0.8 l/s W
SFP for balanced systems	2.0 l/s W
Heat recovery efficiency	66%

6.20.2.3 LPG storage

Note: Oil storage below ground is not recommended if other options are available, as underground tanks are difficult to inspect and leaks may not be immediately obvious.

LPG installations are controlled by legislation enforced by the HSE, which includes the following requirements applicable to dwellings:

The LPG tank should be installed outdoors and not within an open pit.	J 5.15
The tank should be adequately separated from buildings, the boundary and any fixed sources of ignition to enable safe dispersal in the event of venting or leaks and in the event of fire, to reduce the risk of fire spreading (see Figure 6.275).	J 5.15

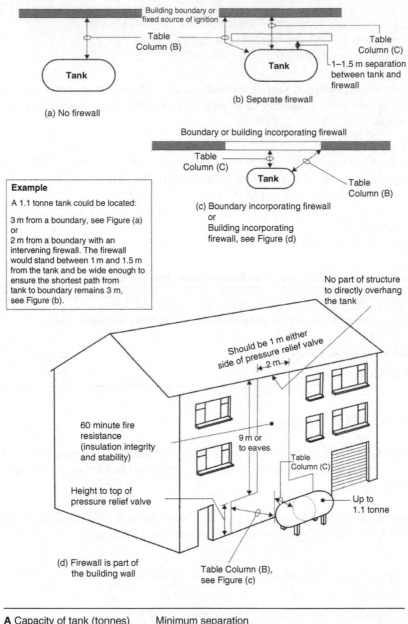

Figure 6.275 Separation or shielding of LPG tanks from the building, boundaries and fixed sources of ignition.

Firewalls may be free-standing, built between the tank and the building, boundary and fixed source of ignition (see Figure 6.275b) or a part of the building, or a fire resistance (insulation, integrity and stability) boundary wall belonging to the property.   J 5.16

Where a firewall is part of the building or a boundary wall, it should be located in accordance with Figure 6.275c.   J 5.16

If the firewall is part of the building, it should be constructed as shown in Figure 6.275d.   J 5.16

Firewalls should be imperforate and of solid masonry, concrete or similar construction.   J 5.17

Firewalls should have a fire resistance (insulation, integrity and stability) of at least 30 minutes.   J 5.17

If firewalls are part of the building as shown in Figure 6.275d, they should have a fire resistance (insulation, integrity and stability) of at least 60 minutes.   J 5.17

To ensure good ventilation, firewalls should not normally be built on more than one side of a tank.   J 5.17

A firewall should be at least as high as the pressure relief valve.   J 5.18

Any pipe carrying natural gas or LPG should be: B2 8.40 (V2)

- of screwed steel or of all welded steel construction;
- installed in accordance with SI 1996 No. 825 and SI 1998 No. 2451.

A protected shaft conveying piped flammable gas should be adequately ventilated direct to the outside air by ventilation openings at high and low level in the shaft.   B2 8.41 (V2)

Where an LPG storage installation consists of a set of cylinders, a way of meeting the requirements is as shown in Figure 6.276.

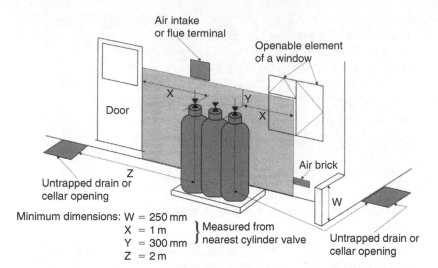

Figure 6.276 Location of LPG cylinders.

Cylinders should stand upright and be secured by straps (or chains) against a wall outside the building, in a well-ventilated position at ground level.	J 5.20
Cylinders should be provided with a firm, level base, such as concrete at least 50 mm thick or paving slabs bedded on mortar.	J 5.20

6.20.2.4 Garages

If a door is provided between a dwelling house and the garage:	B3 5.5 (V1)

- the floor of the garage should be laid so as to allow fuel spills to flow away from the door to the outside; or
- the door opening should be positioned at least 100 mm above garage floor level.

6.21 Kitchens and utility rooms

The Building Act 1984 (Sections 25 and 69) stipulates that plans for proposed buildings will ensure that all occupants of the house will be provided with a supply of *wholesome water, sufficient for their domestic purposes*. This can be achieved by:

- connecting the house to water supplies from the local water authority (normally referred to as the statutory water undertaker); or
- taking water into the house by means of a pipe (e.g. from a local recognized supply); or
- providing a supply of water within a reasonable distance from the house (e.g. such as from a well).

If an occupied house is not within a reasonable distance of a supply of 'wholesome water' or if the local authority is not satisfied that the water supply is capable of supplying 'wholesome water', then it can give notice that the owner of the building must provide water within a specified time. It also has the authority to prohibit the building from being occupied.

6.21.1 Requirements

Cold water supply

(1) There must be a suitable installation for the provision of:

    (a) wholesome water to any place where drinking water is drawn off;

    (b) wholesome water or softened wholesome water to any washbasin or bidet provided in or adjacent to a room containing a sanitary convenience;

    (c) wholesome water or softened wholesome water to any washbasin, bidet, fixed bath or shower in a bathroom; and

    (d) wholesome water to any sink provided in any area where food is prepared.

(2) There must be a suitable installation for the provision of water of suitable quality to any sanitary convenience fitted with a flushing device.

(Approved Document G1)

Water efficiency
Reasonable provision must be made by the installation of fittings and fixed appliances that use water efficiently for the prevention of undue consumption of water.

(Approved Document G2)

Requirement G2 applies only when a dwelling is erected or formed by a material change of use of a building.

Food preparation
A suitable sink must be provided in any area where food is prepared.

(Approved Document G)

Sanitary conveniences
There must be a suitable installation for the provision of water (of suitable quality) to any sanitary convenience fitted with a flushing device.

(Approved Document G1)

Any room containing a sanitary convenience, a bidet, or any facility for washing hands must be separated from any kitchen or any area where food is prepared.

(Approved Document G6)

Fire safety
- *There shall be an early warning fire alarm system for persons in the building.*
- *There shall be sufficient escape routes that are suitably located to enable persons to evacuate the building in the event of a fire.*
- *Safety routes shall be protected from the effects of fire.*
- *In an emergency, the occupants of any part of the building shall be able to escape without any external assistance.*

(Approved Document B1)

- *The spread of flame over the internal linings of the building shall be restricted.*
- *The heat released from the internal linings shall be restricted.*

(Approved Document B2)

- *Buildings shall be sub-divided by elements of fire-resisting construction into compartments.*
- *Openings in fire-separating elements shall be suitably protected in order to maintain the integrity of the element (i.e. the continuity of the fire separation).*

(Approved Document B3)

Ventilation
There shall be adequate means of ventilation provided for people in the building.

(Approved Document F)

Noise
Dwellings shall be designed so that any domestic noise that is generated internally does not interfere with the occupants' ability to sleep, rest and engage in their normal activities in satisfactory conditions.

(Approved Document E2)

6.21.2 Meeting the requirements

6.21.2.1 Cold water supplies

The cold water supply shall:

- be reliable;   G1
- be wholesome;

- have a pressure and flow rate sufficient for the operation of all appliances and locations planned in the building;
- convey wholesome water or softened wholesome water without waste, misuse, undue consumption or contamination of water.

6.21.2.1.1 Cold water storage systems

The cold water storage cistern into which the vent pipe discharges should be supported on a flat, level, rigid platform.

The platform should extend a minimum of 150 mm in all directions beyond the edge of the maximum dimensions of the cistern.	G3.15
The cistern should be accessible for maintenance, cleaning and replacement.	G3.16

Cold water pipes that pass through a compartment wall or compartment floor (unless the pipe is in a protected shaft), or through a cavity barrier, should incorporate one of the following alternatives:

Proprietary seals (any pipe diameter) that maintain the fire resistance of the wall, floor or cavity barrier.	B3 7.6–7.7 (V1) B3 10.6 (V2)
Pipes with a restricted diameter where fire-stopping is used around the pipe, keeping the opening as small as possible.	B3 7.6 and 7.8 (V1) B3 10.5 and 10.7 (V2)
Sleeving – a pipe of lead, aluminium, aluminium alloy, fibre-cement or UPVC, with a maximum nominal internal diameter of 160 mm, may be used with a sleeving of non-combustible pipe as shown in Figure 6.30.	B3 7.6 and 7.9 (V1) B3 10.5 and 10.8 (V2)

6.21.2.2 Water efficiency

The cold water supply shall:

- be reliable;	G1
- be wholesome;	
- have a pressure and flow rate sufficient for the operation of all appliances and locations planned in the building;	

- convey wholesome water or softened wholesome water without waste, misuse, undue consumption or contamination of water.

6.21.2.3 Food preparation

A suitable sink must be provided in any area where food is prepared and:

> this sink should discharge through a grating, a trap and a branch   G6.5
> discharge pipe to an adequate system of drainage.

Note: Where a dishwasher is provided in a separate room that is not the principal place for the preparation of food, an additional sink need not be provided in that room.

6.21.2.4 Storage of food

In accordance with Building Act 1984 Sections 28 and 70, all houses or buildings that have been converted into dwellings **must** provide sufficient and suitable accommodation for storing food or 'sufficient and suitable space for the provision of such accommodation by the occupier'.

6.21.2.5 Refuse chutes and storage

Note: Refuse storage chambers, refuse chutes and refuse hoppers should be sited and constructed in accordance with BS 5906.

> The access to refuse storage chambers should   B1 5.57 (V2)
> not be sited adjacent to escape routes or final
> exits, or near to windows of flats.
>
> Refuse chutes and rooms provided for the   B1 5.55 and 5.56 (V2)
> storage of refuse should:
>
> - be approached either directly from the open air or by way of a protected lobby;
> - be separated from other parts of the building by fire-resisting construction;
> - **not** be located within protected stairways or protected lobbies.
>
> A wall separating a habitable room or kitchen   E 2.28
> and a refuse chute should have mass (including
> any finishes) of at least 1320 kg/m^2.

A wall separating a non-habitable room which is in   E 2.28
a dwelling from a refuse chute should have a mass
(including any finishes) of at least 220 kg/m².

In buildings containing flats, any wall that   B2 8.13 (V2)
encloses a refuse storage chamber should be
constructed as a compartment wall.

6.21.2.6 Sanitary conveniences

For sanitary conveniences the cold water supply shall:

- be reliable;   G1
- be either wholesome, softened wholesome or of suitable quality;
- have a pressure and flow rate sufficient for the operation of
  the sanitary appliances.

In buildings where the Food Hygiene (England) Regulations 2006 (SI 2006114)
and the Food Hygiene (Wales) Regulations 2006 (SI 2006131 W5) apply, in
addition to any hand-washing facilities associated with a water closet or other
sanitary facility, **separate** hand-washing facilities may be needed.

6.21.2.6.1 Alternative sources of water

Wholesome water need not be used for toilet flushing, irrigation, etc. In these
circumstances, water from alternative sources may be used. These include:

- water abstracted from wells, springs, boreholes or watercourses;
- harvested rainwater;
- reclaimed greywater; and
- reclaimed industrial process water.

    In all cases:

the water obtained from an alternative source shall incorporate   G1.7
measures to minimize the impact on water quality from:

- failure of any components;
- failure due to lack of maintenance;
- power failure; and
- any other measures identified in a risk assessment.

Any system or unit used to supply dwellings with water from   G1 .4
alternative sources shall be subject to a risk assessment by the
system designer and manufacturer.

6.21.2.7 Fire safety

In small premises:

• store rooms should be enclosed in a fire-resisting construction;	B1 3.35 (V2)
• clear glazed areas should be provided in any partitioning that separates the kitchen from the open floor area, to enable any person within the kitchen to have an early visual warning of an outbreak of fire.	B1 3.36 (V2)

6.21.2.7.1 Smoke alarms

Smoke alarms should be positioned where fires are most likely to start (e.g. in kitchens and living rooms).	B1 1.11 (V1) B1 1.10 (V2)
Where the kitchen area is not separated from the stairway or circulation space by a door, there should be a compatible interlinked heat detector or heat alarm in the kitchen, in addition to whatever smoke alarms are needed in the circulation space(s).	B1 1.13 (V1) B1 1.12 (V2)

6.21.2.7.2 Escape routes

Any inner room that is a kitchen or utility room that is situated not more than 4.5 m above ground level, whose only escape route is through another room, shall be provided with an emergency egress window.	B1 2.9 (V1) B1 2.5 (V2)

6.21.2.7.3 Ancillary accommodation

Ancillary accommodation in kitchens, such as staff changing rooms, locker rooms and store rooms, should be enclosed by fire-resisting construction.	B1 3.50 (V2)

6.21.2.8 Ventilation

Extract ventilation may be by natural means (e.g. by passive stack ventilation (PSV) or by mechanical means such as an extract fan or central system), but please note:

It is now a **mandatory** requirement that, if you carry out any 'building work' and there is an existing extract fan, PSV or cooker hood extracting to outside in the kitchen, you should either retain or replace it!

In kitchens, any automatic control **must** provide sufficient flow during cooking with fossil fuel (e.g. gas) to avoid build-up of combustion products.	F Table 5.2a

Note: Manual boost controls should also be provided in kitchens to guard against the possibility of a single centrally located switch being left in an incorrect mode of operation.

All kitchens and utility rooms shall be provided with extract ventilation to the outside which is capable of operating either: • intermittently at a minimum extract rate of 30 l/s (adjacent to hob) and 60 l/s elsewhere; or • continuously with a minimum extract rate of 13 l/s.	F 5.5
Common outlet terminals and/or branched ducts shall **not** be used for wet rooms such as kitchens or utility rooms.	F Tables 5.2a–5.2d
PSV devices shall have a minimum: • internal duct diameter of 125 mm for kitchens and 100 mm for utility rooms; and • a cross-sectional area of 12,000 mm^2 for kitchens and 8000 mm^2 for utility rooms.	F Tables 5.2a–5.2d
Where there was no previous ventilation opening, or where the size of the original ventilation opening is not known, replacement window(s) shall have an equivalent area greater than 2500 mm^2.	F 7.6
Wall- or ceiling-mounted centrifugal fans (which are fitted with a 100 mm diameter flexible duct or rectangular duct and which are designed to achieve 60 l/s for kitchens) should not be ducted further than 3 m and should have no more than one 90° bend.	F Tables 5.2a–5.2d
Automatic controls for ventilators that are designed to work continuously in kitchens **must** be capable of providing sufficient flow during cooking with fossil fuels (e.g. gas) to avoid the build-up of combustion products.	F Tables 5.2a–5.2d

> All food and beverage preparation areas shall be   F 6.10
> provided with intermittent air extract ventilation.

In all non-domestic kitchens:

> •   should have separate and independent extraction   B1 5.50 (V2)
>   systems; and
> •   extracted air should not be recirculated.

If any of the work being carried out in the kitchen of an existing building is 'building work' (as defined in Regulation 3 of the Building Regulations) then you will need to comply with the appropriate requirements of this regulation.

6.22 Loft conversions

6.22.1 Requirements

Fire safety
*   *There shall be an early warning fire alarm system for persons in the building.*
*   *There shall be sufficient escape routes that are suitably located to enable persons to evacuate the building in the event of a fire.*
*   *Safety routes shall be protected from the effects of fire.*
*   *In an emergency, the occupants of any part of the building shall be able to escape without any external assistance.*

(Approved Document B1)

*   *The spread of flame over the internal linings of the building shall be restricted.*
*   *The heat released from the internal linings shall be restricted.*

(Approved Document B2)

Ventilation
There shall be adequate means of ventilation provided for people in the building.

(Approved Document F)

Protection from falling
Stairs shall be provided with barriers where it is necessary to protect people in or about the building from falling.

(Approved Document K2

A new issue of Approved Document K came into force in 2013.

Stairs and ladders
Stairs and ladders shall be so designed, constructed and installed as to be safe for
people moving between different levels in or about the building.

(Approved Document K1)

Note: This requirement only applies to stairs that form part of the building.

6.22.2 Meeting the requirements

6.22.2.1 Fire safety

Where the conversion of an existing roof space (e.g. a loft conversion to a two-storey house) means that a new storey is going to be added, then the stairway will need to be protected with fire-resisting doors and partitions.

The floor(s), both old and new, shall have the full 30-minute standard of fire resistance shown in Approved Document B, Appendix A, Table A1, unless: • only one storey is being added; • the new storey contains no more than two habitable rooms; and • the total area of the new storey is less than $50\,\text{m}^2$.	B3 4.7 (V1)
In those places where the floor only separates rooms (and not circulation spaces), a modified 30-minute standard of fire resistance may be applied.	B3 4.7 (V1)
New habitable rooms that are the result of a material alteration and which are above ground-floor level (or at ground-floor level where no final exit has been provided) shall be equipped with: • a fire detection and fire alarm system; • smoke alarms in accordance with BS 5839-6.	B1 2.20a (V1) B1 1.8 (V1)

6.22.2.2 Ventilation

The following guidance (extracted from the 2006 version of Approved Document F) has been included here for information purposes:

Fans and/or ducting placed in or passing through an unheated void or loft space should be insulated to reduce the possibility of condensation forming.	F (2006) App E

The inner radius of any bend should be greater than or equal to the diameter of the ducting being used.	F (2006) App E
Vertical duct rises may need to be fitted with a condensation trap in order to prevent the backflow of any moisture.	F (2006) App E
The circular profile of a flexible duct should be maintained throughout the full length of the duct run (see Figure 6.277).	F (2006) App E

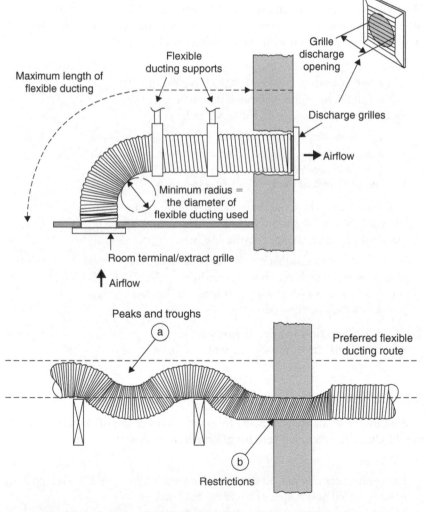

Figure 6.277 Installation of ducting.

If a back-draught device is used it may be incorporated into the fan itself.   F (2006) App E

Flexible ducting should be installed without any peaks or troughs (see Figure 6.277).   F (2006) App E

6.22.2.3 Protection from falling

Pedestrian guarding that is capable of preventing people from being injured by falling from a height of more than 600 mm should be provided.

6.22.2.4 Stairs

The rise of a stair **shall** be between 150 mm and 220 mm, with any going between 220 mm and 400 mm and a maximum pitch of 42° (see Table 1.1 of Approved Document K1).   K1 1.2

Note: The normal relationship between the dimensions of the rise and going is twice the rise plus the going (i.e. 2R + G).

6.22.2.4.1 Spiral and helical stairs

The design of spiral and helical stairs shall be in accordance with BS 5395-2.   K1 1.28

6.22.2.4.2 Handrails for stairs

If the stairs are 1000 mm or wider, there should be a handrail (whose top shall be 900 mm to 1100 mm from the pitch line or floor) on both sides and the handrail may form the top of a guarding if the heights are matched.   K1 1.34

6.22.2.4.3 Construction of steps

Steps should have level treads in accordance with Table 1.1 of Approved Document F1.   K1 1.5

> Steps with open risers (especially those that are likely to be used by children under five years old) should be constructed so that a 100 mm diameter sphere cannot pass through the open risers and overlap treads should be a minimum of 16 mm.
>
> K1 1.7 and 1.8

Steps which have more than 36 risers in consecutive flights should make at least one change of direction, between flights, of at least 30° (see Approved Document B2 for further details).

6.22.2.4.4 Alternating tread stairs

Alternating tread stairs may be used in a loft conversion, but only when as shown in Figure 6.283, the stair is for access to only one habitable room and the construction of the alternate tread stair complies with Figure 6.278 and:

> K1 1.29
>
> K1 1.30
>
> • alternating steps are uniform with parallel nosings;
> • all treads have slip-resistant surfaces;
> • tread sizes over the wider part of the step conform to Approved Document K1, Table 1.1; and
> • a minimum clear headroom of 2 m is provided.

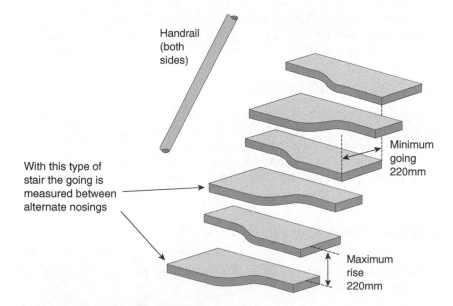

Figure 6.278 Alternating tread stair.

6.22.2.4.5 Tapered treads

For stairs with tapered treads:

- the rise and going shall comply with the recommendations shown in Approved Document H (1.2 and 1.3); K1 1.25
- measured tapered treads shall be in accordance with Figure 6.279.

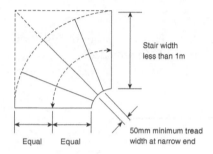
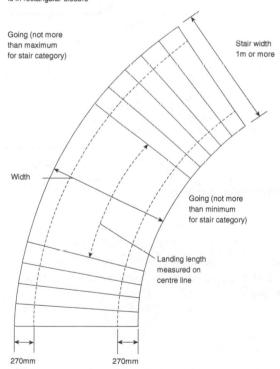

Figure 6.279 Measured tapered treads.

For stairs with tapered treads:

> • consecutive tapered treads shall use the same going. K1 1.26

 Note: If a stair consists of straight and tapered treads, the going of the tapered treads shall not be less than the going of the straight treads.

6.22.2.4.6 Headroom

> On the access between levels, the minimum headroom shown K1 1.11
> in Figure 6.280 should be provided.

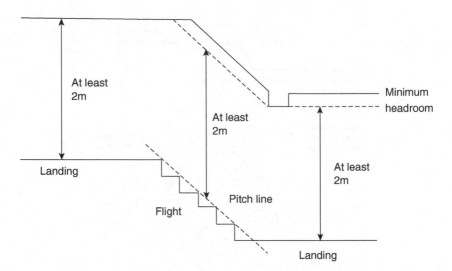

Figure 6.280 Minimum headroom.

> Where there is insufficient space to achieve the height shown K1 1.13
> in Figure 6.280, reduced headroom as shown in Figure 6.281
> should be provided.

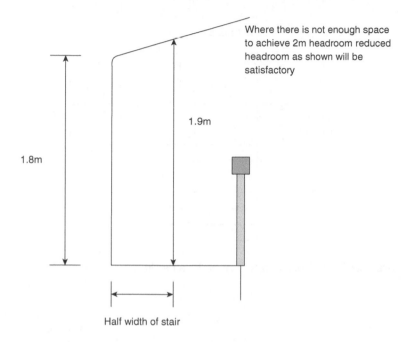

Where there is not enough space to achieve 2m headroom reduced headroom as shown will be satisfactory

1.9m

1.8m

Half width of stair

Figure 6.281 Reduced headroom for loft conversions.

6.22.2.4.7 Landings

Landings shall be provided at the top and bottom of every flight (see Figure 6.198).	K1 1.20
A landing may include part of the floor of the building, but it should be kept clear of permanent obstructions. It may have doors to cupboards and ducts that open over a landing at the top of a flight (see Figure 6.282), but only when they are kept locked or shut when in normal use.	K1 1.21

 Note: Also see Approved Document B.

All landings must be level, unless the landing is at the top (or bottom) of a flight which is formed by the ground, in which case it may have a maximum gradient along the direction of travel of 1:60 provided that the surface is paved ground or otherwise made permanently firm.	K1 1.22
A door may swing across a landing at the bottom of flight, but only as shown in Figure 6.283.	K1 1.24

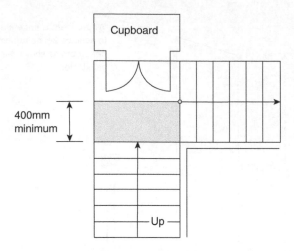

Figure 6.282 Cupboards opening on to a landing.

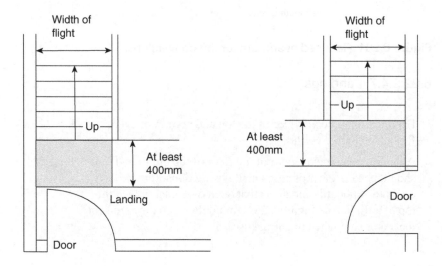

Figure 6.283 Landings next to doors in dwellings.

6.22.2.5 Ladders

A fixed ladder may be used for access in a loft conversion that contains one habitable room, but **only** if it is fitted with fixed handrails on **both** sides.

 Retractable ladders may **not** be used as means of escape.

6.23 Extensions and additions to buildings

Note: Also see the requirements and recommendations for loft conversions (section 6.22) and conservatories (section 6.24) with respect to extensions and additions.

Under Regulation 17D of the Building Regulations, the construction of an extension may trigger the requirement for consequential improvements.

For example, if a building has a total useful floor area greater than $1000\,m^3$ and the proposed building work includes an extension, the initial provision of any fixed building service, or an increase to the installed capacity of any fixed building services, consequential improvements should be made to improve the energy-efficiency of the whole building and:

* thermal units with high U-values should be upgraded;
* existing windows (but not display windows), roof windows, rooflights and doors (excluding high-usage entrance doors) within the area served by the fixed building service with an increased capacity should be replaced;
* heating systems, cooling systems and air-handling systems that are more than 15 years old should either be replaced or be equipped with improved controls;
* any general lighting system serving an area greater than $100\,m^2$ which has an average lamp efficacy of less than 40 lamp-lumens per circuit-watt should be upgraded with new luminaires or improved controls;
* energy metering should be installed if less than 10 per cent of the building's energy demand is provided by a low or zero carbon (LZC) energy system; and
* the building should be upgraded with an additional LZC energy system **provided** that the system would achieve a simple payback within seven years or less.

Note: In 2013, for a period of three years, the size of single-storey rear extensions which can be built under permitted development has been increased and an associated neighbour consultation scheme introduced. Until 30 May 2016, householders may build larger single-storey rear extensions under permitted development; size limits have doubled from $4\,m^2$ to $8\,m^2$ for detached houses, and from $3\,m^2$ to $6\,m^2$ for all other houses. More information on this is contained in Chapters 4 and 5.

6.23.1 Requirements

Ventilation
There shall be adequate means of ventilation provided for people in the building.

(Approved Document F)

Conservation of fuel and power
Reasonable provision shall be made for the conservation of fuel and power in buildings by:

(a) *limiting heat gains and losses*

    (i) *through thermal elements and other parts of the building fabric; and*
    (ii) *from pipes, ducts and vessels used for space heating, space cooling and hot water services;*

(b) *providing fixed building services which*

    (i) *are energy efficient;*
    (ii) *have effective controls; and are commissioned by testing and adjusting as necessary to ensure they use no more fuel and power than is reasonable in the circumstances.*

(Approved Document L)

 A new version of Approved Document L came into force on 6 April 2014.

Responsibility for achieving compliance with the requirements of Part L rests with the person carrying out the work. That person may be, for example, a developer, a main (or sub-) contractor, or a specialist firm directly engaged by a private client.

The person responsible for achieving compliance should either him- or herself provide a certificate or obtain a certificate from the subcontractor that commissioning has been successfully carried out. The certificate should be made available to the client and the building control body.

6.23.1.1 Fire risk analysis

Approved Document B now includes a requirement for the responsible person (i.e. the person carrying out work to a building) to make available to the owner (other than houses occupied as single private dwellings) 'fire safety information' concerning the design and construction of the building or extension, plus details of the services, fittings and equipment that have been provided, in order that he or she (when required under the new Regulatory Reform (Fire Safety) Order 2005 – Statutory Instrument 2005 No. 1541) might complete a fire risk analysis.

The sort of information required must include basic advice on the proper use and maintenance of systems provided in the building such as:

- emergency egress windows;
- fire doors;
- smoke alarms; and
- sprinklers.

6.23.2 Meeting the requirements

6.23.2.1 Changes in the legal requirements

The Building Regulations exempt some extensions from the energy-efficiency requirements. This exemption from the requirements of Part L only applies where the existing doors are retained (or replaced) and:

- the extension is at ground level;
- the extension has a floor area of less than $30\,\mathrm{m}^2$;
- the existing walls, doors and windows of the dwelling have been either retained or, if removed, replaced by walls, windows and doors which meet the energy-efficiency requirements of Part L:2013; and
- the heating system of the dwelling has not been extended into the extension.

If a new extension does not meet all these requirements then it is not exempt and **must** comply with the relevant energy-efficiency requirements of Part L, and reasonable provision should be made to ensure effective thermal separation between the heated area in the existing dwelling and the extension, i.e. the walls, doors and windows between the dwelling and the extension should be insulated and draught-proofed to at least the same extent as in the existing dwelling.

If a building extension is **not** exempt from the energy-efficiency requirements, then:

- there should be effective thermal separation between the heated area in the existing building and the extension, i.e. the walls, doors and windows between the building and the extension should be insulated and draught-proofed to at least the same extent as in the existing building;
- particular attention should be paid to independent temperature and on/off controls to any heating system installed within the extension; and
- glazed elements should meet the standards set out in Table 6.104 and opaque elements should meet the standards set out in Table 6.105.

Removing and not replacing any of the thermal separation between the building and an existing exempt extension, or extending the building's heating system into the extension, mean that the extension ceases to be exempt!

Table 6.104 Standards for controlled fittings

Fitting	Standard (W/m² K)
Windows, roof windows and glazed rooflights	1.8 for the whole unit
Alternative option for windows in buildings that are essentially domestic in character	Window energy rating of B and C
Plastic rooflight	1.8
Curtain walling	<1.8
Pedestrian doors where the door has more than 50% of its internal face area glazed	1.8 for the whole unit
High-usage entrance doors for people	3.5
Vehicle access and similar large doors	1.5
Other doors	1.8
Roof ventilators (including smoke extract ventilators)	3.5

Table 6.105 Standards for new thermal elements

Element	Standard (W/m² K)		
Wall	0.262		
	0.16		
	If a building extension is a conservatory or porch that is *not* exempt from the energy efficiency requirements, then:		
	• the effective thermal separation between the heated area in the existing building (i.e. the walls, doors and windows between the building and the extension) should be insulated and draught proofed to at least the same extent as in the existing building;		
	• particular attention should be paid to independent temperature and ON/OFF controls to any heating system installed within the extension; and		
	• glazed elements should meet the standards set out in Table 6.104 and opaque elements should meet the standards set out in Table 6.105.		
Pitched roof-insulation at ceiling level	Fitting		Standard (W/m² K)
	Windows, roof windows and glazed rooflights		1.8 for the whole unit
	Alternative option for windows in buildings that are essentially domestic in character A window energy rating of Band C		
	Plastic rooflight		1.8
	Curtain walling		< 1.8
	Pedestrian doors where the door has more than 50% of its internal face area glazed		1.8 for the whole unit
	High-usage entrance doors for people		3.5

Table 6.105 (Continued)

Element	Standard (W/m² K)	
	Vehicle access and similar large doors	1.5
	Other doors	1.8
	Roof ventilators (including smoke extract ventilators)	3.5
	Removing and not replacing any of the thermal separation between the building and an existing exempt extension, or extending the building's heating system into the extension, means that the extension ceases to be exempt!	

Element	Standard (W/m² K) (W/m² N
Wall	0.262
Pitched roof – insulation at ceiling level	0.16
Pitched roof – insulation at rafter level	0.18
Flat roof or roof with integral insulation	0.18
Floors	0.224
Swimming pool basin	0.25

Element	Standard
Pitched roof-insulation at rafter level	0.18
Flat roof or roof with integral insulation	0.18
Floors	0.224
Swimming pool basin	0.25

6.23.2.2 Ventilation

If the additional room is connected to an existing habitable room:

- which now has no windows opening to outside, then the ventilation opening (or openings) shall be greater than 8000 mm² equivalent area; F 3.8a(i)
- which still has windows opening to outside, but with a total background ventilator equivalent area of at least 5000 mm² equivalent area, then the ventilation opening (or openings) shall be greater than 8000 mm² equivalent area; F 3.8a(ii)
- which still has windows opening to outside, but with a total background ventilator equivalent area of at least 5000 mm² equivalent area, then there should be background ventilators of at least 8000 mm² equivalent area: F 3.8a(iii)
  ○ between the two rooms; and
  ○ between the additional room and outside.

If the extension is the addition of a habitable room to an existing building, then:

- internal doors between the wet room and the existing   F 3.13
  building should have an undercut of at least minimum area
  7600 mm² (equivalent to an undercut of 10 mm above the
  floor finish for a standard 760 mm width door);
- whole-building and extract ventilation can be provided by: F 3.12

  - an intermittent extract and a background ventilator of
    at least 2500 mm² equivalent area; or
  - a single-room heat recovery ventilator; or
  - a passive stack ventilator; or
  - a continuous extract fan.

6.23.2.3 Historic and traditional buildings

When work is undertaken on any historic or traditional building, the aim should always be to improve energy-efficiency as far as is reasonably practicable without prejudicing the character of the host building or increasing the risk of long-term deterioration of the building fabric or fittings.

In general, new extensions to historic or traditional buildings should   L
comply with the standards of energy-efficiency as set out in Part L. The
only exception is where there is a particular need to match the external
appearance or character of the extension to that of the host building.

Building inspectors normally require the use of 'sympathetic treatment' in the restoration of the historic character of a building that has been subject to previous inappropriate alteration (e.g. replacement doors).

Particular issues that could warrant sympathetic treatment – and where advice from others would probably be beneficial – include:

- restoring the historic character of a building that has   L2B 3.11
  been subject to previous inappropriate alteration (e.g.
  replacement windows, doors and rooflights);
- rebuilding a former historic building (e.g. following a
  fire or filling a gap site in a terrace); and
- renovating the fabric of historic buildings to allow it to
  'breathe' to control moisture and potential long-term
  decay problems.

6.23.2.4 Thermal elements

Extensions to dwellings should either use newly constructed thermal elements (that meet the requirements for the conservation of fuel and power) or use existing (or new) doors, windows, roof windows and rooflights that meet these standards (see Table 6.106).

Table 6.106 Standards for control fittings

Fitting	Standard (W/m² K)
Window, roof window or rooflight	1.6
Doors with >50% of internal face glazed	1.8
Other doors	1.8

In most circumstances the most reasonable provision is to limit the total area of windows, roof windows and doors in extensions so that it does not exceed the sum of:

- 25 per cent of the floor area of the extension; and
- the total area of any windows or doors which, as a result of the extension works, no longer exist or are no longer exposed.

Where the proposed extension has a total useful floor area that is both:

- greater than 1000 m²; and
- greater than 25 per cent of the total useful floor area of the existing building

then the work should be regarded as a **new building**.

One way of complying is to show that the area-weighted U-value of all the elements in the extension is no greater than that of an extension of the same size and shape.

If upgrades are proposed to the existing dwelling, such upgrades should be implemented to a standard that is no worse than that shown in column (b) of Table 6.107.	L1B 4.7

Table 6.107 Upgrading retained thermal elements

Element	(a) Threshold U-value (W/m² K)	(b) Improved U-value (W/m² K)
Wall – cavity insulation	0.70	0.55
Wall – external or cavity insulation	0.70	0.30
Floor	0.70	0.25
Pitched roof – insulation at ceiling level	0.35	0.16
Pitched roof – insulation between rafters	0.35	0.18
Flat roof or roof with integral insulation	0.35	0.18

6.23.2.5 Opening areas of rooflights and windows

The area of windows and rooflights in the extension should   L2B 4.4
generally not exceed the values given in Table 6.108.

Table 6.108 Opening areas in the extension

Building type	Windows and personnel doors as percentage of exposed wall	Rooflights as percentage of area of roof
Residential buildings where people temporarily or permanently reside	30	20
Places of assembly, offices and shops	40	20
Industrial and storage buildings	15	20
Vehicle access doors and display windows and similar glazing	As required	n/a
Smoke vents	n/a	As required

6.24 Conservatories

6.24.1 Requirements

Ventilation
There shall be adequate means of ventilation provided for people in the building.

(Approved Document F)

Conservation of fuel and power
Reasonable provision shall be made for the conservation of fuel and power in buildings by:

(a) limiting heat gains and losses

(i) through thermal elements and other parts of the building fabric; and
(ii) from pipes, ducts and vessels used for space heating, space cooling and hot water services;

(b) providing fixed building services which

(i) are energy efficient;
(ii) have effective controls; and are commissioned by testing and adjusting as necessary to ensure they use no more fuel and power than is reasonable in the circumstances.

(Approved Document L)

A new version of Approved Document L came into force on 6 April 2014.

 Note: In 2013, for a period of three years, the size of single-storey rear extensions including conservatories which can be built under permitted development has been increased and an associated neighbour consultation scheme introduced. Until 30 May 2016, householders may build larger single-storey conservatories under permitted development; size limits have doubled from $4\,m^2$ to $8\,m^2$ for detached houses, and from $3\,m^2$ to $6\,m^2$ for all other houses. More information on this is contained in Chapters 4 and 5.

6.24.2 Meeting the requirements

6.24.2.1 Ventilation

The general ventilation rate for conservatories (and adjoining rooms) with a floor area greater than $30\,m^2$ can be achieved by the use of background ventilators.

> Habitable rooms without an openable window may be F 1.14
> ventilated through a conservatory (see Figure 6.284) provided
> that:
>
> - the conservatory has purge ventilation and an $8000\,mm^2$ background ventilator; and
> - there is a closable opening between the room and the conservatory that also is equipped with purge ventilation and an $8000\,mm^2$ background ventilator.

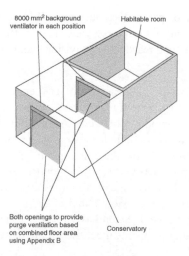

Figure 6.284 A habitable room ventilated through a conservatory.

6.24.2.2 *Conservation of fuel and power*

Under Regulation 17D of the Building Regulations, the construction of a conservatory may trigger the requirement for consequential improvements.

6.24.2.2.1 Changes in the legal requirements

Regulation 9 of the Building Regulations exempts some conservatory and porch extensions from the energy-efficiency requirements. This exemption from the requirements of Approved Document L is only applicable where the existing doors are retained (or replaced) and:

- the conservatory or porch is separated from the house by external-quality walls, doors or windows;
- the conservatory or porch has an independent heating system with separate temperature and on/off controls;
- the conservatory or porch is at ground level;
- the conservatory or porch has a floor area of less than $30\,\text{m}^2$;
- the existing walls, doors and windows of the dwelling have either been retained or, if removed, been replaced by walls, windows and doors which meet the energy-efficiency requirements of Approved Document L; and
- the heating system of the dwelling has **not** been extended into the conservatory or porch.

You are advised not to construct conservatories where they will restrict ladder access to windows serving rooms in roof or loft conversions, particularly if any of the windows are intended to help escape or rescue if there is a fire.

Any new structural opening between the conservatory and the existing house will require Building Regulations approval, even if the conservatory itself is an exempt structure.

If a new conservatory or porch does not meet all these requirements, it is not exempt and **must** comply with the relevant energy-efficiency requirements of Approved Document L, and reasonable provision should be made to ensure that:

• there is effective thermal separation between the heated area in the existing building and the extension, i.e. the walls, doors and windows between the building and the extension should be insulated and draught-proofed to at least the same extent as in the existing building;	L1A L2B 4.12
• particular attention has been paid to independent temperature and on/off controls to any heating system installed within the extension; and	
• glazed elements meet the standards set out in Table 6.109 and opaque elements meet the standards set out in Table 6.110.	

Table 6.109 Standards for controlled fittings

Fitting	Standard (W/m² K)
Windows, roof windows and glazed rooflights	1.8 for the whole unit
Alternative option for windows in buildings that are essentially domestic in character	Window energy rating of B and C
Plastic rooflight	1.8
Curtain walling	<1.8
Pedestrian doors where the door has more than 50% of its internal face area glazed	1.8 for the whole unit
Other doors	1.8
Roof ventilators	3.5

Table 6.110 Standards for new thermal elements

Element	Standard (W/m² K)
Wall	0.28
Pitched roof – insulation at ceiling level	0.16
Pitched roof – insulation at rafter level	0.18
Flat roof or roof with integral insulation	0.18
Floors	0.22

In most circumstances the best reasonable provision is to limit the total area of windows, roof windows and doors in extensions so that it does not exceed the sum of:

- 25 per cent of the floor area of the extension; and
- the total area of any windows or doors which, as a result of the extension works, no longer exist or are no longer exposed.

 One way of complying is to show that the area-weighted U-value of all the elements in the extension is no greater than that of an extension of the same size and shape.

If upgrades are proposed to the existing dwelling, such upgrades should be implemented to a standard that is no worse than that shown in column (b) of Table 6.111.   L1B 4.7

Table 6.111 Upgrading retained thermal elements (existing dwellings)

Element	(a) Threshold U-value (W/m² K)	(b) Improved U-value (W/m² K)
Wall – cavity insulation	0.70	0.55
Wall – external or cavity insulation	0.70	0.30
Floor	0.70	0.25
Pitched roof – insulation at ceiling level	0.35	0.16
Pitched roof – insulation between rafters	0.35	0.18
Flat roof or roof with integral insulation	0.35	0.18

6.24.2.2.2 New dwellings

- If a conservatory or porch is installed at the same time   L1A 2.29
  as the construction of a **new** dwelling and adequate
  thermal separation is provided between the dwelling
  and the conservatory or porch, and the dwelling's
  heating system is not extended into the conservatory or
  porch, then follow the guidance in Approved Document
  L1B.
- If a conservatory or porch is installed at the same   L1A 2.29
  time as the construction of a **new** dwelling and
  no, or inadequate, thermal separation is included
  between the dwelling and the conservatory or porch,
  or the dwelling's heating system is extended into
  the conservatory or porch, follow the guidance from
  Approved Document L1A, including a TER/DER and
  TFEE/DFEE rate calculation.

Acronyms

ABBE	Awarding Body for the Built Environment
ABS	acrylonitrile butadiene styrene
ACE	Amalgamated Chimney Engineers
ach	air changes per hour
AD	Approved Document
ADAHIPP	Association of Home Information Pack Providers
AIBC	approved inspector building control
ATTMA	Air Tightness Testing and Measurement Association
BAFSA	British Automatic Fire Sprinkler Association
BBA	British Board of Agrément
BCB	building control body
BCO	building control officer
BER	building (CO_2) emission rate
BESCA	Building Engineering Services Competence Assessment
BFCMA	British Flue and Chimney Manufacturers Association
BHIF	British Hardware Industry Federation
BRB	building control body
BRE	Building Research Establishment
BREEAM	BRE environmental assessment method
BS	British Standard
BSI	British Standards Institution
BSRIA	Building Services Research and Information Association
CCL	climate change levy
CEA	commercial energy assessor
CIBSE	Chartered Institution of Building Services Engineers
CIC	Construction Industry Council
CIRIA	Construction Industry Research and Information Association
CO	carbon monoxide
CO_2	carbon dioxide
CoPSO	Council of Property Search Organisations
CP	competent person self-certification scheme
CRed	carbon reduction
CSH	Code for Sustainable Homes
DCER	dwellings carbon emission rate
DCLG	Department for Communities and Local Government
DEA	domestic energy assessors
DEC	display energy certificates
DECC	Department of Energy and Climate Change

DER	dwelling (CO_2) emission rate
DETR	Department of the Environment, Transport and the Regions
DFE	decorative fuel effect
DFEE	dwelling fabric energy-efficiency
DfES	Department for Education and Skills
DH	Department of Health
DIAG	Directive Implementation Advisory Group
DQRA	detailed quantitative risk assessment
DSMA	Door and Shutter Manufacturers Association
EEA	European Economic Area
EFBD	Energy Performance of Buildings Directive
EPC	energy performance certificate
EST	Energy Saving Trust
EU	European Union
FAME	fatty acid methyl esters
FPA	Fire Protection Association
GGF	Glass and Glazing Federation
GQRA	generic quantitative risk assessment
H&S	health and safety
HBF	Home Builders Federation
HETAS	Heating Equipment Testing and Approval Scheme
HIP	Home Information Pack
HSE	Health and Safety Executive
HVAC	heating, ventilation and cooling
IACSC	International Association of Cold Storage Contractors
IET	Institution of Engineering and Technology
ILU	Inter-Language Unification
IstructE	Institution of Structural Engineers
LABC	local authority building control
LPG	liquid petroleum gas
LRV	light reflectance value
LUR	logical user requirement
LZC	low or zero carbon
MDD	Medical Devices Directive
MEV	mechanical extract ventilation
MOU	memorandum of understanding
MVHR	continuous mechanical supply and extract with heat recovery
NACE	National Association of Chimney Engineers
NACS	National Association of Chimney Sweeps
NAPIT	National Association of Professional Inspectors and Testers
NBS	National Building Specification
NCM	National Calculation Methodology
NDEPC	non-domestic energy performance certificate
NEF	National Energy Foundation
NES	National Energy Services
NFA	National Fireplace Association

NHER	National Home Energy Rating
NICEIC	National Inspection Council for Electrical Installation Contracting
NIMBY	Not In My Back Yard
ODPM	Office of the Deputy Prime Minister
OFTEC	Oil Firing Technical Association Ltd
PCCB	Property Codes Compliance Board
PCST	pre-completion sound testing
PSA	Property Services Agency
PSV	passive stack ventilation
PVC-U	unplasticized polyvinyl chloride
RCD	residual current device
RCEP	Royal Commission on Environmental Pollution
RdSAP	Reduced data Standard Assessment Procedure
RIBA	RIBA Bookshops
RVA	Residential Ventilation Association
SAP	Standard Assessment Procedure
SBEM	Simplified Building Energy Model
SCI	Steel Construction Institute
SEP	specific fan power
SFA	Solid Fuel Association
SO_2	sulphur dioxide
SRHRV	single room heat recovery ventilator
TEHVA	The Electric Heating and Ventilation Association
TER	target emissions rate
TFEE	target fabric energy-efficiency
TRADA	Timber Research and Development Association
TSO	The Stationery Office
TVOC	total volatile organic compound
UF	urea formaldehyde
UKAEA	United Kingdom Atomic Energy Authority
UPVC	unplasticized polyvinyl chloride
VOC	volatile organic compounds
WAUILF	Workplace Applied Uniform Indicated Low Frequency (application)
YFR	Yearly Forecast Rational

Bibliography

Standards referred to

British Standards

Compliance with a British, European or International Standard does not in itself confer immunity from legal obligations. British Standards can, however, provide a useful source of information that could be used to supplement or provide an alternative to the guidance given in an Approved Document.

When an Approved Document makes reference to a named standard, the relevant version of the standard is the one listed at the end of the publication. However, if this version of the standard has been revised or updated by the issuing standards body, the new version may be used as a source of guidance, provided it continues to address the relevant requirements of the Regulations.

Drafts for Development (DDs) are not British Standards. They are issued in the DD series of publications and are provisional in nature. They are intended to be applied on a provisional basis so that information and experience of their practical application may be obtained and the document developed. Where the recommendations of a DD are adopted, care should be taken to ensure that the requirements of the Building Regulations are adequately met. Any observations that a user may have in relation to any aspect of a DD should be passed onto the British Standards Institution.

Title	Standard
Acoustics. Measurement of sound absorption in a reverberation room	BS EN 20354:1993
Acoustics. Measurement of sound insulation in buildings and of building elements	BS EN ISO 140
Part 1:1980 *Recommendations for laboratories*	
Part 3:1995 *Laboratory measurement of airborne sound insulation of building elements*	
Part 4:1998 *Field measurements of airborne sound insulation between rooms*	
Part 6:1998 *Laboratory measurements of impact sound insulation of floors*	
Part 7:1998 *Field measurements of impact sound insulation of floors*	
Part 8:1998 *Laboratory measurements of the reduction of transmitted impact noise by floor coverings on a heavyweight standard floor*	

Title	Standard
Acoustics. Method for the determination of dynamic stiffness	BS EN 29052-1:1992
Part 1 Materials used under floating floors in dwellings	
Acoustics. Rating of sound insulation in buildings and of building elements	BS EN ISO 717
Part 1:1997 Airborne sound insulation	
Part 2:1997 Impact sound insulation	
Acoustics. Sound absorbers for use in buildings. Rating of sound absorption	BS EN ISO 11654:1997
Aggregates for concrete	BS EN 12620:2002
Anti-flooding devices	BS EN 13564
Application of fire safety engineering principles to the design of buildings. Code of practice	BS 7974:2001
Automatic electrical controls for household and similar use. Particular requirements for temperature sensing controls	BS EN 60730-2-9:2002
Barriers in and about buildings. Code of practice [2011]	BS 6180
Building components and building elements. Thermal resistance and thermal transmittance. Calculation method	BS EN ISO 6946:1997
Building hardware. Panic exit devices operated by a horizontal bar. Requirements and test methods	BS EN 1125:1997
Building materials and products. Hygrothermal proporties. Tabulated design values	BS EN 12524:2000
Building valves. Combined temperature and pressure relief valves. Tests and requirements	BS EN 1490:2000
Building valves. Inline hot water supply tempering valves. Tests and requirements	BS EN 15092:2008
Buildings and structures for agriculture. Various relevant parts including:	BS 5502
Part 33:1991 Guide to the control of odour pollution	
Part 52:1991 Code of practice for design of alarm systems, emergency ventilation and smoke ventilation for livestock housing	
Capillary and compression tube fittings of copper and copper alloy	BS 864
Part 2:1983 Specification for capillary and compression fittings for copper tubes	
Cast iron pipes and fittings, their joints and accessories for the evacuation of water from buildings. Requirements, test methods and quality assurance	BS EN 877:1999
Cement	BS EN 197
Part 1:2000 Composition, specifications and conformity criteria for common elements	
Part 2:2000 Conformity evaluation	
Chimneys. Clay/ceramic flue blocks for single wall chimneys. Requirements and test methods	BS EN 1806:2006

Title	Standard
Chimneys. Clay/ceramic flue liners. Requirements and test methods	BS EN 1457:2009
Chimneys. Components. Concrete flue blocks	BS EN 1858:2003
Chimneys. Components. Concrete flue liners	BS EN 1857:2003
Chimneys. Design installation and commissioning of chimneys	BS EN 15287-1:2007
Chimneys. General requirements	BS EN 1443:2003
Chimneys. Metal chimneys. Test methods	BS EN 1859:2009
Chimneys. Requirements for metal chimneys. Metal liners and connecting flue pipes	BS EN 1856-2:2004
Chimneys. Requirements for metal chimneys. System chimney products	BS EN 1856-1:2003
Chimneys. Thermal and fluid dynamic calculation methods. Chimneys serving one appliance	BS EN 13384-1:2002
Cisterns for domestic use. Cold water storage and combined feed and expansion (thermoplastic) cisterns up to 500 l	BS 4213:2004
Code of practice for accommodation of building services in ducts	BS 8313:1997
Code of practice for assessing exposure of walls to wind-driven rain	BS 8104:1992
Code of practice for building drainage	BS 8301:1985
Code of practice for control of condensation in buildings	BS 5250:2002
Code of practice for daylighting	BS 8206: Part 2
Code of practice for design and installation of damp-proof courses in masonry construction	BS 8215:1991
Code of practice for design and installation of natural stone cladding and lining	BS 8298-1:2010
Code of practice for design and installation of non-load-bearing precast concrete cladding	BS 8297:2000
Code of practice for design and installation of small sewage treatment works and cesspools	BS 6297:1983
Code of practice for design of non-load-bearing external vertical enclosures of buildings	BS 8200:1985
Code of practice for domestic butane- and propane-gas-burning installations. Installations at permanent dwellings, residential park homes and commercial premises with installation pipework sizes not exceeding DN 25 for steel and ON 28 for corrugated stainless steel or copper	BS 5482-1:2005
Code of practice for drainage of roofs and paved areas	BS 6367:1983
Code of practice for earth retaining structures	BS 8002:1994
Code of practice for external renderings	BS 5262:1991
Code of practice for fire door assemblies with non-metallic leaves	BS 8214:1990
Code of practice for flues and flue structures in buildings	BS 5854:1980
Code of practice for foundations	BS 8004:1986

Title	Standard
Code of practice for mechanical ventilation and air-conditioning in buildings	BS 5720:1979
Code of practice for non-automatic fire-fighting systems in buildings	BS 9990:2006
Code of practice for oil firing. Installations of 45 kW and above output capacity for space heating hot water and steam supply services	BS 5410-2:1978
Code of practice for oil firing. Installations up to 44 kW output capacity for space heating and hot water supply purposes. AMD 3637	BS 5410-1:1997
Code of practice for protection of structures against water from the ground	BS 8102:2009
Code of practice for protective barriers in and about buildings	BS 6180:1995
Code of practice for sanitary pipework	BS 5572:1978
Code of practice for sheet roof and wall coverings	BS 5247
Part 14:1975 Corrugated asbestos-cement	
Code of practice for sheet roof and wall coverings	BS CP 143
Code of practice for site investigations	BS 5930:1999
Code of practice for solar heating systems for domestic hot water	BS 5918:1989
Code of practice for stone masonry	BS 5390:1976 (1984)
Code of practice for the audio-frequency induction-loop systems (AFILS)	BS 7594:1993
Code of practice for the control of condensation in buildings	BS 5250:2002
Code of practice for the design of helical and spiral stairs	BS 5395-2:1984
Code of practice for the design of industrial type stairs, permanent ladders and walkways	BS 5395-3:1985
Code of practice for the design of stairs for limited access	BS 5395-4:2011
Code of practice for the operation of fire protection measures	BS 7273
Part 1:1990 Electrical actuation of gaseous total flooding extinguishing systems	
Part 2:1992 Mechanical actuation of gaseous total flooding and local application extinguishing systems	
Part 3:2000 Electrical actuation of pre-action sprinkler systems	
Code of practice for the storage and on-site treatment of solid waste from buildings	BS 5906:1980 (1987)
Code of practice for thermal insulation of cavity walls (with masonry or concrete inner and outer leaves) by filling with urea-formaldehyde (UF) foam systems	BS 5618:1985
Code of practice for use of masonry	BS 5628
Part 1:1992 Structural use of un-reinforced masonry	
Part 2:2000 Structural use of reinforced and prestressed masonry	

Title	Standard
Part 3:2001 *Materials and components, design and workmanship*	
Code of practice for ventilation principles and designing for natural ventilation. AMD 8930 1995	BS 5925:1991
Cold rolled steel flat products with high yield strength for cold forming. Technical delivery conditions	BS EN 10268:2006
Components for residential sprinkler systems. Specification and test methods for residential sprinklers	BS 9295:2011
Components for smoke and heat control systems	BS 7346
Part 2:1990 *Specification for powered smoke and heat exhaust ventilators*	
Part 6:2005 *Components for smoke and heat control systems. Specifications for cable systems*	
Part 7:2006 *Components for smoke and heat control systems. Code of practice on functional recommendations and calculation methods for smoke and heat control systems for covered car parks*	
Concrete	BS 5328
Part 1:1990 *Guide to specifying concrete*	
Part 2:1990 *Method for specifying concrete mixes*	
Part 3:1990 *Specification for the procedures to be used in producing and transporting concrete*	
Part 4:1990 *Specification for the procedures to be used in sampling, testing and assessing compliance of concrete*	
Concrete	BS 8500
Part 1:2002 *Method of specifying and guidance for the specifier*	
Part 2:2002 *Specification for constituent materials and concrete*	
Construction and testing of drains and sewers	BS EN 1610:1998
Copper and copper alloys. Plumbing fittings	BS EN 1254:1998
Part 1 *Fittings with ends for capillary soldering or capillary brazing to copper tubes*	
Part 2 *Fittings with compression ends for use with copper tubes*	
Part 3 *Fittings with compression ends for use with plastic pipes*	
Part 4 *Fittings combining other end connections with capillary or compression ends*	
Part 5 *Fittings with short ends for capillary brazing to copper tubes*	
Copper and copper alloys. Seamless, round copper tubes for water and gas in sanitary and heating applications	BS EN 1057:1996
Copper indirect cylinders for domestic purposes. Open vented copper cylinders. Requirements and test methods	BS 1566-1:2002
Design of aluminium structures	BS EN 1999

Title	Standard
Design of buildings and their approaches to meet the needs of disabled people. Code of practice	BS 8300:2009
Design of buildings and their approaches to meet the needs of disabled people. Code of practice	BS 8300:2001
Design of composite steel and concrete structures	BS EN 1994
Design of concrete structures	BS EN 1992
Design of masonry structures	BS EN 1995
Design of steel structures	BS EN 1993
Design of structures for earthquake resistance	BS EN 1998:2004
Part 1 *General rules, seismic actions and rules for buildings*	
Part 2 *Foundations, retaining structures and geotechnical aspects*	
Design of timber structures	BS EN 1995
Design, installation, testing and maintenance of services supplying water for domestic use within buildings and their curtilages	BS 6700:2006 1
Discharge and ventilating pipes and fittings, sand-cast or spun in cast iron	BS 416
Part 1:1990 *Specification for spigot and socket systems*	
Part 2:1990 *Specification for socketless systems*	
Drain and sewer systems outside buildings	BS EN 752
Part 1:1996 *Generalities and definitions*	
Part 2:1997 *Performance requirements*	
Part 3:1997 *Planning*	
Part 4:1997 *Hydraulic design and environmental aspects*	
Part 5:1997 *Rehabilitation*	
Part 6:1998 *Pumping installations*	
Part 7:1998 *Maintenance and operations*	
Ductile iron pipes, fittings, accessories and their joints for sewerage applications. Requirements and test methods	BS EN 598:1995
Emergency lighting	BS 5266
Part 1:1988 *Code of practice for the emergency lighting of premises other than cinemas and certain other specified premises used for entertainment*	
Emergency lighting code of practice for the emergency lighting of premises	BS 5266-1:2005
Factory-made insulated chimneys	BS 4543-1:1990
Part 1 *Methods of test*	
Part 2 *Specification for chimneys with stainless steel flue linings for use with solid fuel fired appliances*	
Fire classification of construction products and building elements	BS EN 13501
Part 1:2002 *Classification using test data from reaction to fire tests*	

Title	Standard
Part 2:2003 *Classification using data from fire resistance tests, excluding ventilation services*	
Part 3:2005 *Classification using data from fire resistance tests on products and elements used in building service installations: fire resisting ducts and fire dampers*	
Part 4:2005 *Classification using data from fire resistance tests on smoke control systems*	
Part 5:2005 *Classification using data from external fire exposure to roof tests*	
Fire detection and alarm systems for buildings	BS 5839
Part 1:2002 *Code of practice for system design, installation and servicing*	
Part 2:1983 *Specification for manual call points*	
Part 6:1995 *Code of practice for the design and installation of fire detection and alarm systems in dwellings*	
Part 6:2004 *Code of practice for the design, installation and maintenance of fire detection and fire alarm systems in dwellings*	
Part 8:1998 *Code of practice for the design, installation and servicing of voice alarm systems*	
Part 9 *Code of practice for the design, installation, commissioning and maintenance of emergency voice communication systems*	
Fire detection and fire alarm devices for dwellings. Specification for heat alarms	BS 5446-2:2003
Fire detection and fire alarm systems. Manual call points	BS EN 54-11:2001
Fire extinguishing installations and equipment on premises	BS 5306
Part 1:1976 (1988) *Hydrant systems, hose reels and foam inlets*	
Part 2:1990 *Specification for sprinkler systems*	
Fire performance of external cladding systems	BS 8414
Part 1:2002 *Test methods for non-load-bearing external cladding systems applied to the face of a building*	
Part 2:2005 *Test method for non-load-bearing external cladding systems fixed to and supported by a structural steel frame*	
Fire precautions in the design, construction and use of buildings	BS 5588
Part 0:1996 *Guide to fire safety codes of practice for particular premises*	
Part 1:1990 *Code of practice for residential buildings*	
Part 4:1998 *Code of practice for smoke control using pressure differentials*	
Part 5:1991 *Code of practice for fire fighting in stairs and lifts*	

Title	Standard
Part 6:1991 *Code of practice for places of assembly*	
Part 7:1997 *Code of practice for the incorporation of atria in buildings*	
Part 8:1999 *Code of practice for means of escape for disabled people*	
Part 9:1989 *Code of practice for ventilation and air conditioning ductwork*	
Part 10:1991 *Code of practice for shopping complexes*	
Part 11:1997 *Code of practice for shops, offices, industrial, storage and other similar buildings*	
Part 12:2004 *Code of practice for construction and use of buildings. Managing fire safety*	
Fire resistance tests for door and shutter assemblies	BS EN 1634-2:2008
Part 1:2000 *Fire doors and shutters*	
Part 2:2008 *Fire door hardware*	
Part 3:2001 *Smoke control doors and shutters*	
Fire resistance tests for load-bearing elements	BS EN 1365
Part 1:1999 *Walls*	
Part 2:2000 *Floors and roofs*	
Part 3:2000 *Beams*	
Part 4:1999 *Columns*	
Fire resistance tests for non-load-bearing elements	BS EN 1364
Part 1:1999 *Walls*	
Part 2:1999 *Ceilings*	
Part 3:2006 *Curtain walling. Full configuration (complete assembly)*	
Fire resistance tests for service installations	BS EN 1366
Part 1:1992 *Ducts*	
Part 2:1999 *Fire dampers*	
Part 3:2004 *Penetration seals*	
Part 4:2006 *Linear joint seals*	
Part 5:2003 *Service ducts and shafts*	
Part 6:2004 *Raised access and hollow core floors*	
Fire safety signs, notices and graphic symbols	BS 5449
Part 1:1990 *Specification for fire safety signs*	
Fire tests on building materials and structures	BS 476
Part 3:2004 *Classification and method of test for external fire exposure to roofs*	
Part 4:2007 *Non-combustibility test for materials*	
Part 6:1989 *Method for determination of the fire resistance of ventilation ducts*	
Part 11:1982 *Method for assessing the heat emission from building materials*	
Part 20:1987 *Method for determination of the fire resistance of elements of construction (general principles)*	

Title	Standard
Part 21:1987 *Fire tests on building materials and structures. Methods for determination of the fire resistance of load-bearing elements of construction*	
Part 22:1987 *Methods for determination of the fire resistance of non-load-bearing elements of construction*	
Part 23:1987 *Methods for determination of the contribution of components to the fire resistance of a structure*	
Part 24:1989 *Method for determination of the fire resistance of ventilation ducts*	
Part 7:1997 *Method of test to determine the classification of the surface spread of flame of products*	
Fire-resistance tests. Fire dampers for air distribution systems	BS ISO 10294-2:1999
Part 2:1999 *Classification, criteria and field of application of test results*	
Part 5:2005 *Intumescent fire dampers*	
Fixed firefighting systems. Automatic sprinkler systems. Design, installation and maintenance	BS EN 12845:2004
Flat roofs with continuously supported coverings. Code of practice	BS 6229:2003
Flue blocks and masonry terminals for gas appliances	BS 1289-1
Part 1:1986 *Specification for precast concrete flue blocks and terminals*	
Part 2:1989 *Specification for clay flue blocks and terminals*	
Fuel oils for agricultural domestic and industrial engines and boilers. Specification	BS 2869:2006
Fuel oils for non-marine use	BS 2869:1998
Part 2:1988 *Specification for fuel oil for agricultural and industrial engines and burners (classes A2, C1, C2, D, E, F, G and H)*	
Geotechnical work and foundations	BS EN 1997
Part 1 *General rules*	
Part 2 *Ground investigation and testing*	
Glass in building. Pendulum test. Impact test method and classification for flat glass [2002, incorporating corrigendum April 2010]	BS EN 12600
Glass in building. Determination of luminous and solar characteristics of glazing	BS EN 410:1998
Glossary of terms relating to solid fuel burning equipment	BS 1846
Part 1:1994 *Domestic appliances*	
Graphical symbols and signs. Safety signs, including fire safety signs. Specification for geometric shapes, colours and layout	BS 5499-1:2002
Gravity drainage systems inside buildings	BS EN 12056:2000
Part 1 *Scope, definitions, general and performance requirements*	

Title	Standard
Part 2 *Wastewater systems, layout and calculation*	
Part 3 *Roof drainage layout and calculation*	
Part 4 *Effluent lifting plants, layout and calculation*	
Part 5 *Installation, maintenance and user instructions*	
Guide for design, construction and maintenance of single-skin air supported structures	BS 6661:1986
Guide to assessment of suitability of external cavity walls for filling with thermal insulants	BS 8208
Part 1:1985 *Existing traditional cavity construction*	
Guide to development and presentation of fire tests and their use in hazard assessment	BS 6336:1998
Guide to the principles of the conservation of historic buildings	BS 7913:1998
Heating boilers. Heating boilers with forced draught burners. Terminology, general requirements, testing and marking	BS EN 303-1:1999
Heating fuels. Fatty acid methyl esters (FAME). Requirements and test methods	BS EN 14213:2003
Household and similar electrical appliances. Safety. Particular requirements for storage water heaters	BS EN 60335-2-21:2003
Hygrothermal performance of building components and building elements	BS EN ISO 13788:2002
Installation and maintenance of flues and ventilation for gas appliances of rated input not exceeding 70 kW net (1st, 2nd and 3rd family gases)	BS 5440-1
Part 1: 2008 *Specification for installation and maintenance of flues*	
Part 2: 2000 *Specification for installation and maintenance of ventilation for gas appliances*	
Installation of chimneys and flues for domestic appliances burning solid fuel (including wood and peat)	BS 6461
Part 1:1984 (1998) *Code of practice for masonry chimneys and flue pipes*	
Installation of domestic heating and cooking appliances burning solid mineral fuels	BS 8303:1994
Part 1 *Specification for the design of installations*	
Part 2 *Specification for installing and commissioning on site*	
Part 3 *Recommendations for design and on-site installation*	
Installation of factory-made chimneys to BS 4543 for domestic appliances	BS 7566
Part 1:1992 (1998) *Method of specifying installation design information*	
Part 2:1992 (1998) *Specification for installation design*	
Part 3:1992 (1998) *Specification for site installation*	
Part 4:1992 (1998) *Recommendations for installation design and installation*	

Title	Standard
Installations for separation of light liquids (e.g. petrol or oil)	BS EN 858:2001
Part 1 *Principles of design, performance and testing, marking and quality control*	
Internal and external wood doorsets, door leaves and frames	BS 4787
Part 1:1980 (1985) *Specification for dimensional requirements*	
Investigation of potentially contaminated land. Code of practice	BS 10175:2001
Lifts and service lifts	BS 5655
Part 5:1989 *Specifications for dimensions for standard lift arrangements*	
Part 7:1983 *Specification for manual control devices, indicators and additional fittings. Amendment slip*	
Lighting for buildings. Code of practice for daylighting	BS 8206-2:2008
Loading for buildings	BS 6399
Part 1:1996 *Code of practice for dead and imposed loads*	
Part 2:1997 *Code of practice for wind loads*	
Part 3:1988 *Code of practice for imposed roof loads*	
Measurement of sound insulation in buildings and building elements	BS 2750
Part 4:1980 *Field measurement of airborne sound insulation between rooms*	
Part 6:1980 *Laboratory measurement of impact sound insulation of floors*	
Part 7:1980 *Field measurements of impact sound insulation of floors*	
Mechanical thermostats for gas-burning appliances	BS EN 257:1992
Method for specifying thermal insulating materials for pipes, tanks, vessels, ductwork and equipment operating within the temperature range 240° C to 170° C	BS 5422:2001
Method of test for ignitability of fabrics used in the construction of large tented structures	BS 7157:1989
Method of test for resistance to fire of unprotected small cables for use in emergency circuits	BS EN 50200:2006
Methods for rating the sound insulation in building elements	BS 5821
Part 1:1984 *Method for rating the airborne sound insulation in buildings and interior building elements*	
Part 2:1984 *Method for rating the impact sound insulation*	
Methods of test for flammability of textile fabrics when subjected to a small igniting flame applied to the face or bottom edge of vertically oriented specimens, Test 2	BS 5438:1989
Methods of testing plastics	BS 2782
Part 1 *Thermal properties: Methods 120A to 12OE: 1990 Determination of the Vicat softening temperature of thermoplastics*	

Title	Standard
Methods of testing. Plastics. Introduction	BS 2782-0:2004
Oil burning equipment. Specification for oil storage tanks	BS 799-5:1987
Particleboards. Specifications. Requirements for load-bearing boards for use in humid conditions	BS EN 312-5:1997
Performance of windows and doors. Classification for weathertightness and guidance on selection and specification	BS 6375-1:2009
Plastics piping systems for non-pressure underground drainage and sewerage. Structured-wall piping systems of unplasticized poly(vinyl chloride) (PVC-U), polypropylene (PP) and polyethylene (PE). General requirements and performance characteristics	BS EN 13476-1:2007
Plastics piping systems for non-pressure underground drainage and sewerage. Unplasticized polyvinylchloride (PVC-U). Specifications for pipes, fittings and the system	BS EN 1401-1:1998
Plastics piping systems for soil and waste (low and high temperature) within the building structure. Acrylonitrilebutadiene-styrene (ABS). Specifications for pipes, fittings and the system	BS EN 1455-1:2000
Plastics piping systems for soil and waste discharge (low and high temperature) within the building structure. Unplasticized polyvinyl chloride (PVC-U). Specifications for pipes, fittings and the system	BS EN 1329-1:2000
Plastics piping systems for soil and waste discharge (low and high temperature) within the building structure. Polypropylene (PP). Specifications for pipes, fittings and the system	BS EN 1451-1:2000
Plastics piping systems for soil and waste discharge (low and high temperature) within the building structure. Polyethylene (PE). Specifications for pipes, fittings and the system	BS EN 1519-1:2000
Plastics piping systems for soil and waste discharge (low and high temperature) within the building structure	BS EN 1565-1:2000
Plastics. Symbols and abbreviated terms. Basic polymers and their special characteristics	BS EN ISO 1043-1:2002
Plastics. Thermoplastic materials. Determination of Vicat softening temperature (VST)	BS EN ISO 306:2004
Powered vertical lifting platforms having non-enclosed or partially enclosed liftways intended for use by persons with impaired mobility	BS 6440:2011
Precast concrete masonry units	BS 6073
Part 1:1981 *Specification for precast concrete masonry units*	
Precast concrete pipes, fittings and ancillary products	BS 5911
Part 2:1982 *Specification for inspection chambers and street gullies*	
Part 100:1988 *Specification for un-reinforced and reinforced pipes and fittings with flexible joints*	
Part 101:1988 *Specification for glass composite concrete (GCC) pipes and fittings with flexible joints*	

Title	Standard
Part 120:1989 *Specification for reinforced jacking pipes with flexible joints*	
Part 200:1989 *Specification for un-reinforced and reinforced manholes and soakaways of circular cross section*	
Pressure sewerage systems outside buildings	BS EN 1671:1997
Principles of the conservation of historic buildings	BS 7913:1998
Profiled fibre cement. Code of practice	BS 8219:2001
Rainwater harvesting systems. Code of practice	BS 8515:2009
Reaction to fire tests for building products. Building products excluding footings exposed to thermal attack by a single burning item	BS EN 13823:2002
Reaction to fire tests for building products. Conditioning procedures and general rules for selection of substrates	BS EN 13238:2001
Reaction to fire tests for building products. Determination of the heat of combustion	BS EN ISO 1716:2002
Reaction to fire tests for building products. Non-combustibility test	BS EN 150 1182:2002
Reaction to fire tests. ignitability of building products subjected to direct impingement of flame. Single-flame source test	BS EN ISO 11925-2:2002
Recommendations for the storage and exhibition of archival documents	BS 5454:2000
Refrigerating systems and heat pumps. Safety and environmental requirements. Installation site and personal protection	BS EN 378-3:2008
Requirements for electrical installations (IET Wiring Regulations, 17th edition)	BS 7671:2008 Amendment No 1
Safety and control devices for use in hot water systems	BS 6283
Part 2:1991 *Specification for temperature relief valves for pressures from 1 bar to 10 bar*	
Part 3:1991 *Specification for combined temperature and pressure relief valves for pressures from 1 bar to 10 bar*	
Safety rules for the construction and installation of lifts	BS EN 81-1:1998
Part 1 *Electric lifts*	
Part 2 *Hydraulic lifts*	
Safety rules for the construction and installation of lifts. Particular applications for passenger and goods passenger lifts. Accessibility to lifts for persons including persons with disability	BS EN 81-70:2003
Safety rules for the construction and installation of lifts. Particular applications for passenger and goods passenger lifts. Fire fighters lifts	BS EN 81-72:2003
Sanitary installations	BS 6465
Part 1:1984 *Code of practice for scale of provision, selection and installation of sanitary appliances*	
Part 2:1996 *Code of practice for space requirements for sanitary appliances*	
Sanitary installations	BS 6485:2006

Title	Standard
Part 1 *Code of practice for the design of sanitary facilities and scales of provision of sanitary and associated appliances*	
Part 2 *Code of practice for the selection, installation and maintenance of sanitary and associated appliances*	
Sanitary tapware. Low pressure thermostatic mixing valves. General technical specifications	BS EN 1287:1999
Sanitary tapware. Thermostatic mixing valves (PN 10). General technical specification	BS EN 1111:1999
Sanitary tapware. Waste fittings for basins, bidets and baths. General technical specifications	BS EN 274:1993
Small wastewater treatment plants less than 50 PE	BS EN 12566-1:2000
Smoke alarm devices	BSEN 14604:2005
Smoke and heat control systems	BS EN 12101
Part 3:2002 *Specification for powered smoke and heat exhaust ventilators*	
Part 6:2005 *Specification for pressure differential systems. Kits*	
Sound insulation and noise reduction for buildings. Code of practice	BS 8233:1999
Specification for aggregates from natural sources for concrete	BS 882:1983
Specification for ancillary components for masonry	BS EN 845
Part 1:2001 *Ties, tension straps, hangers and brackets*	
Part 2:2001 *Lintels*	
Part 3:2001 *Bed joint reinforcement of steel meshwork*	
Specification for asbestos-cement pipes, joints and fittings for sewerage and drainage	BS 3656:1981 (1990)
Specification for calcium silicate (sandlime and flintlime) bricks	BS 187:1978
Specification for cast iron spigot and socket drain pipes and fittings	BS 437:1978
Specification for cast iron spigot and socket flue or smoke pipes and fittings	BS 41:1973
Specification for clay and calcium silicate modular bricks	BS 6649:1985
Specification for clay bricks	BS 3921:1985
Specification for clay flue linings and flue terminals	BS 1181:1999
Specification for copper and copper alloys. Tubes	BS 2871
Part 1:1971 *Copper tubes for water, gas and sanitation*	
Specification for copper direct cylinders for domestic purposes	BS 699:1984
Specification for copper hot water storage combination units for domestic purposes	BS 3196:1981
Specification for dedicated liquid petroleum gas appliances. Domestic flueless space heaters (including diffusive catalytic combustion heaters)	BS EN 449:2002
Specification for design and construction of fully supported lead sheet roof and wall coverings	BS 6915:2001
Specification for design, installation, testing and maintenance of services supplying water for domestic use within buildings and their curtilages	BS 6700:1987

Title	Standard
Specification for direct surfaced wood chipboard based on thermosetting resins	BS 7331:1990
Specification for domestic appliances. Specification for installation design	BS 4543
Specification for electrical controls for household and similar general purposes	BS 3955:1986
Specification for fabrics for curtains and drapes	BS 5867
Part 2:1980 Flammability requirements	
Specification for fibre boards	BS 1142:1989
Specification for flexible joints for grey or ductile cast iron drain pipes and fittings (BS 437) and for discharge and ventilating pipes and fittings (BS 416)	BS 6087:1990
Specification for galvanized low carbon steel cisterns, cistern lids, tanks and cylinders. Metric units	BS 417-2:1987
Specification for impact performance requirements for flat safety glass and safety plastics for use in buildings	BS 6206:1981
Specification for installation in domestic premises of gas-fired ducted-air heaters of rated input not exceeding 60 kW	BS 5864:2004
Specification for installation of domestic gas cooking appliances (1st, 2nd and 3rd family gases)	BS 6172:2004
Specification for installation of gas fired catering appliances for use in all types of catering establishments (1st, 2nd and 3rd family gases)	BS 6173:2001
Specification for installation of gas fires, convector heaters, fire/back boilers and decorative fuel effect gas appliances	BS 5871:2005
Part 1 Gas fires, convector heaters and fire/back boilers and heating stoves (1st, 2nd and 3rd family gases)	
Part 2 Inset live fuel effect gas fires of heat input not exceeding 15 kW (2nd and 3rd family gases)	
Part 3 Decorative fuel effect gas appliances of heat input not exceeding 20 kW (2nd and 3rd family gases)	
Specification for installation of gas-fired boilers of rated input not exceeding 70 kW	BS 6798:2009
Specification for installation of hot water supplies for domestic purposes using gas fired appliances of rated input not exceeding 70 kW	BS 5546:2000
Specification for ladders for permanent access to chimneys, other high structures, silos and bins	BS 4211:1987
Specification for low-voltage switchgear and control gear assemblies. Particular requirements for low-voltage switchgear and control assemblies intended to be installed in places where unskilled persons have access to their use	BS EN 60439-3:1991
Specification for masonry units	BS EN 771
Part 1:2003 Clay masonry units	
Part 2:2001 Calcium silicate masonry units	
Part 3 Aggregate concrete masonry units	
Part 4:2001 Autoclaved aerated concrete masonry units	
Part 5 Manufactured stone masonry units	
Part 6:2001 Natural stone masonry units	

Title	Standard
Specification for metal flue pipes, fittings, terminals and accessories for gas-fired appliances with a rated input not exceeding 60 kW. AMD 8413	BS 715:2005
Specification for metal ties for cavity wall construction	BS 1243:1978
Specification for modular co-ordination in building	BS 6750:1986
Specification for mortar for masonry	BS EN 998
Part 2:2002 Masonry mortar	
Specification for open fireplace components	BS 1251:1987
Specification for performance requirements for cables required to maintain circuit integrity under fire conditions	BS 6387:1994
Specification for performance requirements for domestic flued oil burning appliances (including test procedures)	BS 4876:1984
Specification for permanently fixed ladders [2005 + AMD Al, Corrigenda C1, C2]	BS 4211
Specification for plastics inspection chambers for drains	BS 7158:2001
Specification for plastics waste traps	BS 3943:1979 (1988)
Specification for Portland cements	BS 12:1989
Specification for powered stairlifts	BS 5776:1996
Specification for prefabricated drainage stack units in galvanized steel	BS 3868:1995
Specification for quality of vitreous china sanitary appliances	BS 3402:1969
Specification for safety aspects in the design, construction and installation of refrigerating appliances and systems	BS 4434:1989
Specification for safety of household and similar electrical appliances	BS EN 60335-2-35:2002
Specification for safety of household and similar electrical appliances. Particular requirements for fixed immersion heaters	BS EN 60335-2-73:2003
Specification for sizes of sawn and processed softwood	BS 4471:1987
Specification for softwood grades for structural use	BS 4978:1988
Specification for stainless and heat-resisting steel plate sheet and strip. AMD 4807, AMD 6646 and AMD 8832	BS 1449-2:1983
Specification for the use of structural steel in building	BS 449
Part 2:1969 Metric units	
Specification for thermoplastics waste pipe and fittings	BS 5255:1989
Specification for thermostats for gas-burning appliances	BS 4201:1979 (1984)
Specification for tongued and grooved softwood flooring	BS 1297:1987
Specification for topsoil	BS 3882:1994
Specification for un-plasticized polyvinyl chloride (PVC-U) pipes and plastics fittings of nominal sizes 110 and 160 for below ground drainage and sewerage	BS 4660:1989
Specification for unplasticized PVC pipe and fittings for gravity sewers	BS 5481:1977 (1989)
Specification for unvented hot water storage units and packages	BS 7206:1990
Specification for urea-formaldehyde (UF) foam systems suitable for thermal insulation of cavity walls with masonry or concrete inner and outer leaves	BS 5617:1985

Title	Standard
Specification for vessels for use in heating systems. Calorifiers and storage vessels for central heating and hot water supply	BS 853-1:1996
Specification for vitreous-enamelled low-carbon-steel fluepipes, other components and accessories for solid-fuel-burning appliances with a maximum rated output of 45 kW	BS 6999:1989 (1996)
Specification for vitrified clay pipes, fittings and ducts, also flexible mechanical joints for use solely with surface water pipes and fittings	BS 65:1991
Specification. Indicator plates for fire hydrants and emergency water supplies	BS 3251:1976
Sprinkler systems for residential and domestic occupancies. Code of practice	BS 9251:2005
Stainless steels. List of stainless steels	BS EN 10088-1:2005
Stairs, ladders and walkways	BS 5395
Part 1:1977 Code of practice for stairs	
Part 2:1984 Code of practice for the design of helical and spiral stairs	
Part 3:1985 Code of practice for the design of industrial type stairs, permanent ladders and walkways	
Steel plate, sheet and strip	BS 1449
Part 2:1983 Specification for stainless and heat-resisting steel plate, sheet and strip	
Steel plate, sheet and strip. Carbon and carbon manganese plate, sheet and strip. General specifications	BS 1449-1:1991
Structural design and loading:	
Actions on structures	BS EN 1990
Part 1.1 General actions – densities, self-weight, imposed loads for buildings	
Part 1.3 General actions – snow loads; with UK National Annex	
Part 1.4 General actions – wind actions; with UK National Annex	
Part 1.6 General actions – actions during execution; with UK National Annex	
Part 1.7 General actions – accidental actions; with UK National Annex	
Part 3 Actions induced by cranes and machinery; with UK National Annex	
Structural design of buried pipelines under various conditions of loading. General requirements	BS EN 1295-1:1998
Structural design of low-rise buildings	BS 8103
Part 1:2011 Code of practice for stability, site investigation, foundations, precast concrete floors and ground floor slabs for housing	
Part 2:2005 Code of practice for masonry walls for housing	
Part 3:1996 Code of practice for timber floors and roofs for housing	
Part 4:1995 Code of practice for suspended concrete floors for housing	

Title	Standard
Structural fixings in concrete and masonry	BS 5080:1993
Part 1:1993 Method of test for tensile loading	
Structural use of aluminium	BS 8118
Part 1:1991 Code of practice for design. Amendment slip	
Part 2:1991 Specification for materials, workmanship and protection	
Structural use of concrete	BS 8110
Part 1:1997 Code of practice for design and construction	
Part 2:1985 Code of practice for special circumstances	
Part 3:1995 Design charts for single reinforced beams, doubly reinforced beams and rectangular columns	
Structural use of steelwork in building	BS 5950
Part 1:2000 Code of practice for design	
Part 2:2001 Specification for materials, fabrication and erection	
Part 3:1990 Design in composite construction	
Part 4:1994 Code of practice for design of composite slabs with profiled steel sheeting	
Part 5:1998 Code of practice for design of cold formed thin gauge sections	
Structural use of timber	BS 5268
Part 2:2002 Code of practice for permissible stress design, materials and workmanship	
Part 3:1998 Code of practice for trussed rafter roofs	
Part 6 Code of practice for timber framed walls	
Part 6.1:1988 Dwellings not exceeding three storeys	
Structural work of composite steel and concrete:	
Design of composite steel and concrete structures, Part 1.1: General rules and rules for buildings	BS EN 1994-1-1
Structural work of masonry:	
Design of masonry structures, Part 1.1: General rules for reinforced and unreinforced masonry structures	BS EN 1996-1-1
Design of masonry structures, Part 2: Design considerations, selection of materials and execution of masonry	BS EN 1996-2
Design of masonry structures, Part 3: Simplified calculation methods for unreinforced masonry structures	BS EN 1996-3
Structural design of low-rise buildings, Part 1: Code of practice for stability, site investigation, foundations, precast concrete floors and ground floor slabs for housing	BS 8103
Structural design of low-rise buildings, Part 2: Code of practice for masonry walls for housing	BS 8103
Structural work of reinforced, pre-stressed or plain concrete:	
Design of concrete structures, Part 1.1: General rules and rules for buildings	BS EN 1992-1-1:2004
Execution of concrete structures	BS EN 13670:2009
Structural work of steel:	
Design of steel structures	BS EN 1993
Part 1.1 General rules and rules for buildings	

Title	Standard
Part 1.3 *General rules – supplementary rules for cold-formed members and sheeting*	
Part 1.4 *General rules – supplementary rules for stainless steels*	
Part 1.5 *Plated structural elements*	
Part 1.6 *Strength and stability of shell structures*	
Part 1.7 *Plated structures subject to out of plane loading*	
Part 1.8 *Design of joints*	
Part 1.9 *Fatigue*	
Part 1.10 *Material toughness and through-thickness properties*	
Part 1.11 *Design of structures with tension components*	
Part 1.12 *Additional rules for the extension of EN 1993 up to steel grades S 700*	
Part 5 *Piling*	
Part 6 *Crane supporting structures*	
Execution of steel structures and aluminium structures, Part 2: Technical requirements for the execution of steel structures	BS EN 1090-2:2008
Structural work of timber:	
Design of timber structures, Part 1.1: General – common rules and rules for building	BS EN 1995-1-1
Recommendations for the design of timber structures to Eurocode 5: Design of timber structures, Part 1: General common rules and rules for buildings	BSI PD 6693-1
Structural design of low-rise buildings, Part 3: Code of practice for timber floors and roofs for housing	BS 8103-3
Test methods for external fire exposure to roofs	DD CEN/TS 1187:21012
Thermal bridges in building construction. Calculation of heat flows and surface temperatures	BS EN ISO 10211-1:1996
Part 1 *General methods*	
Thermal bridges in building construction. Calculation of heat flows and surface temperatures	BS EN ISO 10211-2:2001
Part 2 *Linear thermal bridges*	
Thermal insulation for use in pitched roof spaces in dwellings	BS 5803-5:1985
Part 5:1985 *Specification for installation of man-made mineral fibre and cellulose fibre insulation*	
Thermal insulation of cavity walls by filling with blown man-made mineral fibre	BS 6232
Part 1:1982 *Specification for the performance of installation systems*	
Part 2:1982 *Code of practice for installation of blown man-made mineral fibre in cavity walls with masonry and/or concrete leaves*	
Thermal insulation. Determination of steady-state thermal transmission properties. Calibrated and guarded hot box	BS EN ISO 8990:1996
Thermal performance of building materials and products. Determination of thermal resistance by means of guarded hot plate and heat flow meter methods. Dry and moist products of low and medium thermal resistance	BS EN 12664:2001

Title	Standard
Thermal performance of building materials and products. Determination of thermal resistance by means of guarded hot plate and heat flow meter methods. Products of high and medium thermal resistance	BS EN 12667:2000
Thermal performance of building materials and products. Determination of thermal resistance by means of guarded hot plate and heat flow meter methods. Thick products of high and medium thermal resistance	BS EN 12939:2001
Thermal performance of buildings. Heat transfer via the ground. Calculation methods	BS EN ISO 13370:2007
Thermal performance of windows and doors. Determination of thermal transmittance by hot box method	BS EN ISO 12567-1:2000
Part 1 Complete windows and doors	
Thermal performance of windows, doors and shutters. Calculation of thermal transmittance	BS EN ISO 10077-1:2000
Part 1 Simplified methods	
Thermal solar systems and components	BSEN 12977:2012
Thermoplastics pipes and associated fittings for hot and cold water for domestic purposes and heating installations in buildings	BS 7291-1:2006
Part 1 General requirements	
Part 2 Specification for cross-linked polyethylene (PE-X) pipes and associated fittings	
Unplasticized PVC soil and ventilating pipes of 82.4 mm minimum mean outside diameter, and fittings and accessories of 82.4 mm and of other sizes	BS 4514:2001
Vacuum drainage systems inside buildings	BS EN 12109:1999
Vacuum sewerage systems outside buildings	BS EN 1091:1997
Ventilation for buildings – design criteria for the indoor environment	BSI PD CR 1752:1999
Ventilation for buildings. Performance testing of components/products for residential ventilation	BS EN 13141:2004
Part 1 Externally and internally mounted air transfer devices	
Part 3 Range hoods for residential use	
Part 4 Fans used in residential ventilation systems	
Part 6 Exhaust ventilation system packages used in a single dwelling	
Part 7 Performance testing of a mechanical supply and exhaust ventilation units (including heat recovery) for mechanical ventilation systems intended for single family dwellings	
Part 8 Performance testing of unducted mechanical supply and exhaust ventilation units (including heat recovery) for mechanical ventilation systems intended for a single room	
Ventilation for buildings. Performance testing of components/products for residential ventilation	BS EN 13141
Part 8:2006 Performance testing of unducted mechanical supply and exhaust ventilation units (including heat recovery) for mechanical ventilation systems intended for a single room	

Title	Standard
Part 9:2008 *Externally mounted humidity controlled external air terminal device*	
Part 10:2008 *Humidity controlled extract air terminal device*	
Vitreous china washdown WC pans with horizontal outlet. Specification for WC pans with horizontal outlet for use with 7.5 L maximum flush capacity cisterns	BS 5503-3:1990
Vitrified clay pipes and fittings and pipe joints for drains and sewers	BS EN 295
Part 1:1991 *Test requirements*	
Part 2:1991 *Quality control and sampling*	
Part 3:1991 *Test methods*	
Part 6:1996 *Requirements for vitrified clay manholes*	
Wall hung WC pan. Specification for wall hung WC pans with horizontal outlet for use with 7.5 L maximum flush capacity cisterns	BS 5504-4:1990
Wastewater lifting plants for buildings and sites – principles of construction and testing	BS EN 12050:2001
Part 1 *Lifting plants for wastewater containing faecal matter*	
Part 2 *Lifting plants for faecal-free wastewater*	
Part 3 *Lifting plants for wastewater containing faecal matter for limited application*	
Wastewater lifting plants for buildings and sites. Principles of construction and testing. Lifting plants for wastewater containing faecal matter for limited applications	BS EN 12050-3:2001
Water supply. Specification for indirectly heated unvented (closed) storage water heaters	BS EN 12897:2006
WC pans and WC suites with integral trap	BS EN 997:2012
Windows and doors. Product standard, performance characteristics. Windows and external pedestrian doorsets without resistance to fire and/or smoke leakage characteristics	BS EN 14351-1:2006
Windows, doors and rooflights. Design for safety in use and during cleaning of windows, including door-height windows and roof windows. Code of practice [2004]	BS 8213-1
Wiring Regulations, incorporating Amendment No. 1:2011	BS 7671:2008 (Amendment No. 1)
Wood preservatives. Guidance on choice, use and application	BS 1282:1999
Wood stairs	BS 585
Part 1:1989 *Specification for stairs with closed risers for domestic use, including straight and winder flights and quarter and half landings*	
Wood-based panels for use in construction. Characteristics, evaluation of conformity and marking	BS EN 13986:2004
Workmanship on building sites	BS 8000
Part 6:1990 *Code of practice for slating and tiling of roofs and claddings*	
Part 13:1989 *Code of practice for above ground drainage and sanitary appliances*	

Part 14:1989 *Code of practice for below ground drainage*

Workmanship on building sites. Code of practice for hot   BS 8000-15:1990
and cold water services (domestic scale)

Note: Copies of all British Standards are available from BSI, PO Box 16206, Chiswick, London W4 4ZL (http://www.bsonline.techindex.co.uk).

Other publications

Air Tightness Testing and Measurement Association (ATTMA)
http://www.attma.org

*   *Measuring Air Permeability of Building Envelopes*

Association for Specialist Fire Protection (ASFP)
http://www.asfp.org.uk

*   ASFP Red Book – *Fire Stopping and Penetration Seals for the Construction Industry*
*   ASFP Yellow Book – *Fire Protection for Structural Steel in Buildings*
*   ASFP Grey Book – *Fire and Smoke Resisting Dampers*
*   ASFP Blue Book – *Fire Resisting Ductwork*

British Automatic Fire Sprinkler Association (BAFSA)
http://www.bafsa.org.uk

*   *Sprinklers for Safety: Use and Benefits of Incorporating Sprinklers in Buildings and Structures*

Building and Engineering Services Association (B&ES)
http://www.b-es.org

*   *A Practical Guide to Ductwork Leakage Testing*
*   DW/144, *Specification for Sheet Metal Ductwork*

Building Research Establishment Ltd (BRE)
http://www.bre.co.uk

*   BR 128, *Guidelines for the Construction of Fire Resisting Structural Elements*
*   BR 135, *Fire Performance of External Thermal Insulation for Walls of Multi-Storey Buildings*
*   BR 187, *External Fire Spread: Building Separation and Boundary Distances*
*   BR 208, *Increasing the Fire Resistance of Existing Timber Floors*
*   BR 262, *Thermal Insulation: Avoiding Risks*
*   BR 274, *Fire Safety of PFTE Based Materials Used in Buildings*
*   BR 364, *Solar Shading of Buildings*

- BR 369, *Design Methodologies for Smoke and Exhaust Ventilation*
- BR 437, *Industrial Platform Floors: Mezzanine and Raised Storage*
- BR 443, *Conventions for U-value Calculations*
- BR 498, *Selecting Lighting Controls*
- BR 454, *Multi-Storey Timber Frame Buildings: A Design Guide*
- Information Paper 1P1/06, 'Assessing the effects of thermal bridging at junctions and around openings in the external elements of buildings'
- Information paper 1P14103, 'Preventing hot water scalding in bathrooms: Using TMVs'

Centre for Window and Cladding Technology (CWCT)
http://www.cwct.co.uk

- *Thermal Assessment of Window Assemblies, Curtain Walling and Non-Traditional Building Envelopes*

Chartered Institution of Building Services Engineers (CIBSE)
http://www.cibse.org

- CIBSE Commissioning Code M, *Commissioning Management*
- CIBSE Guide A, *Environmental Design*
- AM 10, *Natural Ventilation in Non-Domestic Buildings*
- *Solar Heating Design and Installation Guide*
- TM 31, *Building Log Book Toolkit*
- TM 36, *Climate Change and the Indoor Environment: Impacts and Adaptation*
- TM 37, *Design for Improved Solar Shading Control*
- TM 39, *Building Energy Metering*

Department for Communities and Local Government (DCLG)
http://www.communities.gov.uk

- *Fire Safety in Adult Placements: A Code of Practice*
- *A Householder's Planning Guide for the Installation of Satellite Television Dishes*
- *Domestic Heating Compliance Guide*
- *Outdoor Advertisements and Signs: A Guide for Advertisers*

Department for Education (DFE)
http://www.education.gov.uk

- Building Bulletin 101, *Ventilation of School Buildings*, School Building and Design Unit

Department for Environment, Food and Rural Affairs (Defra)
http://www.defra.gov.uk

- *The Government's Standard Assessment Procedure for Energy Rating of Dwellings: SAP 2005* (available at: http://www.bre.co.uk/sap2005)
- *Rainwater and Greywater: Technical and Economic Feasibility*
- *Rainwater and Greywater: A Guide for Specifiers*

- *Rainwater and Greywater: Review of Water Quality Standards and Recommendations for the UK*

Department of Health (DH)
http://www.dh.gov.uk

- Health Technical Memorandum 05-02, *Guidance to Support of Functional Provisions in Healthcare Premises*

Door and Hardware Federation (DHF)
http://www.dhfonline.org.uk

- *Code of Practice for Fire-Resisting Metal Doorsets*
- *Hardware for Fire and Escape Doors*

Electrical Contractors' Association (ECA) and National Inspection Council for Electrical Installation Contracting (NICEIC)
http://www.eca.co.uk and http://www.niceic.org.uk

- *ECA Comprehensive Guide to Harmonised Cable Colours*
- *Electrical Installers' Guide to the Building Regulations*
- *New Fixed Wiring Colours: A Practical Guide*

Energy Saving Trust (EST)
http://www.est.org.uk

- CE66, *Windows for New and Existing Housing*
- CE129, *Reducing Overheating: A Designer's Guide*
- GIL20, *Low Energy Domestic Lighting*
- GPG268, *Energy Efficient Ventilation in Dwellings: A Guide for Specifiers*, 2006

English Heritage
http://www.english-heritage.org.uk

- *Building Regulations and Historic Buildings*

Environment Agency
http://www.environment-agency.gov.uk

- Pollution Prevention Guidelines (PPG18), *Managing Fire Water and Major Spillages*

Fire Protection Association (FPA)
http://www.thefpa.co.uk

- *Design Guide*

Food Standards Agency (FSA)
http://www.food.gov.uk

- *Code of Practice: Food Hygiene – a Guide for Businesses*

Glass and Glazing Federation (GGF)
http://www.ggf.org.uk

* *A Guide to Best Practice in the Specification and Use of Fire Resistant Glazed Systems*

Health and Safety Executive (HSE)
http://www.hse.gov.uk

* *Workplace (Health, Safety and Welfare) Regulations*
* *Legionnaires' Disease: Control of Legionella Bacteria in Water Systems*

Institution of Engineering and Technology (IET)
http://www.theiet.org

* *Electrician's Guide to the Building Regulations*
* IEE Guidance Note 1, *Selection and Erection of Equipment*, 4th edition
* IEE Guidance Note 2, *Isolation and Switching*, 4th edition
* IEE Guidance Note 3, *Inspection and Testing*, 4th edition
* IEE Guidance Note 4, *Protection against Fire*, 4th edition
* IEE Guidance Note 5, *Protection against Electric Shock*, 4th edition
* IEE Guidance Note 6, *Protection against Overcurrent* , 4th edition
* IEE Guidance Note 7, *Special Locations*, 2nd edition
* *IEE On-Site Guide* (BS 7671, *IEE Wiring Regulations*, 16th edition)
* *New Wiring Colours*, 2004

Institution of Structural Engineers (IstructE)
http://www.istructe.org

* *Dynamic Performance Requirements for Permanent Grandstands Subject to Crowd Action: Recommendations for Management, Design and Assessment*

Metal Cladding and Roofing Manufacturers Association (MCRMA)
http://www.mcrma.co.uk

* *Guidance for Design of Metal Cladding and Roofing to Comply with Approved Document L2*

Modular and Portable Building Association (MPBA)
http://www.mpba.biz

* *Energy Performance Standards for Modular and Portable Buildings*

National Association of Rooflight Manufacturers (NARM)
http://www.narm.org.uk

* *Use of Rooflights to Satisfy the 2002 Building Regulations for the Conservation of Fuel and Power*

National Health Service (NHS)
http/ /www.nhs.uk

* National Health Service Model Engineering Specifications D 08, *Thermostatic Mixing Valves (Healthcare Premises)*, Revision 1 (archived historical document no longer current), 1997 (http://www.thenbs.com/PublicationIndex/DocumentSummary.aspx?PubID=412&DocID=275997)

Passive Fire Protection Federation (PFPF)
http://www.pfpf.org

- *Ensuring Best Practice for Passive Fire Protection in Buildings*

Planning Portal
http://www.planningportal.gov.uk

- Copies of Approved Documents

Steel Construction Institute (SCI)
http://www.steel-sci.org

- SCI P197, *Designing for Structural Safety: A Handbook for Architects and Engineers*
- SCI P288, *Fire Safe Design: A New Approach to Multi-Storey Steel-Framed Buildings*, 2nd edition
- SCI P313, *Single Storey Steel Framed Buildings in Fire Boundary Conditions*

Thermal Insulation Manufacturers and Suppliers Association (TIMSA)
http://www.timsa.org.uk

- *HVAC Guidance for Achieving Compliance with Part L of the Building Regulations*

Timber Research and Development Association (TRADA)
http://www.trada.co.uk

- *Timber Fire-Resisting Doorsets: Maintaining Performance under the new European Test Standard*

Water Regulations Advisory Scheme (WRAS)
http://www.wras.co.uk

- *Water Regulations Advisory Scheme: Water Regulations Guide*
- WRAS Information and Guidance Note No. 9-02-05, *Marking and Identification of Pipework for Reclaimed (Greywater) Systems*

Wood Protection Association (WPA)
http://www.wood-protection.org

- *Industrial Wood Preservation: Specification and Practice*

Legislation

Standard	Title
SI 1991/1620	Construction Products Regulations 1991
SI 1992/2372	Electromagnetic Compatibility Regulations 1992

SI 1992/3004	The Workplace (Health, Safety and Welfare) Regulations 1992
SI 1994/1886	The Gas Safety (Installation and Use) Regulations 1994
SI 1994/3051	Construction Products (Amendment) Regulations 1994
SI 1994/3080	Electromagnetic Compatibility (Amendment) Regulations 1994
SI 1994/3260	Electrical Equipment (Safety) Regulations 1994
SI 1999/1148	The Water Supply (Water Fittings) Regulations 1999
SI 2000/3184	The Water Supply (Water Quality) Regulations 2000
SI 2001/3335	Building (Amendment) Regulations 2001
SI 2005/1726	Energy Information (Household Air Conditioners) (No. 2) Regulations 2005
SI 2006/14	The Food Hygiene (England) Regulations 2006
SI 2006/31	The Food Hygiene (Wales) Regulations 2006
SI 2006/652	Building and Approved Inspectors (Amendment) Regulations 2006
SI 2009/3101	The Private Water Supplies Regulations 2009
SI 2010/66	The Private Water Supplies (Wales) Regulations 2010

Other

- Commission Decision 2005/823iEC of 22 November 2005 amending Decision 2001/671/EC, regarding the classification of the external fire performance of roofs and roof coverings
- 89/106/EEC, *The Construction Products Directive*
- 93/68/EEC, *The CE Marking Directive*
- Anti-Social Behaviour Act 2003
- Education Act 1996
- Electrical Equipment (Safety) Regulations 1994 (SI 1994 No. 3260)
- Electromagnetic Compatibility (Amendment) Regulations 1994 (SI 1994 No. 3080)
- Electromagnetic Compatibility Regulations 1992 (ISI 1992 No. 2372)
- Equality Act 2010
- Equality Act 2010 (Disability) Regulations 2010
- European Communities Act 1972
- Gas Safety (Installation and Use) Regulations 1998, SI 1998 No. 2451
- Health and Safety (Safety Signs and Signals) Regulations 1996
- Health and Safety at Work etc. Act 1974
- Pipelines Safety Regulations 1996, SI 1996 No. 825
- Water Industry Act 1991
- Workplace (Health, Safety and Welfare) Regulations 1992

References

- Circular 11/95, *The Use of Conditions in Planning Permissions*
- Airports Authority Act 1975
- Ancient Monuments and Archaeological Areas Act 1979
- Atomic Energy Authority Act 1954
- Building Act 1984
- Building (Approved Inspectors etc.) Regulations 2010, SI No. 2215
- Building Regulations 2010, SI No. 2214
- City of London (Various Powers) Act 1977
- Civil Aviation Act 1982
- Clean Air Act 1956
- Climate Change Act 2008
- Construction (Design and Management) Regulations 2007 (CDM 2007)
- Control of Pollution Act 1974
- Criminal Law Act 1977
- Development of Rural Wales Act 1976
- Education Act 1944
- Education Act 1980
- Environmental Protection Act 1990
- Faculty Jurisdiction Measure 1964
- Fire Precautions Act 1971
- Fire Precautions (Workplace) Regulations 1997
- Flood and Water Management Act 2010 (c.29)
- Gas Safety (Installation and Use) Regulations 1998
- Greater London Council (General Powers) Act 1967
- Health and Safety at Work etc. Act 1974
- Highways Act 1980
- Housing Act 1957
- Housing Act 1985
- Housing and Building Control Act 1984
- Interpretation Act 1978
- Local Government (Access to Information) (Variation) Order 2006
- Local Government Act 1972
- Local Government (Miscellaneous Provisions) Act 1976
- Local Government (Miscellaneous Provisions) Act 1982
- Local Government, Planning and Land Act 1980
- Local Land Charges Act 1975
- Localism Act 2011
- London Government Act 1963
- London Local Authorities Act 2007
- Milton Keynes (Urban Area and Planning Functions) Order 2004
- New Towns Act 1981
- Offices, Shops and Railway Premises Act 1963
- Party Wall etc. Act 1996

- Planning Act 2008
- Planning (Listed Buildings and Conservation Areas) Act 1990
- Prisons Act 1952
- Public Health Act 1936
- Public Health Act 1961
- Public Health (Control of Disease) Act 1984
- Radioactive Substances Act 1960
- Restriction of Ribbon Development Act 1935
- Safety of Sports Grounds Act 1975
- Thurrock Development Corporation (Area and Constitution) Order 2003
- Town and Country Planning Act
- Town and Country Planning (Application of Subordinate Legislation to the Crown) Order 2006
- Town and Country Planning (Control of Advertisements) (England) Regulations 2007
- Town and Country Planning (Environmental Impact Assessment) (England and Wales) Regulations 1999
- Town and Country Planning (General Permitted Development) (Amendment) (England) Order 2001
- Town and Country Planning (General Permitted Development) (Amendment) (England) Order 2005
- Town and Country Planning (General Permitted Development) (Amendment) (England) Order 2006
- Town and Country Planning (General Permitted Development) (Amendment) (England) Order 2007
- Town and Country Planning (General Permitted Development) (Amendment) (England) Order 2008, SI 2008 No. 675
- Town and Country Planning (General Permitted Development) (Amendment) (England) Order 2010
- Town and Country Planning (General Permitted Development) (Amendment) (England) Order 2011
- Town and Country Planning (General Permitted Development) (Amendment) (England) Order 2012
- Town and Country Planning (General Permitted Development) (Amendment) (No. 2) (England) Order 2008, SI 2008 No. 2362
- Town and Country Planning (General Permitted Development) (Amendment) (No.2) (England) Order 2010
- Town and Country Planning (General Permitted Development) (Amendment) (No. 2) (England) Order 2012
- Town and Country Planning (General Permitted Development) (Amendment) Order 1996
- Town and Country Planning (General Permitted Development) (Amendment) Order 1998
- Town and Country Planning (General Permitted Development) (Amendment) Order 1999

- Town and Country Planning (General Permitted Development) (England) (Amendment) (No. 2) Order 2005
- Town and Country Planning (General Permitted Development) Order 1995
- Water Act 1945
- Water Act 1973
- Water Act 1981
- Water Industry Act 1991

Useful contact names and addresses

 Authors' hint
Please note that the following lists of contact names and addresses are accurate at the time of publication. It is quite possible, however, that some of them may have changed by the time you come to refer to them.

Building Act

The following professional body is willing to provide general and informal advice about the Act. **However**, any advice given should **not** be seen as being endorsed by the UK government.

Royal Institution of Chartered Surveyors (RICS)
Parliament Square
London SW1P 3AD
Tel: 0870 333 1600
Fax: 020 7334 3811
www.rics.org.uk

Professional advice

The following bodies hold lists of their members who may be willing to provide professional advice or act as 'surveyors' under the Act – again with the proviso that any advice given should **not** be seen as being endorsed by the UK government.

Architecture and Surveying Institute
Register of Party Wall Surveyors
St Mary House
15 St Mary Street
Chippenham
Wiltshire SN15 3WD
Tel: 01249 444505
Fax: 01249 443602

Association of Building Engineers (ABE)
Private Practice Register
Lutyens House
Billing Brook Road
Weston Favell
Northampton NN3 8NW
Tel: 0845 126 1058
Fax: 01604 784220

Pyramus and Thisbe Club
Florence House
53 Acton Lane
London NW10 8UX
Tel: 020 8961 3311
Fax: 020 8963 1689

Royal Institute of British Architects (RIBA)
Clients Advisory Service
66 Portland Place
London W1B 1AD
Tel: 020 7580 5533
Fax: 020 7255 1541

Royal Institution of Chartered Surveyors (RICS)
Parliament Square
London SW1P 3AD
Tel: 0870 333 1600
Fax: 020 7334 3811

Professional contacts

Asbestos specialist

Asbestos Information Centre Ltd
5a The Maltings
Stowupland Road
Stowmarket
Suffolk 1P14 SAG
Tel: 0904 517 0156

Concrete specialist

British Ready Mixed Concrete Association
The Bury
Church Street
Chesham

Buckinghamshire HP5 1JE
Tel: 01494 791050

Damp, rot, infestation

British Wood Preserving and Damp Proofing Association
Building No. 6
The Office Village
4 Romford Road
Stratford
London E15 4EA
Tel: 020 8519 2588

Countryside Council for Wales
Maps y Ffynnon
Penrhosgarnedd
Bangor
Gwynedd LL57 2DW
Tel: 01248 355782

English Nature
Northminster House
Northminster Road
Peterborough PE1 1VA
Tel: 01733 340345

Local department of environmental health
Refer to your local directory.

Scottish Natural Heritage
12 Hope Terrace
Edinburgh EH9 2AS
Tel: 0131 447 4784

Decorators

British Decorators' Association
32 Coton Road
Nuneaton
Warwickshire CV11 5TW
Tel: 0247 635 3776

Scottish Decorators' Federation
Castlecraig
Business Park
Players Road
Stirling FK7 7SH
Tel: 01786 448838

Electricians

National Inspection Council for Electrical Installation Contracting
Warwick House
Houghton Hall Park
Houghton Regis
Dunstable LU5 SZX
Tel: 0870 013 0382

Fencing erectors

Fencing Contractors Association
Warren Rd
Trellech
Monmouthshire NP25 4PQ
Tel: 07000 560722

Glazing specialists

Glass and Glazing Federation
44–48 Borough High Street
London SE1 1XB
Tel: 020 7403 7177

Heating installers

British Coal Corporation
Hobart House
Grosvenor Place
London SW1X 7AE
Tel: 020 7235 2020

British Gas regional office
Refer to your local directory.

Electricity supply company
Refer to your local directory.

Heating and Ventilating Contractors' Association
Esca House
34 Palace Court
London W2 4JG
Tel: 020 7229 2488

National Association of Plumbing, Heating and Mechanical Services Contractors
Ensign House

Ensign Business Centre
Westwood Way
Coventry CV4 8JA
Tel: 01203 470626

Home security

British Security Industry Association
Kirkham House
John Comyn Drive
Worcester WR3 7NS
Tel: 0845 389 3889

Local crime prevention officer
Refer to your local directory.

Local fire prevention officer
Refer to your local directory.

Master Locksmiths Association
5d Great Central Way
Woodford Halse
Daventry
Northants NN1 3PZ
Tel: 01327 262255

National Approval Council for Security Systems
Queensgate House
14 Cookham Road
Maidenhead
Berks S16 8AJ
Tel: 01628 37512

Insulation installers

Draught Proofing Advisory Association Ltd
PO Box 12
Haslemere
Surrey GU27 3AH
Tel: 01428 654011

External Wall Insulation Association
PO Box 12
Haslemere
Surrey GU27 3AH
Tel: 01428 654011

National Association of Loft Insulation Contractors
PO Box 12
Haslemere
Surrey GU27 3AH
Tel: 01428 654011

National Insulation Association
2 Vimy Court
Vimy Road
Leighton Buzzard
Beds LU7 1FG
Tel: 08451 636363

Plasterers

Federation of Master Builders
Gordon Fisher House
14–15 Great James Street
London WC1N 3DP
Tel: 020 7242 7583

Plumbers

National Association of Plumbing, Heating and Mechanical Services Contractors
Ensign House
Ensign Business Centre
Westwood Way
Coventry CV4 8JA
Tel: 01203 470626

Robust Details

Robust Details Ltd
Davy Avenue
Knowlhill
Milton Keynes MK5 8NB
Tel: customer service 0870 240 8210
Tel: technical support 0870 240 8209
Fax: 0870 240 8203
Technical email support: technical@robustdetails.com
Administrative email support: administration@robustdetails.com
Other support: customerservice@robustdetails.com

Roofers

Builders Merchants Federation
15 Soho Square

London W1V 3HL
Tel: 020 7439 1753

National Federation of Roofing Contractors
24 Worship Street
London EC2A 2DY
Tel: 020 7638 7663

Ventilation

Heating and Ventilating Contractors' Association
Esca House
34 Palace Court
London W2 4JG
Tel: 020 7229 2488

Other useful contacts

British Board of Agrément (BBA)
Bucknalls Lane
Garston
Watford WD5 9BA
Tel: 01923 665300
Fax: 01923 665301
Email: contact@bba.star.co.uk
http://www.bbacerts.co.uk

British Standards Institution (BSI)
389 Chiswick High Road
London W4 4AL
Tel: 020 8996 9001
Fax: 020 8996 7001
Email: csenices@bsigroup.com
http://www.bsigroup.com

Chartered Institute of Plumbing and Heating Engineering
64 Station Lane
Hornchurch
Essex RM12 6NH
Tel: 01708 472791
Fax: 01708 448987
Note: Approved contractor person scheme (Building Regulations)

Fenestration Self-Assessment Scheme
Fensa Ltd
54 Ayres Street

London SE1 1EU
Tel: 020 7645 3700
Fax: 020 7407 8307

HETAS Ltd
Orchard Business Centre
Stoke Orchard
Cheltenham
Gloucestershire GL52 7RZ
Tel: 0845 634 5626
Note: Registration scheme for companies and engineers involved in the installation and maintenance of domestic solid-fuel-fired equipment.

HMSO
The Stationery Office
PO Box 29
Norwich NR3 1GN
Tel: orders/general enquiries 0870 600 5522
Fax: orders 0870 600 5533
http://www.tsoshop.co.uk

OFTEC
Oil Firing Registration Scheme
Foxwood House
Dobbs Lane
Kesgrave
Ipswich IPS 2QQ
Tel: 0845 658 5080
Fax: 0845 658 5181

Robust Details Limited
Davy Avenue
Knowlhill
Milton Keynes MK5 8NB
Tel: customer service 0870 240 8210
Tel: technical support 0870 240 8209
Fax: 0870 240 8203

UKAS
United Kingdom Accreditation Service
21–47 High Street
Feltham
Middlesex TW3 4UN
Tel: 020 8917 8400

WIMLAS
WIMLAS Limited
St Peter's House
6–8 High Street
Iver
Buckinghamshire SL0 9NG
Tel: 01753 737744
Fax: 01753 792321
Email: wimlas@compuserve.com

Useful websites

Building Act and Building Regulations: http://www.communities.gov.uk (the new DCLG site)
Approved Documents: http://www.planningportal.gov.uk/buildingregulations /approveddocuments

Air Tightness Testing and Measurement Association (ATTMA)	http://www.attma.org
British Flue and Chimney Manufacturers Association (BFCMA)	http://www.feta.co.uk/bfcma
British Standards Institution (BSI)	http://www.bsigroup.com
Building Research Establishment Ltd (BRE)	http://www.bre.co.uk
Building Services Research and Information Association (BSRIA)	http://www.bsria.co.uk
Centre for Window and Cladding Technology (CWCT)	http://www.cwct.co.uk
Chartered Institution of Building Services Engineers (CIBSE)	http://www.cibse.org
Department for Business, Innovation and Skills	http://www.bis.gov.uk
Department for Education	http://www.education.gov.uk
Department of Energy and Climate Change (DECC)	http://www.decc.gov.uk
Department for Transport (DFT)	http://www.dft.gov.uk
Energy Saving Trust (EST)	http://www.energysavingtrust. org.uk
English Heritage	http://www.english-heritage. org.uk
Environment Agency	http://www.environment- agency.gov.uk
Gas Safe Register	http://www.gassaferegister. co.uk
Health and Safety Executive (HSE)	http://www.hse.gov.uk

Heating and Ventilating Contractors' Association (HVCA)	http://www.hvca.org.uk
Heating Equipment Testing and Approval Scheme (HETAS)	http://www.hetas.co.uk
Institution of Gas Engineers and Managers	http://www.igem.org.uk
Local councils (A–Z list)	http://www.idea.gov.uk/idk/org/la-data.do
Metal Cladding and Roofing Manufacturers Association (MCRMA)	http://www.mcrma.co.uk
Microgeneration Certification Scheme	http://www.microgenerationcertification.org
Modular and Portable Buildings Association (MPBA)	http://www.mpba.biz
My Community Rights	http://mycommnunity rights.org.uk/community-right-to-bid
National Association of Rooflight Manufacturers (NARM)	http://www.narm.org.uk
Oil Firing Technical Association Ltd (OFTEC)	http://www.oftec.org
Solid Fuel Association (SFA)	http://www.solidfuel.co.uk
Thermal Insulation Manufacturers and Suppliers Association (TIMSA)	http://www.timsa.org.uk
TrustMark	http://www.trustmark.org.uk
Assessable thresholds	http://www.tso.co.uk
British Standards	http://www.bsonline.techindex.co.uk
Building near trees	http://www.nhbc.co.uk
Building research	http://www.bre.co.uk
Carbon dioxide from natural sources and mining areas	http://www.bgs.ac.uk
	http://www.tso.co.uk
	http://www.defra.gov.uk
	http://www.ciria.org.uk
Cladding	http://www.mcrma.co.uk
Concrete in aggressive ground	http://www.bre.co.uk
Contaminated land	http://www.defra.gov.uk
	http://www.ciria.org.uk
	http://www.hse.gov.uk
Contamination in disused coal mines	http://www.tso.co.uk
Demolition	http://www.ciria.org.uk
Electrical safety	http://www.theiet.org
Environmental aspects	http://www.arup.com
	http://www.environment-agency.gov.uk
Excavation and disposal	http://www.defra.gov.uk
	http://www.ciria.org.uk

Flood protection	http://www.ciria.org.uk
	http://www.environment-agency.gov.uk
Flooding from sewers	http://www.defra.gov.uk
	http://www.ciria.org.uk
Foundations	http://www.bre.co.uk
Gas-contaminated land	http://www.defra.gov.uk
	http://www.bre.co.uk
	http://www.ciria.org.uk
Geo-environmental and geotechnical investigations	http://www.ags.org.uk
Glass and glazing	http://www.ggf.org.uk
Hardcore	http://www.bre.co.uk
Health and safety	http://www.hse.gov.uk
	http://www.defra.gov.uk
Land quality	http://www.environment-agency.gov.uk
Landfill gas	http://www.ciwm.co.uk
	http://www.defra.gov.uk
	http://www.gassim.co.uk
Laying water pipelines in contaminated ground	http://www.fwr.org
Low-rise buildings	http://www.bre.co.uk
Materials and workmanship	http://www.tso.co.uk
Methane	http://www.bgs.ac.uk
	http://www.tso.co.uk
	http://www.defra.gov.uk
	http://www.ciria.org.uk
Oil seeps from natural sources and mining areas	http://www.bgs.ac.uk
	http://www.tso.co.uk
	http://www.defra.gov.uk
	http://www.ciria.org.uk
Petroleum retail sites	http://www.petroleum.co.uk
Pollution control	http://www.communities.gov.uk
Protection of ancient buildings	http://www.spab.org.uk
Radon	http://www.bre.co.uk
Robust construction details	http://www.tso.co.uk
Roofing design	http://www.mcrma.co.uk
Shrinkable clay soils	http://www.bre.co.uk
Soil sampling	http://www.defra.gov.uk
Soils, sludge and sediment	http://www.ciria.org.uk
Subsidence	http://www.bre.co.uk
Thermal bridging	http://www.bre.co.uk
	http://www.tso.co.uk
Thermal insulation	http://www.bre.co.uk
Timbers	http://www.bre.co.uk

Books by the same authors

Title	Extracts from book reviews	ISBN
Wiring Regulations in Brief (3rd edition) 	• Tired of trawling through the Wiring Regs? • Perplexed by Part P? • Confused by cables, conductors and circuits? Then look no further! This handy guide provides an on-the-job reference source for electricians, designers, service engineers, inspectors, builders, students and DIY enthusiasts. Topic-based chapters link areas of working practice – such as cables, installations, testing and inspection, and special locations – with the specifics of the regulations themselves. This allows quick and easy identification of the official requirements relating to the situation in front of you. The requirements of the regulations, and of related standards, are presented in an informal, easy-to-read style that strips away confusion. Packed with useful hints and tips, and highlighting the most important or mandatory requirements, this book is a concise reference on all aspects of the 17th edition of the *IEE Wiring Regulations* and Part P of the Building Regulations.	Routledge 978-0-415-52687-6
Water Regulations in Brief 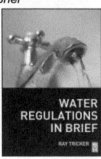	*Water Regulations in Brief* is a unique reference book, providing all the information needed to comply with the regulations, in an easy-to-use, full-colour format. Crucially, unlike other titles on this subject, this book doesn't just cover the Water Regulations; it also clearly shows how they link in with the Building Regulations, Water Bylaws and Wiring Regulations, providing the only available complete reference to the requirements for water fittings and water systems. Structured in the same logical, time-saving way as the author's other bestselling 'In Brief' books, *Water Regulations in Brief* will be a welcome change to anyone tired of wading through complex, jargon-heavy publications in search of the information they need to get the job done.	Routledge 978-1-85617-628-6

Title	Extracts from book reviews	ISBN
Scottish Building Standards in Brief 	*Scottish Building Standards in Brief* takes the highly successful formula of Ray Tricker's previous 'In Brief' series and applies it to the requirements of the Building (Scotland) Regulations 2004. With the same no-nonsense and simple-to-follow guidance – but written specifically for the Scottish Building Standards – it's the ideal book for builders, architects, designers and DIY enthusiasts working in Scotland.	Routledge 978-0-750-68558-0
ISO 9001:2008 for Small Businesses *(5th edition)* 	The new edition of this top-selling quality management book now includes: • relevant examples that put the concepts and requirements of the standard into a real-life context; • down-to-earth explanations to help you determine what you need to work in compliance with and/or achieve certification to ISO 9001:2008; • a brand new chapter on the cost of implementing ISO 9001:2008; • an example of a complete, generic quality management system consisting of a quality manual plus a whole host of quality processes, quality procedures and work instructions; • access to a free, software copy of the generic QMS files (available from the author) to give you a starting-point from which to develop your own documentation.	Routledge 978-0-415-70390-1
ISO 9001:2008 Quality Manual and Audit Checksheets *(5th edition)* 	The publication is made up of the following parts: <u>GENERIC QUALITY MANAGEMENT SYSTEM</u> A complete soft copy of the generic Quality Management System describe in *'ISO 9001:2008 for small Businesses'* (Edition 5) and consisting of a Quality Manual, Quality Processes, Quality Procedures and Work Instructions <u>AUDIT CHECKSHEETS</u> All the major audit checksheets and forms contained in "Auditing Quality Management Systems" that are required to conduct either for a simple internal review of a particular process or an external (e.g. third party) assessment of an organisation against the formal requirements of ISO 9001:2008	Herne European Consultancy Ltd 978-0-9927585-01

Title	Extracts from book reviews	ISBN
ISO 9001:2000 Audit Procedures (2nd edition) 	In order to meet the recommendations, requirements and specifications of ISO 9001:2000, organizations must undertake an audit of their own quality procedures and those of their suppliers. Likewise, when supplying ISO 9001:2000 accredited customers, suppliers must be prepared to undergo a similar audit. Revised, updated and expanded, *ISO 9001:2000 Audit Procedures* describes the methods for completing management reviews and quality audits, and outlines the experiences of working with 9001:2000 since its launch in 2000. It also includes essential new material on process models, generic processes, the requirements for mandatory documented procedures, and detailed coverage of auditors' questionnaires	Routledge 978-0-750-66615-2
Quality Management System for ISO 9001: 2000 (2nd edition) 	*Quality Management System for ISO 9001:2008* and accompanying CD is probably the most comprehensive set of ISO 9001:2008 compliant documents available worldwide. Fully customizable, it can be used as a basic template for any organization wishing to work in compliance with, or gain registration to, ISO 9001.	Herne European Consultancy Ltd 978-0-954-86474-3
ISO 9001:2000 in Brief (2nd edition) 	Revised and expanded, this new edition of an easy-to-understand guide provides practical information on how to set up a cost-effective ISO 9001 compliant quality management system.	Butterworth-Heinemann 978-0-750-66616-9

Title	Extracts from book reviews	ISBN
Auditing Quality Management Systems (2nd edition)  Auditing Quality Management Systems	This publication is the result of over a decade's experience of all major international standards for integrated quality management systems. It is a comprehensive CD containing all the major audit checksheets and forms that are required to conduct either a simple internal audit or an external assessment of an organization against the formal requirements of ISO 9001:2008.

Now fully updated to include checklists for:

• project management;
• health and safety in the workplace.

 Note: Also includes 'background notes for auditors' | Herne European Consultancy Ltd 978-0-954-86477-4 |
| *MDD Compliance Using Quality Management Techniques*  MDD COMPLIANCE using Quality Management Techniques | The Medical Device Directive (MDD) is difficult to understand and interpret, but this book covers the subject superlatively.

In summary, the book is a good reference for understanding the Medical Device Directive's requirements and will aid companies of all sizes in adding these requirements to an existing quality management system. | Butterworth-Heinemann 978-0-750-64441-9 |
| *Quality and Standards in Electronics* Quality & Standards in Electronics | A manufacturer or supplier of electronic equipment or components needs to know the precise requirements for component certification and quality conformance to meet the demands of the customer. This book ensures that the professional is aware of all the UK, European and international necessities, and knows the current status of these regulations and standards and where to obtain them. | Newnes 978-0-750-62531-9 |

Title	Extracts from book reviews	ISBN
Environmental Requirements for Electromechanical and Electronic Equipment	This is the definitive reference containing all of the background guidance, typical ranges, details of recommended test specifications, case studies and regulations covering the environmental requirements on designers and manufacturers of electrical and electromechanical equipment worldwide.	Butterworth-Heinemann 978-0-750-63902-6
CE Conformity Marking	CE marking can be regarded as a product's trade passport for Europe. It is a mandatory European marking for certain product groups to indicate conformity with the essential health and safety requirements set out in the European Directive. This book contains essential information for any manufacturer or distributor wishing to trade in the European Union. Practical and easy to understand.	Butterworth-Heinemann 978-0-750-64813-4
Optoelectronics and Fiber Optic Technology	An introduction to the fascinating technology of fibre optics. Students, technicians and professional readers will benefit from this publication. Simply written in an easily accessible style which does not put the reader off and covers all of the basic topics in an appropriate and logical order. Topical areas such as optoelectronics in LANs and WANs, cable TV systems, and the global fibre-optic highway make this book essential reading for anyone who needs to keep up with the technology of modern data communications.	Butterworth-Heinemann 978-0-750-65370-1
The Cyder Book	A unique combination of a historical overview of cider making through the ages, the cider making process and a collection of recipes using cider and cider apples.	Herne European Consultancy Ltd 978-0-954-86476-7

Index

absorbent materials
  ceilings 402, 528–9, 611
  floors 402
abutting construction, junctions 492
access 677–93
  assembly buildings 684–5, 708 9
  Building Regulations xxi, 41–2,
    108–9
  cleaning windows 94, 641–2, 683–4
  disabled people 709–21
    Building Regulations 2010 xxi,
      41–2, 108–9
    entrances 711
    water closets 727, 728–9, 730–7
  drainage systems 365–6
  entrances 193, 711
    doors 193, 698, 768–70
    signs 686, 687
  fire services 73, 379–80, 676
  guarding 679–80
  handrails for stairs 680–1
  lobbies/corridors 696 9
  for maintenance 588, 624, 683–4
  ramped 681
  requirements 677–8
  route hazards 670, 678 9
  signs, disabled people 719, 721
  stepped 630, 684
  uni-sex toilets 731–5
  vehicles 682–3
  ventilation systems 335–8
  water closets 727
access statements/strategies xxi
acoustics 721
  in schools 77, 386, 408
  see also echoes
acronyms 825–7
additional rooms, ventilation 339
additions to buildings *see* extensions
addresses and contact names 857–67
adjoining dwellings, noise 76, 386
advertisements
  application fees 163
  Building Regulations approval/
    planning permission 120, 121–2,
    151, 163, 167, 238

aerials: Building Regulations approval/
    planning permission 168–9,
    238–9
agricultural applications for planning
    permission 132–3, 162
air
  combustion appliances 775–7
  extract rates 306–7, 325, 327, 333–4
  indoor pollutants 305–6
  intakes 308
  pollutants 304–6
  supplies 333, 775–7
  see also ventilation
air conditioning 333, 656, 702
air flow rate 78
air-handling plant 813
airborne sound
  floors/ceilings 400, 403–8, 528–9
  refuse facilities 481
  stairs 405, 596–7, 610–13
  wall requirements 478–88, 496, 498,
    512
aircrete blocks 498
alarms
  carbon monoxide 90, 564
  disabled people 712, 713–14, 726,
    736–7, 742, 764
  electrical safety 764
  emergency pull cords 713–14, 726,
    747, 764
  fire alarms 671, 742, 761
  heat alarms 525
  toilets 736–7
  see also smoke alarms
alterations 25, 178
  buildings other than dwellings 162,
    178
  electrical safety 751, 759
  roofs 200, 223, 548–9
  structural inside 231, 266
ancillary accommodations 218, 802
annexes
  size/proportion requirements 297–8,
    435
  wall requirements 434, 435–7
appeals 10–11, 146, 149–53